Lecture Notes in Computer Science 16405

Founding Editors

Gerhard Goos
Juris Hartmanis

Editorial Board Members

Elisa Bertino, *Purdue University, West Lafayette, IN, USA*
Wen Gao, *Peking University, Beijing, China*
Bernhard Steffen, *TU Dortmund University, Dortmund, Germany*
Moti Yung, *Columbia University, New York, NY, USA*

The series Lecture Notes in Computer Science (LNCS), including its subseries Lecture Notes in Artificial Intelligence (LNAI) and Lecture Notes in Bioinformatics (LNBI), has established itself as a medium for the publication of new developments in computer science and information technology research, teaching, and education.

LNCS enjoys close cooperation with the computer science R & D community, the series counts many renowned academics among its volume editors and paper authors, and collaborates with prestigious societies. Its mission is to serve this international community by providing an invaluable service, mainly focused on the publication of conference and workshop proceedings and postproceedings. LNCS commenced publication in 1973.

Mario Pavone · Carlos A. Coello Coello ·
Raffaele Cerulli · Salvatore Greco ·
El-Ghazali Talbi
Editors

Decision Sciences

Third Decision Science Alliance
International Summer Conference, DSA ISC 2025
Catania, Italy, June 19–20, 2025
Proceedings

 Springer

Editors
Mario Pavone
University of Catania
Catania, Italy

Raffaele Cerulli
University of Salerno
Fisciano-Salerno, Italy

El-Ghazali Talbi
University of Lille
Lille, France

Carlos A. Coello Coello
Cinvestav - Instituto Politécnico Nacional
Cinvestav, Mexico

Salvatore Greco
University of Catania
Catania, Italy

ISSN 0302-9743   ISSN 1611-3349 (electronic)
Lecture Notes in Computer Science
ISBN 978-3-032-21810-0   ISBN 978-3-032-21811-7 (eBook)
https://doi.org/10.1007/978-3-032-21811-7

This Springer imprint is published by the registered company Springer Nature Switzerland AG
The registered company address is: Gewerbestrasse 11, 6330 Cham, Switzerland

If disposing of this product, please recycle the paper.

Preface

Designing systems capable of learning, reasoning, and therefore making decisions autonomously is becoming increasingly important and decisive in the modern world. Today, there is a growing need to create methodologies capable of interpreting complex contexts from which to choose the optimal actions to be taken, and this often occurs under uncertainty conditions. In addition, the exponential growth of available data, the development of robust and effective learning architectures, and optimization methods have made an integrated vision between *Intelligent Systems* and *Decision Making* essential. Under this vision, artificial intelligence is not only a tool for prediction and/or classification but becomes a cognitive component capable of interpreting dynamic contexts, evaluating trade-offs, and supporting decision-makers in making the right choices.

"Intelligent Systems and Decision Making: Human Insights in the Era of AI" was the theme proposed for this edition of ISC 2025 and the event aims to provide a rigorous and inclusive forum for sharing knowledge, experiences, and outcomes that advance the state-of-the-art in AI-based decision making. The main mission of ISC 2025 was (*i*) to consolidate the theoretical and methodological foundations of intelligent systems and decision optimization in complex and uncertain contexts; (*ii*) to promote advanced research and best business practices, with a focus on the transferability and measurability of their impact; and finally, (*iii*) to bridge the gap between industry and academia by strengthening strategic collaboration focused on value, reliability, and sustainability.

ISC 2025 was the 3*rd* edition of the *Decision Science Alliance International Summer Conference* and was held at the Department of Economics and Business of the University of Catania, Italy, from 19 to 20 June 2025. This edition was focused on the *Intelligent Systems and Decision Making: Human Insights in the Era of AI* theme, with particular focus on the new frontiers of AI-driven Decision Making in academic and industrial research. This edition was promoted by the *Decision Science Alliance*[1] and the *Complex Intelligent Systems research group*[2] of the Department of Mathematics and Computer Science of the University of Catania, Italy, and was held in the beautiful city of Catania, located between fire and sea, suspended between the majesty of Mount Etna and the blue of the Mediterranean. By hosting the event, the city of Catania gave participants the opportunity to live in an atmosphere where history intertwines with everyday life: baroque palaces that speak of ancient rebirths, bustling markets full of voices and colors, traditions that withstand the passage of time.

Over 156 participants, 50 corporate, including several startups, were involved in ISC 2025, and 22 special tracks were offered, with a total of 100 speakers from academia and industry, along with 3 internationally renowned plenary speakers, namely:

- Jafar Rezaei, Delft University of Technology, Netherlands;
- El-Ghazali Talbi, University of Lille, France;

---

[1] https://decisionsciencealliance.org/.

[2] http://cis-lab.dmi.unict.it/.

- Fatos Xhafa, Technical University of Catalonia, Spain.

Their challenging plenary talks not only captivated and attracted the many participants but were a great source of inspiration in both academic and industrial scopes. The real success of ISC 2025 was to provide a unique, friendly and engaging environment that served as bridge between academic and industrial worlds, animated by a genuine desire to exchange knowledge and experiences; create new collaborations; and discuss the new shared challenges, and the future vision of AI-driven Decision Making in support of industry.

As program chairs and editors of this volume and on behalf of all organizing staff, we are proud of how ISC 2025 unfolded and of the excellent quality of each contribution received and included in this volume. Then our first sincere thanks go to all authors and participants who, with their enthusiasm, professionalism, and stimulating discussions, have enriched the scientific quality offered by ISC 2025, along with the three wonderful plenary speakers. We would like to express our sincere gratitude to the more than 60 sponsors who supported and enriched the entire event, without whom we wouldn't have been able to achieve any planned goals.

Finally, a special and big thank you for their important and enormous efforts in the organization and running of ISC 2025 goes to all members of the Decision Science Alliance, all members of the Complex Intelligent Systems research group, and all members of the organizing team in the many and several roles. Without all of them, who donated their time, expertise, enthusiasm, and dedication, it would not have been possible to offer the international community such a successful event as ISC 2025 turned out to be. As last, we would like to express our cordial appreciation to Roberto Cellini, Head of the Department of Economics and Business, and Orazio Muscato, Head of the Department of Mathematics and Computer Science, both at the University of Catania, for supporting and hosting ISC 2025, but primarily for trusting us to organize this successful scientific event.

June 2025

Mario Pavone
Carlos A. Coello Coello
Raffaele Cerulli
Salvatore Greco
El-Ghazali Talbi

Organization

General Chairs

Raffaele Cerulli	University of Salerno, Italy
Salvatore Greco	University of Catania, Italy
Vincenzo Cutello	University of Catania, Italy

Program Chairs

Mario Pavone	University of Catania, Italy
Carlos A. Coello Coello	CINVESTAV-IPN, Mexico
El-Ghazali Talbi	University of Lille, France
Angel A. Juan	Universitat Politècnica de València, Spain
Giovanni Righini	University of Milan, Italy

Proceedings Chair

Salvatore Corrente	University of Catania, Italy

Publicity and Social Chair

Carmine Cerrone	University of Genoa, Italy

Local Organizers

Carolina Crespi	University of Catania, Italy
Claudia Cavallaro	University of Catania, Italy
Francesco Zito	University of Catania, Italy
Alessio Mezzina	University of Catania, Italy
Carmine Sorgente	University of Salerno, Italy
Mario Lepore	University of Salerno, Italy

Program Committee

Alberto Ceselli	University of Milan, Italy
Andrea Schaerf	University of Udine, Italy
Antonio Mauttone	Universidad de la República, Uruguay
Christian Blum	Spanish National Research Council, Spain
Claudia Archetti	University of Brescia, Italy
David Álvarez-Martínez	Universidad de Los Andes, Colombia
David López López	ESADE Business School, Spain
Farouk Yalaoui	University of Technology of Troyes, France
Fatos Xhafa	Universitat Politècnica de Catalunya, Spain
Francesco Carrabs	University of Salerno, Italy
Francisco Salas-Molina	Universitat Politècnica de València, Spain
Fred Gardi	Hexaly, France
Gerarda Fattoruso	University of Foggia, Italy
Gian Luca Di Tanna	SUPSI, Switzerland
Günther Raidl	TU Wien, Austria
J. Alberto Conejero	Universitat Politècnica de València, Spain
Jairo Montoya-Torres	Universidad de La Sabana, Colombia
Javier Faulin	Public University of Navarre, Spain
Jin-Kao Hao	Université d'Angers, France
José-Luis Guisado-Lizar	University of Seville, Spain
Josefa Mula	Universitat Politècnica de València, Spain
Luca Di Gaspero	University of Udine, Italy
Luis Cadarso	University Rey Juan Carlos, Spain
Marc Bara	EAE Business School, Spain
Marc Escoto Gomar	Universitat Politècnica de València, Spain
Massimo Squillante	University of Sannio, Italy
Mauro Dell'Amico	University of Modena and Reggio Emilia, Italy
Paolo Toth	University of Bologna, Italy
Patricia Rodriguez	Universitat Oberta de Catalunya, Spain
Patrick Siarry	Université Paris-Est Créteil, France
Paula Carroll	University College Dublin, Ireland
Pau Fonseca	Universitat Politècnica de València, Spain
Rodrigo Linfati	Universidad del Bío-Bío, Chile
Silvano Martello	University of Bologna, Italy
Simone Martins	Universidade Federal Fluminense, Brazil
Thomas Stützle	Université Libre de Bruxelles, Belgium
Vittorio Maniezzo	University of Bologna, Italy

Contents

On the Acceptability of Computer Tools with Randomised Decision Making

Olivia Erdélyi[1], Gábor Erdélyi[1], and Vladimir Estivill-Castro[2](✉)

[1] University of Canterbury, Christchurch, New Zealand
{olivia.erdelyi,gabor.erdelyi}@canterbury.ac.nz
[2] Universitat Pompeu Fabra, Barcelona, Spain
vladimir.estivill@upf.edu

Abstract. Humans accept technology as a more impartial decision-maker in many environments, even when judgement involve moral decisions on other humans (although in some scenarios, human participation in the decisions is considered essential). Also, humans accept utilitarian justifications as suitable for implementation in software systems making moral decisions (despite utilitarianism is not widely accepted as the sole moral principle). Based on game theory, if machines are to make ethical decisions based on utilitarianism, while at the same time, machines themselves are not moral, optimal strategies would necessarily be mixed strategies. Thus, machines would be making randomised decisions. Randomisation challenges the long-standing notion that equates machines to deterministic artefacts. However, generative AI has propelled the acceptability of unexpected, divergent and randomised behaviour. We performed a study that shows that humans find it quite acceptable that machines could make moral decisions involving randomisation.

Keywords: AI Ethics · Moral decisions · Transparency · Game Theory

1 Introduction

Take sport as an example, judgement and decision-making involving humans are considered complex, involving several psychological issues and potentially highly subjective [5]. It is not surprising that the *video assistant referee system (VARS) for automated soccer decision-making* is considered an improvement to fairness [26]. Some conclude that *"there are no convincing arguments against the use of technological official aids in binary referee situations"* [31]. Major League Baseball in the USA is trialling a ball-strike assisting technology in the 2025 pre-season. People often perceive technology as a neutral and fair decision-maker.

Regarding moral judgements about other humans, machine-assisted decision-making is also perceived as an improvement. In morally complex scenarios, regulations demand that the ultimate ruler be human; but the literature points out the risks of human manipulation of the computer recommendations [16], suggesting even this ultimate human ruler introduces unfair bias.

M. Pavone et al. (Eds.): DSA ISC 2025, LNCS 16405, pp. 1–16, 2026.
https://doi.org/10.1007/978-3-032-21811-7_1

Variants of the trolley dilemma [20,55] are used to elicit humans' perceptions on moral human-robot interaction [3,35]. In scenarios involving autonomous vehicles (AVs), humans prefer machines to take a utilitarian approach to moral dilemmas in selecting the human victims [8,35]. Utilitarian reasoning is widely accepted in AI ethics due to its clear, operationalisable principles [21,30].

However, legislation, regulations and authorities attribute moral responsibility only to humans. For AVs, this position is confirmed by reports [7] recommending that any tragedy involving AV's would be attributed to humans only and not machines. In the case of traditional AI systems, the accountability or the responsibility falls with humans: the developer, the designer and ultimately *the person that uses the tool*" [1]. Even though it is now almost impossible to *"trace back responsibility for actions performed by distributed, hybrid systems of human and artificial agents"* [53], moral responsibility remains with the realm of humans, and any practice to avoid it is *"a moral wrong"* [53].

Therefore, following game theory, if machines make ethical decisions based on utilitarian principles while lacking inherent morality, their optimal approach would involve mixed strategies, leading to randomised decisions [17]. This challenges the traditional view of machines as purely deterministic.

Although the mechanisms by which humans trust and attach credibility to AI-generated information are complex [42], research [45] shows that medical students perceive Generative AI as credible, often *"clearer and more engaging"* [27] than human-generated information. Some claim that *"science written by generative AI is perceived as less intelligent, but more credible and trustworthy than science written by humans"* [38], although this is claim is based on the Natural Language Processing tools and Generative AI creation of summaries [37]. Thus, the rise of generative AI has increased the acceptance of unpredictable, divergent, and randomised behaviour. In this paper, we describe a study that indicates that humans generally find it acceptable for machines to make moral decisions involving randomisation.

We survey different groups of human participants, posing different scenarios where a software system or a human is to make a decision affecting people. We found different levels of conviction among the groups, but machines seem to be a preferable judge in several scenarios compared to humans. Nevertheless, in a few scenarios, humans are preferable. For instance, when judging the level of command of a foreign language, humans are regarded as more reliable than computers. Finally, we explore the issue of machines using randomisation when making their choices. We found that randomisation could be acceptable.

This paper is organised as follows. Section 2 gives more details on the matter of trust and credibility of artificial systems' decisions. Section 3 describes our empirical study and the Subsect. 3.2 focuses on highlighting the most significant results offering a discussion on the observed inclination of the responses. We conclude with final remarks in Sect. 4.

2   Motivation

Non-cooperative game theory studies decision-making by selfish, rational players each aiming to maximise their utility. In the current climate of concern for the socio-economic, moral and legal implications of artificial intelligence, where interpretability, explainability and trustworthiness are sought [52,53] the fundamental concept of game theory, namely the notion of Nash equilibrium, provides an ultimate level of transparency. In a Nash equilibrium, even if the strategy of all other players is fully disclosed to a player P, player P has no incentive to modify its strategy and that happens for all players P. While a Nash equilibrium always exists, it may only exist with mixed strategies.

While how users of a particular software gain trust in that application depends on many factors, it is indisputable that entire societies rely on computer technology for all sorts of activities. Globally, trust in recommendations by computers is high, many times surpassing those of human experts. Models about credibility name the phenomenon *the machine heuristic* [50], and consists of trusting computers' choices over those of human experts. Researchers have long analysed and found that the veracity attributed by humans to news selected by computers is higher than that of editors [51]. Credibility of the news and the media was as shaky on the eve of the US election in November 1952 as it is today. Moreover, the American public had no trust in the so-called "mechanical monsters" and "electronic brains" [11]. Nevertheless, several computer science and technology historians argue that the live broadcast on election night of a computer's remarkable performance with a sample of 5% sparked the fascination with number crunching, data analytics and the so-called informed decision-making we experience today.

This massive ability of our digital technologies to collect even more enormous data sets, combined with the techniques to generate predictive models, have resulted in the growth of the field of machine learning to the point that the general public identifies the applications of deep learning with the parent discipline of artificial intelligence (AI). Perhaps due to the drastic byproduct of avoiding moral responsibility. *"The technical complexity and dynamism of ML algorithms make them prone to concerns of 'agency laundering': a moral wrong which consists in distancing oneself from morally suspect actions, regardless of whether those actions were intended or not, by blaming the algorithm"* [53]. This is dangerously linked to human experts adopting computer output against their understanding. *"Today, the failure to grasp the unintended effects of mass personal data processing and commercialisation, a familiar problem in the history of technology, is coupled with the limited explanations that most ML algorithms provide. This approach risks favouring avoidance of responsibility through 'the computer said so' type of denial. This can lead field experts, such as clinicians, to avoid questioning the suggestion of an algorithm even when it may seem odd to them"* [53].

Naturally, there is a large body of research aimed at improving decisions by humans when incorporating computer assistance in decision making [12,48,

56]. However, when they believe there is AI, novel users attribute computers or applications with almost unconditional, or potentially additional, trust [49].

The trust in systems and applications on board digital computers is closely tied to the capacity of precision, and deterministic computation. Since 1952, computers have demonstrated reliability in obeying their programs and performing precisely the same behaviour, producing the same output for the same input. They function! They perform the mathematical notion of *function*. For each value in the domain, there is one and only one output in the co-domain. The software engineering method of test-driven development is currently a fundamental technique that, with version control systems and continuous integration, has accelerated software productivity while raising trust in computer software, eradicating failures by enabling traceability [46]. At the core of test-driven development is the elaboration of input-output pairs that synthesise large data sets of expected results.

The discipline of machine learning emerged from the *knowledge engineering bottleneck* that seemed to hamper the development of expert systems and recommender systems because human expertise was hard to transfer from humans to machines. The focus became on building such knowledge from data sets. The majority of supervised machine learning (ML) established knowledge views classifier development as the search for a mathematical function [6,23,40,41]. As a result, it is typically implemented using a deterministic algorithm that consistently generates the same output for a given input. *"The objective is to find an effective approximation, $f'(x)$, of the underlying function, $f(x)$, that governs the predictive relationship between inputs and outputs"* [23]. For unsupervised learning (or clustering), the algorithm is optimising a criterion that optimises a learning principle. The formulations are typically intractable and result in heuristics that may be caught in local optima, but conceptually, there is an optimal clustering—that is, one answer for the given input. Similarly, feedback-loop controllers are fundamentally defined as functions, where each specific input corresponds to a single output. In robotics, safety is often defined as a function ensuring robots maintain a safe distance and halt quickly at the stimulus of nearby people.

With the arrival of generative AI, the expectations for the output of systems have radically changed. There is not only one (expected or known) output to each input. Generative AI is expected to produce imaginative and genuine innovation. It is a creative machine, and these software systems are built to deliver never-seen-before text, images, music, code, and even video. Generative AI learns the distribution that fits its training data and uses that to sample such a distribution to generate new, original outputs imitating human constructs. Generative AI's output is mainly judged by how indistinguishable it is from what a human could have produced, and for that, there is the expectation of as much diversity of outputs as potential human authors.

3   Empirical Study

We developed a questionnaire to examine trust in technology versus distrust in human decision-makers without tech support, focusing on scenarios where AI and robotics might replace humans. Inspired by real-world AI use in judicial decisions [33], we anticipated general support for machine-assisted decision-making but expected differing views on full substitution, especially in morally impactful contexts. We surveyed three participant groups with varying familiarity with AI ethics. The study also explored views on deterministic versus random AI decision-making.

3.1   Method

Before the survey, we familiarised them with the *Equal Opportunity Principle* (EOP) [25] illustrating how computer systems could be provided with an ethical principle for their decisions, The EOP introduction (and questions and responses are on-line [15]) offers participants a relatively simple, underlying framework to avoid confrontation with the many fairness definitions against which classifiers' performance may be judged [25].

We recruited participants through four different methods. The first set of participants were Australian full-time employed high-school teachers holding a university degree who participated in a series of professional development workshops to include or expand the use of technology in their pedagogy. They came from 34 different schools with their levels of delivery including junior secondary, year 7 to 9, year 7 to 12, year 10, STEM P-6, year 9 to 12, VET, primary school, year 8 to 12, and K 6. Participation in the survey was an optional activity offered in the workshops. Of 65 invited participants, 53 replied to our questionnaire, and 12 declined. We will refer to this group of participants as *teachers* in Sect. 3.2.

The second means of recruitment was an online survey [35] using Survey-Monkey Inc.[1] through which we collected 124 responses. We limited the group of acceptable respondents since we believe the concepts and issues regarding the potential applications of AI in supporting and replacing human decision-making are not trivial, particularly regarding moral standing and fairness: (1) One excluded respondent class were children, as their views on robots significantly differ from those of adults [28]. (2) We selected respondents from the US who were above the age of 25, held a graduate degree and had a household income of $50K-$200k. Our choice of participant group was motivated by the fact that individuals' engagement in policy-making largely depends on their socialisation and education and we targeted the US because of its culture of deep and genuine citizen participation [4]. Many reports [2] highlight different

---

[1] The data was collected using SurveyMonkey Audience. We chose Age Balancing: Basic, Gender Balancing: Census, Country: United States (USA) - SurveyMonkey, Household income INCOME: $50k-$200k+, age: 25-100+, Education: Graduate degree. Information on how respondents are recruited to SurveyMonkey is available here: www.surveymonkey.com/mp/audience,.

levels of technology adoption in the US, but we assume that the minimum household income and their use of SurveyMonkey represent a homogeneous group. We note that other online experiments have a similar sample size and data collection approach for a generic view by humans [35]. We will refer to this group of participants as *online respondents* in Sect. 3.2.

Third, we performed robotics demonstrations on our campus. Our robots engaged visitors in conversation by randomly picking shorter survey questions. 17 visitors became curious and agreed to complete the survey. We will refer to this group of participants as *Open Day visitors* in Sect. 3.2.

Finally, we collected 13 surveys from IT academics who attended a department's research seminar. We will refer to this group of participants as *academics* in Sect. 3.2. All survey participants were provided with a consent and information sheet as per the conditions of our Research Ethics Committee's approval. Thus, participation was voluntary and anonymous and could stop at any time. For the SurveyMonkey collection, we ensured respondents could not go back and alter their answers for previous questions. The other data collection methods used printed questionnaire versions, so we cannot guarantee that people did not revise their answers. However, this is very unlikely as we found no amendments when capturing the answers.

Most of the questions in our questionnaire can be considered Likert-type scale questions (i.e., we give the survey participants five points they can choose from). We will avoid debating whether the collected data should be considered interval-level or ordered-categorical data whenever possible. As mentioned above, we will use a five-item option with a clear middle neutral point. Questions are symmetrical, and although potentially not equidistant, their presentation aimed for respondents to assume equidistance. We will use diverging stacked bar charts when presenting the descriptive statistic analysis of ordinal data [24] and bar charts in the remaining cases. Since we will be comparing different groups of respondents for the same question, the responses can not be paired (sample sizes are different), and we are not making any assumptions about the normality of the data. We have overlapping samples, which we will analyse for statistical difference of medians using the non-parametric Mann-Whitney U test[2] [13].

Given two random variables X and Y with their cumulative distribution functions f and g, respectively. For ordered samples $x_1, \ldots, x_n$ and $y_1, \ldots, y_m$ of X and Y, respectively, let the statistic U count the number of times a y value precedes an x value. The Mann-Whitney U Test uses the statistic U for testing the hypothesis $f = g$ [36]. The null hypothesis H_0 in our tests is going to be that the two cumulative distribution functions are the same (i.e., two different groups have the same scoring tendency). We will use the significance level $\alpha = 0.05$ in all our tests. Furthermore, we will specify the *p-value*, the probability (assuming that H_0 is true) that we obtain a test statistic that is at least as extreme as the observed value of the test statistic. If the *p*-value

---

[2] We replicated the calculation using the Mann-Whitney U Test Calculator from https://www.socscistatistics.com and R [43].

is less than the significance level, we reject the null hypothesis and accept the alternative hypothesis.

Our survey included several questions to gauge participants' acceptance of technological assistance. We also included questions to explore whether participants consider explanations necessary for the system's decisions. We do not discuss these results as they mostly show a diversity of familiarity with technology. All research instruments are available online (URL withheld because of double blind requirement).

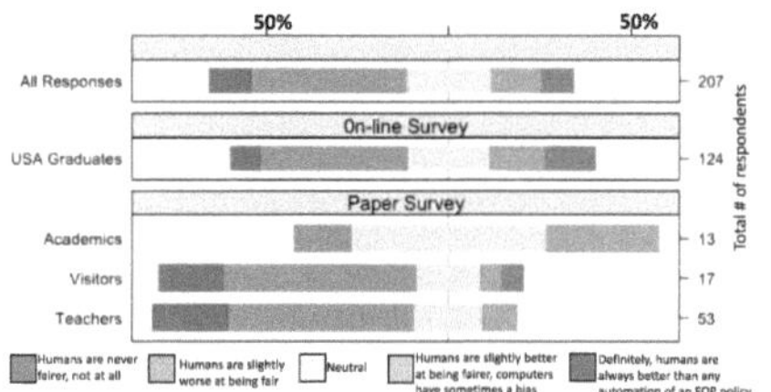

(a) Question 1: Who is fairer, humans or computers, to implement EOP.

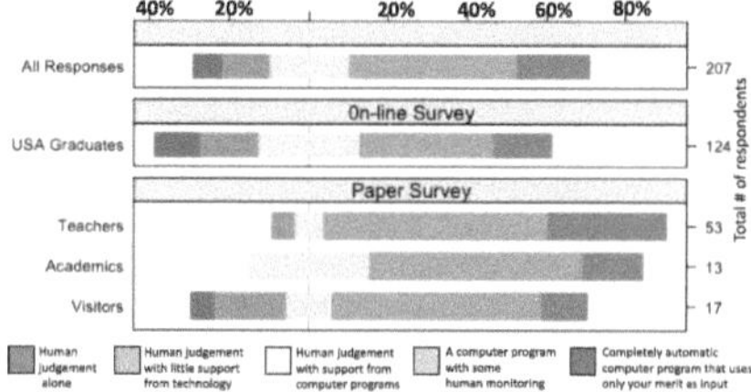

(b) Question 3: How to select which passenger should be left behind from a flight.

Fig. 1. Divergent bar charts of responses to Question 1 and Question 3

3.2 Analysis

Evidence for Preference for Automated Judgements The results here show that some participants' distrust of humans leads them to prefer AI over human decision-makers. Figure 1a shows the divergent bar chart of responses to Question 1 from our groups as percentages. Eleven teachers felt humans are never fairer, and none felt humans are better than automation. A small percentage (13.7%) of SurveyMonkey respondents believed that humans can be fairer, but 40.3% believed humans are slightly worse at being fair. Teachers' and visitors' median and mode indicated that humans were somewhat worse at being fair, while the academic view was neutral. The difference between teachers and academics is statistically significant ($U = 137$, $p = 0.00042$) and so is the difference between visitors and academics ($U = 49.5$, $p = 0.0057$).

Discussion: Figure 1a shows results for Question 1 and assesses individuals' level of trust in human decision-makers who are not supported by technology. The responses accumulate away from the use of human judgement alone. In this case, the results are consistent with the generally accepted view [33] that machines analyse facts without human bias, irrationality, or mistakes. Thus, teachers and Open Day visitors seem to perceive machines as more impartial than humans. Although this view may shift due to the growing recognition of the need for a human-in-the-loop approach [14], these groups of respondents do not seem to be dissuaded from automation.

Question 3 (refer to Fig. 1b), which concerns the automation of decisions on regrouping passengers across flights also shows a preference for AI decisions over human involvement. Teachers' median and mode are again for a computer program with some human monitoring. Not a single academic selected only human judgement. Open Day visitors' median and mode are also for this level of automation, and so is the online survey mode. The teachers' preference for automation significantly differs from that of online respondents ($U = 1921$, $p = 0.0001$) and Open Day visitors ($U = 302.5$, $p = 0.04338$).

The responses for Question 11 (see Fig. 2a) on automating visa applications show an almost identical pattern to Question 3. Responses are clearly in favour of automation and reject human judgement: Teachers' and Open Day visitors' median and mode unambiguously reflect this preference,

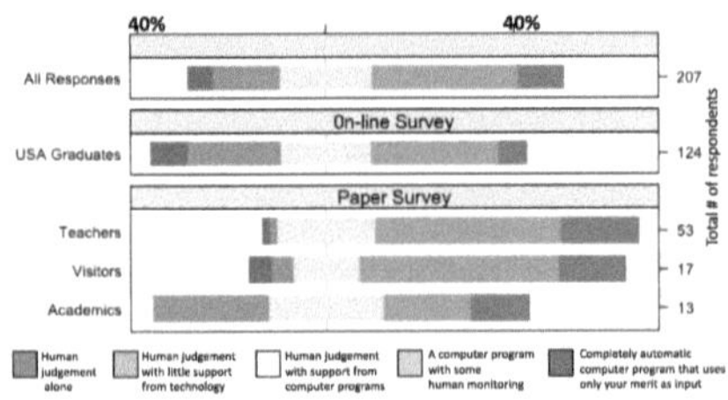

(a) Question 11: Who should evaluate a travel visa application, human or artificial agent.

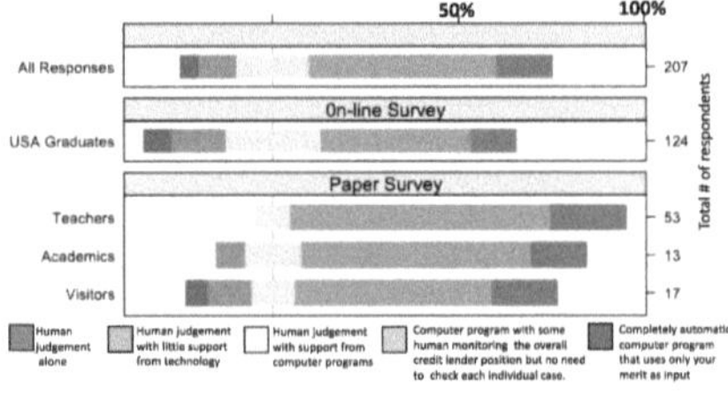

(b) Question 12: Who should decide over awarding a credit-card and its limit, humans or machines.

Fig. 2. Divergent bar charts of responses to Question 11 and Question 12

and the mode of online respondents is also for automation (the median is neutral). To a slightly lesser extent, academics also favour automation. Both Question 3 and Question 11 received a non-trivial number of responses on the extreme values, but mainly advocating fully automated decisions. Albeit in the opposite direction, we will see below the results for Question 2, which generally favours human judgement. But even in Question 2, there are also non-trivial numbers of those who would prefer the selection to be entirely without human intervention. The difference between the responses of teachers and online respondents is again statistically significant ($U = 1968$, $p = 0.0001$). But interestingly, here, Open Day visitors' high preference for an AI-judgement is statistically significant with respect to online respondents ($U = 696$, $p = 0.02382$).

Question 12—concerning decisions on credit-card applications—shows responses even more in favour of automation (refer to Fig. 2b). The median and mode for teachers, academics, and visitors indicate that ruling over credit-card applications should be a matter for computer programs with some human monitoring without checking each case. Almost all teachers favour automation, rejecting human judgement even if supported by technology. Online respondents, academics, and visitors also show a strong tendency towards automation: human judgement is less than 8% of the responses for all participants. All groups' preference for the use of technology; nevertheless, teachers are in favour of automation to such an extent that the difference between their views and those of online respondents is statistically significant ($U = 1967.5$, $p = 0.00001$).

Discussion: The high level of acceptance of automated decision-making over human judgement in this scenario may be due to the massive benefits of automation in the credit-card industry that pushed the early penetration of machine-learning technology: Machine learning has been applied to credit-card application assessments for more than 25 years [10]. In this setting, there have been calls for accepting technology with little questioning [44].

The results for Question 14 show central/neutral responses for Open Day visitors and online respondents (see Fig. 3a). But teachers show a statistically significant higher preference for automation over online respondents ($U = 2276.5$, $p = 0.00124$) and Open Day visitors ($U = 272.5$, $p = 0.0151$).

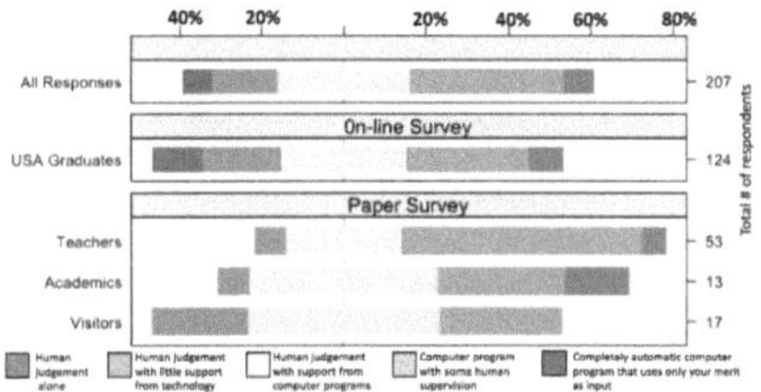

(a) Question 14: Who assesses fairly a funding application from a school, humans or artificial agents.

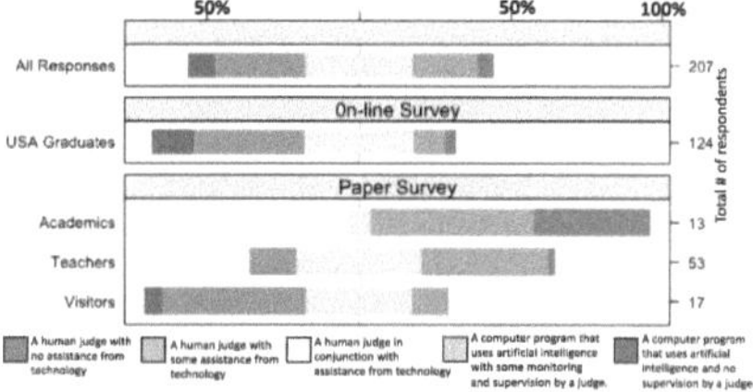

(b) Question 21: Who is fairer at specifying the sentences for a convicted criminal.

Fig. 3. Divergent bar charts of responses to Question 14 and Question 21

Question 21 (see Fig. 3b) shows the public's tendency towards centrality concerning automating sentencing decisions, with a very slight shift in favour of human intervention: Open Day visitors' median and mode, and online respondents' mode indicate a preference for technologically assisted human sentencing decisions. On the other hand, the median and mode of academics and teachers are clearly in favour of AI-enabled, automated decision-making with some monitoring and supervision by a judge. Teachers' preference for an AI-judge is statistically significant over that of online respondents ($U = 1778$, $p = 0.00001$) and Open Day visitors ($U = 228$, $p = 0.00236$). Academics' support for the AI-judge is even stronger: None of them contemplates a position below neutral. Hence, academics' preference for automated decisions in this setting is statistically significant not only concerning online respondents ($U = 123$, $p = 0.00001$), but also Open Day visitors ($U = 12$, $p = 0.00001$) and teachers ($U = 120.5$, $p = 0.00016$).

Discussion: The teachers were surveyed after a professional development day for the inclusion of ICTs in the teaching-learning process. Some of the technologies demonstrated and discussed were virtual-reality, robotics, and cloud tools. Therefore, they may hold high hopes for the automation of judgement. Alternatively, ICT academics and teachers may be more aware of works on bounded rationality—suggesting that cognitive limitations prevent humans from being fully rational [34,47]— and studies indicate that humans suffer from predictable biases that influence judgement [29]. These insights may encourage them to be more cautious and meticulous when evaluating and designing decision-making methods that may affect human beings. The fact that the systematic design of such methods is typically supported by computer programs may provide a further explanation for this group's high preference for automation.

Evidence for Preference for Human Involvement in Judgements. Let us now turn to the results showing that some participants exhibit distrust in AI decision-makers and hence prefer their human counterparts. Figure 4a shows the corresponding divergent bar chart of responses for Question 2, which aimed to determine the public's preferred level of automation in the context of a language test. The accumulated results did not swing either way: Teachers preferred some automation, and academics even more. Their mode and median were for a computer program with some human monitoring. Open Day visitors' views were in line with the online survey. The remarkable point here is the issue at the extremes: In all data collections, there are some non-trivial preferences for exclusively human judgement without technological support. Apart from academics, all other respondents rejected total automation.

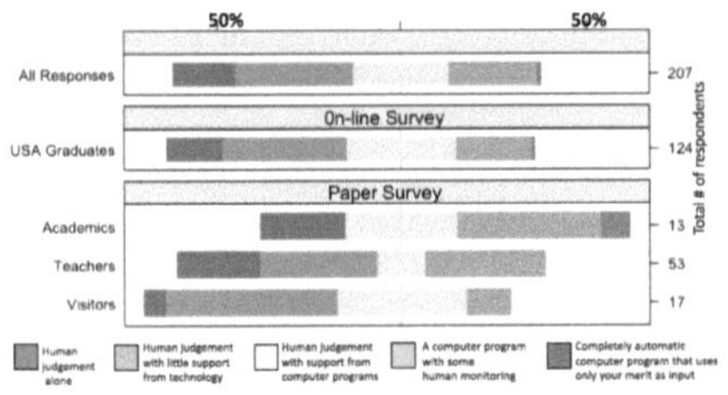

(a) Question 2: Who is fairer at evaluating command of foreign language.

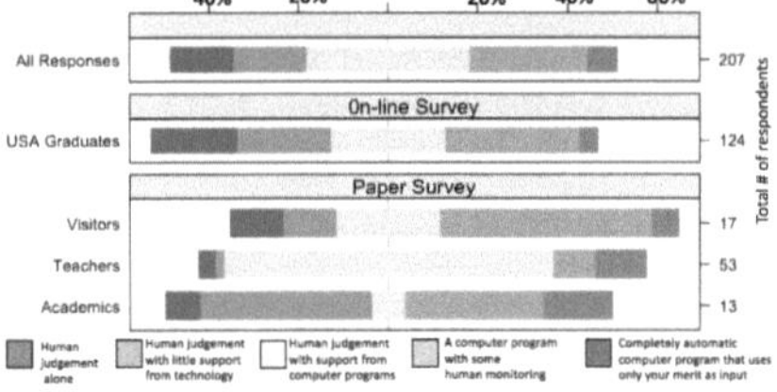

(b) Question 5: Would a minority person be judged fairly by human/artificial security officer.

Fig. 4. Divergent bar charts of responses to Question 2 and Question 5

Although the results we discuss next show a tendency for a neutral response—with some involvement of technology—they also indicate a clear desire by participants to maintain some human involvement, rejecting decisions by fully autonomous AI decision-makers. Question 5 considers a scenario in which a subject, who belongs to a visible minority, needs to request assistance. Some responses here show strong evidence that reliance on human judgement is unacceptable. The median and mode for both online respondents and Open Day visitors are that if they were a minority, they would prefer the involvement of computer programs with some human supervision. Teachers' and academics' responses to this question are remarkably neutral (refer to Fig. 4b).

Analysis as to Whether Randomisation Is Acceptable. For Question 13 and Question 20, we were expecting responses to favour automatic choice since the questions suggest that a sensible choice is a random choice. Another reason for this expectation was that our earlier questions (Question 1 and Question 3) show that our participants do regard computers as effective in calculations. As we argued earlier, a fair choice is random if the case's merit between two humans cannot be determined. These two questions have been worded to suggest that computers would be of valuable assistance even when making random choices. We find responses to Question 13 (see Fig. 5a) are not inclined to a particular level of machine intervention. The results indicate a preference for computer involvement in such randomised decision-making alongside significant human participation. Only for the teachers, the median is to use a computer's random choice and this is statistically significant over online respondents responses ($U = 2,040$ and $p = 0.0003$).

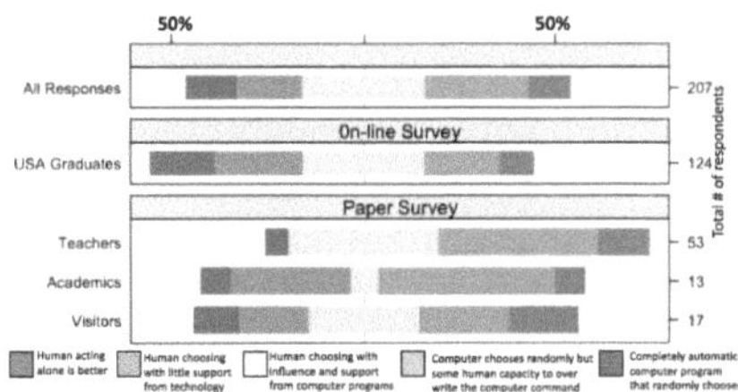

(a) Question 13: If selecting which pedestrian to harm, what provides better randomness.

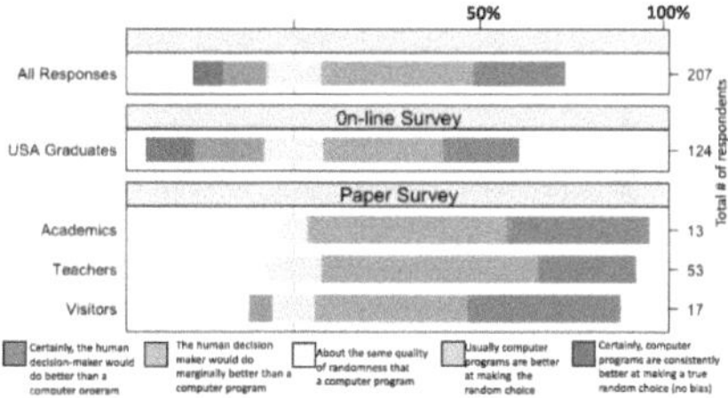

(b) Question 20: If selecting which passenger to eject from a flight, what provides better randomness.

Fig. 5. Divergent bar charts of responses to Question 13 and Question 20

Question 20 recalls the scenario of Question 3, where airline staff eject some passengers from a plane. The scenario now insinuates that a random choice is required, and the responses here favour a strong use of computers in sharp contrast to the results we just described for Question 13. In this setting, only less than 13% of the online respondents chose the human alone option. Teachers, Open Day visitors, and academics decisively favour computer use. The option where the human decision-maker would do marginally better than a computer program received no responses from the teachers, no responses from the academics, and only one from the Open Day visitors. Comparing teachers' responses to Question 20 with those to Question 13, we find they are statistically significant ($U = 923$ and $p = 0.0118$). Similarly, academics' responses to Question 20 and Question 13 ($U = 38$ and $p = 0.0914$) and Open Day visitors' responses to Question 20 and Question 13 ($U = 78$ and $p = 0.116$) are also statistically significant.

Discussion: Question 13 concerns a scenario where a decision results in a human being suffering some physical harm. Under such circumstances, exclusively human and fully automated decisions are the least preferred. However, the three groups (academics, teachers, and visitors) display an absolute and radical shift in favour of computer/machine randomness when the issue (Question 20) does not seem to be a matter of life or death. Human preferences for AI decisions seem to depend on the issues at stake. Question 24 offers a Likert scale with the standard options (Strongly Disagree, Disagree, Neutral, Agree, Strongly Agree) regarding the respondents' belief that computers can obtain fair random numbers for an unbiased choice when equal merit. Figure 6 shows that respondents strongly believe that computers can reproduce a truly random generator.

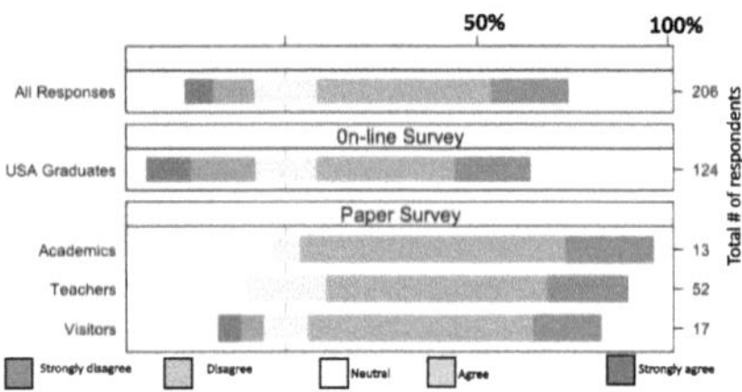

Fig. 6. Responses to Question 24: Computers would be able to generate random numbers to make fair decisions when the merits of two people are equal

Although online respondents have a median and a mode that agree on computers' ability to achieve randomness, from a Mann-Whitney test we get a p-value of 0.2183. Thus, we can reject the null hypothesis that the online respondents and the Open Day visitors have the same scoring tendency at the 5% level. Teachers and academics clearly believe even stronger (see Fig. 6) in achieving fair decisions with random generators in computers. It is not surprising that respondents differ when discussing randomness, seeing as there are several accepted mathematical approaches on the issue [54].

Finally, Question 25 and Question 26 offer only yes/no options. Question 25 aimed to find out if participants are aware of CPUs' capacity to use thermal noise within the silicon as an entropy source. Overwhelmingly, our respondents answered no. Only two of the academics were aware of this issue. Question 26 tested for participants' awareness of the controversy around replacing the random oracle model with cryptographic hash functions. Again, positive responses were minimal (below 1%) except for the academics, where 2 in 13 (i.e., 15%) indicated their knowledge of the random oracle controversy.

Discussion: Overall, it would seem that the public has trust in machines exhibiting non-deterministic and random behaviour with what seems like little understanding of implementation issues. Indeed, people are generally not particularly bothered by how machines work as long as their use is socially considered an improvement: *"Risks do not militate against the introduction of automated driving if the balance of risks is fundamentally positive"* [7].

4   Conclusion

This paper has argued that if AI systems are involved in automating decisions affecting human lives, such decisions would need to be randomised for many scenarios. We also noted that this would revolutionise the notion of machine (automata): Determinism and deductive reasoning would alternate with random choices of carefully learned distributions. AI systems would give the impression of possessing some free spirit despite consistently following the law and making information based decisions. Importantly, their choices would be auditable, explainable and rationalisable (in the sense of game theory) and justifiable (in human interpretability). Most people could grow comfortable with AI decisions, perceiving them as fairer and less biased than human decision-makers. Others would likely find the idea of giving chance such a central role in decisions that so crucially affect their lives highly unsatisfactory.

Our study provides empirical evidence that there is some level of acceptance for using randomness in algorithms for decision-making. While people prefer automated decisions over human ones for many scenarios, they hesitate to rely on randomised decisions in life-or-death situations. Additionally, our research highlights policy and implementation challenges related to transparency requirements and the difficulty of generating truly random sequences, which are essential considerations in AI policy and standardisation discussions.

International AI policy and standardisation initiatives increasingly attempt to guarantee AI systems' highest possible transparency and explainability to ensure accountability. While this is a commendable goal, our key message to policymakers is that maximising transparency is neither the only nor necessarily the best solution. Instead, we need to consider the potential drawbacks of transparency and explore alternative accountability mechanisms that allow us to garner the benefits of randomisation.

The discussion here raises a series of other interesting questions. For instance, do our findings also hold for non-supervised learning tasks? On the game theory side, we note that AI interactions would fall in the realm of repeated games from a game-theoretic perspective, and that human agents would be inclined to game the system, if transparency requirements reveal the AI is deterministic. Human agents would adapt. Should the AI adapt as well? Should it include ML modules for continuous learning? While the examples could continue, we hope to have sparked some interest in the scientific and policymaking communities to engage with these and many more related open problems.

References

1. Acharya, D.B., Kuppan, K., Divya, B.: Agentic AI: autonomous intelligence for complex goals-a comprehensive survey. IEEE Access **13**, 18912–18936 (2025)
2. Anderson, M., Kumar, M.: Digital divide persists even as lower-income Americans make gains in tech adoption. Fact Tank (2019). https://pewrsr.ch/2vK1HIo
3. Angelopoulos, G., Estivill-Castro, V.: Human-robot dialogue that elicits the alignment of moral principles for driverless vehicles. ACM/IEEE International Conference on Human-Robot Interaction, HRI 2024, pp. 196–200. ACM (2024)

4. Australia's Parliament: Citizens' engagement in policymaking and the design of public services. Tech. Paper 1 2011–12, Canberra (2012)
5. Bar-Eli, M., Plessner, H., Raab, M.: Judgment, decision-making and success in sport. Wiley (2011)
6. Bishop, C.M.: Pattern Recognition and Machine Learning (Information Science and Statistics). Springer, Berlin (2006)
7. BMVI: Ethics commission automated and connected driving. Tech. rep., Federal Ministry of Transport and Digital Infrastructure, Germany (2017). www.mbdi.de
8. Bonnefon, J.F., Shariff, A., Rahwan, I.: The social dilemma of autonomous vehicles. Science $\mathbf{352}$(6293), 1573–1576 (2016)
9. Buolamwini, J., Gebru, T.: Gender shades: intersectional accuracy disparities in commercial gender classification. In: Conference on Fairness, Accountability and Transparency, FAT 2018, vol. 81, pp. 77–91. PMLR (2018)
10. Carter, C., Catlett, J.: Assessing credit card applications using machine learning. IEEE Expert $\mathbf{2}$(3), 71–79 (1987)
11. Chinoy, I.: Predicting the Winner: The Untold Story of Election Night 1952 and the Dawn of Computer Forecasting. Potomac Books, U of Nebraska Press (2024)
12. Cummings, M.L.: Automation bias in intelligent time critical decision support systems. Decision making in aviation, pp. 289–294. Routledge (2017)
13. Derrick, B., White, P.: Comparing two samples from an individual Likert question. Int. J. Math. Statist. $\mathbf{18}$(3), 1–13 (2017)
14. EC: Ethics guidelines for trustworthy AI. Tech. rep., European Commission High-Level Expert Group on Artificial Intelligence — AI HLEG, B-1049 Brussels (2019)
15. Erdélyi, G., Erdélyi, O.J., Estivill-Castro, V.: Randomized classifiers vs human decision-makers: trustworthy AI may have to act randomly and society seems to accept this. CoRR abs/ arXiv: 2111.07545 (2021)
16. Esthappan, S.: Assessing the risks of risk assessments: Institutional tensions and data driven judicial decision-making in U.S. pretrial hearings. Social Problems (2024)
17. Estivill-Castro, V.: Game theory formulation for ethical decision making. Robotics and Well-Being. Intelligent Systems, Control and Automation: Science and Engineering, vol. 95, pp. 25–38. Springer (2019)
18. Feldman, M., Friedler, S.A., Moeller, J., Scheidegger, C., Venkatasubramanian, S.: Certifying and removing disparate impact. ACM SIGKDD International Conference on Knowledge Discovery and Data Mining, KDD, pp. 259–268. ACM, NY, USA (2015)
19. Fishkin, J.: Bottlenecks: A New Theory of Equal Opportunity. Oxford University Press, Oxford (2014)
20. Foot, P.: The problem of abortion and the doctrine of the double effect. Oxford Rev. $\mathbf{5}$, 5–15 (1967)
21. Grau, C.: There Is No "I in "Robot: Robots and Utilitarianism. IEEE Intell. Syst. $\mathbf{21}$(04), 52–55 (2006)
22. Hardt, M., Price, E., Srebro, N.: Equality of opportunity in supervised learning. In: International Conference on Neural Information Processing Systems, NIPS , pp. 3323–3331 (2016)
23. Hastie, T., Tibshirani, R., Friedman, J.: The Elements of Statistical Learning: Data Mining, Inference, and Prediction. Springer, 2nd edn. (2017)
24. Heiberger, R., Robbins, N.: Design of diverging stacked bar charts for Likert scales and other applications. J. Statist. Softw. $\mathbf{57}$(5), 1–32 (2014)

25. Heidari, H., Loi, M., Gummadi, K.P., Krause, A.: A moral framework for understanding fair ML through economic models of equality of opportunity. In: Conference on Fairness, Accountability, and Transparency, FAT, pp. 181–190. ACM, NY (2019)
26. Held, J., Cioppa, A., Giancola, S., Hamdi, A., Ghanem, B., Van Droogenbroeck, M.: VARS: video assistant referee system for automated soccer decision making from multiple views. In: IEEE/CVF Conference on Computer Vision and Pattern Recognition, pp. 5086–5097 (2023)
27. Huschens, M., Briesch, M., Sobania, D., Rothlauf, F.: Do you trust ChatGPT? – perceived credibility of human and ai-generated content (2023). https://arxiv.org/abs/2309.02524
28. Kahn, P., Jr., Friedman, B., Hagman, J.: I care about him as a pal: conceptions of robotic pets in online Aibo discussion forum, CHI, pp. 632–633. Poster pp (2002)
29. Kahneman, D., Tversky, A.: Prospect theory: an analysis of decision under risk. Econometrica **42**(2), 263–292 (1979)
30. Karnouskos, S.: The role of utilitarianism, self-safety, and technology in the acceptance of self-driving cars. Cognition Technol. Work **23**(4), 659–667 (2021)
31. Kjærsgaard, T.: Defending technology: a normative defence of technologically assisted officiating in binary referee situations, Sport, Ethics and Philosophy, pp. 1–13 (2024)
32. Kleinberg, J., Mullainathan, S., Raghavan, M.: Inherent trade-offs in the fair determination of risk scores. In: 8th Innovations in Theoretical Computer Science Conference, pp. 43:1–43:23 (2017)
33. Kugler, L.: AI judges and juries. Commun. ACM **61**(12), 19–21 (2018)
34. Lindblom., T., Mavruk, T.S.S.: Decision-making–rational, bounded, or behavioral. Proximity Bias in Investors' Portfolio Choice, pp. 35–60 (2017)
35. Malle, B.F., Scheutz, M., Arnold, T., Voiklis, J., Cusimano, C.: Sacrifice one for the good of many? People apply different moral norms to human and robot agents. In: ACM/IEEE International Conference on Human-Robot Interaction, HRI, pp. 117–124. ACM, NY, USA (2015)
36. Mann, H.B., Whitney, D.R.: On a test of whether one of two random variables is stochastically larger than the other. Ann. Math. Stat. **18**(1), 50–60 (1947)
37. Markowitz, D.M.: From complexity to clarity: How AI enhances perceptions of scientists and the public's understanding of science (2024). https://arxiv.org/abs/2405.00706
38. Markowitz, D.M.: Science written by generative AI is perceived as less intelligent, but more credible and trustworthy than science written by humans (2024). https://hackernoon.com/study-finds-generative-ai-appears-less-intelligent-yet-more-credible-than-humans-in-science-writing
39. McCrudden, C.: Merit principles. Oxford J. Legal Stud. **18**(4), 543–579 (1998)
40. Michie, D., Spiegelhalter, D.J., Taylor, C.C., Campbell, J. (eds.): Machine Learning, Neural and Statistical Classification. Ellis Horwood, USA (1995)
41. Mitchell, T.M.: Machine Learning, 1st edn. McGraw-Hill Inc, USA (1997)
42. Ou, M., Zheng, H., Zeng, Y., Hansen, P.: Trust it or not: understanding users' motivations and strategies for assessing the credibility of AI-generated information, p. 14614448241293154. New Media & Society (2024)
43. R Core Team: R: A Language and Environment for Statistical Computing. R Foundation for Statistical Computing, Vienna (2013)
44. Rosenberg, R.S.: The social impact of intelligent artefacts. AI & Soc. **22**(3), 367–383 (2008)

45. Sami, A., et al.: Medical students' attitudes toward AI in education: perception, effectiveness, and its credibility. BMC Med. Educ. **25**(1), 82 (2025)
46. Sheta, S.V.: The role of test-driven development in enhancing software reliability and maintainability (2023)
47. Simon, H.A.: Models of man, social and rational: mathematical essays on rational human behavior in a social setting. John Wiley & Sons, New York, NY (1957)
48. Skitka, L.J., Mosier, K.L., Burdick, M.: Does automation bias decision-making? Int. J. Hum Comput Stud. **51**(5), 991–1006 (1999)
49. Solans, D., Beretta, A., Portela, M., Castillo, C., Monreale, A.: Human response to an AI-based decision support system: a user study on the effects of accuracy and bias. CoRR abs/ arXiv: 2203.15514 (2022)
50. Sundar, S.S.: The MAIN Model: A Heuristic Approach to Understanding Technology Effects on Credibility, p. 73–100. MIT Press, Cambridge, MA (2008)
51. Sundar, S.S., Nass, C.: Conceptualizing sources in online news. J. Commun. **51**(1), 52–72 (01 2006)
52. Tsamados, A., et al.: The ethics of algorithms: Key problems and solutions. Ethics, Governance, and Policies in Artificial Intelligence, pp. 97–123. Springer, Cham (2021)
53. Tsamados, A., et al.: The ethics of algorithms: key problems and solutions. AI & Soc. **37**(1), 215–230 (2022)
54. Volchan, S.B.: What is a random sequence? Amer. Math. Monthly **109**, 46–63 (2002)
55. Wu, S.S.: Autonomous vehicles, trolley problems, and the law. Sci. Eng. Ethics **22**(1), 1–13 (2020)
56. Zhang, Y., Liao, Q.V., Bellamy, R.K.E.: Effect of confidence and explanation on accuracy and trust calibration in AI-assisted decision making. In: Conf. on Fairness, Accountability, and Transparency, FAT, pp. 295–305. ACM, NY, USA (2020)

Exploring Different Parameters in the Design of Cat-in-a-Grid Tropical Cyclone Parametric Solutions

Marc Escoto[1]($\boxtimes$) , Wenwen Chen[1] , Roberto Guidotti[2] ,
Laura Verderame[2] , and Guillermo Franco[2]

[1] Research Center on Production Management and Engineering, Universitat Politècnica de València, Ferrandiz-Carbonell, 03802 Alcoy, Spain
{mescgom,cwenwen}@upv.es
[2] Guy Carpenter & Company, LLC, 1166 Avenue of the Americas, New York 10036, NY, USA
{roberto.guidotti,guillermo.e.franco}@guycarp.com

Abstract. Parametric solutions emerged in the 1990s as an alternative to the traditional insurance solutions. Parametric solutions offer a faster and more efficient risk transfer approach, not relying on lengthy claims process, but on pre-agreed payouts triggered by specific parameters. This paper explores a regression-based methodology to support the design of cat-in-a-grid tropical cyclone parametric solutions, considering different tropical cyclone characteristics, typically publicly available in the aftermath of the event. The paper specifically investigates different single-metric parametric solutions for Jamaica based on maximum wind speed, minimum central pressure, and radius of maximum wind speed, respectively. However, the methodology is general and can be applied to different regions and hazards. Results show that parametric solutions based on maximum wind speed and minimum central pressure provide good accuracy metrics and a satisfactory agreement with loss estimates for historical events, whereas parametric solutions based on the radius of maximum wind speed have limited predictive power, suggesting a better use of such parameter as a complement of maximum wind speed and/or minimum central pressure in multi-metric parametric solutions.

Keywords: Parametric Tropical Cyclone Risk Transfer · Cat-in-a-grid Parametric Solution · Accuracy Estimation

1 Introduction

Parametric insurance has emerged as a practical and efficient way to manage the financial impact of natural disasters, serving as either an alternative or complement to tradition indemnity-based insurance. Traditional indemnity-based insurance requires detailed damage assessments after an event to determine the payout [15,16]. Parametric insurance, on the other hand, provides fixed compensation when a specific physical parameter, such as wind speed, in the case

M. Pavone et al. (Eds.): DSA ISC 2025, LNCS 16405, pp. 17–32, 2026.
https://doi.org/10.1007/978-3-032-21811-7_2

of tropical cyclones, or peak ground acceleration, in the case of earthquakes, exceeds a predefined threshold [1–3]. These parameters are often available in near real-time, allowing for faster claims processing and simpler operation [14]. Parametric insurance maintains transparency between insured and insurer by defining clear payout criteria based on publicly available data, such that any response could be understood and verified by all parties involved. Transparent mechanisms help to build trust among stakeholders, including insurers, governments and affected communities.

Despite these advantages, designing an effective parametric insurance program remains challenging. A key issue is the basis risk, the difference between the actual compensation received from the parametric policy and the compensation expected on the basis of the sustained losses. The ultimate goal of any design of parametric solutions is to minimize such basis risk. Specifically, in this paper, we explore cat-in-a-grid tropical cyclone risk transfer designs and, to maintain the transparency of the trigger, investigate using parameters such as maximum wind speed, minimum central pressure and radius of maximum wind speed, which are typically publicly available near real-time in the aftermath of an event. Our goal is to explore how using different parameters could improve the predictive accuracy of such parametric solutions, while maintaining transparency and interpretability. Following the formulation described in [4], the main contributions of this work are the following: (i) to propose and compare tropical cyclone parametric design relying on three different parameters; (ii) to explore their predictive power based on data split between training and testing data; and (iii) to apply and validate these approaches to a set of relevant historical events for Jamaica.

The remainder of the paper is organized as follows. Section 2 outlines the problem setting, discussing the methodological aspect of the problem, introducing alternative parameters and accuracy metrics used in the paper. Section 3 discusses the findings for Jamaica: the performance of the parametric solutions is assessed based on accuracy metrics and their predicted power is evaluated using a train-test split. The historical performance of the calibrated parametric solutions is assessed in Sect. 4. Finally, Sect. 5 offers concluding remarks and directions for future research.

2   Tropical Cyclone Parametric Solutions

The "cat-in-a-box", and its variant, the "cat-in-a-circle" are the most widely utilized parametric triggers currently used in the industry. These triggers depend on predefined spatial boundaries—rectangular (box) or circular (circle)—to ascertain whether a qualifying event, such as a tropical cyclone of a certain intensity, has occurred. Common metrics like maximum wind speed (MWS) or minimum central pressure (MCP) are often employed to trigger these solutions due to their strong correlation with cyclone intensity and their availability from standardized observational or modeled data sources [12].

These parametric structures are known as first-generation triggers, which utilize broad event-level parameters to determine payout eligibility. First-generation triggers present several practical benefits, despite their higher exposure to basis risk: they are easier to design, quicker to settle, more transparent, and more amenable to standardization in global markets [4,11]. First-generation triggers can be categorized into single-domain and multi-domain triggers based on their spatial complexity. Single-domain triggers are simple and intuitive, ideal for insuring individual properties or localized assets. They include traditional cat-in-a-box and cat-in-a-circle structures, triggering payouts when hazard parameters exceed defined thresholds within the specified area. However, these triggers lack spatial flexibility and may not capture the full extent of hazards across broader regions.

To address these limitations, as discussed in [4], multi-domain triggers have emerged, featuring more complex spatial configurations—such as multiple boxes or grid-based arrangements—that better represent hazard dynamics over large, varied areas. This work focuses on the cat-in-a-grid method, examining its potential to enhance first-generation parametric insurance products by improving spatial resolution while maintaining simplicity, speed, and transparency.

2.1   Design of Cat-in-a-Grid Tropical Cyclone Parametric Solutions

The parametric design illustrated in this paper is based on the cat-in-a-grid parametric trigger framework proposed by Franco et al. [4]. This approach reduces the basis risk and improves the accuracy of the payments by extending the area of coverage beyond that typically considered for a cat-in-a-box trigger and partitioning the area into a regular grid of cells. Each cell has its own trigger function, which is calibrated using simulated tropical cyclone data, extracted from a catastrophe model (see e.g. Grossi et al. [6]). Simulated tropical cyclone data typically include the physical characteristics of the stochastic tropical cyclone (i.e. the latitude and longitude of its track points), the relevant intensity measures (i.e. the maximum wind speed, the minimum central pressure, and the radius of maximum wind speed simulated at each track point), and the losses it causes to the considered exposure. Once a real event strikes, each cell intersected by the tropical cyclone track provides a loss estimate (or index), based on the cell-specific trigger function; the estimates from each cell are then combined using a weighted average, where better-performing cells (the ones with lower variability) have higher weights. Readers are referred to [4] for a detailed explanation of the implementation. A summary of the main modeling steps is outlined below for completeness, with reference to maximum wind speed as parameter of interest.

  i. For each event, the maximum wind speed within each grid cell crossed by the event is determined and associated with the total loss produced by the event. The simulated events are then binned into wind speed intervals. For each interval, the median wind speed and the corresponding median loss are calculated to form representative training points.
 ii. These median points are used to train the loss function, which is represented by a cubic polynomial. A threshold is applied at a certain value of wind speed, below which the storm is assumed to cause no loss. This helps reduce

the influence of outliers and creates a smoother, more reliable loss estimation curve for each grid cell.

iii. The mean squared error (MSE) between the predicted and modeled losses is calculated. Weights are assigned to each cell based on its relative accuracy, defined as the inverse ratio of the MSE of the predicted results to the MSE of a null prediction.

iv. The loss estimation function is evaluated using only the events where the cyclone track intersects that cell. As a result, only the cells affected by a given event contribute to the aggregated trigger. The local loss estimators are then combined into a global parametric trigger using a weighted average.

v. The payout is calculated based on where the estimated loss falls relative to the policy's attachment and exhaustion level. If the loss estimate comes in below the attachment level, no payout is triggered. When losses exceed the attachment level but remain below the exhaustion, a partial payout is disbursed. In cases where the loss estimate surpasses the exhaustion level, the full limit payout is disbursed.

2.2   Alternative Tropical Cyclone Parameters

This study focuses on exploring different tropical cyclone parameters as possible alternatives to MWS, and to test their predictive power or overfitting risk. While MWS is typically used in parametric trigger systems due to its direct correlation with cyclone-induced losses, other easily accessible metrics can also be used. Among these, minimum central pressure (MCP) and the radius of maximum wind speed (RMW) stand out due to their immediate observational availability and physical relationship to a cyclone's destructive potential.

MCP serves as a robust indicator of cyclone intensity, with lower pressures consistently linked to stronger systems and broader wind fields. The MCP-based design follows a similar grid-based framework as the MWS approach, but with key adaptations to account for MCP's inverse relationship with cyclone intensity. Unlike wind speed, lower pressure values indicate stronger storms, requiring two modifications to the local loss function calibration:

- Inverse-Reflection Transformation: Since loss potential increases as MCP decreases, the training data is transformed via x-axis reflection. For a cell S_k with central pressure p_k, we define the reflected value as $p_k^{\text{ref}} = p_{\max} - p_k$, where $p_{\max}$ is the highest observed pressure in this cell. This ensures the transformed variable p_k^{ref} behaves analogously to wind speed.
- Modified Local Loss Function: The cubic polynomial is fitted to reflected pressures, with thresholds adjusted accordingly.

The RMW measures the distance from a tropical cyclone's center to the location of its strongest winds, generally found at the inner edge of the eyewall in well developed cyclones. Whereas losses can definitely occur outside the RMW, this distance is related to the area extend of extreme storm hazard, including storm surge. Even if the RMW of tropical cyclone is an important factor for tropical cyclone intensity estimation and disaster prevention, the physics mechanisms that associates the RMW to other cyclone's characteristics are still object of research (see, e.g., Ren et al. [13]). As a result, even if -in the extend of this paper- we are treating this parameter in the same way of the other parameters, the correlation between RMW and loss is usually less strong than the correlation between MWS and loss or MCP and loss, making the use of such parameter in isolation more challenging in the design of parametric solutions.

2.3   Accuracy Metrics

In this paper, we consider a number of evaluation metrics, namely the mean absolute error (MAE), the mean squared error (MSE), and the coefficient of determination (R^2). The MAE indicates the average of the residuals in the dataset, i.e., the average of the difference in absolute terms between the predicted and target values. The MSE assesses the variance of the residuals, given by the square root of the average of the squared difference between predicted and target values. Finally, R^2, between 0 and 1, measures the proportion of the variance in the dependent variable that is explained by the adopted model. In general, the lower the MAE and MSE values, the higher the solution's accuracy; higher R^2 values indicate that the independent variables in the model explain well the variability in the dependent variable and are thus more desirable. The following equations illustrate how the aforementioned evaluation metrics are computed for n observations: y_i represents the true value of the target variable, $\hat{y}_i$ is the predicted value, and $\bar{y}$ is the average of value of the target variable.

$$MAE = \frac{\sum_{i=1}^{n} |y_i - \hat{y}_i|}{n} \tag{1}$$

$$MSE = \frac{\sum_{i=1}^{n} (y_i - \hat{y}_i)^2}{n} \tag{2}$$

$$R^2 = 1 - \frac{\sum_{i=1}^{n} (y_i - \hat{y}_i)^2}{\sum_{i=1}^{n} (y_i - \bar{y})^2} \tag{3}$$

In the context of catastrophe models, where the dependent variable is the loss to a particular asset or portfolio induced by a natural event, it is of interest to assess the amount of loss that would be exceeded with a certain return period. It is therefore common the use of exceedance probability (EP) curves to predict the monetary impact of extreme events, such as earthquakes and hurricanes, on a considered portfolio [9]. Therefore, in addition to the three accuracy evaluation metrics introduced above, in this paper, we compare the EP curves obtained with the different parametric solutions against the EP curve obtained with the reference catastrophe model.

3   Parametric Tropical Cyclone Risk Transfer for Jamaica

The tropical cyclone catastrophe risk model for Jamaica used in this paper to calibrate the different parametric solutions has been provided by Risk Engineering + Development (RED). The probabilistic tropical cyclone risk model consists of thousands of synthetic tropical cyclone events representing an equivalent catalog length of 10,000 years. Each event has associated a loss value to the considered exposure in Jamaica, and -at 6-hour time steps- a value of center longitude, center latitude, MWS, MCP and RMW. More details on the RED model used in this paper, and specifically on its modules (hazard, exposure, vulnerability and loss computation) can be found in [4].

Table 1. Correlation matrix of MWS, MCP, and RMW parameters.

Parameters	MWS	MCP	RMW
MWS	1.00	-0.96	-0.23
MCP	-0.96	1.00	0.24
RMW	-0.23	0.24	1.00

We performed the correlation analysis for MWS, MCP, and RMW after aggregating the data by event and grid cell. Correlations were computed across this aggregated dataset to reduce noise from temporal variability and better capture the spatial relationships among the variables. As shown in Table 1, MWS and MCP are highly negatively correlated ($r = -0.96$), consistently with the physical inverse relationship between wind intensity and central pressure. The RMW shows a weak negative correlation ($r = -0.23$) with the MWS and a weak positive correlation ($r = 0.24$) with the MCP, reflecting the complex interaction between this tropical cyclone characteristic and the other two parameters.

3.1   Parametric Design Based on Different Parameters

We designed three different parametric solutions following the methodology described in Sect. 2, one for each tropical cyclone parameter: MWS, MCP and RMW. Table 2 shows a comparison of the accuracy metrics presented in Sect. 2.3 for the three parametric solutions, each one based on a different parameter. We can observe that the performance of the RMW-based parametric solution is lower than the performance of the other two solutions based on MWS and MCP, respectively. This result aligns with the earlier observation that the RMW may pose challenges when used in isolation in the design of tropical cyclone parametric solutions.

When comparing the MWS-based and MCP-based parametric solutions, their performance is relatively close. The MCP solution slightly outperforms the MWS one in terms of MAE. However, the MWS solution demonstrates

Table 2. Accuracy evaluation metrics for the different parametric solutions.

Parameters	MAE	MSE	R^2
MWS	9.196e6	2.678e16	0.595
MCP	9.024e6	2.982e16	0.548
RMW	1.791e7	6.512e16	0.015

slightly better results in both the MSE and the R^2 metric. This combination of a lower MAE but higher MSE suggests that while the solution tends to make smaller errors on average, it occasionally produces larger outliers that significantly impact the squared error. Despite this, the accuracy performance of the MCP solution supports the hypothesis that it could still serve as a viable metric in the design of parametric solutions.

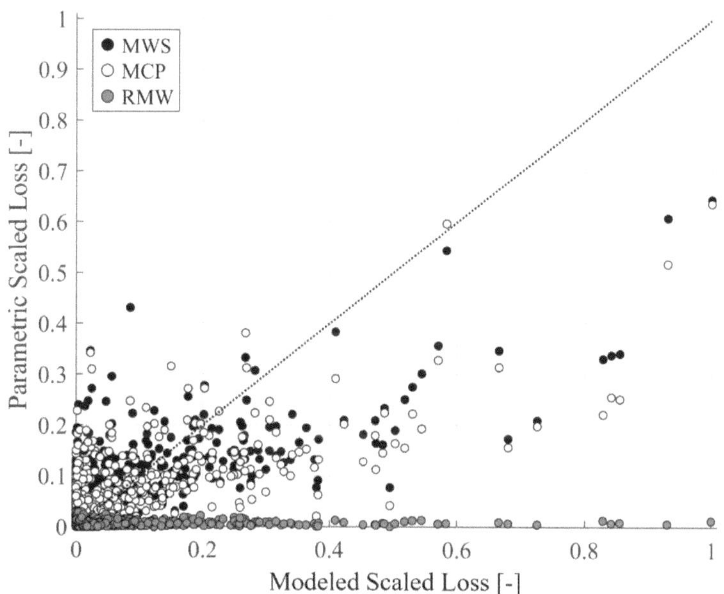

Fig. 1. Scatter plot of the three parametric solutions.

To further compare the performance of the different solutions, a scatter plot and an EP curve were generated to visually assess the distribution of prediction errors and the frequency of extreme deviations. Figure 1 shows the scatter plot of the parametric scaled losses predicted by the three regression models against the modeled scaled losses, where the scaled loss values are expressed as a fraction of the maximum modeled loss (to which is assigned the value 1). The predictions from the MWS-based and MCP-based solutions reflect the overall trend of the

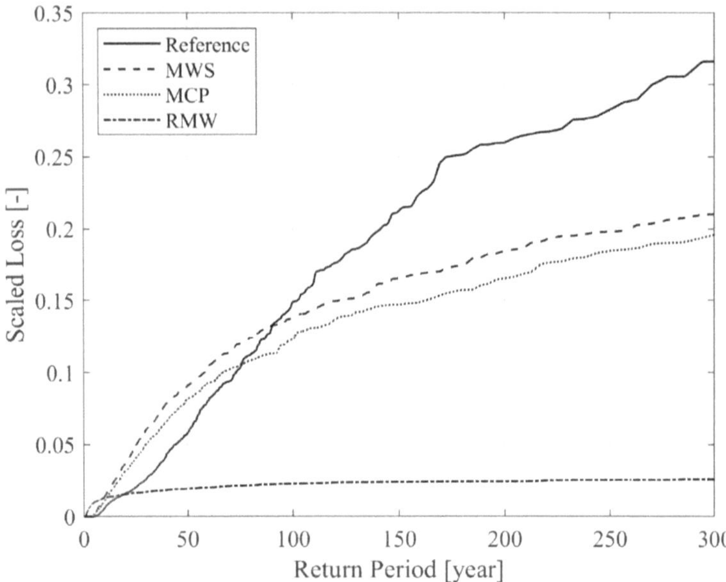

Fig. 2. EP curves of the three parametric solutions, with respect to the reference solution.

modeled losses and are closer to the identity line. However, the predictions from the RMW-based solution are mostly concentrated in the lower range of the y-axis, indicating a systematic tendency to significantly underestimate losses.

Figure 2 presents the EP curves for the different parametric solutions. The solid line represents the reference EP curve based on the modeled loss data. The EP curves generated by the MWS-based and MCP-based parametric solutions generally follow similar trends to the reference curve. Both the MWS and MCP based solutions tend to overestimate the losses for return periods lower than 100 years, with the overestimation slightly lower for the MCP-based solution. In contrast, for return periods exceeding 100 years, both MWS-based and MCP-based solutions underestimate the potential losses, failing to closely capture the tail risk associated with more extreme but less frequent events. The EP curve of the RMW-based solution shows significant deviation across the entire range of return periods of interest, confirming it is the least accurate solution, as anticipated by the overall poor performance metrics in Table 2.

Another aspect to take into account is the distribution of cell weights, which reflect each cell's predictive reliability. Since weights are assigned based on variability in the parametric solution performance, their spatial distribution offers valuable insight into where the solution captures cyclone impacts most effectively. Cells with higher weights would contribute more to the final loss estimate, typically corresponding to areas where the relationship between wind intensity and losses is stable and easier to model. Conversely, lower weights indicate greater uncertainty—often due to heterogeneous exposure or limited cyclone

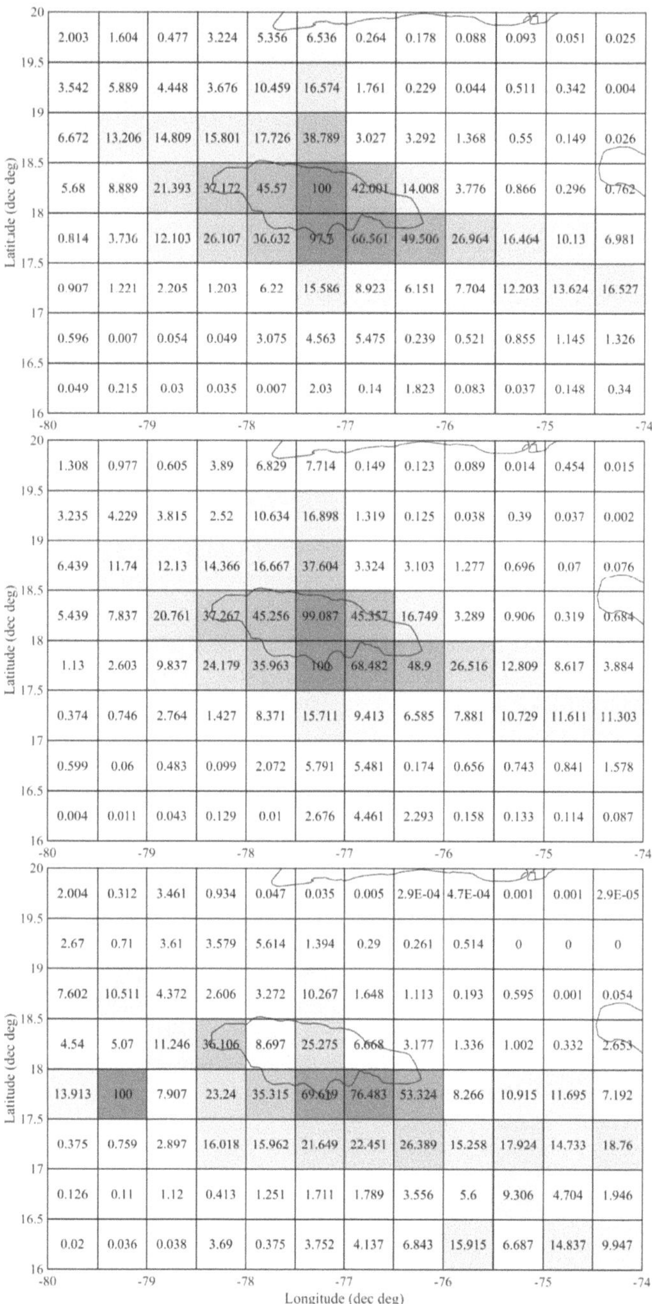

Fig. 3. Weights maps for: (top) the MWS-based solution, (middle) the MCP-based solution, and (bottom) the RMW-based solution.

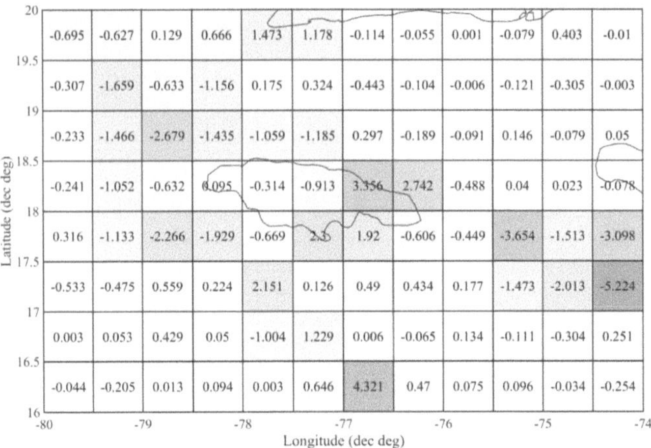

Fig. 4. Difference between the weights in the parametric solutions based on MCP and MWS.

interaction. This weighting mechanism ultimately ensures that the aggregate trigger gives more influence to accurate local predictors, enhancing the robustness of the parametric insurance scheme. Figure 3 showcases the distribution of cell weights for the different parametric solutions.

The analysis of the spatial weight distribution reveals notable differences across the three solutions. As seen in Fig. 4, both the MWS-based and the MCP-based solutions assign the highest weights to grid cells located in the center of the island, which aligns with the regions of highest exposure to damaging winds and likely loss concentration. These two solutions show a remarkably similar weights pattern, which is consistent with their strong correlation and comparable predictive power. Notably, the highest weights differences (still limited to a few units) are located at the boundary of the domain, in cells with generally large variability and low weight, hence with an overall marginal relevance in the parametric loss calculation. In contrast, the RMW-based solution exhibits a less coherent weight distribution, with its most heavily weighted cell located offshore, far from the insured exposure.

3.2 Parametric Design Calibration: 70% Train-30% Test Split

In this section we explore the predictive power of the computed parametric solutions. To evaluate the ability of each solution to generalize to new data, we tested their predictive capabilities using a train-test split. For this purpose, the dataset was divided into 70% for training and 30% for testing. This approach is especially important when developing parametric solutions to predict losses caused by tropical cyclones, as the goal is not just to fit known data, but to more accurately forecast future events that may differ in intensity, location, or impact. By reserving a portion of the data during training, we can more realistically

Table 3. Accuracy evaluation metrics for the parametric solutions, considering a data split between training and test sets.

Parameter	Train			Test		
	MAE	MSE	R^2	MAE	MSE	R^2
MWS	9.205e6	2.379e16	0.592	9.798e6 (+6.44%)	3.664e16 (+54.01%)	0.565 (-4.56%)
MCP	9.211e6	2.688e16	0.539	9.541e6 (+3.58%)	3.965e16 (+47.51%)	0.529 (-1.86%)
RMW	1.006e7	5.829 e16	0.001	1.043e7 (+3.68%)	8.415e16 (+44.36%)	0.001 (0%)

assess how well a parametric solution performs under new, unseen conditions. Table 3 presents the results in terms of accuracy metrics obtained during both the training and testing phases for all three solutions.

As observed in Sect. 3.1, the RMW-based solution continues to under-perform across all the accuracy metrics considered, confirming its limited predictive value. When comparing the MCP-based and the MWS-based solutions, previous results —without a train-test split— showed that the MWS-based solution outperformed the MCP-based solution in terms of MSE and R^2, while the MCP-based solution had a better MAE. This pattern largely holds with the data split. During training, the MWS-based solution achieves better performance across most metrics, with a nearly identical MAE compared to the MCP-based solution, but lower MSE and higher R^2.

In the testing phase the gap between the MWS-based and MCP-based solutions narrows. The MCP-based solution achieves a slightly lower MAE, while the MWS-based solution maintains better results in MSE and R^2, although the differences are less pronounced than in the training phase. For instance, the MAE and R^2 of the MWS-based solution shift from 9.205e6 and 0.592 in training to 9.798e6 (+6.44%) and 0.565 (–4.56%) in testing, respectively. In comparison, the MCP-based solution varies from 9.211e6 and 0.539 to 9.541e6 (+3.58%) and 0.529 (–1.86%), respectively. These results suggest that while the MWS-based solution has a slightly better overall performance, the MCP-based solution is a valid alternative, particularly in terms of average error.

4   Performance Against Historical Events

In this Section we applied the trained solutions to a series of what-if scenarios constructed from historical tropical cyclone events. We selected a set of seven major events for Jamaica in the last 50 years, with their characteristics extracted from the International Best Track Archive for Climate Stewardship (IBTrACS) Catalog [7,8] (see Fig. 5).

We used the three parametric solutions to estimate the potential losses associated with each event. Table 4 and Fig. 6 compare the performance of different parametric solutions across the events set. Estimates for the most recent events, starting with Gilbert (1988), are taken from the report of the Global Facility for Disaster Reduction and Recovery [5], and are based on estimates from

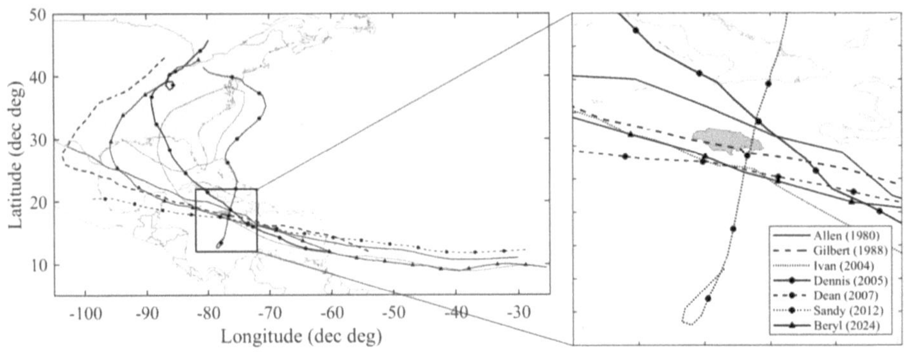

Fig. 5. Set of historical event for Jamaica.

Table 4. Parametric Losses for the three different solution considered, compared with estimated loss [mUSD]

Event	Loss (est.)	Model Optimization		
		MWS	MCP	RMW
Allen (1980)	100	965.4	158.3	–
Gilbert (1988)	1000	1966.5	612.8	–
Ivan (2004)	580	2444.9	2130.9	–
Dennis (2005)	34.5	310.5	101.8	100.3
Dean (2007)	329.34	1817.4	1746.7	327.5
Sandy (2012)	109.1	192	234	413.6
Beryl (2024)	200	1175.3	385.1	403.7

the Planning Institute of Jamaica (PIOJ). Estimates for the oldest event, Allen (1980), are extracted from [10]. Estimated losses values have not been adjusted for inflation. Among the historical events considered, Gilbert (1988), making landfall, and Ivan (2004) and Dean (2007), passing south of the island, were the most damaging for Jamaica, with estimated losses of USD 1.0b, USD 580 m and USD 329.34 m, respectively. Sandy (2012), despite making landfall and crossing the island from south to north, resulted in a lower amount of estimated losses (USD 109.1 m), mainly due to its relatively low wind-speed values. For Beryl (2024) the estimated losses amounted to USD 200 m. It is important to note that losses following catastrophic events are extremely challenging to estimate, as the economic impact of an event on the affected community has long-lasting consequences, therefore it is reasonable to assume that the actual overall impact on the community likely exceeds the estimated loss value.

Whereas the results obtained for each hurricane differ from one parametric solution to another, it is possible to make the following observations:

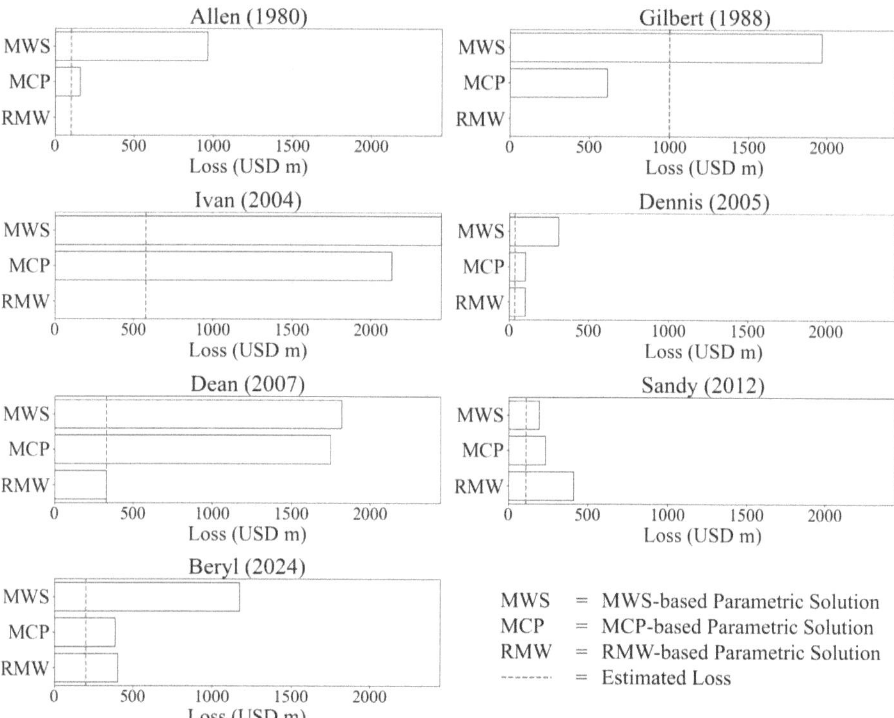

Fig. 6. Performance of the parametric solutions for the set of historical events considered.

i. The parametric predictions tend to consistently overestimate the estimated reference value, with the main exception of hurricane Gilbert (1988) for the MCP-based solution. The highest loss is predicted by the MWS-based solution at USD 2.445b for hurricane Ivan (2004), while the MCP-based solution with the highest estimate is USD 2.131b for the same event.

ii. For the oldest events Allen (1980), Gilbert (1988), and Ivan (2004), the RMW data were not available, therefore the RMW-based prediction is not available for those events.

iii. For the cases where a RMW-based solution is available, for hurricanes Dean (2007) the loss estimates for the MWS-based and MCP-based solutions are aligned and significantly higher than the RMW-based solution's estimate, whereas for hurricanes Dennis (2005) and Beryl (2024) the MWS-based solution has higher loss estimates than the MCP-based and RMW-based solution, which are aligned.

iv. The RMW-based solution has higher losses estimated for hurricane Sandy (2012), which is generally considered a relatively low damage event for Jamaica, with reported losses of USD 109.1 m.

In general, the results indicate that the best-performing solutions (i.e., parametric losses closer to the actual estimated losses) are obtained for the MWS-based and MCP-based solutions, with the MCP-based solution showing overall a slightly better fit. It should be noted, however, that the range of possible values of wind speed is much larger than the range of possible values of central pressure, therefore a small variation in the values of the MCP could lead to large variations in the parametric loss value. This consideration helps explaining the slightly better results of the MWS-based model in the training and test data (see Sect. 3.2). It should not surprise the overall low performance of the RMW-based model: in our analysis we considered the RMW similarly to MWS or MCP, even if this parameter has a lower correlation with losses and therefore it is rarely used in isolation.

A possible reason of the high losses estimates from the MWS-based and MCP-based solution for the offshore events Ivan (2004), Dean (2007), Allen (1980), Dennis (2005) and Beryl (2024) could be attributed to the parametric model resolution. A finer resolution may change the cells crossed by those offshore events, resulting in different, likely lower estimates. Excessively increasing the resolution of the parametric model used in the parametric solutions poses however the risk of overfitting the parametric model to the underlying catastrophe model, increasing the importance of considering data split in test and train, as shown in Sect. 3.2.

Even if the overall performance of the parametric solutions does not exactly match the actual loss estimates for the considered historical events, it is important to realize that whereas the parametric solutions are tested against historical loss estimates (which are affected by large uncertainty), they are not calibrated based on them, rather based on a stochastic catalog simulating thousands of years of tropical cyclones activity. Notwithstanding, parametric losses (or indexes) computed for historical events could provide insights to set appropriate parametric payout thresholds. For example, if the insured would expect to receive a payout for hurricane Gilbert (1988) using a MCP-based parametric solution, then the threshold value of the index to trigger a payout should be lower than 612.8 m, which is the loss value (i.e. index) estimated by that solution for Gilbert (1988). Conversely, if the insured would not expect to receive a payout for hurricane Sandy (2012) using a MCP-based parametric solution, then the threshold value of the index to trigger a payout should be higher than 234 m, which is the loss value (i.e. index) estimated by that solution for Sandy (2012).

5 Conclusions and Future Work

This paper focuses on the development and enhancement of regression-based approaches in the design of tropical cyclone parametric solutions. The paper proposes and compares three parametric designs based on three different parameters, namely MWS, MCP and RMW. The design methodology used in this paper is developed for Jamaica in the context of tropical cyclone risk; however, it is general and can be applied to different regions and to other natural hazards.

Main findings of this paper suggest that the parametric designs based on MWS and MCP provide comparable results, and overall better accuracy metrics than the designs based solely on RMW. It is important to underline that the three considered parametric solutions are limited at one parameter per model. Moreover, we considered the RMW similarly to the other two parameters, MWS and MCP, even if its correlation with the losses is not as well-defined. A possible path to improve the performance of the parametric solutions would be to extend the presented approaches to multiple dimensions, developing multi-metric parametric solutions. Future work could therefore explore the combination of multiple parameters to further improve the robustness and accuracy of the parametric solutions, for example considering RMW as a secondary supporting parameter to complement the information provided by the MWS or MCP.

Additionally, whereas the original cell weight computation reflects prediction variance as a proxy for reliability, it does not automatically ensure optimal loss minimization. Future work could focus on developing alternative weighting methodologies, explicitly targeting the reduction of the prediction error, while maintaining the transparency of the parametric solution.

Finally, the paper explores the predictive power of the parametric solutions considering a data split between training and testing data. Whereas beyond the scope of this work, future research could dive deeper in the importance of conducting a train-test data split in order to minimize overfitting, and to increase the robustness of the parametric solutions, especially in cases where a finer grid resolution would be considered to increase accuracy.

Acknowledgments. The authors are grateful to Risk Engineering + Development (www.redrisk.com), and in particular to Dr. Paolo Bazzurro and Dr. Gianbattista Bussi, for granting access to the modeling results for the case study discussed in this paper. Yuda Li, Juliana Castañeda, Mohammad Peyman, and Ye Yuan are kindly acknowledged for their early contributions in the ideation of the present work. Finally, the authors would like to acknowledge Prof. A. Juan for his valuable comments and insights, functional to the development of the paper.

Disclosure of Interests.   The authors have no competing interests to declare that are relevant to the content of this article.

References

1. Franco, G.: Minimization of trigger error in cat-in-a-box parametric earthquake catastrophe bonds with an application to Costa Rica. Earthq. Spectra **26**, 983–998 (2010)
2. Franco, G.: Construction of customized payment tables for cat-in-a-box earthquake triggers as a basis risk reduction device. In: Proceedings of the International Conference on Structural Safety and Reliability (ICOSSAR), pp. 5455–5462 (2013)
3. Franco, G., Guidotti, R., Field, E., Milner, K., Lee, Y., Stein, R.: An exploration of parametric earthquake risk transfer solutions that dynamically adapt to seismicity changes. In: Proceedings of the 17th World Conference on Earthquake Engineering (2020)

4. Franco, G., et al.: Typology and design of parametric cat-in-a-box and cat-in-a-grid triggers for tropical cyclone risk transfer. Mathematics **12**(11), 1768 (2024)
5. GFDRR: Advancing disaster risk finance in Jamaica (2018). https://www.gfdrr. org/en/publication/advancing-disaster-risk-finance-jamaica
6. Grossi, P., Kunreuther, H., Windeler, D.: An introduction to catastrophe models and insurance. In: Grossi, P., Kunreuther, H. (eds.) Catastrophe Modeling: A New Approach to Managing Risk. Catastrophe Modeling, vol. 25, pp. 23–42. Springer, Boston, MA (2005). https://doi.org/10.1007/0-387-23129-3_2
7. Knapp, K.R., Diamond, H.J., Kossin, J.P., Kruk, M.C., Schreck, C., et al.: International best track archive for climate stewardship (IBTrACS) project, version 4. NOAA National Centers for Environmental Information (2018)
8. Knapp, K.R., Kruk, M.C., Levinson, D.H., Diamond, H.J., Neumann, C.J.: The international best track archive for climate stewardship (IBTrACS) unifying tropical cyclone data. Bull. Am. Meteor. Soc. **91**(3), 363–376 (2010). https://doi.org/10.1175/2009BAMS2755.1
9. Kunreuther, H.: Risk analysis and risk management in an uncertain world. Risk Anal. Int. J. **22**(4), 655–664 (2002)
10. Lawrence, M.B., Pelissier, J.M.: Atlantic hurricane season of 1980. Mon. Weather Rev. **109**(7), 1567–1582 (1981)
11. Lin, X., Kwon, W.J.: Application of parametric insurance in principle-compliant and innovative ways. Risk Manag. Insur. Rev. **23**(2), 121–150 (2020)
12. Powell, M.D., Reinhold, T.A.: Tropical cyclone destructive potential by integrated kinetic energy. Bull. Am. Meteor. Soc. **88**(4), 513–526 (2007). https://doi.org/10.1175/BAMS-88-4-513
13. Ren, H., Dudhia, J., Li, H.: The size characteristics and physical explanation for the radius of maximum wind of hurricanes. Atmos. Res. **277**, 106313 (2022). https://doi.org/10.1016/j.atmosres.2022.106313, https://www.sciencedirect.com/science/article/pii/S016980952200299X
14. Surminski, S., Bouwer, L.M., Linnerooth-Bayer, J.: How insurance can support climate resilience. Nat. Clim. Chang. **6**(4), 333–334 (2016)
15. Wald, D.J., Franco, G.: Financial decision-making based on near-real-time earthquake information. In: 16th World Conference on Earthquake Engineering, pp. 1–13. Santiago de Chile, Chile (2017)
16. Wald, D.J., Franco, G.: Money matters: rapid post-earthquake financial decision-making. Nat. Hazards Observer **40**(7), 24–27 (2016)

Energy-Aware Navigation in Maze-Like Environments: Balancing Exploration and Recharge Strategies in Autonomous Agents

Carolina Crespi[1]([✉]) [iD], Vincenzo Cutello[1] [iD], Attilio Di Natale[1,2], and Mario Pavone[1,2] [iD]

[1] Department of Mathematics and Computer Science, University of Catania, V.le Andrea Doria 6, 95125 Catania, Italy
{carolina.crespi,cutello}@unict.it, mpavone@dmi.unict.it
[2] ANTs Lab: Advanced New Technologies Research Laboratory, Via Rigolato 51, 95024 Acireale, CT, Italy

Abstract. This paper presents an energy-aware framework for autonomous agents navigating maze-like environments characterized by limited visibility and heterogeneous traversal costs. The model extends a previously established navigation framework based on visibility, memory, and exploratory tendency by integrating an internal energy regulation mechanism. Agents are equipped with a recharge threshold that governs when exploration is interrupted to restore energy, allowing them to balance goal pursuit with resource preservation. Simulations were conducted across different maze densities and recharge strategies to evaluate the effects of energy-awareness on navigation performance. The results suggest that the optimal recharge threshold is context-dependent, varying according to environmental complexity and connectivity.

Keywords: Autonomous agents · Energy-aware navigation · Unknown environment exploration

1 Introduction

The study of autonomous agents operating in unknown environments has gained increasing attention in recent years, driven by advances in robotics [1], artificial intelligence, and distributed systems [2]. In both simulated and real-world contexts, agents must interpret local information, coordinate movement, and continuously update internal representations of the environment to achieve individual or collective goals. A wide range of approaches has been proposed to address these challenges. Some studies have focused on the role of memory in navigation under uncertainty, showing how past experiences can be leveraged to improve path efficiency in partially known or dynamically changing environments [3,4]. Other works have explored coordination and interaction mechanisms among agents, analyzing how cooperative and competitive dynamics influence collective performance and adaptability in maze-like environments [5–8].

Beyond agent-based simulations, robotics research has developed adaptive sensing and reactive control strategies that allow robots to dynamically adjust trajectories and avoid obstacles without prior environmental knowledge [9], as well as navigation frameworks designed to operate under uncertainty, such as radiation-informed path planning in dangerous environments [10] and thermal-based decision-making for UAVs in visually degraded conditions [11]. Deep reinforcement learning has further advanced autonomous navigation by integrating perception, learning, and control into unified architectures for adaptive motion [12], while cooperative multi-agent frameworks have demonstrated how memory-based coordination and distributed decision-making enable collective navigation in complex, partially observable domains [13].

Beyond coordination and learning, another emerging dimension of autonomy concerns the management of internal resources. Recent research has increasingly emphasized the importance of resource-awareness in autonomous systems, where energy is treated not merely as an operational limitation but as an active component of the decision-making process. Autonomous agents can regulate their energetic behavior through local sensing and adaptive feedback, balancing resistance and performance in dynamic environments. For instance, self-docking and onboard monitoring mechanisms enable robots to perform self-regulated recharging, transforming energy management into a dynamic behavioral variable [14, 15]. Multi-agent energy management systems integrate forecasting and feedback control to coordinate distributed resources efficiently [16, 17], while energy-aware path planning and optimization allow UAVs and autonomous vehicles to adapt trajectories and control policies to their energetic state [18–20]. Learning-based approaches further extend this principle, showing how proactive control and deep reinforcement learning can jointly optimize energy and task performance in multi-agent systems [21, 22].

Building on this research direction, the present study focuses on the individual dimension of autonomy, investigating how agents regulate their energy resources during navigation. In particular, this work extends a previously developed framework [23], originally designed to analyze how visibility, memory, and exploratory tendency influence agent performance in maze-navigation tasks, by introducing a new self-regulatory component: *energy-awareness*. In the original framework, agents were characterized by visibility range, memory capacity, and exploration bias, and results showed that these factors jointly affected their ability to balance exploration and route efficiency. In the proposed extension, agents are equipped with an internal control parameter, the *recharge threshold*, which determines when an agent interrupts its exploration to search for a charging station. By embedding energy-awareness, this study try to connect cognitive and energetic dimensions of autonomy: visibility and memory define how agents perceive and recall the environment, exploratory tendency shapes how they approach and balance exploration, and the recharge threshold regulates their adaptability under energy constraints. The aim is to evaluate how different recharge thresholds affect agent behavior and system efficiency across maze environments of varying complexity, which offer a structured yet uncertain setting for analyzing adaptive navigation strategies [24,25] and test algorithm performance [26].

2   The Model

We consider a weighted maze modeled as a square grid of size $N \times N$ with a single entrance and a single exit. Each grid element is a cell (i, j) associated with a traversal weight w_{ij}. Traversable cells have weights uniformly distributed in $w_{ij} \in (0, 0.7]$ to capture heterogeneous traversal costs; black cells denote non-traversable walls with $w_{ij} = 1$. The entrance, located at the top-left corner, is assigned an undefined value (NaN) to prevent misidentification as a valid target, whereas the exit, at the bottom-right corner, has weight 0 to guarantee an unobstructed passage to the goal. The maze also considers a density parameter $\rho \in [0, 1]$ expressing the proportion of traversable cells over the total grid: low values yield sparse layouts with few viable routes, while high values produce denser, more connected topologies with multiple alternative paths. A small, fixed fraction p of traversable cells are designated as charging stations and distributed in the maze; these cells act as energy sources that agents can use whenever their residual energy drops below a threshold. Figure 1 illustrates the maze configurations used in the experiments for the three density levels considered.

Agents navigate the maze under limited information and are endowed with three behavioral parameters controlling their decisions: visibility (η), memory (λ), and exploratory tendency (β). Visibility determines the agent's perceptual horizon (maximum number of cells it can sense ahead), while memory encodes the number of previously visited cells. The exploratory tendency $\beta \in [0, 1]$ biases movement toward discovering new routes (*exploration*) or reusing known ones (*exploitation*). Given a current cell i and a candidate move toward cell j, the transition probability is

$$\bar{p}_{j,\beta} = 1 - \beta + (2\beta - 1) \frac{e^{\sigma_j}}{\sum_k e^{\sigma_k}}, \tag{1}$$

where $\sigma_j = \frac{\pi_{\text{overlap}}}{\pi_{\text{tot}}}$ measures the overlap between the candidate path and the set of previously explored cells (π_{overlap}) normalized by the total number of explored cells (π_{tot}). Thus, $\beta = 0$ yields pure exploration (preference for unvisited paths), $\beta = 1$ pure exploitation (preference for known paths), and intermediate values produce an adaptive balance. Energy dynamics couple motion with resource feasibility. Each agent starts with an initial energy level

$$E_0 = c \cdot \sum_{i,j=1}^{N} w_{ij}, \tag{2}$$

where c is an empirically chosen constant that scales E_0 proportionally to the cumulative traversal cost of the maze. Moving across a traversable cell of weight

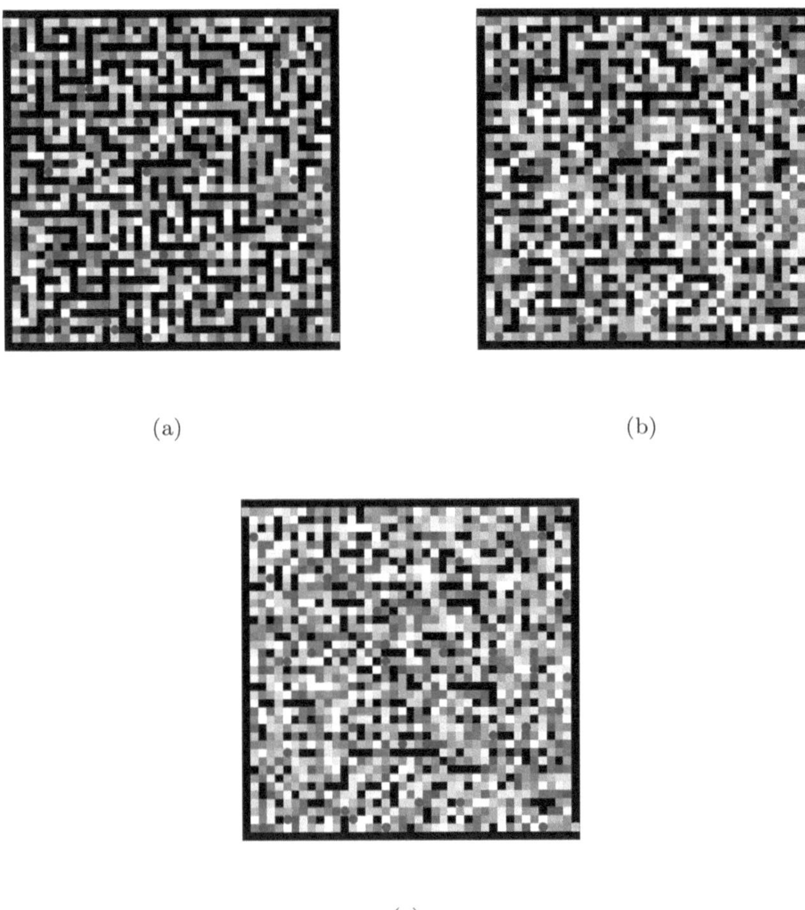

(a)   (b)

(c)

Fig. 1. Maze configurations at different density levels: (a) $\rho = 0.2$, (b) $\rho = 0.5$, and (c) $\rho = 0.8$. Black cells represent non-traversable walls with weight $w_{ij} = 1$, while lighter gray tones indicate lower traversal costs. The entrance (green circle) is placed at the top-left corner, the exit (orange circle) at the bottom-right corner, and blue circles denote the locations of the charging stations. As the density ρ increases, walls become fewer and the number of traversable paths connecting the entrance and the exit increases.

w_{ij} decreases the agent's energy by w_{ij}; an agent that reaches zero energy before finding the exit is considered dead. Charging stations are unknown a priori and can only be discovered via exploration. Upon reaching a station, an agent recharges at a constant per-tick rate r according to the discrete update rule

$$E_i(t+1) \;=\; \min\!\big(E_0,\; E_i(t) + r\big), \tag{3}$$

so that the total time spent recharging is proportional to the current energy deficit. Agents continuously monitor their energy and, when it falls below a *recharge threshold* θ, prioritize the search for a charging station within their visibility range. If both a reachable station and the exit are visible but the residual energy is insufficient to safely reach the exit, the agent opts to recharge. This behavior introduces an explicit trade-off between goal pursuit and energetic viability, shaping route selection, recharging times at stations, and overall success. Simulations are bounded by a maximum time horizon T_{max}; a simulation terminates when an agent reaches the exit or the time limit is met.

3   Experimental Setup

The environment used for the simulations consists of a maze of size 41×41, for a total of 1681 cells, of which $p = 2\%$ are designated as charging stations. This corresponds to approximately 34 charging stations per maze, whose positions remain fixed across all configurations to ensure that performance variations depend exclusively on maze density rather than station placement. Three maze densities were tested ($\rho = 0.2, 0.5, 0.8$), corresponding respectively to sparse, medium, and dense configurations. As the density ρ increases, the number of walls decreases, producing more connected environments with a greater number of traversable paths between the entrance and the exit. Four recharge strategies were evaluated, defined by the threshold parameter $\theta \in \{0, 0.2, 0.5, 0.8\}$, which represents the remaining energy fraction that triggers the search for a charging station. The recharge rate was set to $r = 0.5$ energy units per tick, meaning that agents regain 0.5 energy units at each simulation step while recharging. The initial energy of each agent was computed as in Eq. 2 setting $c = 0.5$. Accordingly, the resulting initial energy values were $E_0 = 170.092$ for $\rho = 0.2$, $E_0 = 196.059$ for $\rho = 0.5$, and $E_0 = 213.718$ for $\rho = 0.8$. The maximum simulation time was fixed at $T_{max} = 10^4$ ticks, expressed in simulated time units. Each agent is characterized by three behavioral parameters inherited from the base framework: *visibility* ($\eta \in [1, \infty)$), *memory size* ($\lambda \in [1, \infty)$), and *exploratory tendency* ($\beta \in [0, 1]$). All combinations of these parameters were systematically tested together with the recharge thresholds θ. Specifically, each agent is defined by a unique triplet (η, λ, β), leading to 330 distinct configurations; for each configuration, 5 agents were simulated to ensure statistical robustness, resulting in a total of $N_{max} = 1650$ agents. All simulation parameters are summarized in Table 1.

Table 1. Model parameters used in the experiments.

Parameter	Symbol	Value/Range
Maze size	$N \times N$	41×41
Maze density	ρ	$\{0.2,\ 0.5,\ 0.8\}$
Recharge threshold	θ	$\{0,\ 0.2,\ 0.5,\ 0.8\}$
Recharge rate	r	0.5 energy units/tick
Initial energy constant	c	0.7
Charging stations	p	2% of total cells (34 per maze)
Visibility	η	$\{1,\ 6,\ 11,\ 16,\ \infty\}$
Memory size	λ	$\{0,\ 10,\ 20,\ 30,\ 40,\ 50,\ 60,\ 70,\ 80,\ 90,\ \infty\}$
Exploratory tendency	β	$\{0.0,\ 0.2,\ 0.4,\ 0.6,\ 0.8,\ 1.0\}$
Maximum simulation time	T_{max}	10^4 ticks
Number of agents per config.	n_a	5

Two complementary analyses were conducted: the *behavioral analysis*, which examines observable agent behaviors such as success rate, exit time, and number of recharges; and the *performance analysis*, which evaluates overall efficiency through three composite metrics, defined as follows. The *Productivity*, denoted as P and formulated in Eq. 4, quantifies the number of agents that successfully reach the exit per unit of recharging time. It expresses how effectively the system converts recharging effort into successful exits, providing a direct measure of energy-to-performance efficiency.

$$P = \frac{N_e}{T_{rc}} \tag{4}$$

The *Efficiency*, denoted as E and defined in Eq. 5, represents the average recharging time required per exited agent. Being the inverse of productivity, it captures the energetic cost associated with each successful exit, where lower values indicate higher efficiency.

$$E = \frac{T_{rc}}{N_e} \tag{5}$$

The *Normalized Efficiency Score*, denoted as F and expressed in Eq. 6 as

$$F = \frac{N_e}{N_{\text{max}}} \left(1 - \frac{T_{rc}}{T_{rc}^{max}} \right) \tag{6}$$

where N_e in the number of exited agents, N_{max} is the total number of agents, T_{rc} is the time spent recharging and T_{rc}^{max} is the maximum recharging time computed empirically as

$$T_{rc}^{max} = \max_{\rho,\theta} \left(N_{rc}^{(\rho,\theta)} \right) \cdot 0.99 \cdot E_0(\rho), \tag{7}$$

where $E_0(\rho)$ denotes the initial energy assigned to agents at a given maze density, and the factor 0.99 is an empirical coefficient. In essence, Eq. 6 combines the exit rate and the energetic cost into a single normalized value. The following section presents the results of these analyses.

4   Results and Discussion

The analyses are organized into two parts: the first focuses on the *behavioral metrics*, describing how agents interact with the environment; the second examines the *performance metrics*, which synthesize the observed behaviors into efficiency indicators at the system level.

4.1   Behavioral Analysis

The behavioral analysis investigates how agents adapt their navigation strategies under different recharge thresholds θ. The three considered metrics, the number of exited agents (N_e), the average time to exit (T), and the average number of recharges (N_{rc}), provide a perspectives on agent success, temporal efficiency, and energy-management behavior.

Figure 2 represents the average number of agents that successfully reached the exit (N_e) for each charging threshold θ and maze density. A general increasing trend is observed: as the recharge threshold rises, the number of exited agents tends to increase. In mazes with lower densities ($\rho = 0.2$ and $\rho = 0.5$), the maximum number of successful exits occurs when the threshold is highest ($\theta = 0.8$), indicating that agents benefit from frequent recharges. Conversely, in the densest maze ($\rho = 0.8$), where multiple paths are available, the highest success rate is reached for $\theta = 0.5$, although the difference with $\theta = 0.8$ remains small. This trend suggests that the number of exited agents is approximately proportional to maze density, since denser environments offer more traversable routes and fewer dead ends, which naturally facilitate exiting. At equal density, higher recharge thresholds appear advantageous, likely because agents that recharge more often maintain higher energy levels throughout navigation and thus have more freedom to explore the maze.

Figure 3 shows the average time to reach the exit (T) for each maze density and recharge threshold. In all three environments ($\rho = 0.2$, 0.5, and 0.8), the average exit time increases with higher recharge thresholds. This trend is consistent with the previous metric: as the threshold increases, agents remain active for longer periods within the maze, leading to longer exit times on average. However, this behavior should not be interpreted as negative, since it corresponds to a greater proportion of agents successfully reaching the exit. Longer exit times thus reflect extended exploration supported by more frequent recharges, which improve the likelihood of survival and task completion. In addition, average exit times are lower in denser mazes, confirming that environments with more traversable paths and fewer obstacles facilitate faster navigation and shorter routes to the exit.

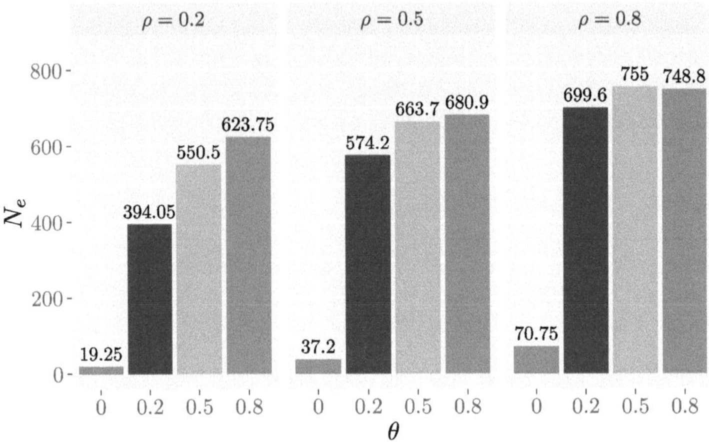

Fig. 2. Average number of agents that successfully reached the exit (N_e) for each maze density and recharge threshold. Higher thresholds generally increase success rates, particularly in low- and medium-density mazes.

Figure 4 represents the average number of recharges (N_{rc}) performed by agents during navigation. The configuration with $\theta = 0$ is excluded, as a zero threshold implies that agents never recharge. As expected, the average number of recharges increases with higher recharge thresholds: when θ is high (e.g., 0.8), agents recharge more frequently, whereas with lower thresholds (e.g., $\theta = 0.2$) they wait until their energy drops close to depletion before recharging. At the same time, the average number of recharges decreases as maze density increases. In less dense environments, where obstacles are more frequent and viable paths are limited, agents perform more recharges on average (up to 23.09 for $\theta = 0.8$). Conversely, in denser mazes, where navigation is easier and multiple routes connect the entrance and exit, the mean number of recharges is significantly lower. This trend is consistent across all threshold values and confirms that as the environment becomes less constrained, agents reach the exit more efficiently and require fewer recharging events.

4.2   Performance Analysis

The performance analysis condenses the behavioral dynamics into aggregate indicators of system efficiency. The three evaluated metrics, *productivity* (P), *efficiency* (E), and *normalized efficiency score* (F), describe how effectively agents convert energy consumption into successful outcomes.

Figure 5 represents the *productivity* (P), which measures the number of successful agents per unit of average recharging time (see Eq. 4). Higher productivity values indicate that agents achieve more exits with less cumulative recharging effort. The configuration with $\theta = 0$ is excluded, since a zero threshold implies that agents never recharge. The observed trend varies according to maze density. In the low-density maze $(\rho = 0.2)$, productivity increases monotonically

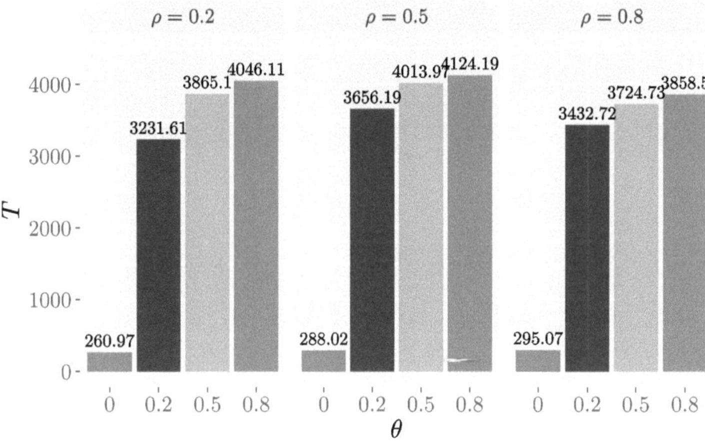

Fig. 3. Average time to reach the exit (T) across different maze densities and recharge thresholds. Higher thresholds lead to longer exit times, while denser mazes result in faster exits due to easier navigation.

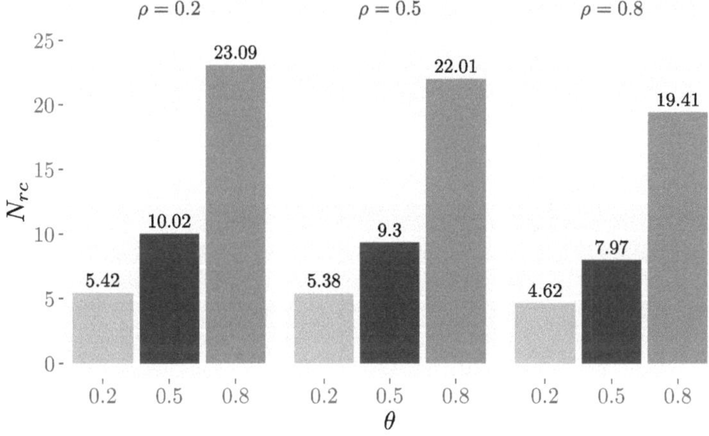

Fig. 4. Average number of recharges (N_{rc}) for each maze density and recharge threshold. Higher thresholds lead to more frequent recharges, while at equal thresholds the number of recharges decreases with maze density.

with the recharge threshold and it is maximum at $\theta = 0.8$. This behavior can be explained by two concurrent effects: first, agents with higher thresholds are better equipped to handle the numerous dead ends typical of sparse mazes, as they rarely risk energy depletion; second, in less connected structures, charging stations are often located along mandatory routes, making recharging events both more probable and more beneficial. In the medium-density maze ($\rho = 0.5$), productivity follows a bell-shaped curve, peaking at $\theta = 0.5$. Here, agents achieve

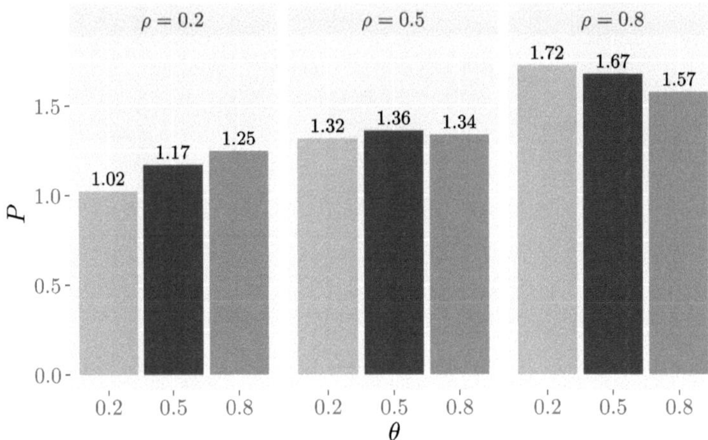

Fig. 5. Productivity ($P = N_e/T_{rc}$) across maze densities and recharge thresholds. Higher productivity indicates more successful agents per unit of recharging time. Frequent recharges enhance performance in sparse mazes, whereas in denser environments excessive recharging reduces overall efficiency.

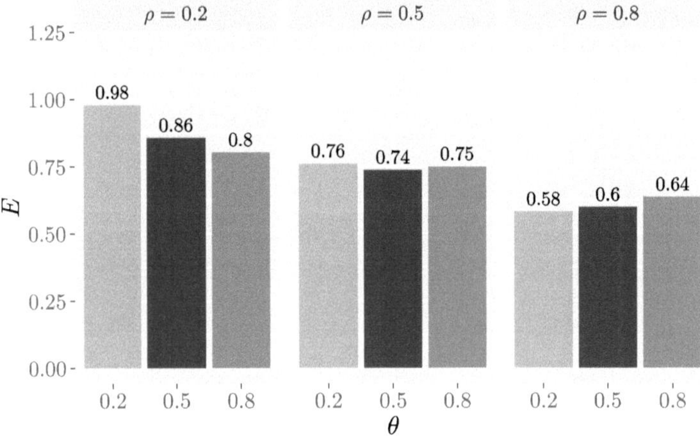

Fig. 6. Efficiency ($E = T_{rc}/N_e$) across maze densities and recharge thresholds. Lower values indicate higher energetic efficiency. Efficiency decreases with θ in sparse mazes, reaches a minimum at $\theta = 0.5$ for medium density, and increases with θ in dense mazes.

an optimal balance between exploration and energy recharging, frequent enough to prevent exhaustion, but not so frequent as to waste time at charging stations. Finally, in the densest maze ($\rho = 0.8$), productivity decreases with increasing threshold values. In this case, the high number of traversable paths makes exploration easier, so frequent recharging becomes unnecessary and may even reduce efficiency by introducing redundant stops.

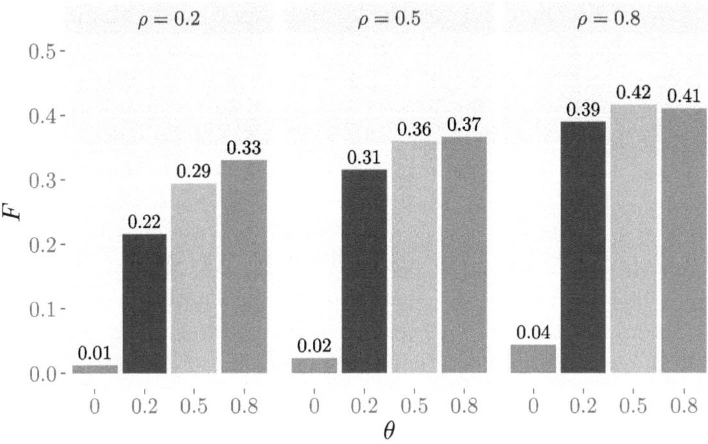

Fig. 7. Normalized Efficiency Score (F) across maze densities and recharge thresholds. Higher values indicate a greater proportion of successful agents with lower recharging time, reflecting higher global efficiency.

Figure 6 shows the *efficiency* (E), defined as the inverse of productivity (see Eq. 5). It represents the average recharging time required per successful agent. Lower E values correspond to higher energetic efficiency, indicating that each agent reaching the exit required little recharging time. Conversely, higher E values reflect lower efficiency, meaning that more time was spent recharging for each successful exit. The behavior of E is specular to that of productivity, exhibiting an inverted bell-shaped trend. In the low-density maze ($\rho = 0.2$), E decreases as the recharge threshold increases, showing that frequent recharges ($\theta = 0.8$) help agents maintain sufficient energy to cope with numerous dead ends and difficult routes. In the medium-density maze ($\rho = 0.5$), efficiency decreases up to $\theta = 0.5$ and then increases again at $\theta = 0.8$, indicating that an intermediate strategy yields the best trade-off between exploration and recharging time. Finally, in the high-density maze ($\rho = 0.8$), efficiency increases with the recharge threshold, meaning that recharging too often becomes counterproductive in open environments where multiple paths already facilitate easier navigation. Overall, agents were most efficient in the densest maze, where the abundance of available routes allowed them to reach the exit with minimal recharging time. In contrast, in the sparsest maze, frequent recharges proved beneficial, as they provided the necessary energy to recover from repeated dead ends and extended exploration attempts. In the intermediate case, the best efficiency was achieved at a moderate threshold, confirming that a balanced recharging strategy performs optimally when the environment is neither too constrained nor too open.

Figure 7 illustrates the *Normalized Efficiency Score* (F), defined in Eq. 6 as the product of the success ratio and the complement of normalized recharging time, i.e., it increases when more agents exit while spending relatively little time recharging. High F values indicate that a large proportion of agents successfully reach the exit while spending relatively little time recharging, thus reflecting high overall system efficiency. Conversely, low F values correspond to configurations where fewer agents complete the task or where excessive time is spent recharging. Across all maze densities, F increases with the recharge threshold θ, indicating that agents benefit from more frequent energy recharge. The score is particularly low for $\theta = 0$, where agents are unable to recharge and thus often fail to complete their exploration. For $\rho = 0.2$ and $\rho = 0.5$, F reaches its maximum at $\theta = 0.8$, while in the densest maze ($\rho = 0.8$) it peaks slightly earlier, at $\theta = 0.5$, with only a marginal decrease at $\theta = 0.8$. This pattern reflects the same balance observed in the previous metrics: in environments with few available paths, frequent recharges sustain exploration and survival, leading to higher efficiency; in dense environments, where there are lot of paths to the exit, further increasing the recharge threshold provides limited benefits.

5   Conclusions

This study explored how autonomous agents regulate their energy resources during navigation in maze-like environments by extending a previously developed framework based on visibility, memory, and exploratory tendency. The extension introduced an energy-awareness component, allowing agents to manage energy through charging stations and a recharge threshold θ, which defines the remaining energy fraction that triggers the search for a recharge. The goal was to assess how different thresholds influence agent behavior and overall system performance under varying environmental densities.

Regarding the *behavioral analysis*, results showed that the recharge threshold strongly affects how agents interact with the environment. Higher thresholds ($\theta = 0.8$) increased the number of agents successfully reaching the exit, particularly in sparse and medium-density mazes, as more frequent recharges helped sustain exploration and prevent premature energy depletion. However, this improvement came with longer exit times, since recharging events extended the total navigation process. The number of recharges per agent decreased with maze density, confirming that denser environments, which offer more connected routes, enable agents to complete the task with fewer energy stops.

In the *performance analysis*, the three metrics of productivity (P), efficiency (E), and normalized efficiency score (F) provided a global view of system behavior. Productivity increased with the recharge threshold in sparse mazes, indicating that frequent recharges are beneficial when navigation is difficult. In medium-density environments, performance peaked at an intermediate threshold ($\theta = 0.5$), which achieved the best trade-off between exploration and recharge time. Since efficiency (E) is defined as the inverse of productivity, its trend mirrored the same pattern in the opposite direction. In contrast, in dense mazes, high thresholds became counterproductive, reducing efficiency by causing unnecessary recharging. The normalized efficiency score confirmed this pattern, showing that the optimal strategy depends on environmental structure: high thresholds favor survival in constrained environments, whereas moderate thresholds are more effective when multiple paths are available.

Overall, the results indicate that optimal performance arises from a context-dependent calibration of the recharge threshold. Specifically, higher thresholds are better in sparse environments where navigation is difficult, intermediate thresholds perform best in medium-density mazes, and lower thresholds are useful in dense environments with lots of paths. Despite these interesting results, some limitations should be acknowledged. The simulations were conducted on a single maze topology, albeit with three different density levels, and with fixed positions of the charging stations. Moreover, parameters such as the recharge rate and the initial energy constant were empirically defined and may influence performance outcomes. Future work should therefore investigate multiple and dynamically changing environments, and explore adaptive or learning-based recharge mechanisms that enable agents to adjust θ in real time according to local conditions.

Acknowledgements. This research is supported by the project PIACERI - PIAno di inCEntivi per la Ricerca di Ateneo 2024/2026—Linea di Intervento i "Progetti di ricerca collaborativa".

References

1. Sakai, T., Nagai, T.: Explainable autonomous robots: a survey and perspective. Adv. Robot. **36**(5–6), 219–238 (2022)
2. Hou, X., Wang, J., Jun, D., Jiang, C., Ren, Y.: Distributed machine learning for autonomous agent swarm: a survey. IEEE Commun. Surv. Tutor. (2025)
3. Crespi, C., Pavone, M. Does a group's size affect the behavior of a crowd? An analysis based on an agent model. In: Elsenbroich, C., Verhagen, H. (eds.) Advances in Social Simulation. ESSA 2023. Springer Proceedings in Complexity, pp. 411–422. Springer, Cham (2024). https://doi.org/10.1007/978-3-031-57785-7_31
4. Cavallaro, C., Crespi, C., Cutello, V., Pavone, M., Zito, F.: Group dynamics in memory-enhanced ant colonies: the influence of colony division on a maze navigation problem. Algorithms **17**(2), 63 (2024)
5. Crespi, C., Fargetta, G., Pavone, M., Scollo, R.A., Scrimali, L.: A game theory approach for crowd evacuation modelling. In: Filipič, B., Minisci, E., Vasile, M. (eds.) Bioinspired Optimization Methods and Their Applications. BIOMA 2020. LNCS, vol.12438, pp. 228–239. Springer, Cham (2020). https://doi.org/10.1007/978-3-030-63710-1_18
6. Crespi, C., Scollo, R.A., Fargetta, G., Pavone, M.: A sensitivity analysis of parameters in an agent-based model for crowd simulations. Appl. Soft Comput. **146**, 110684 (2023)
7. Crespi, C., Fargetta, G., Pavone, M., Scollo, R.A.: An agent-based model to investigate different behaviours in a crowd simulation. In: Mernik, M., Eftimov, T., Crepinšek, M. (eds.) BIOMA 2022. LNCS, vol. 13627, pp. 1–14. Springer, Cham (2022). https://doi.org/10.1007/978-3-031-21094-5_1
8. Crespi, C., Fargetta, G., Pavone, M., Scollo, R.A.: An agent-based model for crowd simulation. In: De Stefano, C., Fontanella, F., Vanneschi, L. (eds.) WIVACE 2022. CCIS, vol. 1780, pp. 1–14. Springer, Cham (2023). https://doi.org/10.1007/978-3-031-31183-3_2

9. Hacene, N., Mendil, B.: Behavior-based autonomous navigation and formation control of mobile robots in unknown cluttered dynamic environments with dynamic target tracking. Int. J. Autom. Comput. **18**(5), 766–786 (2021)
10. Groves, K., Hernandez, E., West, A., Wright, T., Lennox, B.: Robotic exploration of an unknown nuclear environment using radiation informed autonomous navigation. Robotics **10**(2), 78 (2021)
11. Boiteau, S., Vanegas, F., Gonzalez, F.: Framework for autonomous uav navigation and target detection in global-navigation-satellite-system-denied and visually degraded environments. Remote Sensing **16**(3), 471 (2024)
12. Bai, Z., Pang, H., He, Z., Zhao, B., Wang, T.: Path planning of autonomous mobile robot in comprehensive unknown environment using deep reinforcement learning. IEEE Internet Things J. **11**(12), 22153–22166 (2024)
13. Xue, Y., Chen, W.: Multi-agent deep reinforcement learning for uavs navigation in unknown complex environment. IEEE Trans. Intell. Veh. **9**(1), 2290–2303 (2023)
14. Sreenivas Rao, M.V., Shivakumar, M.: IR based auto-recharging system for autonomous mobile robot. J. Robot. Control (JRC) **2**(4), 244–251 (2021)
15. Harik, E.H.C.: Design and implementation of an autonomous charging station for agricultural electrical vehicles. Appl. Sci. **11**(13), 6168 (2021)
16. Areekkara, S., Kumar, R., Bansal, R.C.: An intelligent multi agent based approach for autonomous energy management in a microgrid. Electr. Power Components and Syst. **49**(1-2), 18–31 (2021)
17. Grosset, J., Fougères, A.J., Djoko-Kouam, M., Bonnin, J.M.: Fuzzy agent-based simulation for managing battery recharging for a fleet of autonomous industrial vehicles. In: ASPAI 2024: 6th International Conference on Advances in Signal Processing and Artificial Intelligence, 2024
18. Alyassi, R., Khonji, M., Karapetyan, A., Chau, S.C.K., Elbassioni, K., Tseng, C.M.: Autonomous recharging and flight mission planning for battery-operated autonomous drones. IEEE Trans. Autom. Sci. Eng. **20**(2), 1034–1046 (2022)
19. Liu, S., Li, X., Meng, M., Gong, X.: Energy-aware coverage path planning of multi-uav based on relative distance scaling cluster method. In: 2023 International Conference on Advanced Robotics and Mechatronics (ICARM), pp. 709–714. IEEE, 2023
20. Zhang, Y., Ai, Z., Chen, J., You, T., Chenglie, D., Deng, L.: Energy-saving optimization and control of autonomous electric vehicles with considering multiconstraints. IEEE Trans. Cybern. **52**(10), 10869–10881 (2021)
21. Fouad, H., Beltrame, G.: Energy autonomy for robot systems with constrained resources. IEEE Trans. Rob. **38**(6), 3675–3693 (2022)
22. hen, X., Chen, T., Zhao, Z., Zhang, H., Bennis, M., Ji, Y.: Resource awareness in unmanned aerial vehicle-assisted mobile-edge computing systems. In: 2020 IEEE 91st Vehicular Technology Conference (VTC2020-Spring), pp. 1–6. IEEE, 2020
23. Crespi, C., Cutello, V., Pavone, M., Zito, F.: An agent framework to explore pathfinding strategies in maze navigation problem. Matematiche (Catania) **79**(2), 555–583 (2024)
24. Khalil Al-Rahman Youssefi and Modjtaba Rouhani: Swarm intelligence based robotic search in unknown maze-like environments. Expert Syst. Appl. **178**, 114907 (2021)
25. Husain, Z., Al Zaabi, A., Hildmann, H., Saffre, F., Ruta, D., Isakovic, A.F.: Search and rescue in a maze-like environment with ant and dijkstra algorithms. Drones **6**(10), 273 (2022)

26. Crespi, C., Scollo, R.A., Pavone, M.: Effects of different dynamics in an ant colony optimization algorithm. In: 2020 7th International Conference on Soft Computing & Machine Intelligence (ISCMI), pp. 8–11. IEEE, 2020

Integrating Proximal Policy Optimization Algorithm for the Study of Synchronization Suppression in fNIRS Visibility Networks

Xhilda Dhamo[1] [iD], Eglantina Kalluçi[1] [iD], and Fatos Xhafa[2]([✉]) [iD]

[1] Faculty of Natural Sciences, University of Tirana, Tirana, Albania
`{xhilda.merkaj,eglantina.kalluci}@fshn.edu.al`
[2] Department of Computer Science, Universitat Politècnica de Catalunya, Barcelona, Spain
`fatos@cs.upc.edu`

Abstract. Synchronization in complex networks is an impactful phenomena with applications in numerous disciplines that have attracted the researcher's attention for decades. Recently, there have been efforts to investigate brain synchronization by representing brain signals as complex networks and analyzing the dynamic interactions between neural regions to uncover patterns of connectivity and coordination. Focusing on functional Near-Infrared Spectroscopy signals converted to visibility networks, this paper incorporates reinforcement learning into the exploration of synchronization suppression in the visibility networks constructed, enhancing interventions to mitigate excessive neural synchronization. In this study, we extend the Kuramoto model by adding input signals to pinned nodes of the networks generated *via* the Reinforced Learning, more precisely the Proximal Policy Optimization algorithm, and analyze the synchronization suppression conditions. Comparison to results from other existing models in the literature is done via an experimental study in a realistic setting.

Keywords: Reinforcement learning · complex networks · synchronization suppression · functional Near- infrared spectroscopy

1 Introduction

Collective synchronization phenomena are widely recognized across numerous scientific disciplines and have been observed in biology, physics, social sciences, technology, and neuroscience for centuries [16, 29–32]. This complex phenomenon involves large populations of coupled oscillators with similar natural frequencies that self-organize into coherent collective modes of motion. Synchronization plays an important role in maintaining stability in real- world systems such as power grids [15], cardiac networks [13, 14], but on the other hand excessive synchronization can lead to system failures [10–12]. In neuroscience abnormal neuronal synchronization has been linked to disorders like Pankirson's disease where excessive rhythmic firing of brain circuits leads to tremors. Understanding and controlling undesired synchronization is crucial to ensure optimal functioning in both natural and engineering systems.

© The Author(s), under exclusive license to Springer Nature Switzerland AG 2026
M. Pavone et al. (Eds.): DSA ISC 2025, LNCS 16405, pp. 48–61, 2026.
https://doi.org/10.1007/978-3-032-21811-7_4

A classical approach to studying synchronization involves modelling the components of the population studied as phase oscillators using the Kuramoto model (KM), which describes a wide variety of synchronization processes like phase synchronization, cluster synchronization, explosive synchronization, chimera states [18–23]. Recent studies have suggested combining the phenomenon of synchronization in complex networks with machine learning algorithms to suppress undesired synchronization [33]. Reinforcement Learning (RL) techniques are integrated into the original synchronization models to automatically generate and modify control signals based on the system's real- time state, resulting in efficient desynchronization [1, 7, 9, 33]. Various strategies have been proposed in the literature in this context, including those that apply control signals to all nodes in a complex networks, a randomly selected subset of nodes or a targeted subset of nodes based on their structural importance, referring as pinning control strategies [4–7, 33].

In our previous studies [2, 17], we investigated the conditions under which synchronization emerges in networks constructed from brain activity data. In particular, [17] modelled the autonomous dynamics of an isolated node using the Rössler system, enabling the characterization of network synchronization through a limited set of parameters. In the second study, [2], we employed the Kuramoto model to evaluate synchronization in the same brain activity networks by estimating global and local order parameters to compare the synchronization dynamics across different states and among individual- specific and hemisphere- specific brain networks. In comparison with these prior works focused on identifying and quantifying synchronization, the current study shifts the focus on the analysis of desynchronization. Specifically, we aim to quantify desynchronization by integrating in the original Kuramoto model input signals obtained from the Reinforcement Learning (RL), specifically the Proximal Policy Optimization (PPO) algorithm, into the KM model. The PPO algorithm belongs to RL methods. It updates the decision-making policy using small batches of experiences obtained from interacting with the environment. The purpose in this study is then to analyze the effects of the input signals on the synchronization and desynchronization in the case when all the nodes of the networks or part of them receive the above input signals focusing on networks obtained from data collected using the Functional Near-Infrared Spectroscopy (fNIRS) optical brain monitoring technique.

fNIRS technology is a recent technology to measure the brain activity of individuals during social interactions. During the time duration of the experiment analyzed in this study, the participants passed three different events: (i) rest event, where participants were doing nothing; (ii) before disturbance (five minutes after the beginning of the experiment a disturbance was given to the participants); (iii) after disturbance. The fNIRS technology measured the brain activity [27] by capturing the oxyhemoglobin (HbO) signals using the optodes positioned in the left and right hemispheres of the prefrontal cortices (PFC) of the participants in the experiment. Our approach consists of three consecutive steps: (i) firstly, we obtain the brain activity time series from fNIRS technology as described before; (ii) secondly, we model the time series describing the brain activity of one PFC in one specific event as a network by employing the visibility graph technique [28]. This means that each participant will result in two networks; (iii) thirdly, we use the classical and extended Kuramoto model to analyze synchronization suppression in the networks constructed in step (ii).

The rest of this paper is organized as follows: Sect. 2 introduces the mathematical background for the visibility networks, the original Kuramoto model, the extended Kuramoto model incorporating input signals and the PPO algorithm. Section 3 introduces the experimental data, analyzes and interprets the computational results, while Sect. 4 concludes the study and outlines future work.

2 Notations, Background and Models

2.1 Notations

The following notations are used throughout the paper (Table 1).

Table 1. Mathematical Notations

Variable name	Definition
N	Data series size (Total number of oscillators)
x_i	Time series value at time t_i
A_{ij}	Entry values of the adjacency matrix A
θ_i	Phase of the i-th oscillator (function of time)
$\frac{d\theta_i}{dt}$	First derivative of the phase of the i-th oscillator
$sin(\theta_j - \theta_i)$	Interaction term
ω_i	Natural frequency of the i-th oscillator
λ	Coupling strength
$r(t)$	Magnitude of the average phase vector
$e^{i\theta_j(t)}$	Unit vector representing the phase of oscillator j on the unit circle
$\psi(t)$	Average phase angle of the system
A_{total}	Total input energy
$A_i(t)$	Action applied only to pinned nodes.
R_t	Reward function
$R_{Before\ (After)}$	Complex order parameter values before (after) the application of input stimuli
S	Synchronization suppression coefficient
D	Distance
L	Total number of λ-s used in computing KM
I_i	Input signals applied to the oscillator i

2.2 Visibility Networks

The construction of the visibility graph is described in detail in [25, 26, 28] maps a time series into a network, that is, it converts a time series into a graph, which inherits several properties of the series in its structure. Let's consider a time series with N data measured at times t_i, $i = 1, 2, \ldots, N$ with values x_i, $i = 1, 2, \ldots, N$ and consecutive time points (t_i, x_i), (t_k, x_k) and (t_j, x_j). Time points (t_i, x_i) and (t_j, x_j) are visible and consequently will become two connected nodes in the visibility graph if for any point (t_k, x_k) between them, they fulfill the following inequation (Eq. (1)):

$$x_k < x_j + \left(x_i - x_j\right)\frac{t_j - t_k}{t_j - t_i} \tag{1}$$

The network, whose nodes fulfill the above condition has four main properties: it is connected, undirected, invariant under affine transformations of the series data and it can be applied to any kind of time series [28].

2.3 Extension of the Kuramoto Model Based on Reinforcement Learning

The Kuramoto model is a mathematical framework that describes the synchronization dynamics of coupled oscillators, providing insights into collective behaviours in complex systems. Here we consider an unweighted and undirected network, composed of N coupled phase-oscillators whose connections are described by the adjacency matrix A ($A_{ij} = 1$ if the oscillators i and j are connected and 0 otherwise, $i, j = 1, 2, \ldots N$).

Then, the Kuramoto model in complex networks is introduced in Eq. (2):

$$\frac{d\theta_i}{dt} = \omega_i + \lambda \sum_{j=1}^{N} A_{ij}\sin\left(\theta_j - \theta_i\right), i, j = 1, 2, \ldots N \tag{2}$$

where θ_i (t) stands for the phase of oscillator i at time t; ω_i is the natural frequency of oscillator i, which defines how the oscillator i would oscillate independently from the other oscillators and we consider it to have a normal distribution with zero mean and unit variance; λ is the coupling strength that defines how strongly oscillators influence each other and here it is considered identical for all connections.

In this study, we extend the original KM described by Eq. (2) by incorporating input signals I_i to the oscillator i:

$$\frac{d\theta_i}{dt} = \omega_i + \lambda \sum_{j=1}^{N} A_{ij}\sin\left(\theta_j - \theta_i\right) + I_i, i, j = 1, 2, \ldots N \tag{3}$$

Nodes in the networks will receive input signals I_i generated by the PPO algorithm. We consider input signals I_i to be continuous and bounded in the interval [-2;2]. We train a PPO agent to apply action to some part of the oscillators with the purpose of suppressing synchronization. The complex order parameter which describes the phase transition is used to measure the degree of synchronization among N oscillators for increasing values of the coupling strength λ (Eq. (4)):

$$r(t)e^{i\psi(t)} = \frac{1}{N}\sum_{j=1}^{N} e^{i\theta_j(t)} \tag{4}$$

where $\psi(t)$ stands for the average phase of the collective dynamics of the system and the degree of synchronization is measured as the modulus of the above order parameter and it takes values in the interval [0; 1], where 0 refers to an incoherent solution and 1 refers to a fully synchronized solution of the network.

2.4 Reinforcement Learning Model: Proximal Policy Optimization Algorithm

Reinforcement Learning (RL) is a powerful machine learning algorithm, which solves complex decision-making problems by enabling an agent to interact with its environment to obtain rewards. The agent performs actions that affect the state of the environment and lead to a either positive or negative reward signal. In addition, the agent performs actions guided by a policy by taking into account its current state and the reward it receives. The overall objective of the RL algorithm is to maximize the cumulative reward gained through the agent's actions [1, 8].

Here, we are focused on the Proximal Policy Optimization (PPO) algorithm [3], a policy gradient method for RL whose aim is to maximise a surrogate objective function using stochastic gradient ascent. The reason for choosing PPO/RL is based on the advantages offered by the RL and PPO within RL. Additionally, various works in the literature [1, 7–9, 33] emphasised the importance of RL to obtain effective desynchronization performance. Furthermore, it has been shown that PPO algorithm is very effective to modulate phase interactions in neural populations, which disrupt pathological synchrony [33].

Next, we briefly emphasize the key components of the PPO (environment, action, reward and agent) in our problem as follows:

Environment. The oscillatory system defined by the Kuramoto model presented in Subsect. 2.2 constitutes the environment in our setting.

Action. The action and state is referred to the input to and the resulting output from the oscillatory system. Here, we consider idealistic δ-shaped pulse actions with constant interpulse interval Δ and amplitude that does not exceed A_{max}. Additionally, the actions are bounded within the interval:

$$[-A_{max} \leq A(t_n) \leq A_{max}], t_n = n\Delta, n = 1, 2, \ldots.$$

Finally, the purpose is to minimize the total input energy defined as $A_{total} = \sum_{t_n} |A(t_n)|$ during the training phase of RL. The state represents the system's configuration at a given time step.

Reward. The reward function guides the RL agent to achieve desynchronization while minimizing energy consumption. We considered as reward function $R_t = -\sum_{i \in P} |A_i(t)|$, where $A_i(t)$ is the action applied only to pinned nodes P.

Agent. The agent learns a policy that maps states to actions. Here we consider the Multilayer Perceptron (ML) Policy and PPO algorithm to update the policy based on interactions with the environment.

2.5 Desynchronization Performance Measure

The aim in this study is to compare the degree of synchronization before and after applying the input signals generated by RL. We consider the synchronization suppression coefficient which is defined in Eq. (5):

$$S = \frac{std\left(R_{Before}\right) - std\left(R_{After}\right)}{std\left(R_{Before}\right)} \tag{5}$$

where R_{Before} denotes the complex order parameter values before the application of input stimuli, while R_{After} denotes the values after the stimuli application.

3 Experimental Study

3.1 Experimental Setting

The experiment carried out in this study follows a study from our recent articles [2, 17], considering three events that happened during the entire experiment. The 18 participants who took part in the cooperative task called "MapTask", were grouped in 9 dyads. The task is described in Fig. 1.

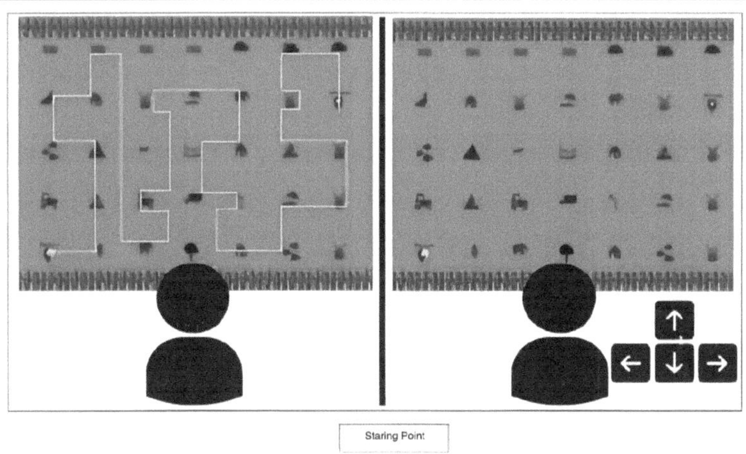

Fig. 1. "MapTask", pA (left) and pB (right)

Both participants in one dyad had the same icons on their screen, but one of them had a path drawn on its screen, whereas the other one did not have that path. The idea was that the second participants had to draw the same path as the first one, based only on the instruction given by the first participant. From now on, we refer to the participant *pA* the one who had the path drawn on his screen and *pB* the one who had to draw the same path [2, 17].

The technology used to measure brain activity is Functional Near-Infrared Spectroscopy Data Acquisition and Pre-processing (fNIRS). Each of the participants had two optods in their prefrontal cortices (PFC): one positioned in the left hemisphere (hL) and the other in the right hemisphere (hR). These optodes captured the oxyhemoglobin (HbO) and deoxyhemoglobin (HbR) signals. Taking into account that the HbO signal is more sensitive to changes in cerebral blood flow than the HbR signal, we focused on the HbO signal [24].

The experimental design takes into account psychological parameters from the literature in the field. The experiment is divided into different events, of different time durations such as one minute, five minutes, which are taken from reference values in the literature. More precisely, here we are interested in the before disturb event. These events are illustrated in Fig. 2. The first event is referred to as a rest event. Participants did nothing during this event. The time duration of the Rest event slightly differs for different participants, but here we consider the first minute of the event. Five minutes after the participants started to communicate and draw the path, a disturbance was given to them. Their screen disappeared immediately for a few seconds and then appeared again. Signals are considered in the time interval before and after the disturbance. The second event corresponds to the one minute signal before the disturbance happens, referred to as *before disturb* event; finally, the third event, referred to as after *disturb* event corresponds to the one minute signal after the disturbance happens. In the figure, *Beg Rest* and *End Rest* stand for the time when the *Rest* event starts and ends, resp. *Beg Exp* and *End Exp* stands for the start and end time of the *Experiment*. *Disturb* is the time when the disturbance happens.

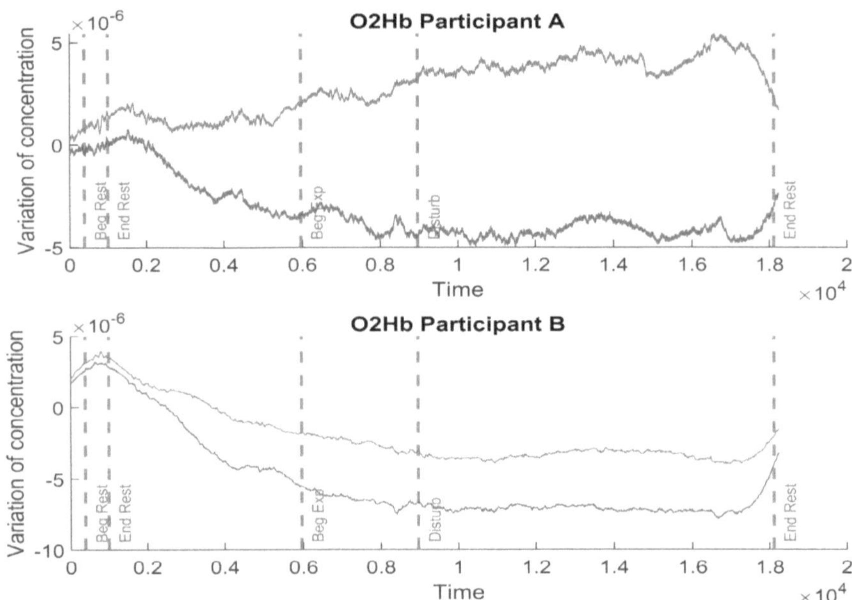

Fig. 2. Rest, Train and Experiment events. Green vertical lines stand for the start and end of each event (Dhamo et al. 2024b).

3.2 Computational Results and Evaluation

Here we present and evaluate the computational results obtained for the original KM and the extended KM models. The oxyhemoglobine signals described in Subsec. 3.1 were mapped into visibility networks with the same number of nodes (613 nodes) of one minute time duration for each event. For each participant we constructed one network corresponding to the *before disturb* event. We refer as lPFC (rPFC) to signals measured in the left (right) prefrontal cortex hemispheres; pAhL (pAhR) the signals measured at the left (right) prefrontal cortex hemisphere of the participant pA; pBhL (pBhR) the signals measured at the left (right) cortex hemisphere of the participant pB. The original KM and the extended KM models are executed for all the constructed networks. The evolution of the phases θ_i, $i = 1, 2, ..., 613$ are obtained for all the networks for different values of the coupling strengths. Different networks reached the stationary state for different values of the coupling strength λ. In Rest networks in both models the values of the coupling strength are taken from interval [0;1] with a very small step of 0.01. The complex order parameter is computed using Eq. (2) within the stationary state for both models before and after applying the input signals I_i. Since the natural frequencies are taken at random, we repeat the execution of the Kuramoto model 10 times for each network and then compute the final complex order parameter as the average of the order parameters obtained in all executions.

For the experiment, we used the open source library Stable Baselines 3 in Python to implement our above mentioned scenarios. The values of the parameters used in the executions are shown in Table 2.

Table 2. The accuracy of piecewise linear approximation according to different values of L.

Learning rate	N° of training mini batches *per* update	N° of steps to run for each environment *per* update	Discount factor	Factor for the bias trade-off	Entropy coefficient for the loss calculation
0.0003	64	2048	0.99	0.95	0

The external signals I_i can be applied to all the nodes in the network or to a part of them. Firstly, we applied the signals in randomly selected nodes starting from 5% up to 100% of the nodes with a step 5. Based on the results, we fixed the ratio of pinned nodes to 50% and continued the experiments by computing the complex order parameter for different values of the coupling strength. Comparisons between evolution of complex order parameters before and after stimuli are provided. In addition, in the second case, the synchronization suppression coefficient is introduced.

Figure 3 visualizes the synchronization suppression coefficient for increasing ratio of the randomly pinned nodes. It indicates an abrupt transition from the case when 0% of the nodes were subject to external signals to when 5% of the nodes received such signals.

There can be seen an instant growth of the suppression coefficient when the PPO signals are added up to 5% of the nodes and after this threshold the suppression coefficient does not change much from each-other, indicating that the more nodes are subjected to the PPO signals I_i the more synchronization is hindered. Taking as reference results in Fig. 3 we fix the ratio of pinned nodes 50% since it corresponds to the highest value of synchronization suppression.

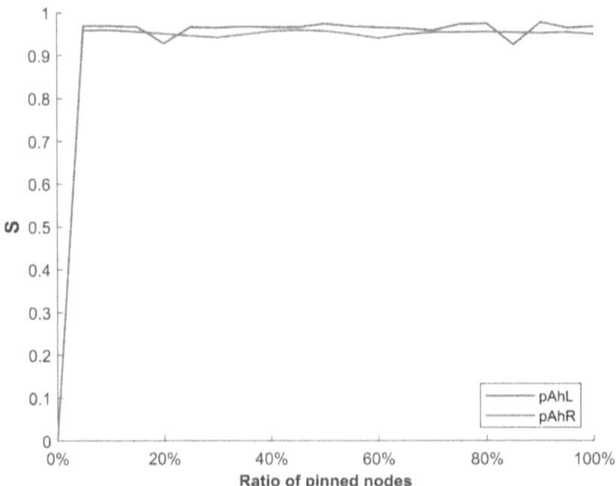

Fig. 3. Synchronization suppression coefficient for randomly selected pinned nodes.

Figures 4, 5, 6, 7 illustrates the results of the complex order parameter computed respectively for pAhL, pAhR, pBhL and pBhR networks for all dyads before and after applying the external signals generated via PPO algorithm. The main plot and the inset visualize respectively the complex order parameter before and after applying the external signal I_i.

Initially, the dyads demonstrate a smooth, second-order phase transition from asynchronous to synchronized states as coupling strength increases. However, upon introducing external signals generated by the PPO algorithm, the insets clearly reveal disruption in this orderly synchronization progression. Specifically, the smooth transitional behavior becomes irregular and less predictable, departing notably from the characteristic second-order phase transition typically observed in Kuramoto systems. This indicates that adding external signals suppresses synchronization in all the networks studied visualized in Figs. 4, 5, 6, 7. Furthermore, the incorporation of these signals into our model disrupts the initial phase transition from the incoherent state up to the coherent state.

From some of the networks, we noticed that the order parameter has an increasing tendency despite that the values are diffused (Fig. 4 Dyas 1–6, 8, Fig. 5 Dyad 1–2, 4–5, 8, Fig. 6 Dyad 1–5, Fig. 7 Dyad 1–4) while there are network cases in which the increase is not clear (Fig. 4 Dyad 9, Fig. 5 Dyad 3,7, Fig. 6 Dyad 7, 8).

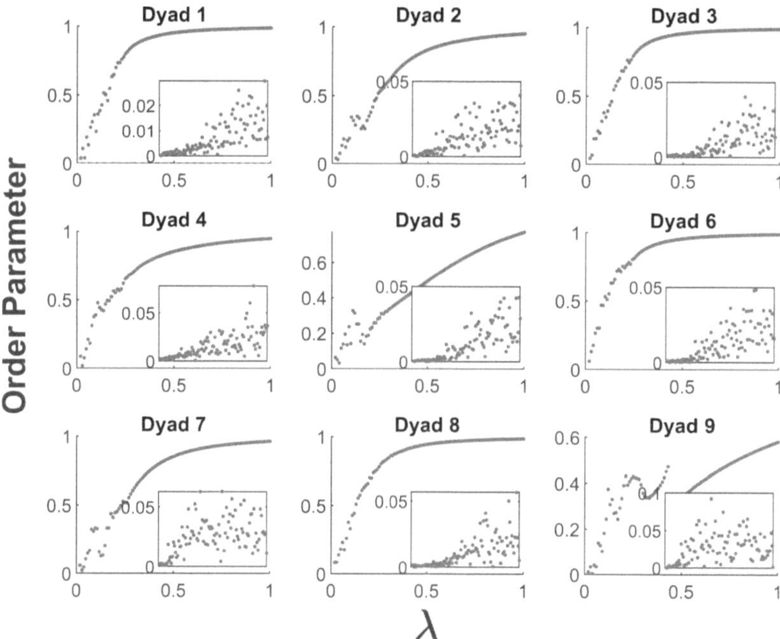

Fig. 4. Complex order parameter before and after applying the external signal generated via PPO algorithm for **pAhL networks**.

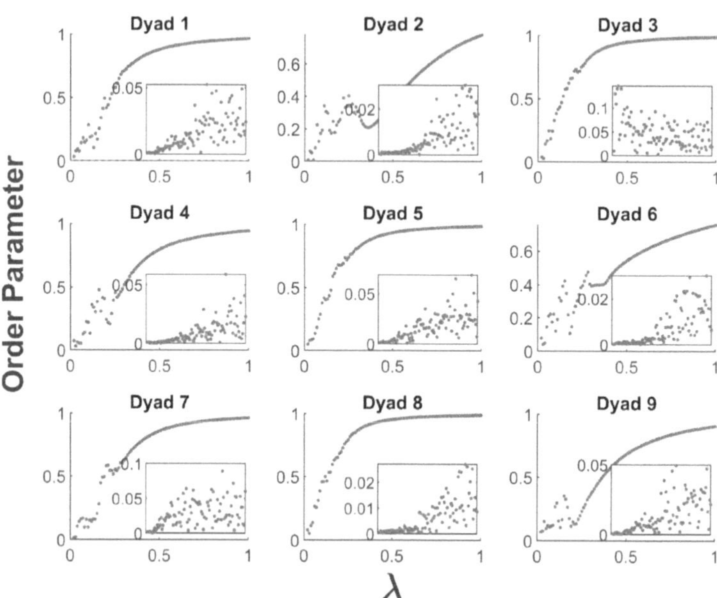

Fig. 5. Complex order parameter before and after applying the external signal generated via PPO algorithm for **pAhR networks**.

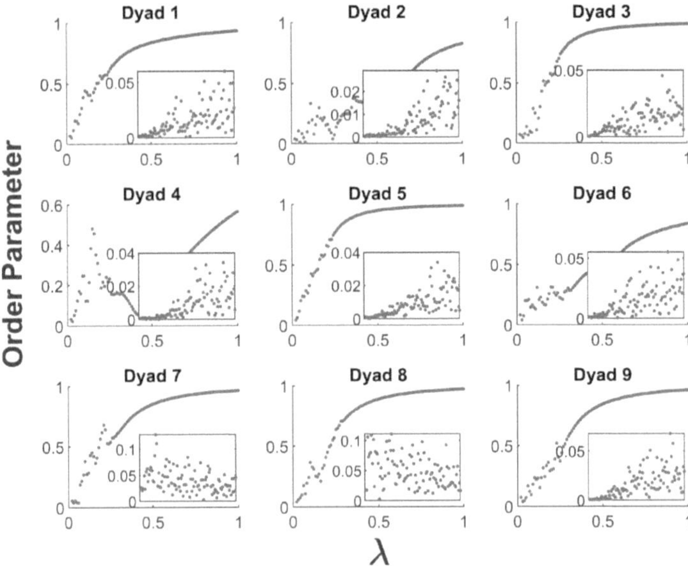

Fig. 6. Complex order parameter before and after applying the external signal generated via PPO algorithm for **pBhL networks**.

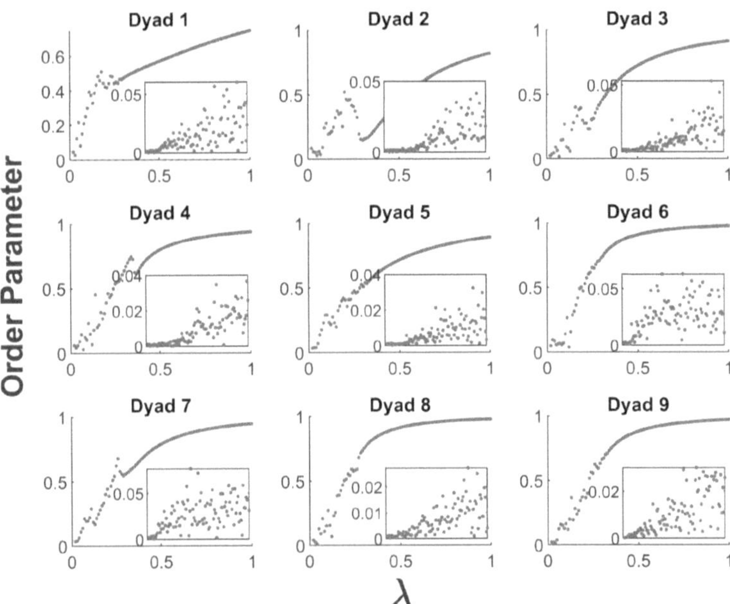

Fig. 7. Complex order parameter before and after applying the external signal generated via PPO algorithm for **pBhR networks**.

In order to numerically quantify the difference between the complex order parameter values before and after stimuli, we compute the following distances:

$$D = \frac{1}{L}\sum_{i=1}^{L}\left(R_{i_{before}} - R_{i_{after}}\right)^2 \tag{6}$$

where L stands for the total number of λ used in computing KM. The distances presented in Table 3 provide evidence of the possible similarities with respect to the order parameters before and after stimuli for all the dyads and all the cortical hemispheres considered. Larger values of the distance indicate less similarity and smallest values indicate more similarity between the curves.

Table 3. Distance between complex order parameter before and after applying stimuli

	Dyad 1	Dyad 2	Dyad 3	Dyad 4	Dyad 5	Dyad 6	Dyad 7	Dyad 8	Dyad 9
pAhL	0.738	0.553	0.748	0.595	0.257	0.757	0.556	0.725	0.151
pAhR	0.601	0.226	0.678	0.517	0.691	0.288	0.562	0.748	0.405
pBhL	0.569	0.226	0.683	0.093	0.721	0.277	0.559	0.575	0.547
pBhR	0.305	0.271	0.442	0.529	0.449	0.618	0.502	0.656	0.602

4 Conclusions and Future Work

In this work, our focus was to explore more in the synchronization of brain signals by representing them as complex networks and incorporating reinforcement learning into the analysis of synchronization suppression. We have broadened the Kuramoto model by introducing input signals to the pinned nodes of the visibility networks fNIRS signals generated using Proximal Policy Optimization algorithm and analyzed the synchronization suppression coefficient before and after the stimuli for 18 patients, which are grouped in 9 dyads. From the analysis, we conclude that external signals have the effect of suppressing synchronization in all the networks studied, and also in our model, which disrupts the initial phase transition from the incoherent state to the coherent state. The values of the synchronization suppression coefficient obtained for all the networks when considering different percentages of the randomly selected nodes indicated an abrupt change when the ratio of pinned nodes increased from 0% up to 5% and after that, these values became stable. As an overall conclusion, we conclude that desynchronization is guaranteed when external signals generated via PPO algorithm are incorporated into the original KM.

In future work, we would like to study potential convergence issues with the Proximal Policy Optimization (PPO) algorithm, namely, the early convergence, which has been reported in other works in the literature. The study of the learning rate (LR) and the use of a dynamic annealing LR would be of interest to study as well.

Acknowledgements. This research work has been supported by the READ Project of the *Albanian-American Development Foundation (AADF)*, Albania.

References

1. Krylov, D., Tachet, R., Laroche, R., Rosenblum, M., Dylov, D.V.: Reinforcement learning framework for deep brain stimulation study. In: Twenty-Ninth International Joint Conference on Artificial Intelligence, Yokohama (2021)
2. Dhamo, X., et al.: Synchronization processes in fNIRS visibility networks. Appl. Netw. Sci. **9** (2024)
3. Schulman, J., Filip, W., Dhariwal, P., Radford, A., Klimov, O.: Proximal policy optimization algorithms. Mach. Learn. (2017)
4. Qiu, X., Yang, L., Guan, C., Leng, S.: Closed-loop control of higher-order complex networks: finite-time and pinning strategies. Chaos, Solitons Fractals. **173** (2023)
5. Tang, Y., Zhou, L., Tang, J., Rao, Y., Fan, H., Zhu, J.: Hybrid impulsive pinning control for mean square synchronization of uncertain multi-link complex networks with stochastic characteristics and hybrid delays. Mathematics. **11** (2023)
6. Yi, C., Xu, C., Feng, J., Wang, J., Yi, Z.: Pinning synchronization for reaction-diffusion neural networks with delays by mixed impulsive control. Neurocomputing. **339**, 270–278 (2019)
7. Chen, G.: Pinning control of complex dynamical networks. IEEE Trans. Consum. Electron. **68**, 336–343 (2022)
8. Krylov, D., Dylov, D.V., Rosenblum, M.: Reinforcement learning for suppression of collective activity in oscillatory ensembles. Chaos. **30** (2020)
9. Naros, G., Naros, I., Grimm, F., Ziemann, U., Gharabaghi, A.: Reinforcement learning of self-regulated sensorimotor β-oscillations improves motor performance. NeuroImage. **134** (2016)
10. Tang, E., Bassett, D.S.: Colloquium: control of dynamics in brain networks. Rev. Mod. Phys. **90** (2018)
11. Breakspear, M., Heitmann, S., Daffertshofer, A.: Generative models of cortical oscillations: neurobiological implications of the Kuramoto model. Front. Hum. Neurosci (2010).
12. Strogatz, S.H., Abrams, D.M., McRobie, A., Eckhardt, B., Ott, E.: Crowd synchrony on the millennium bridge. Nature. **438** (2005)
13. Jalife, L.: Mutual entrainment and electrical coupling as mechanisms for synchronous firing of rabbit sino-atrial pace-maker cells. J. Physiol. **356** (1984)
14. Winfree, A.T.: The Geometry of Biological Time. Springer, New York, NY (2001)
15. Dörfler, F., Bullo, F.: Synchronization and transient stability in power networks and nonuniform Kuramoto oscillators. SIAM J. Control. Optim. **50** (2012)
16. Osipov, G.H., et al.: Synchronization phenomena in networks of oscillatory and excitable Luo-Rudy cells. In: Complex Dynamics in Physiological Systems: from Heart to Brain. Understanding Complex Systems. Springer, Dordrecht (2009)
17. Dhamo, X., et al.: Global synchronization measure applied to brain signals data. In: Studies in Computational Intelligence. Springer, France (2024)
18. Abrams, D.M., Strogatz, S.H.: Chimera states for coupled oscillators. Phys. Rev. Lett. **93** (2004)
19. Boccaletti, S., et al.: Explosive transitions in complex networks' structure and dynamics: percolation and synchronization. Phys. Rep. **660**, 1–94 (2016)
20. Gómez-Gardeñes, J., Gómez, S., Arenas, A., Moreno, Y.: Explosive synchronization transitions in scale-free networks. Phys. Rev. Lett. **106** (2011)
21. Sorrentino, F., Pecora, L.M., Hagerstrom, A.M., Murphy, T.E., Roy, R.: Complete characterization of the stability of cluster synchronization in complex dynamical networks. Sci. Adv. **2** (2016)
22. Pecora, L.M., Sorrentino, F., Hagerstrom, A.M., Murphy, T.E., Roy, R.: Cluster synchronization and isolated desynchronization in complex networks with symmetries. Nat. Commun. **5** (2014)

23. Arenas, A., Díaz-Guilera, A., Pérez-Vicente, C.J.: Synchronization processes in complex networks. Phys. D Nonlinear Phenom. **224**, 27–34 (2006)
24. Wang, X., Zhang, Y., He, Y., Lu, K., Hao, N.: Dynamic inter-brain networks correspond with specific communication behaviors: using functional near-infrared spectroscopy Hyperscanning during creative and non-creative communication. Front. Hum. Neurosci. **16** (2022)
25. Lacasa, L., Nuñez, A., Roldán, É., Parrondo, J.M.R., Luque, B.: Time series irreversibility: a visibility graph approach. Eur. Phys. J. B. **85** (2012)
26. Lacasa, L., Luque, B., Luque, J., Nuño, J.C.: The visibility graph: a new method for estimating the Hurst exponent of fractional Brownian motion. Europhys. Lett. **86** (2009)
27. Li, R., Mayseless, N., Balters, S., Reiss, A.L.: Dynamic inter-brain synchrony in real-life inter-personal cooperation: a functional near-infrared spectroscopy hyperscanning study. NeuroImage. **238** (2021)
28. Lacasa, L., Luque, B., Ballesteros, F., Luque, J., Nuño, J.C.: From time series to complex networks: the visibility graph. Appl. Math. **105**, 4972–4975 (2008)
29. Jiruska, P., de Curtis, M., Jefferys, J.G.R., Schevon, C.A., Schiff, S.J., Schindler, K.: Synchronization and desynchronization in epilepsy: controversies and hypotheses. J. Physiol. **591**, 787–797 (2013)
30. Dörfler, F., Francesco, B.: Synchronization in complex networks of phase oscillators: a survey. Automatica. **50**, 1539–1564 (2014)
31. Pikovsky, A., Rosenblum, M., Kurths, J.: Synchronization. Cambridge University Press (2001)
32. Arenas, A., Díaz-Guilera, A., Kurths, J., Moreno, Y., Zhou, C.: Synchronization in complex networks. Phys. Rep. **469**, 93–153 (2008)
33. Li, K., Yang, L., Guan, C., Leng, S.: Reinforcement learning-based pinning control for synchronization suppression in complex networks. Heliyon. **10** (2024)

Self-Adaptation and Big Data Analytics to Enhance Decision-Making in Smart City Platforms: A Systematic Literature Review

Mubashir Ali[1], Fabio Moretti[2], Cristiano Novelli[2], and Patrizia Scandurra[3(✉)]

[1] School of Computer Science, University of Birmingham, Birmingham, UK
`m.ali.16@bham.ac.uk`
[2] Smart Cities and Communities (TERIN-ICER-SCC) lab., Roma, Italy
`{fabio.moretti,cristiano.novelli}@enea.it`
[3] DIGIP, University of Bergamo, Bergamo, Italy
`patrizia.scandurra@unibg.it`

Abstract. In recent years, engineering software platforms aimed at supporting the development and integration of applications and services for smart cities has become increasingly significant and some initiatives have gained traction to pursue environmental, economic and social goals. Decision science plays a crucial role and needs to be incorporated into such platforms to inform and make decision-making more effective. To address these challenges, self-adaptation (SA) and big data analytics (BDA) are two prominent approaches for decision science that have not yet been sufficiently explored in the context of smart city.

This paper presents the results of a systematic literature review on software platforms for smart cities, with a focus on SA and BDA, and their interrelation. This review can benefit both researchers and practitioners in the field, who can use it as a reference for conceiving smart city platforms that enable seamless integration and cooperation of diverse and competing city services, leveraging SA and BDA for decision science.

1 Introduction

Recently, there is an increasing trend toward adopting a *citizen-centric* concept of Smart City and software platforms [2,11,13], which can aggregate as much information and "polymerize" the city services. This is giving rise to the concept of *smart city as a platform* where service providers can easily and uniformly access to a variety of information thanks to one source of aggregated information and end-to-end APIs. Such platforms typically make use of Internet-of-Things (IoT), cloud computing, and big data management [14]. IoT sensors and smart city apps ("citizens as sensors") generate large amounts of heterogeneous data in real time and space. These urban data streams are vital information that can be processed to provide novel services and manage or optimize different and competing concerns of decision science in the context of a smart city [22].

© The Author(s), under exclusive license to Springer Nature Switzerland AG 2026
M. Pavone et al. (Eds.): DSA ISC 2025, LNCS 16405, pp. 62–77, 2026.
https://doi.org/10.1007/978-3-032-21811-7_5

However, over time various engineering challenges have arisen in the design, development, and deployment of smart city applications. Traditionally, city services are implemented via heterogeneous siloed ICT-based solutions by public-private partnerships that operate in specific business domains (such as street lighting, waste management, city surveillance, parking, public transport, etc.). Moreover, cities and their citizens have different characteristics and needs, and sector-specific digital platforms for urban services need to be configured and connected properly for a specific city instance. Other concerns include integrability, scalability, extensibility and context-awareness issues due to the high number and diversity of application domains and city infrastructures [1,23] and to socio-technical aspects [15].

To this purpose, *Self-Adaptation* (SA) [8,36] and *Big Data Analytics* (BDA) [9] are two prominent approaches to handle in part such issues and develop smart city platforms with more effective services and decision-making processes [13]. On the one hand, big data management is mandatory for the aggregation and mining of large volumes of changing and unstructured data, and BDA solutions are essential for enabling decision science. Integrating BDA into smart city platforms together with human-comprehensible dashboards and control rooms allows city managers to effectively extract emergent characteristics, accurate predictions, and hence have a more effective decision-making process for resource monitoring and management [13,17]. On the other end, SA is now regarded as a key capability for modern software systems to manage run-time uncertainties. It ensures that systems meet their functional requirements and expected Quality of Service (QoS) by adjusting themselves at run-time. SA may help smart cities platforms in adapting to dynamic requirements: self-configuring in response to situations and changing needs of the city and its citizens, self-optimizing based on key performance indicators, and also, if guided by BDA, transforming volumes of data into actionable insights to anticipate changes. For example, a city traffic management system can collect data from various sources, perform near-real-time data correlation analysis, and optimize traffic guidance accordingly.

This paper aims at identifying and reviewing the main academic articles in the literature from 2010 to mid-2024 about software platforms for smart cities, by highlighting the aspects of SA and BDA. As promoted by the IBM vision of autonomic computing [18], we rely on the MAPE-K (Monitor, Analyse, Plan and Execute over a shared Knowledge) feedback control loop model as the mainstream mechanism for SA. We conducted a systematic literature review by searching major scientific databases, resulting in 268 primary studies after applying stringent exclusion and inclusion criteria. We rigorously defined a framework for categorizing such primary studies and synthesized the obtained data to produce a clear overview of the state of the art about SA and BDA in smart city platforms. In particular, we used the taxonomy by Krupitzer et al. [20] as assessment framework for SA, which we applied to all selected studies.

This paper is organized as follows. Section 2 describes some other related surveys and reviews. Section 3 summarizes the research method used for this review. Section 4 presents the main result, while Sect. 5 sums up the key findings. Finally, Sect. 6 concludes the paper and outlines future steps.

2   Related Work

Only a few surveys have so far attempted to map the existing proposals of software platforms for smart city, and their scopes have been rather limited to the fundamental concept of smart city and BDA methods for the development of applications. We here summarize those most relevant to our review.

Habibzadeh Hadi et al. [13] presented a multi-faceted survey of machine intelligence in modern smart city applications. The work presents a layered view of smart city infrastructure, emphasizing the data plane as central to computation and storage, and evaluates data processing, analytics, machine learning, and visualization for specific applications.

The survey of Santana et al. [31] considers 23 platforms to analyze enabling technologies and functional/non-functional requirements. They also proposed a reference architecture to guide the development of a smart city platform.

ChuanTao Yin et al. [37] conducted a literature survey on smart cities to understand its definition and application domains. They also examined enabling technologies for the development of smart city platforms and also proposed a smart city reference architecture, which consists of four different layers: data acquisition, data vitalization, common data and services, and domain application.

Ruben et al. [30] presented a survey to analyze the work done so far in the domain of smart city. They provide an overview of the smart city definitions discussed in the literature, technologies and methodologies used in the development of smart city software platforms, smart city application domains, and at the end they highlight the research challenges and future opportunities for smart cities.

Eiman Al Nuaimi et al. [4] conducted a review to analyze the application of big data to support smart cities. Yosra Hajjaji et al. [14] reviewed big data and IoT-based applications in smart environments to identify key areas of application, current trends, data architectures, and ongoing challenges in these fields.

We have also found fewer studies focusing on self-adaptation in the context of different niche application domains, such as in cyber-physical systems [25], and in IoT Systems for Smart Cities [26]. A recent concept of smart, sustainable, and resilient city utilizes knowledge systems, AI algorithms, and adaptation mechanisms to ensure urban efficiency, quality of life, competitiveness, and inherent resiliency. Although this concept is demonstrated through a web-based platform and virtual reality simulations, a concrete solution has yet to be developed.

SA and BDA remains relatively underexplored in the current literature. BDA has received a lot of attention in the last few years in the smart city context, while we did not find any survey or review article focusing on the link between SA and BDA (the intended goal of our review). In relation to the preceding reviews, our contribution is the first comprehensive study that investigates into prior studies about the intersection of SA and BDA in smart city platforms.

3   Study Design

We applied the conventional three-stage process for conducting systematic literature reviews [19][1]. This section summarizes the activities in the stages *Planning*,
Conducting, and *Reporting*. A package (including the protocol, the selected studies, and the collected and analyzed data) is also publicly available in [3].

3.1   Planning Review

First, we defined the protocol. This included identifying the research questions,
the search strategy, the screening strategy, and the data extraction strategy.

Research Questions. The following questions shaped the whole investigation
study:

- **RQ1**: *How is self-adaptation implemented in software platforms for smart
  cities?* Rationale: the answer to this question will help understand the extent
  to which current software platforms for smart cities target (self-)adaptation,
  and which runtime adaptation mechanism they adopt.
- **RQ2**: *Which concerns (goals) of adaptation are handled in software platforms
  for smart cities?* Rationale: this question will help to determine which types
  of adaptive features, such as self-* capabilities associated with autonomic
  computing [18], are pursued in the context of a smart city. As a sub-question,
  we aim to identify real concerns in one or more interrelated domains, possibly
  supported by BDA technologies for decision making.
- **RQ3**: *What types of services are developed for smart cities using big data
  analytics?* Rationale: this will categorize application domains requiring BDA.
- **RQ4**: *Which big data technologies are used in services for smart cities?* Rationale: this will identify suitable big data technologies for various domains.

Search Strategy. The search process was designed as a multi-stage approach
to include all potentially relevant studies. It combines manual and automatic
searches. The manual search aimed to identify pilot studies on well-known software platforms for Smart Cities. The automatic search utilized various widely
adopted scientific digital libraries. The following two search strings were used
to search into the digital libraries according to the two groups of formulated
research questions, namely {RQ1,RQ2} and {RQ3,RQ4}, respectively.

First search string(s): s_1 ***AND*** s_2 where:
 - $s_1 \equiv$ (*"smart city"*) ***AND*** (*platform **OR** software **OR** system **OR**
   framework*)
 - $s_2 \equiv$ (*adaptation **OR** adaptive **OR** "self-adaptation" **OR** "self-adaptive"
   OR "MAPE-K"*)

Second search string: (*"smart city"*) ***AND*** (*application **OR** service*) ***AND***
 (*"urban data analytic" **OR** "big data analytic" **OR** "machine learning"*)

---

[1] We did not adopt AI/ML automation tools for data extraction, also because they
were not off-the-shelf when we carried out most of the work.

Screening Strategy. As preliminary screening, the search strategy targeted peer-reviewed journal and conference papers, and was applied mainly on the title, keywords, and abstract. The considered time frame is the decade 2010–2020 (pre-pandemic period), plus the additional period from 2021 to mid-2024. Of course, duplicates of studies have been removed as well.

In order to further filter the studies resulting from the this preliminary screening, we defined the following inclusion and exclusion criteria.

Inclusion Criteria: studies on software platforms for smart cities adopting self-adaptation mechanisms; studies in which smart city services are developed by using machine learning techniques and BDA.

Exclusion Criteria: studies not written in English; secondary or tertiary studies (e.g., systematic reviews, surveys, etc.); studies in the form of editorials, tutorial, and posters, because they do not provide enough information; commercial online brochures by ICT companies (such as IBM, Microsoft, Huawei, etc.) since we found very little technical information about them.

When going through each selected study for starting with data extraction, some other studies that were semantically out of scope were excluded. Moreover, since we have divided our main search string into two different sub-strings, as a final cross-check, papers resulting from the first sub-string which were on the scope encompassed by the second sub-string were included in the final sample of studies for the RQ group {RQ3,RQ4}, and vice-versa.

Data Extraction Strategy. First, common information were extracted directly from the documents such as title, authors, a brief summary, publication year, venue, etc. Then, different data were extracted to answer the research questions depending on the RQ set. For the set {RQ1,RQ2} (SA mechanisms for smart city platforms) data features extracted include: adaptation mechanism (MAPE-K loop, self-organizing agents, etc.) and other dimensions of adaptation of the classification framework by Krupitzer et al. [20] (for RQ1); and concerns of adaptation, application domain, and BDA adoption for adapting (for RQ2). Data features for the set {RQ3,RQ4} (concerning smart city services and BDA) are the application domain (for RQ3), and big data technologies (for RQ4).

3.2 Conducting Review

In this phase we executed the previously defined protocol.

Manual Search and Selection. We started with some pilot studies about well-known smart city platforms that we identified based on our prior knowledge using Google Scholar. This initial set of studies was then expanded by manually reviewing additional papers cited in each study. The initial screening of these additional studies was based on their title and abstract, whereas the final decision about their inclusion or not was based on their content. We obtained 15 potentially relevant studies. Details of these selected papers are reported in Appendix A, available online in our replication package [3].

Automatic Search and Selection. The results of the overall automatic search and selection process for the two different groups of search strings is shown in Fig. 1 (the number of the studies selected during the stages is also depicted). We used five widely adopted digital libraries: Scopus, ACM Digital Library, IEEE Xplore, ScienceDirect, and SpringerLink. We obtained two final study sets from 2010 to mid-2024: 56 studies for the first search string and 197 studies for the second search string. The list of these last studies are also available in [3]. Data was extracted by exploiting and synthesizing data features from the research questions. This involved manually searching data features from each selected study. A combination of content analysis and narrative synthesis was used to interpret the findings correctly. The results this data synthesis are presented in Sect. 4.

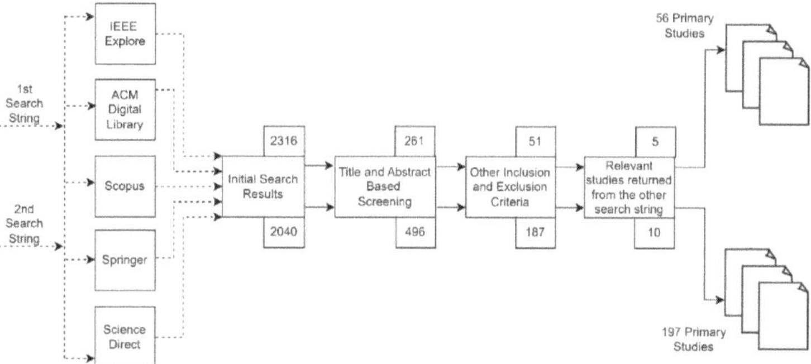

Fig. 1. Multi-staged automatic search and selection process.

3.3   Reporting Review

Data extracted from the studies was collected in two separate spreadsheets for the RQ sets {RQ1,RQ2} and {RQ3,RQ4}. These spreadsheets are available online in the data package [3], while the main findings are described in Sect. 4.

4   Results

4.1   Answers to Research Questions {RQ1,RQ2}

From the manual and automatic search, 71 primary studies (from both manual and automatic searches) were considered totally for data extraction.

Mechanism and Dimensions of Adaptation (RQ1). To answer research question RQ1, we examined whether the collected software platforms perform self-adaptation to support decision-making in smart cities, and how this is engineered. We discovered that only a small number of case studies adopt the MAPE-K feedback control loop as architecture style [18], but (as better explained below) there are other studies that implement self-adaptation differently. To the end of extracting and classifying the adaptation mechanisms from all these studies, we adopted the dimensions of the taxonomy by Krupitzer et al. [20] for describing self-adaptation in software systems, namely: *Adaptation Control, Time, Reason, Level,* and *Technique.* We choose such a taxonomy among others since it reasonably answers the *5W + 1H* questions introduced by M. Salehie et al. in [29][2] for eliciting adaptation requirements, and it includes context adaptation.

Adaptation Control. Based on the analyzed studies, we first investigated whether self-adaptation is supported or not trying to understand the *Adaptation Control* mechanism ("How to adapt?"). From the decade set 2010–2020, only four software platforms [7,12,17,33] (papers[3] P11, P12, P17, and P15 in Appendix A) adopt the MAPE-K feedback control loop model as adaptation mechanism. Two additional papers [5,34] come from the three-year (post-pandemic) period 2021–23. With this limited number of studies (six studies in total), we therefore investigated whether the MAPE-K control loop model is supported "implicitly", i.e., by cause-effect dependencies of interactive software components that actually realize closed feedback control loops for monitoring and actuation (though the study does not explicitly states their role of MAPE components), or if another type of adaptation mechanism is used (e.g., self-organizing agents).

Figure 2 shows the main adaptation approaches identified in the studies. Remarkably, most of the platforms adopt MAPE-K loops implicitly (62.9%), while 15.7% of the studies do not detail (in the text or in the software architecture) any type of adaptation. Only 10% use an agent-based mechanism to realize adaptation, also combined with explicit (2.9%) MAPE-K loops, and 2.9% apply a nature-inspired approach. By the obtained percentages, it is worth to note that control architectures based on the MAPE-K loop model are less popular w.r.t. an agent-based approach. A possible reason for this could be the major flexibility and scalability provided by existing ready-to-use agent-based simulation tools and services, which in the context of smart city come in handy for simulating use cases and conducting experiments with real data and virtual reality.

For the studies using MAPE-K control loops implicitly, we proceeded with a fine-grained analysis to get more insight on the "completeness of control" by measuring it in terms of supported MAPE functions. The results are shown in Fig. 3) by denoting, for example, with MA solutions supporting only sensing and analysis, with PE only planning and actuation, etc. Many approaches cover all

---

[2] As in [29], the question: "Who has to perform the adaptation?" is not investigated, since SA refers to the ability of a system to automatically adapt to changes.

[3] We use the notation Px to refer to a surveyed paper listed in Appendix A.

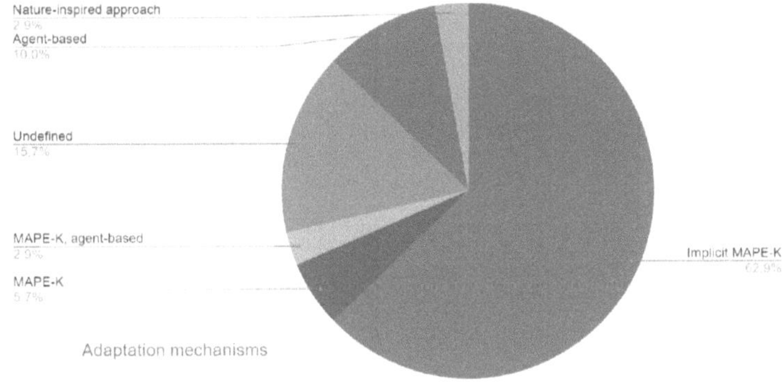

Fig. 2. About the adaptation mechanisms

MAPE functions, closing the feedback loop. Some other approaches support only one MAPE function, limiting their architectures' scope as standalone solutions.

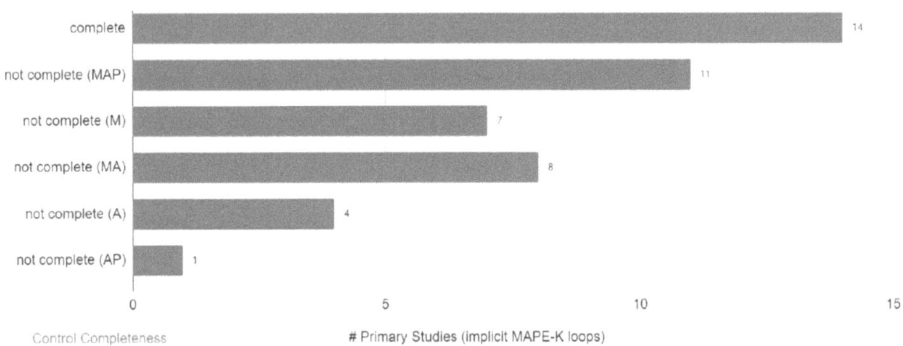

Fig. 3. About control completeness of studies using implicit MAPE-K control loops.

Time Dimension. This aspect is related to the when-question ("When should we adapt?") and distinguish two time dimensions *before* or *after* the need for adaptation for describing the temporal aspect of adaptation [20, 24].

The simplest form is *reactive adaptation*; usually, self-adaptive systems tend to be reactive, i.e. they adapt in response to changes without anticipating what the next adaptation needs will be. *Proactive adaptation* addresses the limitations of reactive adaptation taking into account not only the current conditions, but how they are estimated to evolve when deciding to adapt. So, based on predicted events, proactive adaptation adjust the system in advance. With proactive adaptation, the monitored data is used to forecast system behavior or environmental state. Proactive adaptation requires computationally intensive reasoning that may be realized as part of the Analysis and Plan phases of a MAPE-K loop with the help of (big)data analytics and machine learning algorithms [28].

In smart city scenarios, reactive adaptation could lead to resource waste, transient unavailability of services and behavioral fluctuations. In contrast, proactive adaptation can help in preventing the occurrence of problems and/or in mitigating the effects of upcoming problems by dynamically re-planning adaptation actions in advance. As an example, if during the execution of a freight transport process a delay is predicted, faster transport services (such as air delivery instead of road delivery) can be scheduled to prevent the delay. This review distinguishes between proactive and reactive adaptation to describe the temporal aspect of adaptation (i.e., when to adapt, before or after the need for adaptation) in the considered studies. As expected, the reactive nature resulted to be the most popular form (66,7%); instead, proactive adaptation in the context of smart city is not explored enough (proactive only 8,3%, mixed proactive/reactive 25%), though it could play a central role for critical services in domains such as energy consumption, transportation, and epidemic containment, to name a few.

Reason. This dimension is related to the why-question ("Why do we have to adapt?"). Adaptation can be triggered by changes in the managed technical resources (e.g., a defect of a hardware component, a software fault, or the availability of an alternative network connection or service), in the environment or context (e.g., the state change of a context variable like air quality), or in the users (e.g., in the user preferences). Answering this question is essential for software architects as it impacts the design of the adaptation logic; the reasons determine the elements that have to be monitored and the adaptation/control actions to actuate. The left side of Fig. 4 shows the distribution of the primary studies among the main categories of changes (*context, technical resource, user,* and *mixed*) as identified in [20]. In most of the examined studies, the trigger for the adaptation process are changes in almost all categories of observed elements (the technical resources, the context, and the users). This is an expected result since all categories of changes are common in the urban landscape of data and services. For some studies, it was not possible to determine the reason for adaptation, hence we kept them in as a separate category (*undefined*).

Level. This aspect is related to the where-question ("Where do we have to implement change?"), i.e. identifying the levels, where the adaptation should take place including: technical managed resources (such as hardware, application, middleware, physical network infrastructure, logical communication, etc.), the system environment and the users. Due to the lack of technical details in the evaluated studies and platform descriptions, we did not report any significant results; however, we were able to identify (see next paragraph) the general type of adaptation that the evaluated platforms are able to actuate.

Technique. This dimension is related to the what-question ("What kind of change is needed?"). In this review, we considered the distribution of the primary

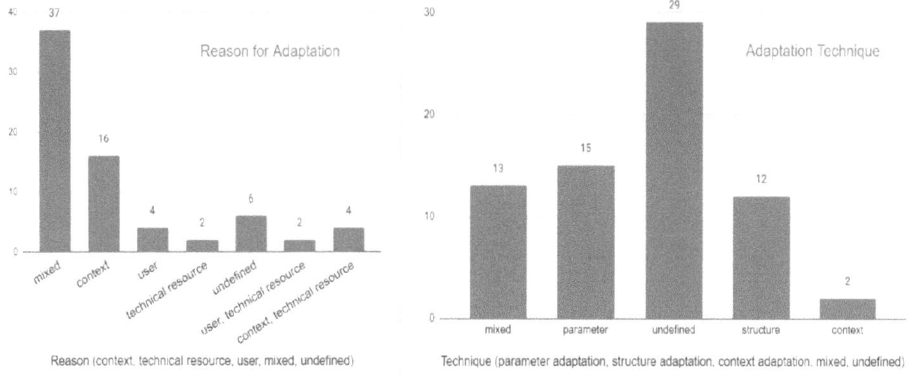

Fig. 4. Reason and Technique dimensions for adaptation

works (see the right side of Fig. 4) along the categories of techniques for adaptation as in [20], namely: *parameter adaptation, structure adaptation,* and *context adaptation.* Parameter refers to adaptation through the change of parameters. Structural adaptation enables the exchange of algorithms or system components dynamically at runtime. Context adaptation refers to any changes in the context. Combinations of techniques (*mixed*) in one adaptation plan is also possible, e.g., changing parameters of one component and adding further ones. We observed that studies are evenly distributed among all categories, except for modifying the context variables. For this last category, we found only a couple of works making pure context adaptation. One is the ingestion and analytics architecture (P1) described in [35] that supports both real-time and historical data analytics in the transportation and energy domain to extract and spread actionable knowledge (e.g., generating a complex event representing an anomaly, which can then in turn be used to notify the user as well). A high number of studies were unclassified (*undefined*), mainly because of the incomplete adaptation control.

Adaptation Concerns (RQ2). When analyzing the primary studies, we also traced the application domains and concerns of adaptation in smart city platforms. In order to relate the studies that adopt BDA with those performing self-adaptation, we specifically searched for those using machine intelligence (big data and AI together) for adapting [32]. We here report only the results of those platforms that feature a complete adaptation control (30 in total) and employ BDA to process data and extract useful knowledge for planning and guiding adaptation (10 out of 30). For these ten studies, Fig. 5 shows their application domain together with their concerns of adaptation. In this way we unveil those sectors that are more demanding of proactive adaptation. The dominant application domains are those related to transportation, surveillance, and traffic management. Concerns of adaptation are QoSs (response time, reliability, privacy and security) and the provision of *context-aware* services such as deliv-

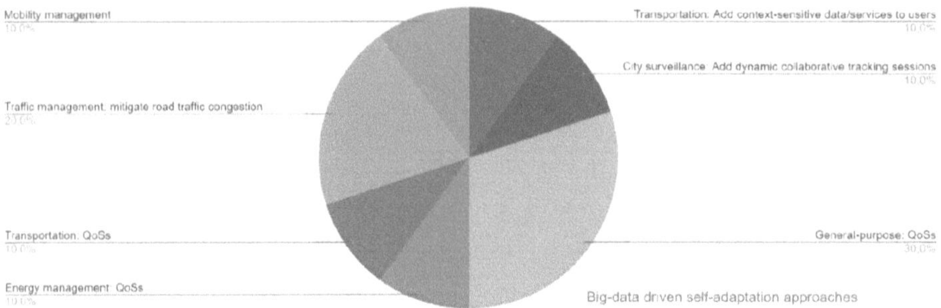

Fig. 5. About studies using BDA to guide self-adaptation

ering traffic- and transport- related data/services to users depending on their context, and activating tracking/monitoring surveillance sessions dynamically. All these application-specific studies involve, as expected, proactive adaptation. Some studies do not mention any specific domain (such as papers P32 [23] and P2 [27] in Appendix A) since their adaptation mechanisms are broadly applied to several domains, but operate with limited autonomy and primarily feature simple reactive adaptations, such as alert messages or recommendations.

4.2   Answers to Research Questions {RQ3,RQ4}

We reviewed 197 studies from the automatic search and 2 from the manual.

Smart City Services (RQ3). We examined the application domains from our primary studies to identify smart services developed using big data technologies. Figure 6) shows that the majority of smart services are offered in the transportation domain 35.03%, hence the development of intelligent transportation systems is an active research area. Other (almost equally covered) application domains include: parking, healthcare, energy optimization, emergency management, road-surface monitoring, public safety and security, environmental monitoring, water, and mobility. Other minor appearances (such as tourism, smart farming, smart education, citizens behavior prediction, street cleaning monitoring, and location identification for real estate investment, to name a few) are also reported (10.15%). In view of the increasing importance of big data, there could be new opportunities for the provision of smart and intelligent services in these least represented domains and for identifying new areas for improvement.

Big Data Technologies and Smart City Services (RQ4). We categorized smart city services into three classes: (i) services that collect and use urban data sets without employing big data technologies, (ii) self-contained services that collect and aggregate urban big data (as structured and unstructured data) and also analyze and extract value by means of AI and analytic frameworks, and (iii) services that apply BDA on urban data provided by third parties.

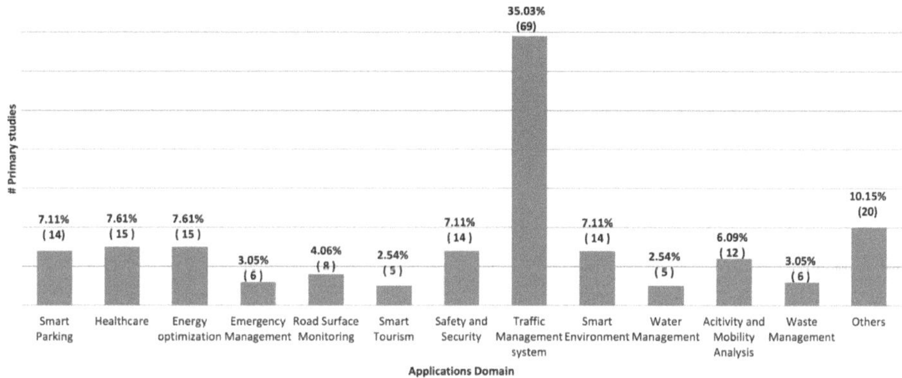

Fig. 6. Distribution of application domains among the studies using big data

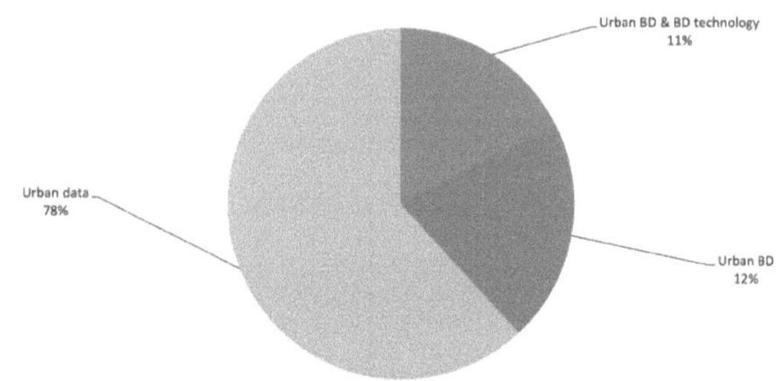

Fig. 7. Smart city services and big data adoption levels

We observed (see Fig. 7) that the majority of services (78%) do not adopt big data (i), hence not big data technology helped the smart city sector to grow. Only 21 studies (11%) device solutions using big data technologies (mostly opensource, like NoSQL databases and parallel data processing tools such as Apache Hadoop and Spark combined) to both collect large urban data sets and then make analytics on them (ii), and 23 (12%) services perform analytics to analyze urban big data collected by third party platforms or data provided by the city administration authorities (iii). This result may be due to several challenges that companies, in general, face in implementing BDA. These include the lack of data scientist skills, and the ever-present security-governance. Data quality and its consistency also presents a big challenge to the management [16].

We finally identified the application domains of the smart services adopting BDA. We found that traffic management is the most prominent domain in which 43% services are developed by using BDA. The second common application domain is energy optimization, in which 13% of the services are based on BDA. We also identified the least common application domains (7%) such as smart education, street cleaning monitoring, and social services diagnosis.

5   Discussion and Outlook

Based on our findings, we list the following key points and research opportunities. *SA mechanisms*: A few number of software platforms expose adaptive features. Most of the architectures that support self-adaptation for city services are agent-based and do not explicitly adopt a MAPE-K control loop architecture. A variety of mechanisms are used to enact limited forms of self-adaptation implicitly with the help of ready-to-use tools (mostly related to auto-scaling or auto-tuning of parameters) and/or component frameworks that do not support complete MAPE loop functions (i.e. they focus only on data monitoring/analysis). This indicates that the field is young, and more advances will emerge in the next years. In particular, *decentralized self-adaptation* and AI agents [21] can unveil new horizons and opportunities for smart cities for the integration of Cyber-Physical-Systems, Social-Technical Systems, IoTs, mobile apps, and digital twins, which are areas of growing scientific and practical interest [10,25].

BDA and machine intelligence: As digitization becomes perfectly integrated to our daily lives, enormous heterogeneous amounts of data collected everyday can be fruitfully used in several application domains [4]. By employing the applications of statistic analysis, predicting modeling and many other BDA techniques, business decisions and city services can be enhanced and made really "smart". Self-adaptation in smart city software platforms can be driven by big data analytics through continuous stream processing and analysis of urban data. Additionally, integrating AI agents can further optimize these processes by automating decision-making, improving efficiency, and providing real-time insights.

6   Future Directions

Based on our findings we want to define a reference software architecture for smart cities that want to move from a reactive logic of urban management to a proactive logic in relation to the time dimension of adaptation [20], by leveraging *prescriptive analytics* through the combined application of BDA, SA and AI. An initial concept for structuring this machine intelligence module is outlined in our earlier work [6]. We also want to examine *gray literature* to look for initiatives from leading ICT companies to determine if academia and industry are aligned.

References

1. European project Sharing Cities. Deliverable D4.2 Urban Sharing Platform Reference Model (2016). http://www.sharingcities.eu/
2. European innovation partnership on smart-cities and communication (2018). http://ec.europa.eu/eip/smartcities/indexen.htm. Accessed 30 Mar 2018
3. Data package for: Self-Adaptation and Big Data Analytics to enhance decision-making in Smart City Platforms (2024). https://zenodo.org/records/15274683
4. Al Nuaimi, E., Al Neyadi, H., Mohamed, N., Al-Jaroodi, J.: Applications of big data to smart cities. J. Internet Serv. Appl. **6**(1), 25 (2015)

5. Alavi, S., Jalili, S., Eslami, M.: Decentralized control architecture for multi-authoring microgrids. Computing 1–26 (2023). https://doi.org/10.1007/s00607-023-01201-w
6. Ali, M.: Big data and machine intelligence in software platforms for smart cities. In: Software Architecture - 14th European Conference, ECSA 2020 Tracks and Workshops, L'Aquila, Italy, September 14-18, 2020, Proceedings, vol. 1269, pp. 17–26. Springer, Cham (2020). https://doi.org/10.1007/978-3-030-59155-7_2
7. Auger, A., Exposito, E., Lochin, E.: iQAS: an integration platform for qoi assessment as a service for smart cities. In: 2016 IEEE 3rd World Forum on Internet of Things (WF-IoT), pp. 88–93 (2016)
8. De Lemos, R., et al.: Software engineering for self-adaptive systems: a second research roadmap. In: Cheng, B.H.C., de Lemos, R., Giese, H., Inverardi, P., Magee, J. (eds.), Software Engineering for Self-Adaptive Systems II: International Seminar, Dagstuhl Castle, Germany, October 24-29, 2010 Revised Selected and Invited Papers, LNCS, pp. 1–32. Springer, Cham (2013). https://doi.org/10.1007/978-3-642-02161-9_1
9. Dobre, C., Xhafa, F.: Intelligent services for big data science. Futur. Gener. Comput. Syst. **37**, 267–281 (2014)
10. D'Angelo, M.: Decentralized self-adaptive computing at the edge. In: Proceedings of the 13th International Conference on Software Engineering for Adaptive and Self-Managing Systems, pp. 144–148. SEAMS '18, ACM, New York, NY, USA (2018). https://doi.org/10.1145/3194133.3194160
11. Gil-Garcia, J.R., Pardo, T.A., Nam, T.: What makes a city smart? Identifying core components and proposing an integrative and comprehensive conceptualization. Information Polity **20**(1), 61–87 (2015)
12. Gurgen, L., Gunalp, O., Benazzouz, Y., Gallissot, M.: Self-aware cyber-physical systems and applications in smart buildings and cities. In: 2013 Design, Automation Test in Europe Conference Exhibition (DATE), pp. 1149–1154 (2013)
13. Habibzadeh, H., Kaptan, C., Soyata, T., Kantarci, B., Boukerche, A.: Smart city system design: a comprehensive study of the application and data planes. ACM Comput. Surv. (CSUR) **52**(2), 1–38 (2019)
14. Hajjaji, Y., Boulila, W., Farah, I.R., Romdhani, I., Hussain, A.: Big data and iot-based applications in smart environments: a systematic review. Comput. Sci. Rev. **39**, 100318 (2021)
15. Huang, Y., Poderi, G., Šćepanović, S., Hasselqvist, H., Warnier, M., Brazier, F.: Embedding internet-of-things in large-scale socio-technical systems: a community-oriented design in future smart grids. The Internet of Things for Smart Urban Ecosystems, pp. 125–150 (2019)
16. Ji, S., Li, Q., Cao, W., Zhang, P., Muccini, H.: Quality assurance technologies of big data applications: a systematic literature review. Appl. Sci. **10**(22) (2020). https://doi.org/10.3390/app10228052
17. Juan, Y.K., Wang, L., Wang, J., Leckie, J.O., Li, K.M.: A decision-support system for smarter city planning and management. IBM J. Res. Dev. **55**(1.2), 3–1 (2011)
18. Kephart, J.O., Chess, D.M.: The vision of autonomic computing **36**, 41–50 (2003)
19. Kitchenham, B.A., Charters, S.: Guidelines for performing systematic literature reviews in software engineering. Technical report. EBSE 2007-001, Keele University and Durham University Joint Report (07)
20. Krupitzer, C., Roth, F.M., VanSyckel, S., Schiele, G., Becker, C.: A survey on engineering approaches for self-adaptive systems. Pervasive Mob. Comput. **17**(PB), 184–206 (2015)

21. Li, J., Zhang, M., Li, N., Weyns, D., Jin, Z., Tei, K.: Generative AI for self-adaptive systems: state of the art and research roadmap. ACM Trans. Auton. Adapt. Syst. **19**(3), 13:1–13:60 (2024). https://doi.org/10.1145/3686803
22. Ma, M., Preum, S.M., Ahmed, M.Y., Tärneberg, W., Hendawi, A., Stankovic, J.A.: Data sets, modeling, and decision making in smart cities: a survey. ACM Trans. Cyber-Phys. Syst. **4**(2) (2019). https://doi.org/10.1145/3355283
23. Martins, P., Albuquerque, D., Wanzeller, C., Caldeira, F., Tomé, P., Sá, F.: City-Action a smart-city platform architecture. In: Arai, K., Bhatia, R. (eds.) Advances in Information and Communication. FICC 2019. LNNS, vol. 69, pp. 217–236. Springer, Cham (2020). https://doi.org/10.1007/978-3-030-12388-8_16
24. Moreno, G.A., Cámara, J., Garlan, D., Schmerl, B.: Proactive self-adaptation under uncertainty: a probabilistic model checking approach. In: Proceedings of the 2015 10th Joint Meeting on Foundations of Software Engineering, pp. 1–12. ESEC/FSE 2015, ACM, New York, NY, USA (2015).https://doi.org/10.1145/2786805.2786853
25. Muccini, H., Sharaf, M., Weyns, D.: Self-adaptation for cyber-physical systems: a systematic literature review. In: Proceedings of the 11th International Symposium on Software Engineering for Adaptive and Self-managing Systems, pp. 75–81 (2016)
26. Nogueira, B.C., Motta, R.C., Delicato, F.C., Batista, T.V.: Self-adaptation in iot systems for smart cities. In: 2023 Symposium on Internet of Things (SIoT), pp. 1–5 (2023). https://doi.org/10.1109/SIoT60039.2023.10390083
27. Puiu, D., et al.: Citypulse: large scale data analytics framework for smart cities. IEEE Access **4**, 1086–1108 (2016)
28. Quin, F., Weyns, D., Bamelis, T., Buttar, S.S., Michiels, S.: Efficient analysis of large adaptation spaces in self-adaptive systems using machine learning. In: 2019 IEEE/ACM 14th International Symposium on Software Engineering for Adaptive and Self-Managing Systems (SEAMS), pp. 1–12. IEEE (2019)
29. Salehie, M., Tahvildari, L.: Self-adaptive software: landscape and research challenges. ACM Trans. Auton. Adapt. Syst. **4**(2) (2009). https://doi.org/10.1145/1516533.1516538
30. Sánchez-Corcuera, R., et al.: Smart cities survey: technologies, application domains and challenges for the cities of the future. Int. J. Distrib. Sens. Netw. **15**(6), 1550147719853984 (2019)
31. Santana, E.F.Z., Chaves, A.P., Gerosa, M.A., Kon, F., Milojicic, D.S.: Software platforms for smart cities: concepts, requirements, challenges, and a unified reference architecture. ACM Comput. Surv. **50**(6) (2017). https://doi.org/10.1145/3124391
32. Schmid, S., Gerostathopoulos, I., Prehofer, C., Bures, T.: Self-adaptation based on big data analytics: a model problem and tool. In: 2017 IEEE/ACM 12th International Symposium on Software Engineering for Adaptive and Self-Managing Systems (SEAMS), pp. 102–108 (2017)
33. Sivrikaya, F., Ben-Sassi, N., Dang, X., Görür, O.C., Kuster, C.: Internet of smart city objects: a distributed framework for service discovery and composition. IEEE Access **7**, 14434–14454 (2019)
34. Solino, A., Batista, T., Cavalcante, E.: Decision-making support to auto-scale smart city platform infrastructures. In: 2023 18th Iberian Conference on Information Systems and Technologies (CISTI), pp. 1–6 (2023). https://doi.org/10.23919/CISTI58278.2023.10212058
35. Ta-Shma, P., Akbar, A., Gerson-Golan, G., Hadash, G., Carrez, F., Moessner, K.: An ingestion and analytics architecture for iot applied to smart city use cases. IEEE Internet Things J. **5**(2), 765–774 (2018). https://doi.org/10.1109/JIOT.2017.2722378

36. Weyns, D.: An Introduction to Self-adaptive Systems: A Contemporary Software Engineering Perspective. (IEEE Press) Wiley, Hoboken (2020)
37. Yin, C., Xiong, Z., Chen, H., Wang, J., Cooper, D., David, B.: A literature survey on smart cities. SCIENCE CHINA Inf. Sci. **58**(10), 1–18 (2015)

Towards Explainable Decision Support Using Hybrid Neural Models for Logistic Terminal Automation

Riccardo D'Elia[1]([✉]) [iD], Alberto Termine[1] [iD], and Francesco Flammini[1,2] [iD]

[1] University of Applied Sciences and Arts of Southern Switzerland,
Dalle Molle Institute for Artificial Intelligence, Lugano, Switzerland
{riccardo.delia,alberto.termine}@supsi.ch, francesco.flammini@unifi.it
[2] Department of Mathematics and Computer Science Ulisse Dini,
University of Florence, Florence, Italy

Abstract. The integration of Deep Learning (DL) in System Dynamics (SD) modeling for transportation logistics offers significant advantages in scalability and predictive accuracy. However, these gains are often offset by the loss of explainability and causal reliability — key requirements in critical decision-making systems. This paper presents a novel framework for interpretable-by-design neural system dynamics modeling that synergizes DL with techniques from Concept-Based Interpretability, Mechanistic Interpretability, and Causal Machine Learning. The proposed hybrid approach enables the construction of neural network models that operate on semantically meaningful and actionable variables, while retaining the causal grounding and transparency typical of traditional SD models. The framework is conceived to be applied to real-world case-studies from the EU-funded project AutoMoTIF (https://automotif-project.eu), focusing on data-driven decision support, automation, and optimization of multimodal logistic terminals. We aim at showing how neuro-symbolic methods can bridge the gap between black-box predictive models and the need for critical decision support in complex dynamical environments within cyber-physical systems enabled by the industrial Internet-of-Things.

Keywords: artificial intelligence · XAI · IoT · trustworthy autonomous systems · transportation · cognitive digital twins

1 Introduction

Intermodal terminals represent one of the most complex and critical environments within the modern logistics ecosystem. Functioning as key nodes where rail, road, and maritime systems intersect, these hubs must manage immense flows of containers, trailers, and vehicles in real time. With operations affected by volatile shipping schedules, fluctuating hinterland traffic, equipment availability, and infrastructure constraints, even minor disruptions can propagate through the system and lead to cascading delays and significant economic losses. In such

© The Author(s), under exclusive license to Springer Nature Switzerland AG 2026
M. Pavone et al. (Eds.): DSA ISC 2025, LNCS 16405, pp. 78–93, 2026.
https://doi.org/10.1007/978-3-032-21811-7_6

dynamic settings, intelligent decision-support tools are essential − not only to forecast and optimize system performance but also to provide explanations that are comprehensible to support trustworthy decisions and ease assessment by terminal operators, engineers, and regulators.

System Dynamics (SD) has long been employed as a foundational methodology to model and manage the complexity of logistics networks. SD models capture the evolution of systems over time through feedback loops, differential equations, and causal relationships grounded in domain expertise. Within terminal operations, SD has been applied to scenarios such as crane scheduling, gate throughput optimization, and yard congestion modeling, providing transparency and accountability through interpretable causal pathways [2,38,41]. These models have the advantage of semantic clarity, aligning with human mental models of causality and system behavior. However, traditional SD approaches suffer from key limitations: they rely heavily on expert-defined rules and assumptions, are constrained by oversimplified functional relationships, and become difficult to scale or calibrate as the number of interacting components and variables increases.

In recent years, Deep Learning (DL) has emerged as a powerful alternative, particularly for handling high-dimensional, nonlinear, and data-rich environments. DL has demonstrated considerable success in logistics for predictive tasks such as berth occupancy forecasting, crane movement duration estimation, or anomaly detection in terminal workflows [9,43,48]. Leveraging historical operational data, DL models can autonomously extract latent patterns without explicit programming. However, this strength is also a weakness: DL architectures function as opaque "black-boxes" [55], offering little to no insight into how decisions are made or what drives their predictions. More critically, these models often infer behavior based on statistical correlations rather than true causal mechanisms − a problem referred to as *causal reliability* [50]. In contexts such as intermodal logistics, where safety, compliance, and operator accountability are paramount, the lack of interpretability[1] and causal transparency undermines the trustworthiness of purely data-driven models.

Related Works. Within the paradigm of trustworthy artificial intelligence (AI), the field of Explainable AI (XAI) has arisen with the objective of understanding the internal rationale of deep learning models. Post-hoc XAI methods such as feature attribution, saliency maps, and surrogate modeling have been applied in logistics for use cases including predictive maintenance, risk analysis, and routing decisions [37,45]. While these methods offer some visibility into input-output relationships, they remain fundamentally limited. First, most

---

[1] In this paper, we use the terms *explainability* and *interpretability* interchangeably to refer to the ability of AI systems to make their operations and decisions understandable to human stakeholders. While explainability typically emphasizes tailored explanations for users, and interpretability refers to the transparency of the model itself, both are treated here as complementary aspects of making AI behavior intelligible, trustworthy, and auditable [13,33].

XAI techniques are applied after model training and do not alter the underlying architecture. Second, the explanations produced often lack semantic clarity or domain alignment, making them difficult for non-technical users to interpret or act upon. As noted in recent reviews of XAI in supply chain and logistics [15,26,54], this gap between technical insight and human interpretability remains a major barrier to adoption. The widespread use of DL methods in the modeling of dynamical systems calls for a paradigm shift to jointly address the challenges of interpretability and causal reliability. Although techniques exist for semantic interpretability, causal discovery, or mechanistic modeling in isolation, no one framework brings together (i) high-level, human-readable variables, (ii) robust interventional reasoning, and (iii) interpretable dynamic equations – leaving a critical integration gap in complex, data-driven environments. This paper advocates for precisely such an integration, bringing together advances from multiple fields to propose a methodological foundation for trustworthy optimization in logistics. In an increasingly digitalized world, such an approach for explainable decision support can be extended and generalized to additional classes of autonomous cyber-physical systems (CPS), in transportation and other domains [20]. From an infrastructural perspective, the approach is enabled by emerging technologies and paradigms, such as the Industrial Internet-of-Things (IIoT), interconnecting logistic terminal components, and Digital Twins (DTs), leveraging on the potential of cloud computing for model-based run-time monitoring, data analytics, and what-if projections [14,19].

Novel Contribution. The approach is positioned as both a theoretical contribution and a practical roadmap for developing robust models of complex, high-risk environments such as intermodal terminals. Rather than focusing on the development and implementation of post-hoc explainability techniques, the proposed approach focuses on the construction of an **interpretable-by-design** neural system dynamics framework. This modeling framework combines the expressiveness of data-driven learning with the interpretability of symbolic reasoning. Rather than retrofitting explanations *post-hoc*, this paradigm seeks to build models that are inherently explainable, causally reliable, and semantically aligned with real-world systems from the outset. A particularly compelling direction is the integration of Neuro-Symbolic AI, which combines the learning capabilities of neural networks with the reasoning structures of symbolic logic [8].

The new research direction proposed in this paper builds on these foundations to achieve a unified methodological approach for explainable decision making in intermodal logistics. At its core, this approach introduces an Interpretable Neural System Dynamics (INSD) pipeline designed to concurrently overcome the challenges of interpretability, scalability, and causal reliability in high-dimensional terminal operations optimization. This pipeline encompasses three main components. As a first step, Concept-Based Interpretability techniques are employed to extract semantically meaningful high-level variables from raw operational data. These "concepts" are aligned with domain-relevant metrics and behaviors, enabling human users to understand the factors driving system behavior [39]. Causal Machine Learning (CML) methods are then used to identify causal dependencies among the extracted concepts. Unlike conventional DL models that rely on statistical association, CML establishes cause–and–effect relationships, anchoring the learned model to the true generative processes of the system [23,44]. Finally, mechanistically interpretable modeling techniques will be leveraged to infer a set of interpretable structural dynamic equations that govern the system's behavior. These equations retain the mathematical transparency of SD while being learned from data, forming a hybrid structure that is both scalable and interpretable [7].

The resulting architecture is envisioned as the backbone of a new generation of Cognitive Digital Twins (CDTs) for intermodal terminals. These CDTs are not merely simulations of physical assets; rather, they are enriched with models that offer support for real-time monitoring, scenario analysis, and adaptive decision-making. Early research on CDTs has demonstrated potential in applications such as disruption recovery, predictive maintenance, and freight parking optimization [3,11,52], but most existing systems rely on black-box learning modules and lack semantic or causal integration.

The paper is organized as follows: Sect. 2 reviews interpretable modeling concepts; Sect. 3 introduces the INSD pipeline; Sect. 4 outlines research challenges; the final section concludes with key implications for decision support in intermodal logistics.

2 Interpretable Modeling in Dynamic Logistics Systems

The interpretability of models in logistics decision support systems is not just a matter of ease of use; it is an essential prerequisite for trust, accountability, and operational safety. In high-stakes environments like intermodal terminals, operators must be able to audit system behavior, trace recommendations back to intelligible causes, and simulate interventions under uncertainty. This section outlines three key components of an integrated interpretability framework for logistics optimization: System Dynamics, Explainable AI, and their hybridization into an appropriate modeling pipeline.

2.1 System Dynamics: Modeling Causality in Complex Logistics Systems

System Dynamics (SD) is a well-established modeling paradigm that represents the evolution of complex systems through interconnected feedback loops, stock-and-flow structures, and time delays. Originally developed for analyzing industrial processes and socio-economic systems, SD has proven especially effective in logistics, where resource flows and operational bottlenecks emerge from nonlinear interactions among interdependent components [2,38,41]. At the heart of SD lies the *causal loop diagram* (CLD), which provides a graphical language to describe how changes in one part of a system propagate to affect others. For example, an increase in crane utilization may reduce yard dwell time, thereby influencing gate throughput − a feedback structure that can be formally captured in a CLD and translated into system equations. These diagrams enable both *semantic interpretability* (clarity about what variables represent) and *mechanistic interpretability* (how they interact). This explicit encoding of causality makes SD a reliable tool in modeling dynamic behaviors in terminal operations, from congestion buildup to equipment scheduling. Despite these strengths, SD models are not without limitations. They are time-consuming to build, requiring deep domain knowledge and manual tuning. Moreover, they struggle to capture fine-grained nonlinearities and real-time adaptation, which are critical in modern, sensor-rich logistics environments. As terminal complexity grows, SD models may become insufficiently expressive or too coarse-grained, necessitating a complementary data-driven approach.

2.2   Explainable Artificial Intelligence and the Interpretability Challenge

Opacity in Deep Learning for Logistics. In contrast to rule-based System Dynamics (SD) models, Deep Learning (DL) approaches provide unmatched scalability and predictive performance by learning complex input-output mappings directly from operational data. These models have been increasingly applied to a range of logistics tasks − including berth occupancy prediction, crane sequencing, and dwell time estimation [4, 40, 46]− enabling real-time decision support across ports, intermodal terminals, and supply chains. However, this predictive power comes at a cost: DL models are notoriously *opaque*, making it difficult for users to audit their reasoning or justify their outputs. This opacity is particularly problematic in logistics, where high-stakes decisions demand transparency, accountability, and regulatory compliance.

Explainable AI Approaches. The field of *Explainable Artificial Intelligence* (XAI) has emerged to overcome these black-box limitations by producing human-readable explanations for automated forecasts and recommendations [25, 47]. Within critical logistics domains, XAI plays a growing role in enhancing human trust, operational safety, and system reliability [35]. For instance, in the transport logistics domain, a hybrid explainable AI framework has been developed to analyze traffic accident data, enabling stakeholders to identify and interpret risk factors associated with different injury severity levels, thereby improving decision-making processes in road safety management [1].

Post-hoc vs. By-Design Explainability. Post-hoc XAI methods − such as SHAP [31] or LIME [42] − generate explanations after a model is fully trained. They offer the advantage of being model-agnostic and easy to apply, providing local, instance-level feature attributions. However, these explanations do not affect the underlying architecture; they may be misleading in the case of feature correlations or non-linear interactions, and do not guarantee that they reflect true causal pathways. By contrast, explainable-by-design approaches build transparency directly into the model's structure [39]. Each component or parameter is explicitly tied to a human-understandable concept (e.g., *yard congestion, gate throughput*), and the architecture enforces modular, often sparse or additive forms that mirror causal loops and feedback mechanisms from system dynamics [16]. Moreover, causal constraints can be embedded during training to ensure that learned relationships correspond to genuine cause-effect dependencies, thus supporting reliable interventional and counterfactual reasoning [36,44]. While such designs may impose limitations on expressive power or require careful adaptation to high-dimensional, time-dependent data, they deliver *global* auditability and *interventional* transparency − properties essential for trust and accountability in safety-critical logistics applications.

Semantic and Mechanistic Opacity. Despite this progress, interpretability remains an open and multi-faceted challenge. As outlined in recent literature [16], two major forms of opacity dominate discussions around XAI:

- *Semantic Opacity* refers to the difficulty of mapping a model's learned internal representations to high-level, human-understandable concepts. Neural networks typically operate in high-dimensional feature spaces where latent variables do not correspond directly to domain-relevant categories. This discrepancy is particularly limiting in decision-making contexts, where well-founded explanations of operational terms are needed to gain insight into action.
- *Mechanistic Opacity* involves the challenge of deciphering how various components of the model − layers, neurons, weights − interact to produce specific outputs [28]. This issue is particularly significant in large-scale neural networks with millions or even billions of parameters, where computations are spread across numerous layers and involve several nonlinear transformations.

Most existing XAI methods fail to simultaneously address both semantic and mechanistic opacity, often focusing on one at the expense of the other [16].

Causal Reliability. A third, often overlooked, dimension of the interpretability challenge is *causal reliability* [44,50]. This refers to a model's ability to reflect the underlying causal mechanisms behind observed data, rather than merely exploiting statistical associations. DL models excel at identifying correlations but lack the structure to reason about cause and effect. This is a critical limitation in logistics operations, where decision-makers must evaluate the effects of critical management decisions (e.g., rerouting a shipment, modifying crane schedules). Without support for interventional and counterfactual reasoning [36], such models fall short of enabling robust planning and adaptation to disruptions.

Integrated Explainability Landscape. These three forms of opacity − semantic, mechanistic, and causal − are deeply interconnected. A model that is mechanistically transparent but semantically unintelligible offers limited utility; likewise, semantically aligned models without causal grounding cannot reliably support robust or fair decision-making. Current efforts in XAI and Causal Machine Learning (CML) have often developed in isolation, creating a fragmented research landscape [12]. Their integration is crucial for modeling complex, dynamic systems like logistics networks, where interpretability must span across various dimensions. Recent surveys in logistics and supply chain research highlight the urgency of these challenges. For instance, Neuro-symbolic paradigms – combining statistical learning with symbolic reasoning – have been proposed to integrate domain knowledge into demand forecasting and inventory control [26]. However, most implementations still rely on post-hoc explanation techniques that fail to guarantee causal reliability. Even when integrated into routing and scheduling algorithms, DL systems struggle to explain the logic of complex, time-dependent decisions [52]. While these models may reduce costs, they often lack the transparency needed for intervention analysis and long-term planning.

This work responds to these limitations by proposing a cohesive framework that unifies the three domains of interpretability: concept-based (semantic) and mechanistic interpretability, and causal reliability. By integrating these perspectives within the modeling of dynamic logistic systems, the project seeks to bridge the gap between black-box predictive performance and human-aligned reasoning. This integrated approach will enable more reliable predictions and the ability to simulate interventions, analyze counterfactuals, and support human oversight in critical decision-making contexts.

3   Toward Interpretable and Causally-Reliable Models

To unify the aforementioned threads, in this section we propose an *Interpretable Neural System Dynamics (INSD)* pipeline – a hybrid approach that integrates concept-based interpretability, causal discovery, and mechanistic modeling into a coherent framework (Fig. 1). The 3-step INSD pipeline addresses both semantic and mechanistic opacity while ensuring causal reliability across time-evolving systems:

1. **Concept Learning**: Leveraging Concept-Based Interpretability (CBI), raw operational data is transformed into high-level variables (e.g., yard congestion, crane idleness) that align with human understanding [39].
2. **Causal Learning**: Causal Machine Learning and Causal Discovery methods are used to uncover dependencies among these concepts, producing a causal graph that encodes plausible cause-and-effect relations [23].
3. **Equation Learning**: Finally, dynamic equations are learned from this causal graph using techniques from neuro-symbolic AI [10,27]. This stage produces interpretable models capable of describing the temporal evolution of the system and enabling long-term intervention planning [7].

Unlike standard neural networks, the resulting model can be queried, audited, and adapted in real time – offering a balance of flexibility, performance, and explainability.

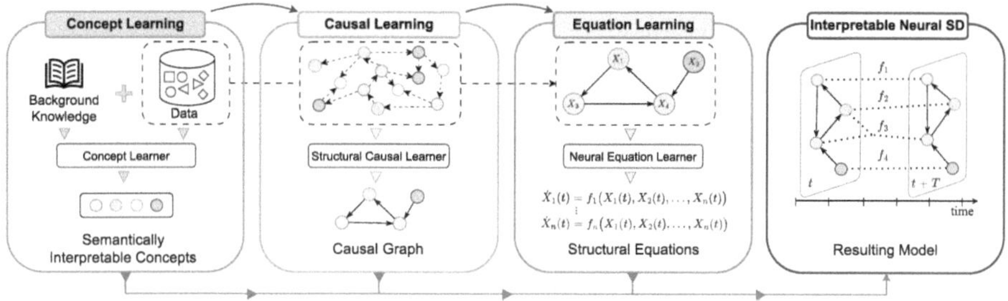

Fig. 1. Overview of the INSD pipeline, from concept learning to causal and equation learning, ensuring interpretability and causal reliability in the resulting model.

3.1   Application of INSD Pipeline to Intermodal Terminals

To illustrate how the proposed pipeline operates in practice, let us consider a digital twin simulation of an intermodal freight terminal. Here, a deep learning model forecasts recurring train delays during peak hours. Initial analyses show a correlation between these delays and increased truck traffic. However, without further interpretability, the underlying causes remain obscure. By integrating *concept-based interpretability*, we can represent domain-relevant variables such as terminal workload, crane handling efficiency, and cross-modal wait times as explicit components in the model. This alignment allows predictions to be traced back to real-world operational concepts, ensuring that decision-makers understand not only what the model predicts but why. To uncover whether truck congestion truly causes the train delays, *causal reasoning* is applied. The model simulates counterfactual scenarios such as, "What if crane availability were increased during peak periods?" or "What if trucks were rerouted to a secondary access road?" This enables planners to test interventions and proactively identify leverage points within terminal operations. Finally, the structural dynamic equations are used to track the flow of effects across the model's internal structure. The interpretable model reveals that the recommendation to reroute trucks stems from predicted access road bottlenecks and not from capacity shortfalls at the gates, as initially assumed. By disentangling the roles of modular components within the model, terminal operators can gain a structured, auditable understanding of how decisions are generated. These forms of interpretability collectively enhance transparency, support trustworthy simulation-based decision-making, and empower human oversight in the operation of complex, dynamic logistics environments such as those addressed in the *AutoMoTIF* project.

3.2   Limitations of Existing XAI Approaches in Logistics

Despite recent advances in Explainable AI for logistics, most existing approaches remain limited in their ability to support meaningful, actionable insight in dynamic operational settings. A key shortcoming is the lack of causal grounding. For example, methods like those proposed in [32] apply post-hoc explanation techniques to pre-trained deep models, providing insights into feature importance without distinguishing between mere statistical associations and true cause-and-effect relationships. These explanations are inherently descriptive rather than interventional, offering no guidance for counterfactual reasoning or scenario planning, essential tasks in high-risk domains such as intermodal logistics. Moreover, most existing models treat interpretability and predictive performance as a tradeoff rather than a design imperative. Additionally, explanations generated from latent neural activations often fail to align with semantically meaningful variables recognizable to domain experts, reinforcing the semantic opacity challenge discussed earlier. In short, current explainable systems either prioritize interpretability at the cost of accuracy or deliver high-performance predictions without ensuring that the outputs are causally valid or human-aligned. They fall short of enabling robust planning, adaptive decision-making, or regulatory accountability.

3.3   INSD Pipeline Novelty and Impact

The unified modeling framework advances the state of the art in both interpretability and causal reasoning for logistics decision-support systems. The proposed Interpretable Neural System Dynamics (INSD) pipeline is novel in three key dimensions:

- **Interpretability-by-Design:** Rather than relying on post-hoc explanation, the INSD framework is designed from the ground up to be semantically and mechanistically interpretable. High-level concepts are learned explicitly from operational data and aligned with domain-relevant abstractions, facilitating intuitive understanding and traceability;
- **Causally Reliable Modeling:** By integrating causal discovery and causal machine learning into the modeling pipeline, the INSD approach distinguishes correlation from causation, enabling robust interventional and counterfactual reasoning. This supports what-if analysis and decision-making under uncertainty;
- **Structural Dynamic Equations Learning:** The final stage of the pipeline involves learning structural dynamic equations that retain the mechanistic interpretability of traditional SD models while leveraging the flexibility of neural architectures.

This hybridization supports structural auditability, bridging the gap between symbolic modeling and deep learning. Moreover, the broader impact of this approach lies in its potential to serve as the core of next-generation Cognitive Digital Twins in logistics. By embedding causally reliable, interpretable models within digital replicas of intermodal terminals, the INSD framework can drive adaptive control, disruption response, and strategic planning with unprecedented transparency and trustworthiness.

4 Research Agenda and Open Challenges

While the integration of System Dynamics and Deep Learning holds great promise for intermodal terminal decision support, realizing this vision demands advances across multiple fronts. Below, we present the key research challenges that must be addressed to build transparent, trustworthy, and operationally robust systems.

Semantic Modeling of Operational Concepts. A foundational challenge lies in defining and extracting high-level, semantically rich concepts (e.g., yard congestion, crane idle time) from heterogeneous, multimodal data streams typical of modern terminals. Concept-based interpretability has shown promise in aligning model internals with human-understandable abstractions through techniques such as Concept Activation Vectors [24] and hybrid causal-concept representation learning [21]. However, most work focuses on static domains or language models; extending these methods to dynamic, streaming sensor data (RFID logs, camera feeds, IoT telemetry) demands novel architectures that can handle time-series invariance and contextual shifts across different terminal layouts. Furthermore, learned concepts must remain robust under operational variability, such as equipment maintenance, procedural changes, or seasonal traffic fluctuations, while preserving alignment with operator mental models to facilitate rapid trust and adoption. This calls for adaptive concept learners that integrate domain ontologies and human-in-the-loop feedback, ensuring that representations not only capture statistical structure but also retain semantic fidelity in real-world settings [49].

Learning Causal Structures in Complex, Dynamic Environments. Intermodal operations involve feedback loops and delayed effects (e.g., a crane breakdown propagates to yard congestion hours later). Neural causal discovery methods have shown promise in uncovering static DAGs from observational data [23, 44]. Recent advances in spatio-temporal causal discovery, such as the SPACY framework, leverage variational inference to infer latent time-series causal graphs, offering a promising blueprint for terminal operations [51]. Yet, these methods typically assume dense observability and stationary processes, whereas terminals are characterized by partial data observation, ad-hoc interventions, and streaming constraints. Adapting causal discovery to this domain thus entails (i) robust handling of sensor noise and data gaps, (ii) support for cyclic feedback between subsystems, and (iii) online updating mechanisms that can incorporate new evidence without retraining from scratch. Embedding interventional calculus (do-calculus) will enable counterfactual reasoning critical for "what-if" scenario planning, such as quantifying the effect of deploying an extra crane during peak hours.

Interpretable Modeling of System Dynamics. While System Dynamics (SD) has long offered white-box causal loop diagrams and differential equation models, it struggles with high-dimensional nonlinearity and real-time calibration. Mechanistic equation-learning methods – such as Sparse Identification of Nonlinear Dynamics (SINDy) and physics-informed neural ODEs – provide data-driven alternatives that can recover transparent governing equations from observations [10]. Yet, applying these methods to the inherently discontinuous and multimodal dynamics of logistics terminals remains an open frontier. Key research tasks include devising modular equation-learning architectures that can isolate subsystem dynamics (e.g., crane cycles, truck arrival patterns) while preserving end-to-end predictive fidelity [53]. Moreover, embedding background domain knowledge, such as physical constraints and workflow rules, can regularize equation learners, yielding models that are both accurate and audit-friendly.

Regulatory and Safety Considerations. As AI systems become increasingly embedded in critical infrastructure, ensuring their transparency, auditability, and compliance with emerging regulatory and ethical standards has become imperative. The complexity and high-risk nature of these domains demand not only effective but also trustworthy and accountable AI systems [6]. *Transparency by design* is a foundational principle that must guide the development of decision-support models in these contexts. As emphasized by Felzmann et al., transparency should not be treated as a post-hoc feature but integrated throughout the system's lifecycle, from data preprocessing to interface design [17]. This ensures that models are intelligible and usable for domain operators, auditors, and policy stakeholders. *Auditability* is crucial where decisions may trigger operational or safety consequences. Fernsel et al. propose a framework to assess AI systems' auditability, focusing on the availability of verifiable claims, structured evidential access, and tooling to support technical and procedural audits [18]. Their work outlines how auditability can be operationalized through traceable pipelines and documentation protocols. To align these technical dimensions with ethical imperatives, *ethics-based auditing* has emerged as a constructive practice for AI governance. Mokander and Floridi argue for continuous auditing processes that evaluate not only system outputs but also development practices, incentive structures, and the ethical alignment of goals [34]. Such frameworks can function as a bridge between abstract ethical principles and deployable technical guidelines. In parallel, the role of *standardization* in governing AI transparency is gaining traction. Högberg shows that transparency can be stabilized and institutionalized through technical standards that codify explanation practices and communication formats [22]. This is particularly relevant in logistics, where operational transparency must translate into consistent and interpretable outputs. Finally, explainability remains central to both technical and organizational transparency. A recent analysis of explainability practices across AI ethics guidelines confirms that explainability is essential for understanding model behavior, building trust, and ensuring responsible deployment [5]. The study underscores that effective implementation often requires multidisciplinary design teams that explicitly link explainability goals to stakeholder needs and system impacts.

Towards an Integrated Evaluation Framework. To ensure that these principles are not only aspirational but also operational, there is a growing need for integrated evaluation frameworks that can assess AI systems along regulatory, ethical, and technical dimensions. For the INSD pipeline, this implies establishing metrics and traceability mechanisms for each layer of the architecture, from causal validity and semantic alignment to user comprehension and audit readiness. Such a framework would not only support internal quality assurance but also serve as a compliance interface for external audits and certification processes, bridging the gap between explainability, governance, and deployment in real-world logistics contexts. A comprehensive framework categorizing XAI design goals and corresponding evaluation methods has been proposed, facilitating iterative design and evaluation cycles in multidisciplinary teams [33]. This work underscores the importance of aligning XAI system design with specific user needs and evaluation strategies, which is crucial for developing transparent and accountable AI systems. Complementing this, a systematic literature review on ethics-based AI auditing highlights the necessity of integrating ethical principles − such as fairness, transparency, and autonomy − into AI system evaluations [29]. The review emphasizes the role of standardized methodologies in operationalizing these principles, ensuring that AI systems meet both ethical standards and stakeholder expectations. Incorporating these frameworks into the INSD pipeline can enhance its capacity to deliver AI decision-support systems that are not only functionally effective but also ethically sound and compliant with emerging regulatory standards.

Conclusion

This paper proposes a roadmap for combining semantic and mechanistic interpretability, causal reliability, and regulatory compliance into a cohesive framework within reference industrial applications. By charting a research agenda across concept identification, causal discovery, and equation learning, we shed light on the technical and organizational requirements for the operationalization of the Interpretable Neural System Dynamics (INSD) pipeline. In doing so, we envision a new generation of Cognitive Digital Twins for intermodal terminals: digital replications enabled by emerging CPS and IIoT technologies that not only forecast and optimize, but also reveal their reasoning in terms that operators can understand and regulators can audit [14,19,30]. Beyond logistics, this approach exemplifies a broader paradigm shift toward interpretability-by-design in AI systems deployed in critical infrastructures, where trust, safety, and accountability are essential [20]. Realizing this vision will demand concerted, multidisciplinary collaboration among AI researchers, system dynamics experts, logistics practitioners, and policymakers. Rewards include enhanced resilience, informed decision-making, and societal trust, justifying this ambitious undertaking. The INSD framework offers not just a path forward for logistics but a template for embedding transparency and causality at the heart of AI-driven decision support across domains.

Acknowledgments.. This work was partly supported by the Swiss State Secretariat for Education, Research and Innovation (SERI) under contract no. 24.00184 (AutoMoTIF project). The project has been selected within the EU Horizon Europe programme under grant agreement no. 101147693. Views and opinions expressed are however those of the authors only and do not necessarily reflect those of the funding agencies, which cannot be held responsible for them.

References

1. Abdulrashid, I., Zanjirani Farahani, R., Mammadov, S., Khalafalla, M., Chiang, W.C.: Explainable artificial intelligence in transport logistics: risk analysis for road accidents. Transport. Res. Part E: Logist. Transport. Rev. **186**, 103563 (2024). https://doi.org/10.1016/j.tre.2024.103563
2. Aschauer, G., Gronalt, M., Mandl, C.: Modelling interrelationships between logistics and transportation operations – a system dynamics approach. Manag. Res. Rev. **38**(5), 505–539 (2015). https://doi.org/10.1108/MRR-11-2013-0271
3. Ashraf, M., Eltawil, A., Ali, I.: Disruption detection for a cognitive digital supply chain twin using hybrid deep learning. Oper Res. Int. J. **24**(2), 23 (2024). https://doi.org/10.1007/s12351-024-00831-y
4. Awasthi, A., Krpalkova, L., Walsh, J.: Deep learning-based Boolean, time series, error detection, and predictive analysis in container crane operations. Algorithms **17**(8), 333 (2024). https://doi.org/10.3390/a17080333
5. Balasubramaniam, N., Kauppinen, M., Rannisto, A., Hiekkanen, K., Kujala, S.: Transparency and explainability of AI systems: from ethical guidelines to requirements. Inf. Softw. Technol. **159**, 107197 (2023). https://doi.org/10.1016/j.infsof.2023.107197
6. Bellogín, A., Grau, O., Larsson, S., Schimpf, G., Sengupta, B., Solmaz, G.: The EU AI act and the wager on trustworthy AI. Commun. ACM **67**(12), 58–65 (2024). https://doi.org/10.1145/3665322
7. Bereska, L., Gavves, E.: Mechanistic interpretability for AI safety – a review (2024). https://doi.org/10.48550/arXiv.2404.14082, arXiv:2404.14082
8. Bhuyan, B.P., Ramdane-Cherif, A., Tomar, R., Singh, T.P.: Neuro-symbolic artificial intelligence: a survey. Neural Comput. Appl. **36**(21), 12809–12844 (2024). https://doi.org/10.1007/s00521-024-09960-z
9. Boute, R.N., Udenio, M.: AI in logistics and supply chain management. In: Merkert, R., Hoberg, K. (eds.) Global Logistics and Supply Chain Strategies for the 2020s, pp. 49–65. Springer International Publishing, Cham (2023). https://doi.org/10.1007/978-3-030-95764-3_3
10. Brunton, S.L., Proctor, J.L., Kutz, J.N.: Discovering governing equations from data by sparse identification of nonlinear dynamical systems. Proc. Natl. Acad. Sci. U.S.A. **113**(15), 3932–3937 (2016). https://doi.org/10.1073/pnas.1517384113
11. Busse, A., Gerlach, B., Lengeling, J.C., Poschmann, P., Werner, J., Zarnitz, S.: Towards digital twins of multimodal supply chains. Logistics **5**(2), 25 (2021). https://doi.org/10.3390/logistics5020025

12. Carloni, G., Berti, A., Colantonio, S.: The role of causality in explainable artificial intelligence (2023). https://doi.org/10.48550/arXiv.2309.09901, arXiv:2309.09901
13. Chazette, L., Brunotte, W., Speith, T.: Exploring explainability: a definition, a model, and a knowledge catalogue (2021). https://doi.org/10.48550/arXiv.2108.03012, arXiv:2108.03012
14. De Benedictis, A., Flammini, F., Mazzocca, N., Somma, A., Vitale, F.: Digital twins for anomaly detection in the industrial internet of things: conceptual architecture and proof-of-concept. IEEE Trans. Industr. Inf. **19**(12), 11553–11563 (2023). https://doi.org/10.1109/TII.2023.3246983
15. Elkhawaga, G., Elzeki, O.M., Abu-Elkheir, M., Reichert, M.: Why should i trust your explanation? An evaluation approach for XAI methods applied to predictive process monitoring results. IEEE Trans. Artif. Intell. **5**(4), 1458–1472 (2024). https://doi.org/10.1109/TAI.2024.3357041
16. Facchini, A., Termine, A.: Towards a taxonomy for the opacity of AI systems. In: Müller, V.C. (ed.) Philosophy and Theory of Artificial Intelligence 2021, vol. 63, pp. 73–89. Springer International Publishing, Cham (2022). https://doi.org/10.1007/978-3-031-09153-7_7, series Title: Studies in Applied Philosophy, Epistemology and Rational Ethics
17. Felzmann, H., Fosch-Villaronga, E., Lutz, C., Tamò-Larrieux, A.: Towards transparency by design for artificial intelligence. Sci. Eng. Ethics **26**(6), 3333–3361 (2020). https://doi.org/10.1007/s11948-020-00276-4
18. Fernsel, L., Kalff, Y., Simbeck, K.: Assessing the auditability of AI-integrating systems: a framework and learning analytics case study (2024). https://doi.org/10.48550/arXiv.2411.08906, arXiv:2411.08906
19. Flammini, F.: Digital twins as run-time predictive models for the resilience of cyber-physical systems: a conceptual framework. Phil. Trans. R. Soc. A **379**(2207), 20200369 (2021). https://doi.org/10.1098/rsta.2020.0369
20. Flammini, F., Alcaraz, C., Bellini, E., Marrone, S., Lopez, J., Bondavalli, A.: Towards trustworthy autonomous systems: taxonomies and future perspectives. IEEE Trans. Emerg. Topics Comput. **12**(2), 601–614 (2024). https://doi.org/10.1109/TETC.2022.3227113
21. Goyal, Y., Feder, A., Shalit, U., Kim, B.: Explaining classifiers with causal concept effect (CaCE) (2020). https://doi.org/10.48550/arXiv.1907.07165, arXiv:1907.07165
22. Högberg, C.: Stabilizing translucencies: governing AI transparency by standardization. Big Data Soc. **11**(1), 20539517241234296 (2024). https://doi.org/10.1177/20539517241234298
23. Kaddour, J., Lynch, A., Liu, Q., Kusner, M.J., Silva, R.: Causal machine learning: a survey and open problems (2022). https://doi.org/10.48550/ARXIV.2206.15475
24. Kim, B., et al.: Interpretability beyond feature attribution: quantitative testing with concept activation vectors (TCAV) (2018). https://doi.org/10.48550/arXiv.1711.11279, arXiv:1711.11279
25. Kobayashi, K., Alam, S.B.: Explainable, interpretable & trustworthy AI for intelligent digital twin: case study on remaining useful life. Eng. Appl. Artif. Intell. **129**, 107620 (2024). https://doi.org/10.1016/j.engappai.2023.107620, arXiv:2301.06676
26. Kosasih, E.E., Papadakis, E., Baryannis, G., Brintrup, A.: A review of explainable artificial intelligence in supply chain management using neurosymbolic approaches. Int. J. Prod. Res. **62**(4), 1510–1540 (2024). https://doi.org/10.1080/00207543.2023.2281663

27. Kramer, S., Cerrato, M., Džeroski, S., King, R.: Automated scientific discovery: from equation discovery to autonomous discovery systems (2023). https://doi.org/10.48550/arXiv.2305.02251, arXiv:2305.02251
28. Kästner, L., Crook, B.: Explaining AI through mechanistic interpretability. Euro. J. Phil. Sci. **14**(4), 52 (2024). https://doi.org/10.1007/s13194-024-00614-4
29. Laine, J., Minkkinen, M., Mäntymäki, M.: Ethics-based AI auditing: a systematic literature review on conceptualizations of ethical principles and knowledge contributions to stakeholders. Inf. Manage. **61**(5), 103969 (2024). https://doi.org/10.1016/j.im.2024.103969
30. Le, T.V., Fan, R.: Digital twins for logistics and supply chain systems: literature review, conceptual framework, research potential, and practical challenges. Comput. Ind. Eng. **187**, 109768 (2024). https://doi.org/10.1016/j.cie.2023.109768
31. Lundberg, S., Lee, S.I.: A unified approach to interpreting model predictions (2017). https://doi.org/10.48550/arXiv.1705.07874, arXiv:1705.07874
32. Luo, C., Li, A.J., Xiao, J., Li, M., Li, Y.: Explainable and generalizable AI-driven multiscale informatics for dynamic system modelling. Sci. Rep. **14**(1), 18219 (2024). https://doi.org/10.1038/s41598-024-67259-4
33. Mohseni, S., Zarei, N., Ragan, E.D.: A multidisciplinary survey and framework for design and evaluation of explainable AI systems. ACM Trans. Interact. Intell. Syst. **11**(3–4), 1–45 (2021). https://doi.org/10.1145/3387166
34. Mokander, J., Floridi, L.: Ethics-based auditing to develop trustworthy AI. Minds Mach. **31**(2), 323–327 (2021). https://doi.org/10.1007/s11023-021-09557-8, arXiv:2105.00002
35. Mugurusi, G., Oluka, P.N.: Towards explainable artificial intelligence (XAI) in supply chain management: a typology and research agenda. In: Dolgui, A., Bernard, A., Lemoine, D., Von Cieminski, G., Romero, D. (eds.) Advances in Production Management Systems. Artificial Intelligence for Sustainable and Resilient Production Systems, vol. 633, pp. 32–38. Springer International Publishing, Cham (2021). https://doi.org/10.1007/978-3-030-85910-7_4, series Title: IFIP Advances in Information and Communication Technology
36. Pearl, J.: Causality: Models, Reasoning, and Inference. Cambridge University Press, 2nd edn. (2009). https://doi.org/10.1017/CBO9780511803161
37. Phan, T.L.J., Gehrhardt, I., Heik, D., Bahrpeyma, F., Reichelt, D.: A systematic mapping study on machine learning techniques applied for condition monitoring and predictive maintenance in the manufacturing sector. Logistics **6**(2), 35 (2022). https://doi.org/10.3390/logistics6020035
38. Pirouzrahi, Z., Vanelslander, T., Nassiri Aghdam, A.: Applying system dynamics modelling to modal shift: a systematic review. Sustain. Futures **9**, 100526 (2025). https://doi.org/10.1016/j.sftr.2025.100526
39. Poeta, E., Ciravegna, G., Pastor, E., Cerquitelli, T., Baralis, E.: Concept-based explainable artificial intelligence: a survey (2023). https://doi.org/10.48550/arXiv.2312.12936, arXiv:2312.12936
40. Puli, V.O.R., Yasmeen, Z.: Predictive analytics and deep learning for logistics optimization in supply chain management. JAIBD **1**(1), 139–150 (2021). https://doi.org/10.31586/jaibd.2021.1187
41. Quaranta, G., Lacarbonara, W., Masri, S.F.: A review on computational intelligence for identification of nonlinear dynamical systems. Nonlinear Dyn. **99**(2), 1709–1761 (2020). https://doi.org/10.1007/s11071-019-05430-7
42. Ribeiro, M.T., Singh, S., Guestrin, C.: Why should i trust you?: explaining the predictions of any classifier (2016). https://doi.org/10.48550/arXiv.1602.04938, arXiv:1602.04938

43. Richey, R.G., Chowdhury, S., Davis-Sramek, B., Giannakis, M., Dwivedi, Y.K.: Artificial intelligence in logistics and supply chain management: a primer and roadmap for research. J. Bus. Logist. **44**(4), 532–549 (2023). https://doi.org/10.1111/jbl.12364
44. Scholkopf, B., et al.: Toward causal representation learning. Proc. IEEE **109**(5), 612–634 (2021). https://doi.org/10.1109/JPROC.2021.3058954
45. Sharma, J., Lal Mittal, M., Soni, G., Keprate, A.: Explainable Artificial Intelligence (XAI) approaches in predictive maintenance: a review. Eng **18**(5), e170423215860 (2024). https://doi.org/10.2174/0187221211866623041708423
46. Sobrie, L., Verschelde, M., Hennebel, V., Roets, B.: Capturing complexity over space and time via deep learning: an application to real-time delay prediction in railways. Eur. J. Oper. Res. **310**(3), 1201–1217 (2023). https://doi.org/10.1016/j.ejor.2023.03.040
47. Sokol, K., Flach, P.: Explainability is in the mind of the beholder: establishing the foundations of explainable artificial intelligence (2022). https://doi.org/10.48550/arXiv.2112.14466, arXiv:2112.14466
48. Soumpenioti, V., Panagopoulos, A.: AI Technology in the field of logistics. In: 2023 18th International Workshop on Semantic and Social Media Adaptation & Personalization (SMAP)18th International Workshop on Semantic and Social Media Adaptation & Personalization (SMAP 2023), pp. 1–6. IEEE, Limassol, Cyprus (2023). https://doi.org/10.1109/SMAP59435.2023.10255203
49. Tao, M.: Semantic ontology enabled modeling, retrieval and inference for incomplete mobile trajectory data. Futur. Gener. Comput. Syst. **145**, 1–11 (2023). https://doi.org/10.1016/j.future.2023.03.012
50. Termine, A., Primiero, G.: Causality problems in machine learning systems. In: The Routledge Handbook of Causality and Causal Methods, pp. 325–341. Routledge, New York, 1 edn. (2024). https://doi.org/10.4324/9781003528937-37
51. Wang, K., Varambally, S., Watson-Parris, D., Ma, Y.A., Yu, R.: Discovering latent structural causal models from spatio-temporal data (2024). https://doi.org/10.48550/arXiv.2411.05331, arXiv:2411.05331
52. Wasi, A.T., Anik, M.A., Rahman, A., Hoque, M.I., Islam, M.S., Ahsan, M.M.: A theoretical framework for graph-based digital twins for supply chain management and optimization (2025). https://doi.org/10.48550/arXiv.2504.03692, arXiv:2504.03692
53. Wei, B.: Sparse dynamical system identification with simultaneous structural parameters and initial condition estimation. Chaos, Solitons Fractals **165**, 112866 (2022). https://doi.org/10.1016/j.chaos.2022.112866
54. Yang, W., et al.: Survey on explainable AI: from approaches, limitations and applications aspects. Hum-Cent. Intell. Syst. **3**(3), 161–188 (2023). https://doi.org/10.1007/s44230-023-00038-y
55. ŞAHiN, E., Arslan, N.N., Özdemir, D.: Unlocking the black box: an in-depth review on interpretability, explainability, and reliability in deep learning. Neural Comput. Appl. **37**(2), 859–965 (2025). https://doi.org/10.1007/s00521-024-10437-2

Joint Optimization of Active Storage Lots, Layout Design, Storage Location Assignment and Order Picking in a Warehouse

Giovanni Righini[✉]

Department of Computer Science, University of Milan, Milan, Italy
`giovanni.righini@unimi.it`

Abstract. Optimal management of large warehouses or regional distribution centers requires to optimally solve a number of interrelated sub-problems such as deciding the layout of the warehouse, i.e. the organization of aisles and shelves, the number of items to be made available for each stock keeping unit (SKU), the location of each SKU in the warehouse and the routing policies to be followed by pickers. These sub-problems are usually analyzed separately in the scientific literature. Moreover, real cases are often characterized by additional constraints and complicating features that are seldom considered in academic papers. We present an attempt to optimize them jointly by means of suitable mathematical optimization models and solvers. The method has been used to optimize the operations in a warehouse with an irregular layout and precedence constraints between different materials. Relevant improvements in the productivity of pickers were observed when the suggestions stemming from this analysis and optimization were compared with the company current practice.

Keywords: Inventory management · Order picking · Storage location assignment

1 Introduction

In business-to-business (B2B) warehouses items are stored to fulfill orders coming from regional distribution centers and sale points. Items in the inventory are cataloged into Stock Keeping Units (SKU), i.e. product types identified by a unique code. Incoming orders are processed to produce picking lists; pickers equipped with a trolley visit the warehouse, pick-up the required amount of items for each SKU following a specified sequence, and bring them to the assembly point, where the collected items are packaged in boxes to be shipped. The time taken by pickers is typically the bottleneck of the whole system. Both the layout of the warehouse and the location of the SKUs determine the time needed by pickers, even when optimal routes are followed. Furthermore, routes of minimum length may not be followed either because pickers do not optimize their duty or because of additional constraints.

© The Author(s), under exclusive license to Springer Nature Switzerland AG 2026
M. Pavone et al. (Eds.): DSA ISC 2025, LNCS 16405, pp. 94–109, 2026.
https://doi.org/10.1007/978-3-032-21811-7_7

All these aspects are interconnected and it is very challenging to optimize them jointly. The case we consider has the following characteristics:

- Picker-to-parts: pickers equipped with trolleys traverse the aisles, visit the locations, pick-up the required items and bring them out to a packing and shipping shop.
- Pick-by-order policy: each picker is assigned a picking list, corresponding to part of a customer order; picking lists are assumed to be given.
- As opposed to Automatic Storage/Retrieval Systems (AS/RS), the layout of warehouses operated by human pickers can be more complex than those studied in the literature. A real example is shown in Fig. 1: two lateral corridors connect parallel aisles along which items are accessible. Furthermore, an additional rack is located along one of the corridors.
- Items of special size may originate special constraints: for instance, some items that are especially long are required to be stored in deep shelves, that are available in a limited amount and must be coupled, that is they must be located back-to-back in adjacent racks, as shown in Fig. 1.
- Precedence constraints can be imposed on the pick-up sequence: when items are made of different materials they must be picked-up according to a predefined sequence (heavier materials first). Only a few papers in the literature consider this kind of constraints [2,8,10].

Main Sub-problems and Related Literature. To introduce the problem, we first briefly review the main sub-problems involved and the related literature.

A sub-problem that must be addressed before tackling the optimization of the pick-up operations is the definition of optimal active stocks, i.e. the number of items to be made available on the shelf for each SKU when a replenishment operation is executed. This problem can be addressed as a variant of the classical economic order quantity problem, a continuous non-linear but convex optimization problem mentioned in all textbooks on inventory optimization (see for instance [11]), with the main difference that holding costs are often replaced by capacity constraints.

The Storage Location Assignment Problem (SLAP) is the problem of assigning each SKU a position. The goal is to minimize the average time needed to access the SKUs that are required to satisfy an expected set of orders. The SLAP has been studied in the literature in its most basic and simplified versions in the last fifteen years. Some useful reviews are [5,6,9]. From a mathematical optimization point of view, the SLAP is a computationally difficult problem even in the case of a single aisle [3].

Strictly related to the SLAP is the optimization of order picking, that is the minimization of the expected distance to be covered by pickers. This problem can be treated under the assumption that the storage locations are given, as in Daniels et al. [4], or jointly optimized with the SLAP, as in Renaud and Ruiz [7] and Bolaños et al. [2]. The latter paper contains a useful classification of the related literature and a rich bibliography and it is one of the few papers in which precedence constraints between SKUs are taken into account. However, the mathematical programming model it proposes cannot be solved by general purpose mathematical programming solvers for instances of realistic size. Furthermore, the layout design sub-problem is not addressed in these references.

To the best of our knowledge, no contribution in the literature deals with all these features simultaneously, a fortiori when the layout itself is not given but can be decided within some geometric and topological constraints. The original contribution of this paper is to make an attempt to develop a method to solve these sub-problems, taking into account that all of them are related to one another. The method is general but its description is illustrated by a specific and successful case study, where its implementation allowed to obtain significant savings.

2   Models and Solutions

Operations and Objectives. The efficiency of the warehouse is measured by the number of customer orders satisfied per unit of time. To maximize efficiency, several decisions and operations must be jointly optimized, since they are strictly related to one another:

- the optimization of replenishment operations;
- the optimization of the layout;
- the optimization of the storage location assignment;
- the optimization of picker routes.

Correspondingly, our method for the joint optimization of the warehouse works in four main steps. Each of the following subsections provides a description of each sub-problem its data, decision variables, constraints and objectives, the proposed solution procedure, its underlying assumptions and their justifications as well as the results obtained in the specific case study.

2.1   Step 1: Optimization of the Active Stock

In non-automated warehouses each rack is divided into several levels, but usually only the bottom-most level contains the active stock, i.e. the stock available to pickers. The upper levels are filled with buffer stock. Therefore, replenishment operations are time-consuming. An operator must access a shelf at a certain level of a rack, pick-up a pallet with a forklift, move it to the floor, open one or more boxes on the pallet, transfer a given number of items from them to the bin on the bottom-level shelf, close the palletized boxes and put back the pallet in its original position. It can be safely assumed that each replenishment operation takes about the same time weakly depending on the amount to be replenished.

Ideally, one wants a low replenishment frequency for all SKUs, but this would imply large active stocks which is in conflict with the capacity constraint, since the shelf capacity that is available for active stock is limited.

The problem of optimally exploiting the available capacity of the bottom-level shelves with the goal of minimizing the overall frequency of replenishment operations is similar to the classical economic order quantity problem mentioned in all textbooks on inventory optimization (see for instance [11]), with the main difference that there are no holding costs to penalize the amounts of inventory; instead, they are replaced by the capacity constraint.

The data needed to formulate this sub-problem can be taken from historial series of pick-up operations recorded in the information system of the warehouse. By pick-up we mean a single access to a shelf to pick-up a certain number of items of an SKU. From these data one can obtain the number of pick-up operations and the total number of items picked-up for each SKU. This yields the total demand satisfied by the warehouse and the total demand for each SKU. In spite of being non-linear, this sub-problem is not computationally challenging, owing to the convexity of the objective function, and it can be solved by the analytical procedure outlined in Appendix A.

Safety Stocks. Since the demand for each SKU is not known a priori, the inventory problem should be classified as non-deterministic, implying the determination of a suitable safety stock. It is well-known that the safety stock can be optimally sized in a slightly different way according to the replenishment policy adopted: the fixed reorder point policy triggers replenishment when the inventory level reaches a suitably tuned threshold; the fixed reorder period policy triggers replenishment when a suitably tuned period of time has passed. The former policy allows for a smaller safety stock compared with the latter one. In both cases, the safety stock depends on the lead time, the mean value and the variance of the demand as well as a parameter depending on the accepted stockout probability.

It can be safely assumed that the scarcest SKUs can be always replenished every day. This assumption is well-justified in a non-optimized scenario, because the satisfied demand is relatively low. However, in a well-optimized scenario it may happen that the replenishment step becomes the bottleneck of the whole system: since an increased level of efficiency of the warehouse implies that larger amounts of demand can be satisfied. If this translates into employing fewer pickers or the same number of pickers for shorter duties, additional workforce can be moved from order picking to replenishment, while the total demand satisfied per day does not change. In this case the replenishment step does not become critical and safety stocks are less and less needed. On the contrary, if the increased efficiency translates into a development of the business and an enlargement of the customer portfolio for the warehouse, then the total demand increases while the resources available for replenishment do not and therefore safety stocks become more and more needed. In our procedure we initially assume that safety stocks can be neglected; should such an assumption not be verified by the results, the analysis can be corrected by increasing the size of the active stock of the most highly-moving SKUs by a suitable safety stock.

In our case study, the lead time was equal to one day, since replenishment operations are typically done in the evening, when pickers are not on duty. Moreover, the decision about the replenishment of scarce SKUs could be taken every day on the basis of the residual stock (fixed reorder point policy) and there was no tight bound on the number of replenishment operations that can be done in one day. Since no change in the customer portfolio was in sight, we assumed that safety stocks could be neglected. Some limited sensitivity analysis confirmed our assumption.

Upper and Lower Bounds. Two limits bound the amount of stock that can be assigned to a shelf for each SKU: the upper bound is given by whole shelf capacity; the lower bound is the capacity of the smallest bin that can be produced in a shelf by suitably placing horizontal and vertical separators in it.

Because of these bounds, the optimal values computed with the procedure of Appendix A may be infeasible in practice: active lot sizes larger than the capacity of a single shelf must be reduced to it, while active lot sizes smaller than the minimum allowed amount must be increased to it. When some upper or lower bounds are not satisfied, the values of the infeasible lots are fixed to their upper or lower bounds, the corresponding volumes are subtracted from the total capacity of the warehouse and the values of the other lots are re-computed.

Capacity Constraints. From the optimal size of active stock, one can obtain the corresponding optimal number of shelves to assign to each type of material. For this purpose, SKUs with special characteristics, such as long items requiring special shelves, are treated as a distinct material.

However, owing to the discrete structure of the racks, subdivided into identical shelves, combined with the constraint that it is not allowed to store the same SKU across different shelves, the optimal values of the lots of active stock computed as shown in Appendix A, although reconciled with upper and lower bounds, are not necessarily feasible, because it is not guaranteed that there exists an assignment of such amounts to the shelves such that the available volume in each shelf is saturated exactly. Therefore, the amounts of active stock obtained by our procedure must be considered as ideal values to tend to, while satisfying packing constraints for each shelf.

Our case study concerned a warehouse containing long and thin items, such as tubes and profiles, stored horizontally with their longest dimension orthogonal to the aisle. The two smallest dimensions of the items determine the consumption of the accessible shelf area facing its aisle and this generates the capacity constraint. Owing to the very small value of the two smallest dimensions of the items, it was not necessary to formulate a two-dimensional packing problem which is computationally hard to solve. On the contrary, items could be treated as if they were deformable, assuming that any fraction of the total area of a shelf can be assigned to each SKU. This originates continuous packing problems, whose solution can be rounded to integer values without introducing significant errors.

A dozen distinct material types were considered, while the number of shelves was approximately 150. The variance in the demand of different types was very large (two orders of magnitude difference between the fastest and the slowest-rotating SKUs) and therefore the number of shelves assigned to each type of material covered a wide range, from a minimum of 0.5 shelves (for long brass SKUs) to a maximum of 50 shelves (for aluminium SKUs).

2.2   Step 2: Optimization of the Layout

The second step in our method concerns the layout and the order picking policy. These two aspects need to be considered jointly, as illustrated hereafter.

Warehouse Zones. In general a warehouse can be set up to serve a large number of small customers (B2C), a small number of large customers (B2B) or a mix of the two types. In a B2B scenario, it is possible that the ground set of SKUs can be (exactly or approximately) partitioned into subsets each one corresponding to a different customer. When this happens, it is intuitively convenient to decompose the whole warehouse optimization problem into almost independent sub-problems and to assign SKUs of different customers to different zones of the warehouse. This allows pickers to visit only one zone for each trip, when their picking lists are obtained from partitioning customer orders without batch picking, i.e. without mixing orders coming from different customers. If some SKU is shared by two or more customers, it may be convenient to duplicate its location in two or more corresponding zones, as if they were different SKUs.

Our case study concerned a B2B warehouse partitioned into three zones related to three main customers. We concentrated on one of these zones, shown in Fig. 1, that was perceived as the most critical one.

Fig. 1. The layout of the warehouse zone in the case study. The input/output position is indicated by the black circle in the bottom left corner. Large rectangles correspond to racks made of deep shelves.

Shelves and SKU Special Features. In warehouses operated by human pickers, the storage is located on parallel racks separated by aisles where pickers can access the shelves. The aisles between parallel racks are connected by one or more corridors running orthogonal to them.

Along its horizontal dimension each rack is partitioned into shelves; they have the same size and each shelf can be partitioned into several bins of variable size, each one with its own SKU. For each SKU, several items can be placed in the corresponding bin. For each zone of the warehouse it is not allowed to store the same SKU in different bins on different shelves.

In some cases specific requirements constrain the layout further. For instance, in our case study some items are longer than the depth of a single shelf. For this reason, they must be stored in special deep shelves obtained by placing three shelves next to one another, so that the central one acts as a prolongation of the depth of the other two. In this way, the depth of the accessible shelves is increased by 50%; however, this introduces the constraint that deep shelves must be located back-to-back in adjacent racks.

Layout and Pickers' Routes. There exist two main policies for pickers to visit the shelves of a rack along an aisle: one is to completely traverse the aisle from an end to the other, which allows accessing all shelves (on both sides) along it; the other is to enter the aisle from an end, visit it up to a certain position and then exit from the same end, which allows accessing only the shelves closest to the input/output end. The former policy is typical of bustrofedic trips, while the latter is called "out-and-back". They are shown in Fig. 2.

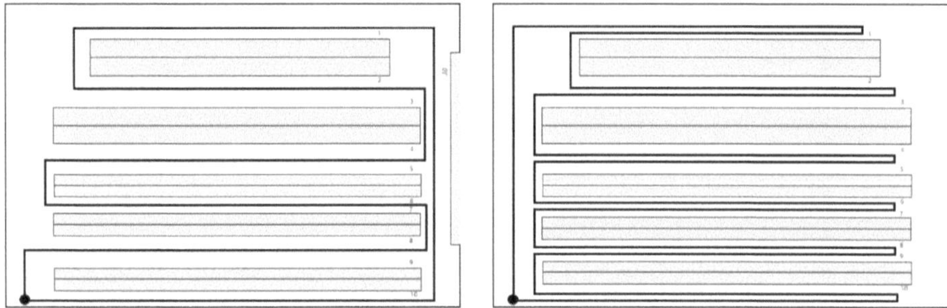

Fig. 2. Left: a bustrofedic trip visiting all shelves in all aisles. Right: a trip using an out-and-back policy.

The two policies can also be mixed along the trip of each picker. In general, slow-moving SKUs (i.e. SKUs with a low level of demand) can be conveniently located at the end of aisles visited according to an out-and-back policy from the other end.

Assuming that an aisle allows visiting s shelves and each shelf has the same probability p or requiring a visit, one can compute the critical value of the probability p for which the expected distance travelled by a picker adopting an out-and-back policy is the same as in a bustrofedic trip, as explained in Appendix B. For instance, when racks are made by $s = 15$ shelves, as in our case study, the corresponding critical value of the probability is $\overline{p} = 9\%$.

In our case study, from the analysis of a data-set of $1,283,879$ pick-ups, the fastest-moving SKU has been picked-up 6653 times. Interpreting frequencies observed in the past as a reliable proxy of probabilities of future events, the pick-up probability of that SKU is about 0.52%. Since SKUs are about 2500, the average number of pick-ups for each SKU in the same period is about 510, which gives an average pick-up probability of about 0.04%, much smaller than $\bar{p}$.

To obtain the probability of a visit associated with a whole shelf containing several SKUs, one should sum the probabilities of the SKUs and subtract the joint probabilities. However, when the probability values are very low, the joint probabilities can be neglected; this allows simplifying the computation of the visit probability of each shelf at the price of slightly overestimating it. For instance, even placing 20 different SKUs in a same shelf (the average number of SKU per shelf is about 16 in our case study), the average visit probability for the shelf would be less than 1%. To further justify the above simplifying assumption, it should be noted that fast-rotating SKUs tend to have relatively large amounts of active stock and therefore the shelves containing them are likely to contain relatively few different items, which makes the approximation error smaller. Even when two shelves can be accessed from each position along an aisle (on the two sides of the aisle), the total probability assigned to each position remains well below the $\bar{p}$ threshold computed above.

Furthermore, the threshold value $\bar{p}$ corresponds to the worst possible case from the point of view of an out-and-back policy, i.e. when all shelves have the same probability of requiring a visit. However, the efficiency of the out-and-back policy is improved by ordering the shelves on each side of each aisle according to their probability of visit, locating the shelves with higher probability closer to the end where the picker enters and leaves and the shelves with lower probability closer to the other end. When the variance of pick-up frequencies is large, as in our case study, this ordering makes the out-and-back policy even more efficient.

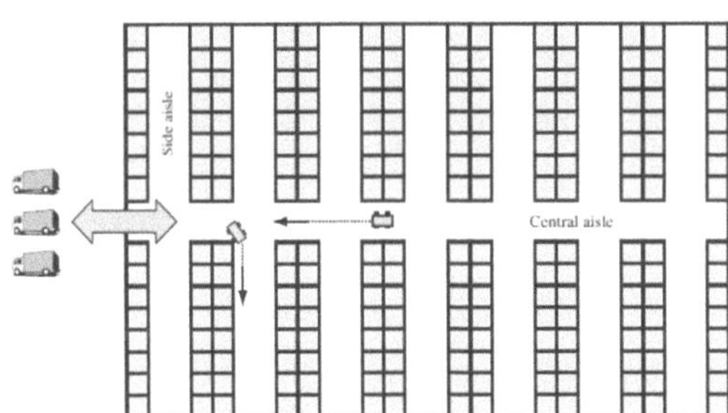

Fig. 3. A layout that allows visiting all aisles with an out-and-back policy (taken from Ghiani, Laporte and Musmanno textbook [11].)

These observations justify the common assumption that warehouses should be organized as shown in Fig. 3, which is taken from a logistics textbook.

However, sometimes the layout can be constrained or given in a different way. This is also the case in the warehouse of our case study, as shown in Fig. 1, where aisles are connected by two lateral corridors instead of a central one. Note that the input/output position for the warehouse zone is at an endpoint of one of the corridors, while a whole rack is located along the other corridor. If both corridors must be used to access the aisles, at least two aisles must be completely traversed as in bustrofedic trips. Intuitively, it is optimal to select them as close as possible to the ends of the corridors as shown in Fig. 4.

Fig. 4. A layout where two aisles (racks $2-3$ and rack 10) as well as the lateral corridors (one of them facing rack 20) are completely traversed, while all the others are visited with out-and-back policy.

To fully exploit the complete traversal of the two selected aisles, it is profitable to select two aisles sided by two racks (like racks $2 - 3$ in Fig. 4), thus avoiding the very first and the very last aisle that are sided by only one rack (like rack 10 in Fig. 4). All these decisions originate different scenarios for which optimal storage locations and picker routes must be determined. However, there is no combinatorial explosion associated with the number of scenarios to consider and their number remains limited and well-manageable.

Hence, different layouts can be tested, where different aisles are traversed with bustrofedic or out-and-back policy and the analysis described in the remainder can be repeated for each of them.

2.3 Step 3: Storage Location Assignment

For each assumed layout and order picking policy, the storage location assignment problem consists of assigning each SKU a position in the warehouse. This must be done while satisfying some additional constraints, such as precedence constraints between different material types.

Precedence Constraint. Since all items are put in the same boxes on picker trolleys, heavier material items must be picked-up before lighter ones. Hence, a remarkable complicating feature of the problem is the fixed sequence of materials: in our case study, it was given by steel, iron, brass, aluminium, PVC.

A notable exception were items of sheet metal. They have two large dimensions and a small one. They must be placed in a separate position on the trolleys and therefore they are not subject to any precedence constraint. Another exception was due to items catalogued as accessories: they are small and non-deformable. Therefore, in spite of being relatively light, they can be picked-up at any point in the sequence.

Two-Level Assignment. Owing to the precedence constraints, before considering each SKU, shelves must be assigned to material types, complying with the precedence constraints and with the constraint on the back-to-back location of deep shelves. In a second-level assignment individual SKU must be suitably located.

The material precedence constraint suggests to group SKUs of the same material in the same shelves. This is an additional restriction to the solution of the bin packing problem mentioned in Step 1, which decomposes into as many independent bin packing problem instances as the number of types of materials.

In Step 1 of our method, we force the definition of homogeneous shelves, still allowing for a small number of shelves to be shared by different materials (consecutive in the precedence sequence), so that half-shelves can be assigned to a material. This is allowed to avoid too large deviations from the ideal values of the active stocks.

Figure 5 shows an example of the outcome of the first-level assignment, i.e. the assignment of materials to shelves.

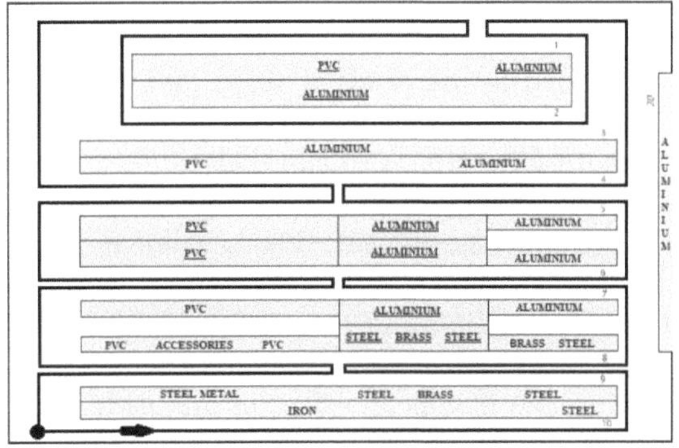

Fig. 5. An example of a scenario with back-to-back deep shelves and a corresponding shelves-material assignment complying with the precedence constraint. Deep SKUs are indicated with underlined font.

From the output of Step 2, for each material type, one can observe the number ω of shelves to be visited with an out-and-back policy. The objective of the assignment is to partition the whole set of SKUs of each given material type in a number of subsets equal to the number of shelves determined in Step 1. Each SKU has an associated ideal active stock and the volume assigned to each SKU in its shelf must deviate as little as possible from it.

As a secondary objective of the optimization, ω shelves in the solution should be assigned slow-moving SKUs in order to maximize the efficiency of the out-and-back policy.

Both these problems can be easily formulated in mathematical programming terms and solved by available software solvers.

For instance, the material type "long PVC" includes 375 SKUs. In order to assign $15 - 16$ SKU to each shelf, 24 shelves are needed. Owing to the layout, $\omega = 8$ shelves must be visited with out-and-back policy. Therefore all long PVC SKUs have been sorted by non-increasing pick-up frequency and the 120 SKUs with minimum frequency have been grouped into 8 subsets of 15 SKUs each to be assigned to the long PVC shelves visited with out-and-back policy. The range of total frequencies of these subsets goes from 127 to 0 pick-ups per year. The remaining 255 long PVC SKUs have been grouped into 16 subsets, with 15 or 16 SKUs each, balancing their total frequency the most. The range of frequencies of the 255 fastest-moving SKUs goes from 3497 to 10 pick-ups per year; the range of total frequencies of the optimally balanced subsets goes from 7474 to 7321 pick-ups per year.

2.4 Step 4: Optimal Picker Routes

The order picking optimization assumes that each picker visits the needed locations according to the predefined sequence associated with the layout. In principle, it is still possible that for some picking lists better trips can be found. However, this is rather unlikely to happen when material precedence constraints must be satisfied.

Since the sequence of the materials is fixed and the material location has been determined in Step 3, the optimization of pickers' routes is pursued by a second-level assignment of the SKUs to the shelves.

For this purpose, the groups of SKUs formed in Step 1 are sorted according to their expected frequency of visit. Then, a linear assignment problem is solved between the set of SKUs groups determined in Step 1 and the set of shelves assigned to the corresponding material. The aim is to place those with lowest probability far from the input/output end for each out-and-back aisle. This allows maximizing the beneficial effect of adopting the out-and-back policy. From a computational viewpoint it requires to solve a non-linear integer optimization problem of small size. In our case study, the optimal solution was produced by a spreadsheet solver in negligible computing time.

It is worth remarking that the out-and-back policy has the further advantage to make it easier to satisfy the material precedence constraint, since the sequence of visits to the shelves of aisles is less constrained compared with the bustrofedic policy: the picker has two chances of visiting each shelf, instead of one.

3 Conclusions

Success and Impact. After analyzing several alternative scenarios, one of them was selected for some simulations. Some picking lists were generated at random, exploiting the probabilities taken from the visit frequencies observed in the past. Pickers were instructed to visit the locations corresponding to the SKUs of the picking lists in the selected scenario and their travel time was measured and compared with that needed for the same picking lists in the current scenario. Astonishing time reductions of about $20\% - 25\%$ were observed, confirming the outcome of several computer-based simulations.

This translates into an increased productivity level for the warehouse, where the time taken by pickers was clearly the bottleneck of the whole process. Either more orders can be processed in the same amount of time by the same number of pickers or the same number of orders can be processed by a reduced number of pickers, devoting the newly available workforce to other functions that are likely to become bottlenecks, like packing and shipping, for example.

Besides the optimization of the current scenario, the method developed can be re-used for any future scenario, allowing to re-allocate SKUs to shelves and aisles or to redefine the layout of pickers' routes on a quantitative basis rather than by manual inspection led by the criterion of selecting the least impacting change, as it was common practice in the warehouse where the case study has been carried out.

The method described in this paper was tailored to the specific case of the warehouse considered. Since each warehouse is likely to have specific features, constraints and objectives, there is no guarantee that the same approach can be transferred to other case studies without any major change. What is generally valid is the decision science method, based on the definition of a mathematical model of the problem at hand, possibly allowing for its decomposition into interdependent sub-problems, each of them being effectively solvable with mathematical programming techniques.

Extensions. While Steps 1, 3 and 4 can be solved at optimality with the aid of a mathematical programming solver, in our case study Step 2 was solved manually in a heuristic way. Hence, a natural development consists of designing an algorithm to automatically generate feasible layouts in Step 2. The computing time we observed in Steps 1, 2 and 4 was very small. Therefore, the described approach is likely to be significantly scaled up.

A further possible extension concerns the optimal composition of picking lists, which implies extending the joint optimization method to include also the so-called order batching problem (see for instance Aerts et al. [1] for a recent reference), i.e. the decomposition-recomposition of orders coming from different customers into picking lists allowing pickers to visit a small portion of the warehouse, following short routes. This brings advantages in the picking step, but on the other side it forces the packing and shipping section to keep several shipments open and incomplete at each point in time. Besides requiring more room and possibly more operators, this may also lead to an increased probability of mistakes in the composition of shipments.

Acknowledgments. This work has been carried out in collaboration with Gianluca Distratis and Luca Buoninconti (KPMG Advisory Group, Milan).

Disclosure of Interests. The author has been responsible for a consulting contract signed by KPMG Advisory Group, Milan, and the University of Milan.

A   Appendix A: Optimal Sizing of Active Storage Lots

Let n be the number of SKUs in the warehouse zone and let $N = \{1, \ldots, n\}$ be the set of their indices. Let v_i be the number of pick-ups for each SKU $i \in N$. Let m_i be the total number of items picked-up for each $i \in N$. Let V the total number of registered pick-ups ($V = \sum_{i \in N} v_i$) and let M be the total number of picked-up items ($M = \sum_{i \in N} m_i$). Let T be the duration of the time period in which the recording has been done (one year). From these data we computed the total demand satisfied by the warehouse, $D = M/T$, expressed in number of items per unit of time. For each SKU $i \in N$, the corresponding demand is $d_i = m_i/T$.

For each SKU $i \in N$, let q_i be size of the active stock, i.e. the optimal number of items that must be available in the active storage, on the bottom-level shelf.

We now introduce a measure of the amount of space taken by the items for each SKU. We use the term "volume" to indicate it, but it can be a measure of length or area, according to the type of items. In our case study, it is a measure of the accessible area of the shelf. If w_i is the unit volume of items for SKU $i \in N$, we define $\bar{q}_i = q_i w_i$ as the amount of SKU $i \in N$ allocated to its shelf, expressed in units of volume. Analogously, we indicate by $\bar{d}_i = d_i w_i$ the demand of SKU $i \in N$ expressed in units of volume per unit of time. Let Q be the overall capacity of the warehouse zone, expressed in the same units of volume as the variables $\bar{q}$.

The frequency of replenishment operations for SKU $i \in N$ is given by $d_i/q_i = \bar{d}_i/\bar{q}_i$ and hence it is inversely proportional to the decision variable $\bar{q}_i$, while $\bar{d}_i$ is given.

The problem of optimally sizing the active storage lots for each SKU $i \in N$ can be formulated with continuous non-negative variables $\bar{q}_i \geq 0 \ \forall i \in N$, a linear constraint

$$\sum_{i \in N} \bar{q}_i \leq Q \tag{1}$$

and a non-linear but convex objective function

$$\text{minimize } z = \sum_{i \in N} \bar{d}_i/\bar{q}_i.$$

Owing to the presence of a single inequality constraint, the optimal solution can be immediately determined from the analytical optimality conditions. The components of the gradient vector of the constraint $g(x) \geq 0$ with $g(x) = Q - \sum_{i \in N} \bar{q}_i$, are unitary negative: $\nabla g_i = -1 \ \forall i \in N$. The gradient of the objective function has components $\nabla z_i = -\bar{d}_i/\bar{q}_i^2 \ \forall i \in N$. Since the inequality constraint

is certainly active at optimality, the first order necessary condition for optimality
($\exists k \neq 0 : \nabla z = k \nabla g$) translates into

$$\exists k \neq 0 : \frac{\overline{d_i}}{\overline{q_i}^2} = k \quad \forall i \in N.$$

Therefore

$$\overline{q_i}^2 = \frac{\overline{d_i}}{k} \quad \forall i \in N$$

for some suitable $k \neq 0$, i.e.

$$\overline{q_i} = \sqrt{\frac{\overline{d_i}}{k}} \quad \forall i \in N.$$

The value of k is obtained from the capacity constraint (1):

$$\left(\sum_{i \in N} \frac{\sqrt{\overline{d_i}}}{Q} \right)^2 = k.$$

Then, one obtains

$$\overline{q_i} = Q \frac{\sqrt{\overline{d_i}}}{\sum_{i \in N} \sqrt{\overline{d_i}}} \quad \forall i \in N.$$

B Appendix B: Critical Value of the Probability in an Out-and-Back Policy

Assume an aisle allows visiting s shelves and each shelf has the same probability
p or requiring a visit. Then, the expected value of the distance to be traveled by
the picker following an out-and-back policy is

$$L = 2 \sum_{i=1}^{s} i p (1 - p)^{s-i}.$$

The probability p such that this value L is equal to s, i.e. the same distance a
picker would travel along the aisle in a bustrofedic trip, is shown in Fig. 6.

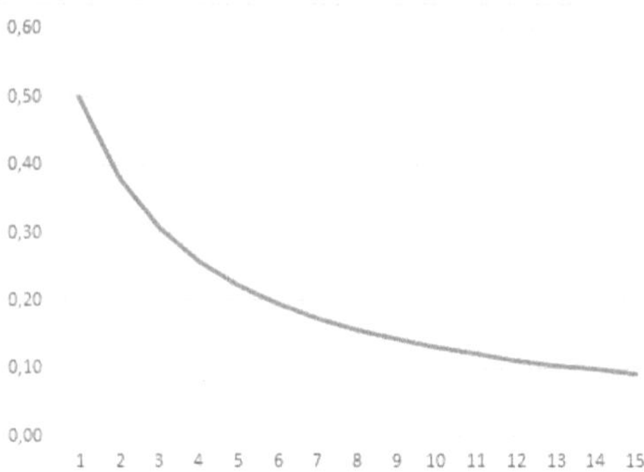

Fig. 6. The value of the probability p for which the out-and-back policy and the bustrofedic policy are equivalent, as a function of the length of the aisle (number of shelves).

When p is low, L turns out to be smaller than s and hence the out-and-back policy is the most efficient. The values of p from which the out-and-back policy is optimal decrease with the length of the aisle.

References

1. Aerts, B., Cornelissens, T., Sörensen, K.: The joint order batching and picker routing problem: modelled and solved as a clustered vehicle routing problem. Comput. Oper. Res. **129**, 105–168 (2021)
2. Bolaños, Z.J., Saucedo Martínez, J.A., Salais Fierro, T.E., Marmolejo Saucedo, J.A.: Optimization of the storage location assignment and the picker-routing problem by using mathematical programming. Appl. Sci. **10**(534), 1–15 (2020)
3. Boysen, N., Stephan, K.: The deterministic product location problem under a pick-by-order policy. Discret. Appl. Math. **161**, 2862–2875 (2013)
4. Daniels, R.L., Rummel, J.L., Schantz, R.: A model for warehouse order picking. Eur. J. Oper. Res. **105**, 1–17 (1998)
5. Gu, J., Goetschalckx, M., McGinnis, L.F.: Research on warehouse operation: a comprehensive review. Eur. J. Oper. Res. **177**, 1–21 (2007)
6. Gu, J., Goetschalckx, M., McGinnis, L.F.: Research on warehouse design and performance evaluation: a comprehensive review. Eur. J. Oper. Res. **203**, 539–549 (2010)
7. Renaud, J., Ruiz, A.: Improving product location and order picking activities in a distribution centre. J. Oper. Res. Soc. **59**, 1603–1613 (2008)
8. Trindade, M.A.M., Sousa, P.S.A., Moreira, M.R.A.: Ramping up a heuristic procedure for storage location assignment problem with precedence constraints. Flex. Serv. Manuf. J. **34**, 646–669 (2022)
9. Van Gils, T., Ramaekers, K., Caris, A., De Koster, R.B.M.: Designing efficient order picking systems by combining planning problems: state-of-the-art classification and review. Eur. J. Oper. Res. **267**, 1–15 (2018)

10. Žulj, I., Glock, C.H., Grosse, E.H., Schneider, M.: Picker routing and storage-assignment strategies for precedence-constrained order picking. Comput. Ind. Eng. **123**, 338–347 (2018)
11. Ghiani, G., Laporte, G., Musmanno, R.: Introduction to Logistic Systems Planning and Control. Wiley (2004)

Bio-Inspired Intelligence Paradigm: The New Era of Smart Industry

Galina Samigulina[1] , Zarina Samigulina[1,2(✉)] , Daulet Bekeshev[1,2] ,
and Diana Butakova[2]

[1] Institute of Information and Computing Technologies, Almaty, Kazakhstan
`samigulinaresearch@gmail.com`
[2] Kazakh-British Technical University, Almaty, Kazakhstan

Abstract. Currently, the development of intelligent production systems based on modern artificial intelligence methods is an urgent problem. Industrial automation has its own specific limitations in the application of artificial intelligence at the level of software and hardware implementation. The creation of a new modified endocrine-immune algorithm (EAIS) for complex objects control and diagnosing the state of equipment is a promising and relevant task. The proposed EAIS algorithm is adapted for work with industrial controllers of the Modicon series (Schneider Electric) of the IEC 61131–3 standard. The results of modeling and experiments were carried out in the Industrial Automation Lab (KBTU JSC). The developed algorithm was tested on the engineering database of the Tengizchevroil oil refinery and modeled using the Modicon M340 controller. The advantages of EAIS include the ability to be placed directly in the controller without using external servers for training. For this purpose, a user-defined functional block DFB (Derived Functional Block) is created with a modified endocrine-immune algorithm EAIS_FBD, implemented in the ST (Structured Text) language. The organization of an intellectual structural program unit (Intellectual Program Organization Unit) is proposed for placing the functional block EAIS_FBD and multiple use in the development of process control systems. Comparative analysis of the results of modeling with the support vector method, decision trees and nearest neighbors showed the superiority of EAIS.

Keywords: Smart Industry · bioinspired algorithms · diagnostic of industrial equipment · modified endocrine-immune algorithm · programmable logic controller

1 Introduction

Currently, a new generation of industrial automation systems is built on complex industrial equipment with modern software. New automation platforms allow flexible integration of artificial intelligence (AI) methods for process control. However, the production environment imposes its own limitations. For example, the industrial programmable logic controller (PLC) Modicon M340 supporting the international standard IEC 61131–3 with a P342010 series processor has an internal user memory of maximum 4096 KB.

M. Pavone et al. (Eds.): DSA ISC 2025, LNCS 16405, pp. 110–123, 2026.
https://doi.org/10.1007/978-3-032-21811-7_8

The maximum number of discrete points is 1024, analog 256 [1]. This amount of memory does not allow the implementation of artificial intelligence built on neural networks within the controller itself. Currently, leading international manufacturers of industrial automation systems offer their solutions for the use of artificial intelligence. For example, AVEVA offers the software product AVEVA Predictive Analytics based on the patented data clustering algorithm OPTiCS (Ordering Points to Identify Clustering Structure) [2], which allows predicting equipment failure time, setting maintenance priorities and issuing recommendations. The system is trained on the unique profile of the installation, taking into account the operating conditions. Historical data is compared with current indicators to identify anomalies. The largest company in the field of industrial solutions Siemens offers a software product "AI Anomaly Assistant", which predicts anomalies in the production process using artificial intelligence [3]. At the level of programmable logic controllers, Siemens has released controllers with support for neural networks of the SIMATIC S7 and SIMATIC Open Controller series [3]. To solve the problem of insufficient memory, part of the calculations is transferred to the local SIMATIC IPC server.

Currently, the creation of smart production systems is carried out through the implementation of AI in solving the following problems: predictive analytics, forecasting and preventing failures (predictive maintenance) [4, 5], optimization of production processes [6], construction of digital twins of real production [7, 8], optimization of robotic complexes [9, 10], development of adaptive control systems for technological processes [11].

However, not all AI algorithms are applicable to industry. For example, the widely known AI method based on neural networks has a limitation in industrial systems due to the insufficient interpretability of the obtained results. In [12], a new neural network (Physics-Guided Neural Networks, PGNNs) is used, which solves this problem by embedding physical values directly into the network structure to improve the interpretability and reliability of the system in the production of seamless pipes. The network is trained on a data set of an industrial hydraulic system, and the data is validated on a neural processing unit (NPU). Research [13] is devoted to the use of Internet of Things (IoT) technology for industrial defect detection based on the Deep Learning (DL) method. The paper discusses the problem associated with DL's requirements for large computing resources, as well as the difficulty in integrating it into IoT devices due to limitations in computing power and memory resources. Digital Signal Processor (DSP) is an important part of IoT. Various optimization strategies have been proposed to detect defects on DSP, as well as a parallel scheme for scaling the model and implementing it on multiple cores.

Interesting from the point of view of solving industry problems are algorithms built on the basis of an artificial immune system (AIS) and their improved varieties that use the principles of operation of the human biological immune system [14]. The basis of the AIS are mechanisms for recognizing anomalies when pathogenic microorganisms and foreign substances enter the body. The immune system has unique properties, such as memory, and the AIS as a biological prototype can remember pathogens for a faster response in case of re-infection, and it can also model the interaction between various components of the system for a collective response, forming an immune network, etc.

A great contribution to the development of this direction was made by such scientists as: James A. Forrest [15], Leandro Nunes de Castro [16], Jon Timmis [17], Vincenzo Cutello [18], Giuseppe Nicosia [19–21], Mario Pavone [22–24] and others.

Smart manufacturing systems built on AIS allow for effective control of technological processes, reducing the frequency of false positives. For example, in [25] a recursive control system based on an immune working mechanism for designing a framework network of multi-agent manufacturing systems is shown. The negative selection algorithm is used for the antibody learning system, which takes into account the disturbance problems. The results show that the false alarm rate has decreased by 4%, and the production balance optimization rate has exceeded 90%.

Another promising AI algorithm is the Artificial Endocrine Algorithm (AEA) [26], which is based on the endocrine system, which regulates processes in the body using hormones. In the biological prototype, the hormonal system is also involved in regulating immunity, it can either enhance or suppress the immune response [27]. The work [28] presents an example of the use of an artificial hormonal system as a mechanism that responds to changes in terrain and allows the robot to move with the least energy expenditure.

Thus, close interaction in a biological prototype of the endocrine and immune systems allows developing a modified artificial intelligence algorithm that has the necessary properties to successfully solve production problems and be integrated into a programmable logic controller for technological processes control based on various industrial platforms that support the IEC 61131–3 standard.

The following structure of the article is proposed. The second section presents the formulation of the research problem, the third section develops a modified endocrine-immune algorithm for diagnosing the state of industrial equipment, the fourth section presents the results of modeling and experiments based on the Industrial Automation Lab (Schneider Electric) laboratory using Modicon series industrial controllers.

This study examines the possibility of developing a modified artificial intelligence algorithm based on artificial immune and endocrine systems (Modified Endocrine-Immune Algorithm, EAIS) for diagnostics and adaptive control of industrial equipment. Particular attention is paid to ensuring characteristics for working with industrial programmable logic controllers: minimization of computational costs due to the controller memory limitation; ensuring self-organization properties for working in a dynamically changing production environment; the presence of self-diagnostic mechanisms for responding to emergency situations; compatibility with programming languages of the IEC 61131–3 standard; minimal delays in data processing. The implementation of the research results into the architecture of programmable logic controllers is possible through the development of an intelligent structural block, the Intellectual Program Organization Unit (IPOU), based on the standard organizational programming unit, the Program Organization Unit (POU) [1, 2].

The research problem statement is formulated as follows: it is necessary to develop a modified artificial intelligence algorithm based on artificial endocrine and immune systems for industrial equipment diagnostics (Modified Endocrine-Immune Algorithm, EAIS). The algorithm must have the following properties for implementation on a PLC: low computational costs, since the controller memory is limited, have the properties of reliability and self-organization, have self-diagnostic mechanisms to eliminate emergency situations, compatibility with the programming languages of controllers of the IEC 61131–3 standard, be well interpreted, have a minimum delay in data processing for work in time close to real time. Integration with the controller is achieved through the development of the concept of an intelligent structural unit Intellectual Program Organization Unit (IPOU) based on the Program Organization Unit (POU) [1, 2].

2 Development of a Modified Endocrine-Immune Algorithm

The developed EAIS algorithm is based on the concept of an artificial endocrine system, which is a bioinspired method in the field of artificial intelligence. The central element of this approach is the term "artificial hormone", which imitates the mechanisms of regulation and impact of biological hormones on various body systems. In nature, the hormonal and immune systems interact with each other through complex and interconnected mechanisms that ensure the maintenance of homeostasis and an adequate response to changes in both the external and internal environment. These systems operate in a unified regulatory context, ensuring the harmonious functioning of the body. Endocrine glands secrete hormones that influence the immune response, while immune system cells are able to modulate the activity of endocrine organs [29, 30].

The hormonal system regulates the immune response through a variety of mechanisms, including both activation and suppression of immune cells such as T cells and macrophages. In turn, immune cells and cytokines can influence the activity of endocrine glands. For example, cytokines can stimulate the production of hormones such as cortisol in response to inflammation [26–30]. Hormones, in turn, can regulate the activity of immune cells, maintaining a balance between activating the immune response and preventing excessive inflammatory activity [26–30]. Thus, the interaction between the hormonal and immune systems is a two-way process in which hormones not only regulate the immune response, but can also be modified in response to immune system activity. This interaction is important for maintaining the body's homeostasis, regulating its response to infections, inflammatory processes, and stressful situations [26].

In the developed modified algorithm, the endocrine mechanism is responsible for identifying informative features, which allows for increasing the accuracy and quality of data processing, and the immune algorithm solves the classification problem, ensuring the system's response based on the principles of the biological immune response. The modified algorithm is presented as a block diagram in Fig. 1.

The metaparameters of the modified hormonal-immune EAIS algorithm are:

- the number of agents *num_agents*;
- the numsber of iterations *num_iterations*;
- the learning rate *learning_rate*;
- the influence of hormones *hormone_influence*, this parameter controls the strength of the influence of the best agents on the rest;
- the number of clones *n_clones*;
- the degree of mutation *mutation_rate*;
- the number of iterations *iterations*;
- the type of distribution for the mutation procedure.

The artificial endocrine algorithm allows using the SHAP (SHapley Additive exPlanations) mechanism [31] to show the degree of significance of each feature for the purpose of data reduction. Then, to solve the classification problem, the most informative features identified by the hormonal system are sent to the immune network. To implement this algorithm in an industrial controller, it is necessary to create a user-defined function block (Derived Functional Block, DFB).

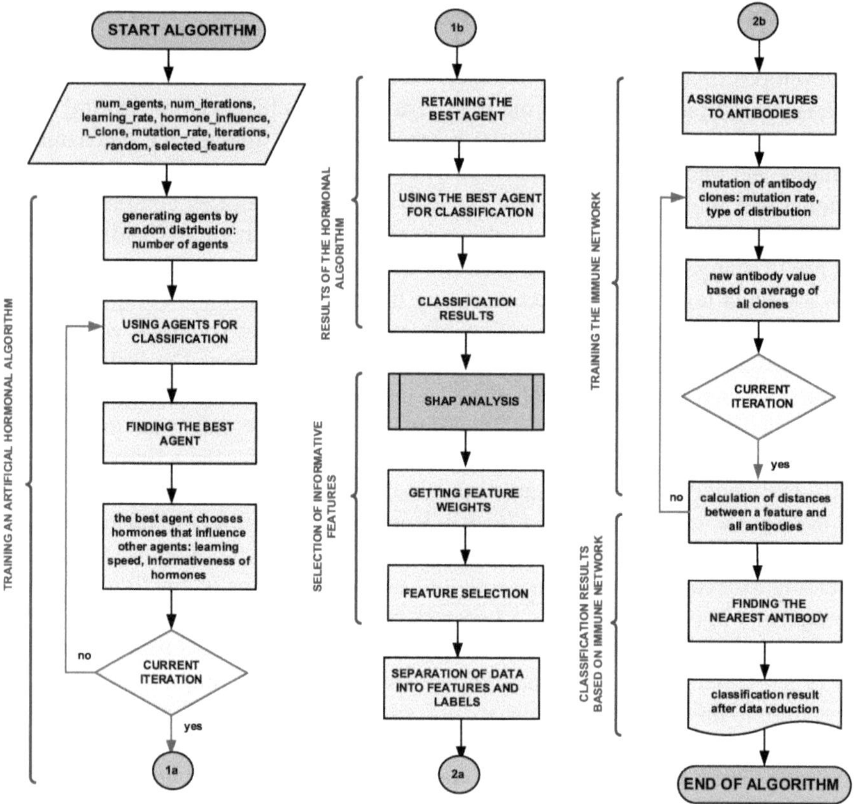

Fig. 1. Modified endocrine-immune algorithm for industrial data processing

User-defined function blocks can be placed in a Program Organization Unit (POU).

2.1 Development of a User-Defined Functional Block of the Modified EAIS Algorithm

The developed EAIS algorithm can be implemented on the EcoStruxture control Expert platform (Schneider Electric) in the form of an intellectual structural unit (Intellectual Program Organization Unit, IPOU). Figure 2 shows the IPOU organization architecture for programming industrial controllers that support the international standard IEC 61131–3.

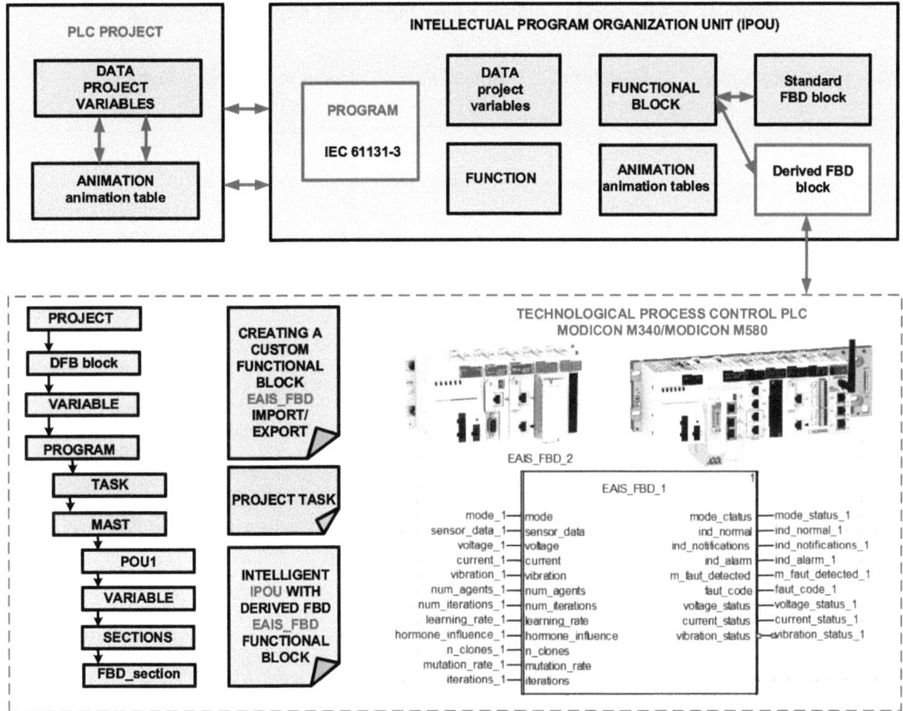

Fig. 2. Structure of the Intellectual Program Organization Unit (IPOU) for IEC 61131–3 Controllers

For technological process control, a working project for PLC programming is created, which consists of a table of variables, device configuration, main program and animation table for tracking the status of the variables.

Within a simple structural unit, there is the possibility of creating a user functional block in the FBD (Functional Block Diagram) language in the DFB (Derived Functional Block) section. Thus, the organization of an intelligent IPOU is possible when creating a user-defined functional block EAIS_FBD for diagnostics of industrial equipment based on the modified EAIS algorithm. The functional block consists of configurable inputs and outputs that signal equipment failure (Table 1).

It is stored in the program library and can be called multiple times. The EAIS algorithm code is implemented in the ST (Structured Text) language inside the functional block. In PLCs programmed on the EcoStruxture Control Expert platform from Schneider Electric, the "TASK" tasks of the project are divided into the following parts: "MAST" - the main task, "FAST" - a fast task, "Event" - rare events, "AUX" - additional tasks.

The call of the developed EAIS_FBD functional block is available from any section of the main task "TASK" of the "MAST" sect [1, 2].

Table 1. EAIS_FBD Function Block Input and Output Specification

No.	Variable name	Data type	Input/Output (I/O)	Discrete/Analog signal (D/A), PLC memory variable (M)	Purpose
1	*mode*	ebool	I	D	operating mode (manual, automatic, stop)
2	*sensor_data*	array of real	I	A	industrial data array
3	*voltage*	real	I	A	voltage value
4	*current*	real	I	A	current value
5	*mode_status*	string	Q	M	text message about the equipment operating mode
6	*ind_normal*	ebool	Q	D	indicator of normal equipment operation
7	*ind_notification*	ebool	Q	D	indicator warning operator
8	*ind_alarm*	ebool	Q	D	indicator of emergency mode of equipment operation
9	*m_fault_detected*	ebool	Q	M	failure detection flag
10	*fault_code*	integer	Q	M	error code
11	*voltage_status*	string	Q	M	equipment voltage status
12	*current_status*	string	Q	M	equipment current status
13	*vibration_status*	string	Q	M	equipment vibration status

3 Results of Modeling and Experiments

Let us consider the results of modeling and experiments of the algorithm operation for analyzing production data. As an example, the D-304 unit with an amine regeneration unit designed to remove sulfur-containing components from the D-304 column cube of the TenggizChevroil oil and gas enterprise was used [32]. The system is deployed on Schneider Electric Modicon M340 and Modicon M241 series controllers (Fig. 3). Temperature and pressure sensors were used to model the operation of the technological process.

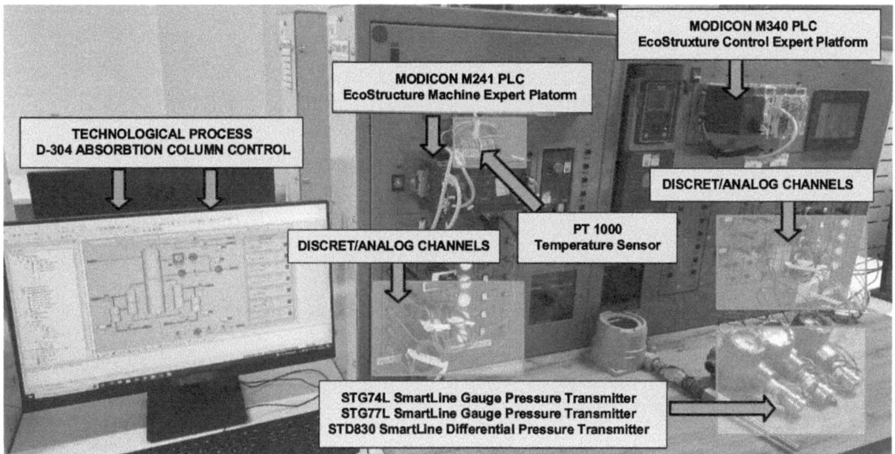

Fig. 3. Modeling of the technological process of the D-304 unit of the TenggizChevroil enterprise

Table 2 shows a fragment of the database of the operation of the D-304 process unit, where PT301027 is a pressure sensor, TE301056 is a temperature sensor, etc.

Table 2. Fragment of the D-304 unit database of the TengizChevroil enterprise

PT301027	PDT31007	TE301056	TE301020	PDT31008	...	PIC31009
1.102	71.383	100.033	106.832	133.233	...	1.089
1.099	71.226	100.101	107.122	131.772	...	1.087
1.097	71.345	100.206	105.456	130.345	...	1.045
1.094	72.456	100.354	106.456	130.234	...	1.043
...	...	...	...	...	...	...
1.101	70.967	100.047	106.532	131.2881813	...	1.087

The main task of the program is to classify objects, defining their "class" - for example, "1" for normal equipment condition and "0" for failure or breakdown. The developed modified algorithm scans the data readings from the sensors. At the first stage, the endocrine component reduces the dimensionality of the data, while preserving the most significant features. Next, the immune part of the algorithm performs classification on the already optimized database. The following parameter values were selected for modeling: number of agents *num_agents* = *100*; number of iterations *num_iterations* = *100*; learning rate *learning_rate* = 2; hormone influence *hormone_influence* = *0.5*;

118   G. Samigulina et al.

n_clones = *50*; mutation rate *mutation_rate* = *0.1*; number of iterations *iterations* = *200*; distribution type for the mutation procedure *random* = *normal*.

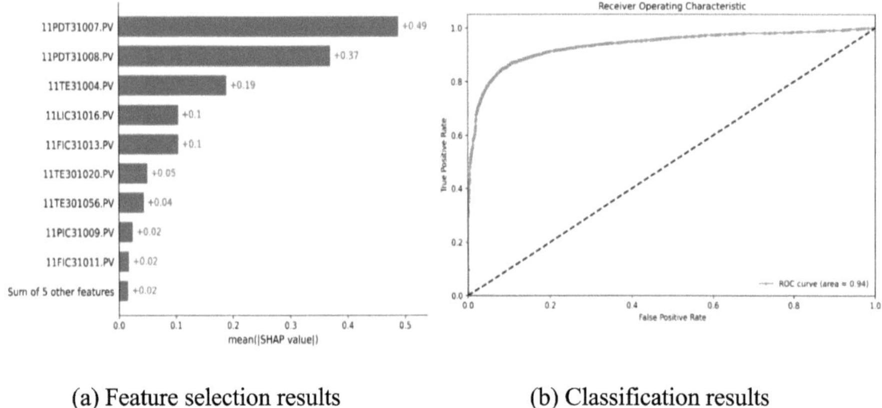

(a) Feature selection results   (b) Classification results

Fig. 4. The result of the modified EAIS algorithm on industrial data: (a) identification of informative features based on the endocrine component of the algorithm; (b) ROC analysis of the solution to the classification problem based on the immune component of the algorithm

Figure 4 shows a visualization of the solution to the classification problem on industrial data based on the EAIS algorithm (Fig. 5).

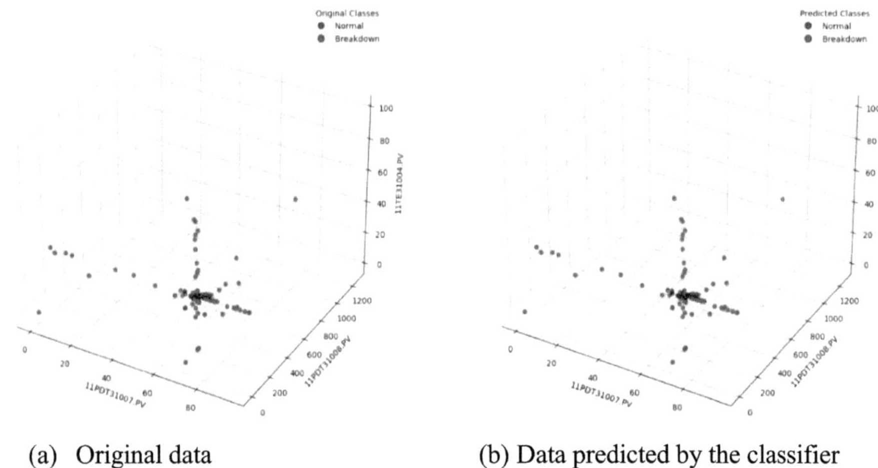

(a) Original data   (b) Data predicted by the classifier

Fig. 5. Visualization of the classification problem solution on industrial data based on the EAIS algorithm: (a) original data; (b) data predicted by the classifier

The efficiency of the classifier is assessed using the following set of metrics: precision reflects the proportion of correctly classified positive examples among all predicted positive ones; recall measures the proportion of correct positive ones among all real positive ones; F-score is the harmonic mean between precision and recall, providing a balanced assessment of these indicators; accuracy expresses the proportion of correctly classified objects among all data, but may be less informative in the case of unbalanced classes; weighted averaging takes into account the importance of different classes in the data; macro-averaging calculates the average value of metrics across all classes, ignoring their frequencies; logarithmic loss (Log Loss) evaluates the accuracy of probabilistic predictions; ROC curve visualizes the relationship between sensitivity and specificity at different classification thresholds [33–36]. Table 3 shows the result of the EAIS algorithm.

Table 3. Efficiency of the EAIS algorithm

Matthews Correlation Coefficient	specificity	balanced accuracy	precision (class 1)	Recall (class 1)	f1-score (class 1)	precision (class 0)	Recall (class 0)	f1-score (class 0)	accuracy	macro avg	weighted avg	ROC curve	evaluation time
0.855	0.880	0.925	0.97	0.88	0.92	0.89	0.97	0.93	0.93	0.93	0.93	0.936	10–50 ms

The obtained modeling results prove the efficiency of the algorithm. The data processing speed depends on the type of devices used. The Modicon M340 controller is equipped with a P342020 processor with a clock frequency of 66 MHz and RAM up to 4 MB. To implement the algorithm, the following minimum computer requirements are required: a processor with 4–8 cores (Intel Core i7/i9 or AMD Ryzen 7/9), with a clock frequency of 3.5 GHz to ensure fast processing of numerical operations.

The recommended RAM is 32 GB, the minimum is 16 GB, a graphics processor with CUDA support (NVIDIA GTX 1660 or RTX 3060), with a memory capacity of 4–6 GB, for data storage it is recommended to use an SSD with a capacity of at least 512 GB. The operating system must be 64-bit, Windows 10/11.

The developed algorithm was compared with other classifiers that are close to EAIS in terms of computational complexity. The following methods were selected as comparison algorithms: support vector machines (SVM) [37], decision trees (DT) [38], and the nearest neighbor method (k-means) [39] (Fig. 6).

For processing engineering data, the support vector method showed an AUC (Area under ROC curve) metric result of 0.28, which indicates incorrect operation of the classifier and overtraining of the data. The SVM method is effective in the case of clearly separable classes, and the data obtained from the controller does not meet such requirements. The decision tree-based classifier showed an AUC result of 0.87. However, the disadvantage of this method is the ability to remember noise in the data, instead of identifying general patterns. Decision trees can be interpretable, but require more computing memory. The algorithm also overfits on training data. Analysis of the database using the nearest neighbor method showed an AUC of 0.84. This algorithm can show good results, but is susceptible to noise in the data.

The classifier is effective on small amounts of data. Thus, the developed modified EAIS algorithm outperforms the considered algorithms (AUC = 0.94), and also processes large data without increasing the memory requirements of the computing device.

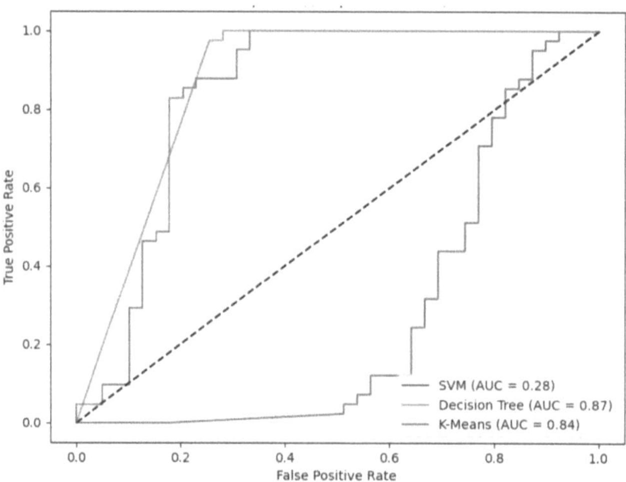

Fig. 6. Comparison of the efficiency of classification algorithms using the Receiver Operating Characteristic (ROC) metric on industrial data

4   Conclusion

The developed modified EAIS algorithm is adaptive for working with industrial data and integration with industrial controllers that support the international standard IEC 61131–3. Unlike neural networks, which require large computing power, this algorithm has higher performance and takes up less space in the memory of devices. The neural network is difficult to interpret, while EAIS uses clear rules based on hormonal regulation, which simplifies diagnostics and debugging of the algorithm. Also, the neural network requires preliminary training outside the controller using external servers for data storage, the modified EAIS algorithm does not need such preliminary training. A feature of the operation of programmable logic controllers is a scanning cycle lasting from 10 to 50 ms, on average about 20 ms. The execution of an algorithm based on a simple neural network without optimization may exceed the duration of the controller scanning cycle, which will lead to a failure of the entire process control program.

These studies are a continuation of a large cycle of work on the development of intelligent technology for the self-assembly of a unified artificial immune system for complex objects control [40, 41].

The research is carried out with the support of the Science Committee of the Ministry of Science and Higher Education of the Republic of Kazakhstan under grant No. AP23486386 (2024–2026).

References

1. Schneider Electric: Modicon M340 Automation Platform. Schneider Electric (2019)
2. Schneider Electric https://www.se.com/
3. Siemens. https://www.siemens.com/

4. Prabu, S., Senthilraja, R., Ali, A.M., Arun, M., et al.: AI-driven predictive maintenance for smart manufacturing systems using digital twin technology. Int. J. Comput. Exp. Sci. Eng. **11**(1) (2025). https://doi.org/10.22399/ijcesen.1099

5. Bharot, N., Verma, P., Soderi, M., Breslin, J.G., et al.: DQ-DeepLearn: data quality driven deep learning approach for enhanced predictive maintenance in smart manufacturing. Procedia Comput. Sci. **232**(19), 574–583 (2024). https://doi.org/10.1016/j.procs.2024.01.057

6. Pasupuleti, M.K.: Smart Manufacturing with Digital Twins: Real-Time Optimization and Process Innovation. In Digital Twin Technology in. Tools for Real-Time Process Optimization. National Education Services, Manufacturing (2024). https://doi.org/10.62311/nesx/905773

7. Besigomwe, K.: Closed-loop manufacturing with AI-enabled digital twin systems. Cognizance J. Multi. Stud. **5**(1), 18–38 (2025). https://doi.org/10.47760/cognizance.2025.v05i01.002

8. Mata, O., Ponce, P., Perez, C., Molina, A., et al.: Digital twin designs with generative AI: crafting a comprehensive framework for manufacturing systems. J. Intell. Manuf. (2025). https://doi.org/10.1007/s10845-025-02583-8

9. Hayat, R.Q., Hussain, F., Masood, U.: Artificial intelligence (AI)-powered line follower robot with hurdle detection and voice control. Int. J. Latest Technol. Eng. Manag. Appl. Sci. **13**(11), 79–83 (2024). https://doi.org/10.51583/IJLTEMAS.2024.131109

10. Chauhan, V.D., Maziarz, G., Dos Santos, S.C.: AI-powered robot-assisted automated assembly inspection system for industry 4.0. In: ASME 2024 International Mechanical Engineering Congress and Exposition (2025). https://doi.org/10.1115/IMECE2024-142813

11. Zohuri, B.: Artificial intelligence and machine learning driven adaptive control applications. J. Mater. Sci. Eng. Technol. **2**(4), 1–4 (2024). https://doi.org/10.61440/JMSET.2024.v2.27

12. Filipović, L., Miličić, L., Ristanović, M., Jovanović, P., et al.: Physics-guided neural network-based feedforward control for seamless pipe manufacturing process. Appl. Sci. **15**(4), 2229 (2025). https://doi.org/10.3390/app15042229

13. Yue, H., Wang, R., Gao, Y., Zhang, J., et al.: Optimising digital signal processor-based defect detection in smart manufacturing with lightweight convolutional neural networks. IET Collaborative Intell. Manufact. **6**(1) (2024). https://doi.org/10.1049/cim2.12092

14. Myakala, P.K., Bura, C., Jonnalagadda, A.K.: Artificial immune systems: a bio-inspired paradigm for computational intelligence. J. Artif. Intell. Big Data. **5**(1), 1–13 (2025). https://doi.org/10.31586/jaibd.2025.1233

15. Hofmeyr, S.A., Forrest, S.: Architecture for an artificial immune system. In: Lecture Notes in Computer Science, vol. 1805, pp. 128–134. Springer (2000)

16. De Castro, L.N., Von Zuben, F.J.: The clonal selection algorithm with engineering applications. In: Proceedings of the Genetic and Evolutionary Computation Conference (GECCO'00), Workshop on Artificial Immune Systems and their Applications, p. 37 (2000)

17. Castro, L.N., de Castro, L.N., Timmis, J.: Artificial Immune Systems: a New Computational Intelligence Approach. Springer Science & Business Media (2002)

18. Cutello, V., Nicosia, G.: Multiple learning using immune algorithms. In: Proceedings of 4th International Conference on Recent Advances in Soft Computing, pp. 102–107. RASC (2022)

19. Ciccazzo, A., Conca, P., Nicosia, G., Stracquadanio, G.: An advanced clonal selection algorithm with ad-hoc network-based hypermutation operators for synthesis of topology and sizing of analog electrical circuits. In: International Conference on Artificial Immune Systems, pp. 60–70 (2008)

20. Conca, P., Nicosia, G., Stracquadanio, G., Timmis, J.: Nominal-yield-area tradeoff in automatic synthesis of analog circuits: a genetic programming approach using immune-inspired operators. In: NASA/ESA Conference on Adaptive Hardware and Systems, pp. 399–406 (2009)

21. Cutello, V., Nicosia, G.: A clonal selection algorithm for coloring, hitting set and satisfiability problems. Neural Nets. **324-337** (2005)

22. Cutello, V., Narzisi, G., Nicosia, G., Pavone, M.: Clonal selection algorithms: a comparative case study using effective mutation potentials. In: International Conference on Artificial Immune Systems, pp. 13–28 (2004)
23. Cutello, V., Nicosia, G., Pavone, M., Timmis, J.: An immune algorithm for protein structure prediction on lattice models. IEEE Trans. Evol. Comput. **11**(1), 101–117 (2007). https://doi.org/10.1109/TEVC.2006.885928
24. Cutello, V., Lee, D., Nicosia, G., Pavone, V., Prizzi, I.: Aligning multiple protein sequences by hybrid clonal selection algorithm with insert-remove-gaps and blockshuffling. In: International Conference on Artificial Immune Systems, pp. 321–334 (2006)
25. Teng, B.: An intelligent manufacturing system based on a recursive control structure. Front. Mech. Eng. **10** (2025). https://doi.org/10.3389/fmech.2024.1437198
26. Jones, T., Thomas, S., Wilson, M.: The role of hormones in immune regulation. J. Immunol. **23**(2), 105–114 (2019)
27. Nunes, C., Sucena, E., Koyama, T.: Endocrine regulation of immunity in insects. FEBS J. **288**(13), 4219–4234 (2020). https://doi.org/10.1111/febs.15581
28. Haomachai, W., Teerakittikul, P.: An artificial hormone system for adaptable locomotion in a sea turtle-inspired robot. In: Proceedings of the 4th International Conference on Control and Robotics Engineering (ICCRE), pp. 136–141. IEEE (2019). https://doi.org/10.1109/ICCRE.2019.8724369
29. Young, L., Smith, J., Goodman, R.: The Influence of Hormones on the Immune System: Mechanisms and Consequences, p. 240. Nauka, Moscow (2020)
30. Belousova, I.L., Kuznetsova, N.A., Ivanova, O.V.: Immunology and Endocrinology, p. 320. Saint Petersburg University Press, Saint Petersburg (2018)
31. Hancock, J., Khoshgoftaar, T., Liang, Q.: A problem-agnostic approach to feature selection and analysis using SHAP. J. Big Data. **12**, 1 (2025). https://doi.org/10.1186/s40537-024-010 41-1
32. Persistent Technological Regulations for the Process of LPG Extraction at U-700. Tengiz Chevroil, TP-ZVP-700-11 (2017)
33. Jackson, M., Harris, T.: Evaluation of classifier effectiveness: issues and solutions. Algorithms Syst. **33**(4), 55–61 (2021)
34. O'Connor, R., Miller, L.: Weighted averaging and macro-averaging in classification tasks. Math. Meth. Comput. Sci. **21**(2), 99–105 (2018)
35. Li, S., Ho, L.: Logarithmic loss and ROC curves: evaluation metrics in classification tasks. Mach. Learn. Stat. **22**(1), 98–107 (2020)
36. Lundberg, S.M., Lee, S.I., Yang, D.: A unified approach to interpreting model predictions. In: Proceedings of the 31st International Conference on Neural Information Processing Systems, pp. 4765–4774 (2017). https://doi.org/10.1145/3294996.3295033
37. Wang, R., Yu, N., An, B.: Research on power equipment fault diagnosis based on improved SVM algorithm. J. Electr. Syst. **20**(5s), 112–125 (2024). https://doi.org/10.52783/jes.1836
38. Tan, P., Gong, L.: Fault detection and failure rate analysis of new energy vehicles based on decision tree algorithm. Appl. Math. Nonlinear Sci. **9**(1) (2024). https://doi.org/10.2478/amns-2024-0803
39. Dhakar, A., Singh, B., Gupta, P.: Comparative performance analysis of different types of k-nearest neighbor (k-NN) classifiers for fault diagnosis of air compressor setup. Eng. Res. Express. **6**(2) (2024). https://doi.org/10.1088/2631-8695/ad5497
40. Samigulina G.A., Samigulina Z.I. Development of intelligent technology for complex objects control based on a unified artificial immune system and principles of immunological homeostasis for industrial automation using modern microprocessor equipment: monograph. p. 196. Science Book Publishing Hous, Yelm, WA, USA (2023). ISBN 978-1-62174-150-3. SAN 920-3230. https://www.elibrary.ru/item.asp?id=50443830.

41. Samigulina, G., Samigulina, Z.: Diagnostics of industrial equipment and faults prediction based on modified algorithms of artificial immune systems. J. Intell. Manuf. **33**(1), 1–18 (2022). https://doi.org/10.1007/s10845-020-01732-5

Energy Transition and Sustainable Water Management in Catalonia: Towards a Holistic Simulation Model

Victor Garcia$^{(\boxtimes)}$ ⓘ and Pau Fonseca i Casas ⓘ

Universitat Politècnica de Catalunya, Barcelona 08034, Spain
`victor.garcia-carrasco@upc.edu`

Abstract. The urgency to transition to a fossil-free energy system has been underscored by ecological, economic, and geopolitical factors. The escalating global warming crisis necessitates complete decarbonisation, with renewables being the only technology capable of scaling up sufficiently. Economically, renewables have become the most cost-effective power generation technologies with significant growth potential. The recent gas crisis in Europe has highlighted the geopolitical implications of energy production. A successful energy transition would not only significantly reduce CO2 emissions and contain global warming but also provide cheaper energy and resilience against geopolitical crises by decentralizing energy production. However, the current pace of transition is insufficient to maintain global warming within the IPCC's limit of 1.5°. Current projections suggest a likely exceedance of 2° by 2050, necessitating accelerated global efforts to mitigate this trend. This research project aims to develop a methodology for creating collaborative models to formulate policies for a decarbonized and sustainable society, generating a framework for the so-called Society 5.0. These models should provide technically and economically viable solutions for a fossil-free energy mix, tailored to Catalonia's specific situation.

Keywords: Water · Energy · Sustainability · Climate change · Catalonia

1 Introduction

The need to achieve an energy system free of fossil energy has been accentuated in recent years by multiple factors. Firstly, the ecological factor due to the emergency situation due to the acceleration of global warming. There is an urgent need for complete decarbonisation and for now the only technology that seems capable of sufficiently scaling up is renewables. Secondly, we have the economic factor, renewable energies have become the cheapest power generation technologies on the market and with enormous growth potential. And thirdly, the geopolitical factor, a factor that has gained importance as a result of the gas crisis in Europe. All this implies that a successful energy transition would allow us to greatly reduce carbon dioxide emissions into the atmosphere and contain warming within a certain margin of security, added to the fact that we could have more energy at a lower cost and be much more resilient to geopolitical crises

© The Author(s), under exclusive license to Springer Nature Switzerland AG 2026
M. Pavone et al. (Eds.): DSA ISC 2025, LNCS 16405, pp. 124–138, 2026.
https://doi.org/10.1007/978-3-032-21811-7_9

since energy production would no longer be monopolized by a few producing countries but would be distributed everywhere following the planet's natural renewable flows. Unfortunately, the pace of transition is also insufficient to keep us within the IPCC's publicly set limit of 1.5°. The most likely scenarios at the current rate of emissions and taking into account the planned measures would involve exceeding 2° of global warming by 2050 (Renewable Energy Agency, 2023). The consensus, then, is that global efforts to mitigate this trend as much more as possible must be accelerated further (Fig. 1).

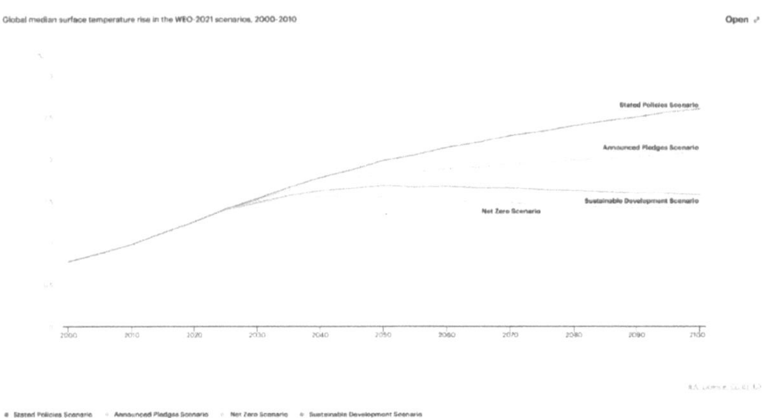

Fig. 1. Increase in average temperature depending on decarbonization trajectories. The current trajectory (blue and orange lines) is clearly not aligned with the 1.5° target

Thus, in this research project we aim to create a work methodology that allows us to generate models that can collaborate with each other, to make policies in favor of reaching a decarbonized and sustainable system. The models we want to build should allow us to obtain technically and economically viable solutions to have an energy mix free of the use of fossil fuels adapted to the situation of Catalonia. These models eventually will have an impact on the decision process at political level becoming the core of the Digital Twins that rules the Society 5.0 decision-making process.

In on operative way, these models should also establish relationships between the generation of electricity from renewable sources and the management of water resources and agricultural waste and should serve to create a standardized methodology to analyze, study and classify energy systems according to the different transition paths. Our project also attacks several points of the 17 Sustainable Development Goals. Specifically, by improving access to drinking water (6), by promoting diversification of supply sources and seeking synergies with the production of renewable energy to lower its cost. We also tackle the problem of obtaining affordable and clean energy (7) that must come with a combination of electrification and transition to cheaper renewable sources. Points (11), (12) and (13) are also objectives of the project as we will try to make the model a circular system as far as possible and with the aim of accelerating as much as possible the decarbonization of the economy. At least in the sectors that affect the system.

2 System Analysis

The energy transition is a complex process, crucial for societal functions like food production, water supply, and industry. We aim to study this system, focusing on Catalonia, to create a detailed, yet generalizable model. This holistic model will consider renewable electricity generation, consumption, and ecological aspects like waste management. It will also explore the interconnections between energy, water, and agriculture, considering new energy-efficient processes. The transition's speed largely depends on reducing electricity and water costs. We'll use the Specification Description Language (SDL) for continuous validation of digital twins, facilitating understanding and early error detection. The model will consider several key factors.

2.1 Water

We know that droughts will become more intense, more frequent and sustained over time. Thus, a vital aspect of the ecological transition is also to ensure the availability of fresh water. Therefore, we will have to take into account in the model variables such as consumption and availability as well as establish links between water and energy. The model should help us to find solutions that keep consumption below resource availability whenever possible so as not to enter into a structural deficit. Currently, unfortunately, we already have the structural deficit situation, so it may also be necessary to contemplate restrictions for quite some time before we can normalize our supply.

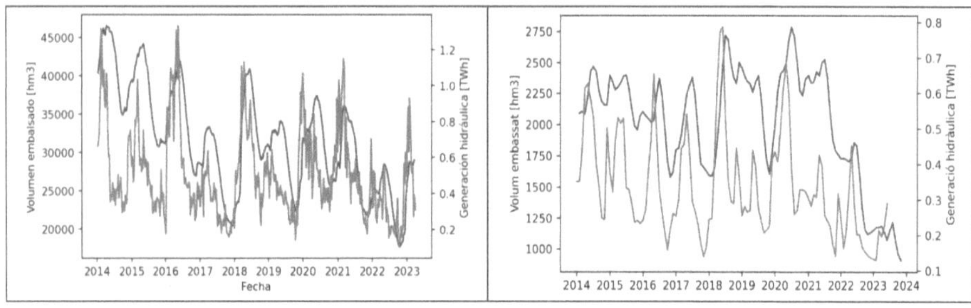

Fig. 2. In blue the level of the reservoirs and in red the hydraulic generation. As expected, there is a strong correlation between water availability and hydraulic generation (Left for Spain, right for Catalonia). In the case of Catalonia, we can see how the level of the reservoirs is in critical condition, at less than 20% capacity (*El Portal de La Sequera*, n.d.) at the time of writing this paper.

It is needed to analyze the sources of consumption and the availability of water, like (i) Residential consumption, (ii) Industrial consumption, (iii) Agricultural and livestock consumption (livestock, and irrigation), (iv) Reservoir capacity and historical availability, (v) Desalination capacity, (vi) Purification, treatment and repurification and (vii) Water capture from air humidity. Figure 2 come from crossing the data obtained from the MITECO website (*El Boletín Hidrológico Semanal*, n.d.)(*REData - Estructura Generacion | Red Eléctrica*, n.d.) on reservoir availability and the REE website on hydraulic generation. In this case, they allow us to analyze the relationship between this source of energy and the availability of water resources. We also observe how, in the case of Catalonia, the level is currently critical and requires urgent intervention. One of the theses of our proposal is that the water problem is also an energy problem and that they must be addressed in an interrelated way in order to find synergies since they are closely linked systems.

2.2 Occupied Area

Agricultural land could be used for renewable energy, including photovoltaics on urban surfaces. Any model should quantify the area needed for wind and solar power, considering the competition for land use. The goal of Catalonia's energy transition policies is not just decarbonization, but also energy self-sufficiency. Electrical interconnections are important for grid balance and cost reduction, but a generation deficit would lead to less competitiveness and energy dependence. Calculating the space needed for self-sufficiency is a key output of our energy and ecological transition analysis.

2.3 Electricity

We will use the generation and installed power data for each technology and we will analyze the growth possibilities of renewables (wind, photovoltaic and hydraulic) to replace the retreating, gas and nuclear sources. We will consider (i) Photovoltaic (parks and self-consumption): Stronger in summer although it requires storage for night production. It is the cheapest source (*Registre d'Autoconsum a Catalunya (RAC). Medi Ambient i Sostenibilitat*, n.d.). (ii) Wind: Stronger in winter and statistically with more night than daytime production. It complements very well with solar energy since both sources are very uncorrelated both seasonally and daily. (iii) Hydraulic: Manageable and strong spring in spring. The perfect renewable complement for wind and solar. In order to have a greater hydraulic use, however, it is necessary to definitively solve the drought problems by resorting to Catalonia. (iv) Gas (combined cycle): Very flexible source but subject to fluctuations in international prices. In principle, non-renewable and dirty, although a small part could be produced locally from biogas digesters or the methanization of hydrogen of electrolytic origin. The goal would be to minimize its use as much as possible. (v) Nuclear: Non-manageable and non-renewable source but without direct emissions. Too expensive in new construction. The three Catalan reactors are expected to close according to the state plan between 2030 and 2035. This means that we will also have to replace all the energy they produced with renewable sources.

2.4 Accumulation

The accumulation of electricity can be direct from reservoirs, batteries or other energy storage systems. Or indirect if it is taken into account that an increase in wind and solar availability results in a lower turbine of the reservoirs or a lower use of them thanks to the greater availability of desalination. There is also the possibility of storing energy in chemical form through processes such as electrolysis and methanization or biomethane fermentation while being able to reuse part of the fossil gas infrastructure for green gas.

2.5 Heat

Heat needs, both industrial and residential, could be included in the model, since in some cases direct conversion from biomass or gas can be more efficient, also allowing renewable surpluses to be released and accumulated in the form of heat.

2.6 Fertilizers

Fertilizers today are produced from fossils, but in the future, they can be made with green ammonia from the electrolysis of water, establishing a direct link between electricity and fertilizer production.

3 State of the Art

As we see in (Breyer et al., 2022), there are more and more studies and reports that support that it is feasible to reach a 100% renewable energy system. Research in this field is a relatively recent phenomenon, which began in the 1970s due to rising oil prices. Since the mid-2000s, research on 100% renewable energy systems has evolved rapidly and has become a prominent field of research encompassing an increasing number of research groups and organizations around the world. Currently, most studies conclude that 100% renewable energy systems are feasible worldwide at low cost and that advanced concepts and methods allow tracing realistic and cost- and resource efficient transition paths towards a future without the use of fossil fuels. In all of them, solar energy and wind energy emerge as the central pillars of a sustainable energy system combined with energy efficiency measures in most transition trajectories. Research has focused on challenges and opportunities related to grid congestion, energy storage, interconnections, electrification of transport and industry and the inclusion of carbon capture and removal approaches. Many of these studies show that reaching between 90 and 95% penetration of renewables in the electricity mix is affordable both technically and economically. David Osmond, an Australian engineer with a wind farm installation company, has done several simulations in the style of those proposed in the project (David Osmond, 2022, 2023). In this case, it has done so based on historical data from the Australian power grid, reaching the conclusion that it is possible to achieve high levels of penetration without installing too much storage. It therefore seems quite accepted that, to obtain high levels of integration of renewable energies, a comprehensive "power-to-X" approach will be necessary, also emphasizing the production of energy vectors such as green hydrogen, synthetic fuels and biogas. In turn, from the report of the think tank on technological disruption Rethink it has been postulated that to reach 100% renewable it will be cheaper and more convenient to the overgeneration electricity than to exceed us with storage (Dorr & Seba, 2020). His thesis is that we are at the beginning of an era of profound and rapid change driven by the technological convergence of photovoltaic solar energy, wind energy and batteries. They estimate that according to the cost reduction curves observed by the different technologies, there is an exponential adoption of the SWB system (Solar, Wind, Battery) which should still suffer a cost reduction of around an additional 70% by 2030. The report argues that it is possible to generate 100% of our electricity with a SWB system and that, in fact, by 2030, these systems will be the most economical option for new energy generation in most countries. SWB systems will produce excess clean energy at almost zero marginal cost, which the authors call "Super Power." This superabundance of clean energy will drive major transformations in a system where surplus energy will not be a problem to be solved, but an opportunity to generate even more economic and social benefits. These excesses would enhance new business models to take advantage of opportunities and create value within the system's new architecture. For example, the abundant generation of drinking water from desalination.

Based on a detailed analysis of the capacity of solar and wind resources to meet electricity demand in 42 countries, the study estimated the following. Depending on several factors, including the mixture of solar and wind generation, annual generating capacity, and energy storage capacity, the reliability of the electricity system based on these energy sources can vary substantially. Thus, with an annual generation equal to annual demand and without energy storage, the most reliable systems can meet between 72% and 91% of electricity demand (Tong et al., 2021). The addition of energy storage and/or surplus power generation can significantly improve system reliability. For example, with 1.5 times annual generation and 12 h of energy storage, reliability can increase to 89–100%. The study also analyses the advantages of increasing interconnection on the largest possible scale. Specifically, results in several countries indicate that the reliability of solar and wind energy systems, excluding energy storage, increases by 7.2% for every factor of 10 in increasing available land area. Thus, we see how coverage improves as we increase interconnections on a continental and global scale.

In summary, the current consensus is that the last 5 to 10% will require alternative solutions to the mere installation of non-hydraulic renewables and batteries (Mai et al., 2022). The manufacture and accumulation of green gas or the construction of robust interconnections seem like practical solutions that could cover the latter 5%. Likewise, countries with more hydraulics, more surface area and better interconnected will have an easier time getting closer to 100%. According to the report, by the think tank Carbon Tracker (Kingsmill Bond et al., 2021), the greatest renewable potential is found in emerging economies given their large available area and low energy demand. But also developed regions such as the USA, Australia, Canada and southern European countries have the potential to generate an overabundance of energy. In the study of the global potential in an electrified economy to cover the demand, they establish that, with photovoltaic solar energy, up to 27 times the current global primary energy consumption could be covered and with wind energy up to 5 times (Perez & Perez, 2022). And all while considering reasonable land use and conversion efficiency restrictions.

3.1 Superiority in Costs of Renewables Compared to Fossil Technologies

Reviewing we see that the EROI (Energy Return On Investment) is a popular metric to evaluate the profitability of energy extraction processes. An EROI greater than 1 indicates that more energy is delivered to society than is used in the extraction process. The authors argue that it is more relevant to use EROI at the point of use (where energy is effectively used) rather than EROI at the extraction point (where energy is extracted), since the processes to bring thermal fuels from extraction to the point of use drastically reduce their EROI (Murphy et al., 2022). The results indicate that renewable energies such as photovoltaic (PV), wind and hydroelectric have EROIs at or above ten, while EROIs for thermal fuels vary significantly, with oil notably below ten. The study proposes a life cycle assessment (LCA) as the methodological framework for calculating the EROI. The article conducts a review of the EROI literature of different energy sources harmonizing the values to allow accurate comparisons. The harmonized results show that most thermal fuels, including biofuels, oil and natural gas, have EROIs well below ten after taking into account the entire production chain to the point of use (Fig. 3).

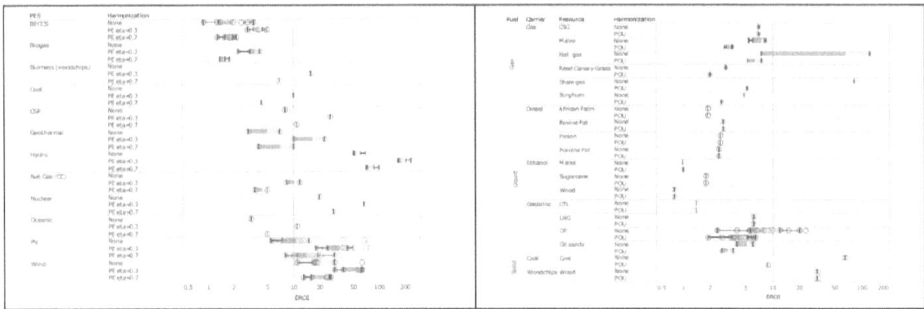

Fig. 3. EROI (TRE) for electricity (left) and thermal fuels (right) (Murphy et al., 2022).

These results are confirmed by the sharp decrease in costs reflected in reports such as de or Fraunhofer, where you can see the enormous decrease in costs of wind and especially photovoltaic technologies in recent years. This decrease in costs is also proof that the EROI of these technologies has increased greatly in recent years making a system based on clean energies viable (Lazard, 2023) (Philipps et al., 2024).

3.2   Studies Analyzing the Possibilities of the Use of Desalination on a Large Scale

Desalination is experiencing a downward trend due to several factors. The cost of seawater desalination varied between $0.35 and $1.87/m3, while the cost of desalination of brackish water ranged from $0.35 to $1.53/m3 through 2019 according to (Eke et al., 2020). Reverse osmosis (OR) technology contributes most of desalinated water production today, indicating that the technology is well developed and can be used to desalinate all types of water sources. These innovations in desalination technology have led by extension to a reduction in the environmental footprint of desalination in recent years (Nassrullah et al., 2020).

During the period between 2010 and 2019, desalination has increased by an average of 7% per year. The ability to obtain cheap energy from renewable sources could accelerate its use, especially in arid countries such as ours.

It will also be necessary to take into account the cost prospects for the coming years of emerging technologies for obtaining drinking water also from air humidity.

3.3   Holistic Models of Energy and Ecological Systems

In the literature we find a variety of models that perform tasks similar to those we want to achieve in ours. A good example could be ENECO 2Calc. This model allows the search for ways of energy transition towards climate neutrality within municipalities. The model is capable of calculating the current final energy demand within various sectors by municipalities, as well as making predictions for future scenarios up to 2050. In addition, it can calculate the ecological and economic footprint based on energy demand (Hammerschmid et al., 2022). The tool has an interdisciplinary approach already between the energy transition and spatial planning, which allows a holistic assessment of the ecological and economic footprint of a region.

But we also find in the literature other much better known and general models. There are integrated assessment models based on system dynamics and that go beyond energy aspects such as MEDEAS that are used to have a more macro vision of transition paths at EU-wide level. This model is divided into the 7 submodules, see Fig. 4.

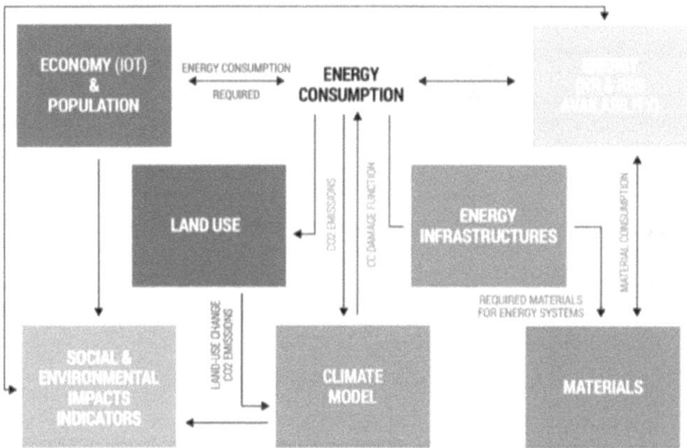

Fig. 4. Diagram of the MEDEAS model (Samsó Roger & Solé Ollé Jordi, 2020).

Or the MARKAL model, more focused on the optimization of energy systems and also well known. This is a linear programming model of bottom-up optimization. It looks for the combination of technologies that minimizes the total cost of the constrained energy system. It represents energy supply chains in detail, including resources, conversion processes, infrastructure and demand and covers long time horizons (typically 40–50 years) with time steps of 5–10 years. It is worth noting that the use of the similar but more detailed and precise TIMES successor model is now being promoted.

4 Methodology

We will build a digital twin using methodologies analogous to those of industry 4.0 projects. A digital twin is a virtual replica of a physical system or process used to simulate its behavior and predict its performance in different situations. Digital twins require continuous input of real-time data that is constantly updated to reflect any changes to the physical system. They are necessary to validate simulation systems because they allow modelers to test and adjust their models before implementing them into the physical system. Finally, digital twins can be used to simulate different scenarios and optimize system performance.

The first part of the work will be to define a conceptual model defining its assumptions, limits, variables and parameters. According to there are three types of assumptions to take into account, systemic data, referring to the characteristics of the data with which we are going to work, statistical distribution, accuracy and quality of the same; The structural systemic ones, referring to the structure of the system, topology, connectivity and interaction between the components and those of simplification, referring to the simplifications that we will make to reduce the complexity of the system, such as the elimination of irrelevant variables or the aggregation of similar variables. All these assumptions are what will allow us to generate the "Digital Master" (Fig. 5).

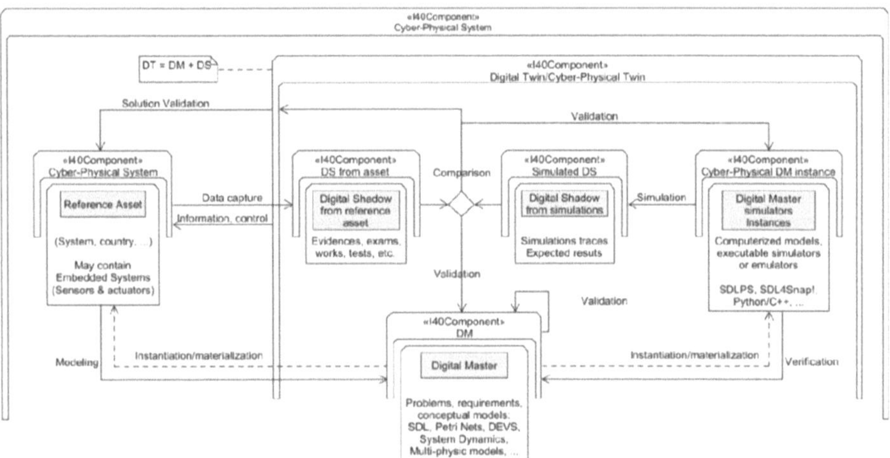

Fig. 5. In this case our conceptual model would be the Digital Master and the Digital Shadow from reference asset would be the system data and the Digital Shadow from simulations would be the synthetic data, from (Fonseca i Casas, 2023).

4.1 Systemic Data Assumptions

One of the challenges we will have will be to collect, clean and structure the data necessary to use it in the model defining the systemic data assumptions. It will also be necessary to build regionalized datasets in the study area, in our case Catalonia, presumably. We must also try to automate as far as possible the obtaining of this data to facilitate the management and maintenance of the model. Some of these data would be the following: (i) Power installed by technology, (ii) Electricity generated by technology, (iii) Capacity and levels of reservoirs, (iv) Desalination capacity, (v) Water purification capacity, (vi) Production and potential of biomass from slurry and other livestock and agricultural waste (vii) Fertilizer consumption, (viii) Demand for water for irrigation and human consumption, and (ix) Costs of different technologies, wind, solar, storage, desalination, among others.

4.2 Systemic Structural Assumptions

Once we have all the required data, we will have to build and approach a model to represent the system with all its variables and interrelationships. From this model it will be necessary to analyze optimization techniques. We want to explore the phase space that would give us to have a model with different key variables to maximize or minimize. For example, the desired solutions should initially meet the following characteristics: (i) Minimize the occupied area, (ii) Maximize the clean energy generated, (iii) Minimize system costs and (iv) Maximize water availability in reservoirs.

It should be borne in mind that there are several variables that will change in the coming years due in part to electrification and global warming. There is always uncertainty when estimating the future, but certain foreseeable changes must be foreseen and taken into account in the models in the form of scenarios in order to make the system robustly resilient. As an example, (i) Increased demand for electricity (residential and mobility electrification), (ii) Increased demand for water (more frequent droughts), (iii) Decreased water availability in reservoirs, (iv) Variability in energy prices and (v) Increase in extreme weather events and their impact on the grid.

4.3 The Validation Processes.

In parallel with the construction of the model, it will have to be subjected to a validation process that may involve several strategies to follow to ensure that the model is accurate and reliable. These are the main validation milestones of a model (i) Validation of the conceptual model, (ii) Data validation, (iii) Model code verification (iv) Operational validation (v) Experimental validation, (vi) Accreditation and (vii) Solution validation (Fonseca i Casas, 2023). We applied this methodology in a COVID pandemic modeling project from 2020 to 2021(Fonseca i Casas et al., 2020, 2021). In this project we tackled the simulation problem through a cross-verification strategy of comparing multiple system dynamics models programmed with InsightMaker, Python and SDLPS, with operational validation with the historical data that fed our models and with solution validation comparing our forecasts with the new data that came out. This approach allowed us to gain insights into the effects of non-pharmaceutical interventions on the evolution of the pandemic. In this case, the digital twin was intended to serve as a tool to inform about the effectiveness and convenience in the application of different public health policies. For the case at hand, the idea of creating a digital twin should allow political decision-making with a better understanding of the capabilities and limitations of the ecological and energy transition. It is also necessary to be a model that can adapt to a rapidly changing reality due to technological innovation and the variation in electricity consumption and generation patterns.

5 The Preliminary Model

We currently have a prototype of the model programmed in Python that allows us to obtain results on the peninsular system from historical data of the installed power and the energy generated by the different generation sources. It would be a piece of the more general and holistic model we want to implement.

For our model we use data on electricity generation with hourly granularity and installed power for different energy sources and in the most extensive interval possible. For now, we have used data from the entire peninsular system, in the future the objective would be to prepare the data for Catalonia.

The base model generates scenarios based on the scaling of historical data based on different proposals for the target installed power for each technology. Being able to get diagrams that shows the share of renewable energies under various wind and solar installation scenarios, see Fig. 6.

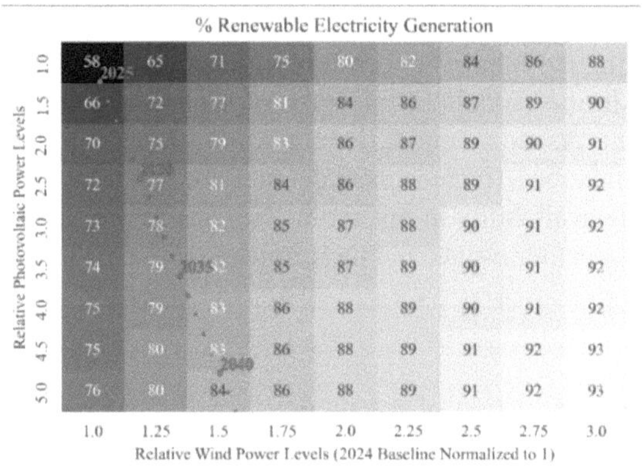

% Renewable Electricity Generation

Relative Photovoltaic Power Levels	1.0	1.25	1.5	1.75	2.0	2.25	2.5	2.75	3.0
1.0	58	65	71	75	80	82	84	86	88
1.5	66	72	77	81	84	86	87	89	90
2.0	70	75	79	83	86	87	89	90	91
2.5	72	77	81	84	86	88	89	91	92
3.0	73	78	82	85	87	88	90	91	92
3.5	74	79	82	85	87	89	90	91	92
4.0	75	79	83	86	88	89	90	91	92
4.5	75	80	83	86	88	89	91	92	93
5.0	76	80	84	86	88	89	91	92	93

Relative Wind Power Levels (2024 Baseline Normalized to 1)

Fig. 6. Table of the share of renewable energies under various wind and solar installation scenarios. In red is shown the (*Plan Nacional Integrado de Energía y Clima*, 2023) planned path.

It also incorporates functions to be able to add the presence of different types of energy storage to the scenarios. Short and long life, in batteries, hydro management or with green hydrogen, etc. With these data we can make hour-by-hour simulations of how a given scenario would have behaved in a specific time period, see Fig. 7.

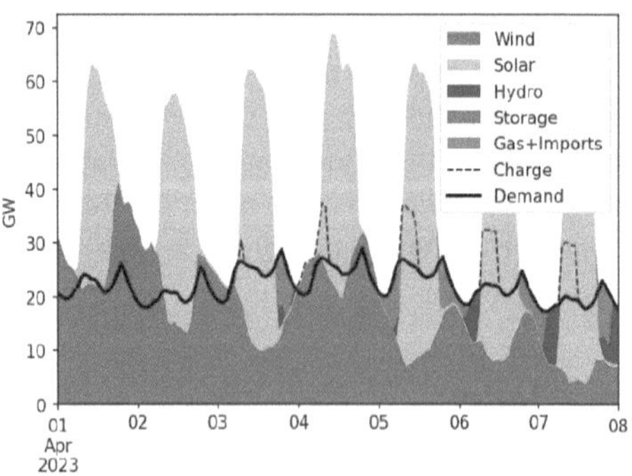

Fig. 7. Graph of electricity generation by various energy sources in a hypothetical 100% renewable scenario.

The analysis of the historical data of the model allows us to analyze the evolution of the number of hours of high renewable penetration. We can see how, since 2020, growth in the peninsular electricity system has been exponential, see Fig. 8 and Fig. 9.

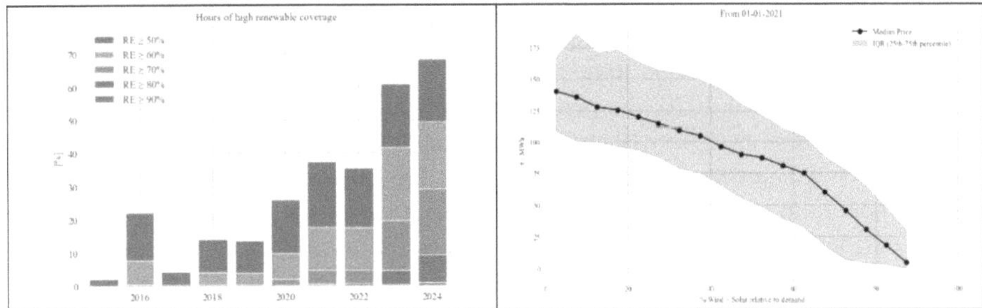

Fig. 8. Evolution of the volume of hours of high penetration in the Iberian electricity system (left) and relationship between the wholesale price of electricity and the amount of solar and wind energy available (right).

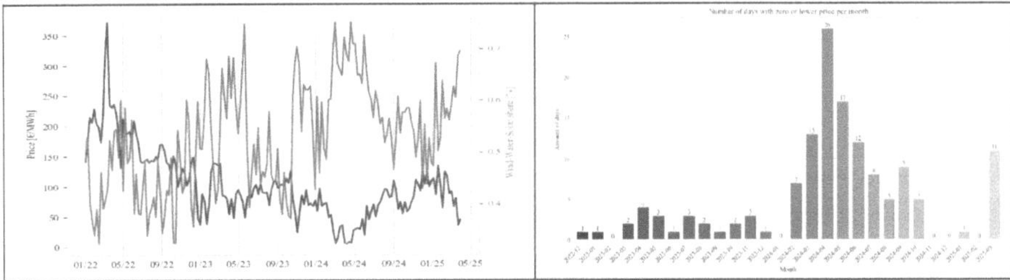

Fig. 9. Price and renewable share dynamics in the Iberian electricity system. (Left) Daily electricity price (€/MWh, blue) and the corresponding share of wind, hydro, and solar generation (%, green) from January 2022 to May 2025, often showing an inverse correlation. (Right) The increasing occurrence of zero or lower price days, shown as the number of days per month from December 2022 to March 2025.

Crossing these data with wholesale electricity prices reveals a strong link between high renewable penetration and low electricity prices, driven by low operating costs (OPEX) yet inflexible sources. Note that these, figures shown refer to the Iberian system. For Catalonia, the values of renewable penetration would be much worse since we are lagging far behind. So, for the Catalan case, the problem of harnessing the renewable surplus would not be so critical yet.

6 Discussion

We have developed a prototype model using Python to analyze the peninsular energy system. This model leverages historical data on installed power and energy generation from diverse sources. Our ultimate goal is to create a comprehensive and holistic model that provides valuable insights for policymakers, energy companies, and researchers. Regarding the data Sources and Granularity, our model relies on hourly granularity data for both electricity generation and installed power across various energy sources. The base model generates scenarios by scaling historical data based on different proposals for target installed power for each technology. The resulting scenarios can be visualized through diagrams, providing a clear picture of their evolution over time.

Our initial analysis reveals some interesting insights regarding, renewable penetration and growth: Since 2020, the peninsular electricity system has experienced exponential growth. However, it's essential to note that this data represents the entire Iberian system. When we narrow our focus to Catalonia, the situation appears less favorable.

Our model aims to provide a comprehensive view of the peninsular energy landscape, considering factors such as renewable penetration, energy storage, and regional nuances. The specific focus on Catalonia will be crucial for addressing localized challenges. Additionally, integrating electricity prices into our analysis adds another layer of understanding.

As a future work, our first objective will be to obtain data from the different parts of the system to be analyzed and regionalize them for Catalonia. This involves building a homogenized data set with all the variables we want to study.

To generate the scenarios, we will scale the generation according to different installed power proposals for each technology and we will analyze the different relationships between the parts of the system. Specially, the link between the level of the reservoirs and hydraulic generation in order to prevent water scarcity caused by severe droughts.

We will build features to incorporate different types of energy storage into the model. Short and long lasting. Again, analyzing here, the role of reservoirs as large energy stores capable of displacing the use of surpluses up to several months after their generation. We will also consider the possibility of using hydrogen gas from electrolysis. Gas can have various uses, seasonal energy storage or fertilizer manufacturing, for example. All these elements must be considered in the model and combined to evaluate which are the scenarios and transition paths with the lowest cost and highest return.

Part of the research project will be not only to perfect the starting model but also to expand it to incorporate the fields mentioned in the document and others that may arise on the fly. The important thing will be to arrive at a general model such as those mentioned in the state of the art but dedicated to Catalonia and perhaps providing some improved or new functionality on existing models. Which should also serve as a reference, starting point and inspiration to develop a better model. Ultimately, developing such sophisticated, data-driven energy system models aligns with the vision of Society 5.0, where complex societal challenges, including sustainable energy provision and resource management, are addressed through the deep integration of cyber and physical systems to enhance human well-being and economic progress.

Disclosure of Interests. The authors have no competing interests to declare that are relevant to the content of this article.

References

David Osmond: 100% Renewables Is Hard - But 96% Is Easy: Windlab's David Osmond (2023). https://www.solarquotes.com.au/blog/renewable-energy-osmond/

Lazard: Levelized cost of Energy report (2023). https://www.lazard.com/research-insights/2023-levelized-cost-of-energyplus/

David Osmond: A near 100 per cent renewables grid is well within reach, and with little storage. RenewEconomy (2022). https://reneweconomy.com.au/a-near-100-per-cent-renewables-grid-is-well within-reach-and-with-little-storage/

Breyer, C., et al.: On the history and future of 100% renewable energy systems research. IEEE Access. **10**, 78176–78218 (2022). https://doi.org/10.1109/ACCESS.2022.3193402

Dorr, A., Seba, T.: Disruption, Implications, and Choices Rethinking Energy (2020)

Eke, J., Yusuf, A., Giwa, A., Sodiq, A.: The global status of desalination: an assessment of current desalination technologies, plants and capacity. Desalination. **495** (2020). https://doi.org/10.1016/j.desal.2020.114633

El Boletín Hidrológico semanal (n.d.). Retrieved 23 Nov 2023. from https://www.miteco.gob.es/es/agua/temas/evaluacion-de-los-recursos-hidricos/boletin-hidrologico.html

El portal de la sequera (n.d.). Retrieved 23 Nov 2023. from https://sequera.gencat.cat/ca/inici/

Pau, F.i.C.: A continuous process for validation, verification, and accreditation of simulation models. Mathematics. **11**(4) (2023). https://doi.org/10.3390/math11040845

Pau, F.i.C., Victor, G.i.C., Joan, G.i.S.: SEIRD COVID-19 formal characterization and model comparison validation. Appl. Sci. (Switz.). **10**(15) (2020). https://doi.org/10.3390/app10155162

Pau, F.i.C., Joan, G.i.S., Victor, G.i.C., Palomes, X.P.i.: Sars-cov-2 spread forecast dynamic model validation thorough digital twin approach, catalonia case study. Mathematics. **9**(14) (2021). https://doi.org/10.3390/math9141660

Hammerschmid, M., Konrad, J., Werner, A., Popov, T., Müller, S.: ENECO2Calc—a modeling tool for the investigation of energy transition paths toward climate neutrality within municipalities. Energies. **15**(19) (2022). https://doi.org/10.3390/en15197162

Mai, T., et al.: Getting to 100%: six strategies for the challenging last 10%. Joule. **6**(9), 1981–1994 (2022). https://doi.org/10.1016/j.joule.2022.08.004

Murphy, D.J., Raugei, M., Carbajales-Dale, M., Estrada, B.R.: Energy return on Investment of Major Energy Carriers: review and harmonization. Sustain. (Switz.). **14**(12) (2022). https://doi.org/10.3390/su14127098

Nassrullah, H., Anis, S.F., Hashaikeh, R., Hilal, N.: Energy for desalination: a state-of-the-art review. In: Desalination, vol. 491. Elsevier B.V (2020). https://doi.org/10.1016/j.desal.2020.114569

Perez, M., Perez, R.: Update 2022 – a fundamental look at supply side energy reserves for the planet. Solar Energ. Adv. **2**, 100014 (2022). https://doi.org/10.1016/j.seja.2022.100014

Philipps, S., Ise, F., Warmuth, W., Projects GmbH, P.: Photovoltaics report (2024). www.ise.fraunhofer.de

Plan Nacional Integrado de Energía y Clima (2023). https://www.miteco.gob.es/es/energia/estrategia-normativa/pniec-23-30.html.

REData: Estructura generacion. Red Eléctrica (n.d.). Retrieved 23 Nov 2023. From https://www.ree.es/es/datos/generacion/estructura-generacion

Registre d'Autoconsum a Catalunya (RAC). Medi Ambient i Sostenibilitat (n.d.). Retrieved 23 Nov 2023. From https://mediambient.gencat.cat/ca/05_ambits_dactuacio/energia/installacions-domestiques/autoconsum/registre-autoconsum-catalunya/

Renewable Energy Agency, I. World Energy Transitions Outlook 2023: 1.5°C Pathway (2023). www.irena.org

Samsó Roger, Solé Ollé Jordi: Medeas Model (2020). https://www.medeas.eu/model/medeas-model
Tong, D., et al.: Geophysical constraints on the reliability of solar and wind power worldwide. Nature. Communications. **12**(1) (2021). https://doi.org/10.1038/s41467-021-26355-z

Long Short-Term Memory Models
for Improving Route Planning Acceptance

Samuele Simone[1]($\boxtimes$) , Alberto Ceselli[1] , and Andrea Bettinelli[2]

[1] Dipartimento di Informatica, Università degli Studi di Milano,
Via Celoria 18, 20133 Milan, Italy
`{samuele.simone,alberto.ceselli}@unimi.it`
[2] OPTIT S.r.l., Via Mazzini, 82, 40138 Bologna, BO, Italy
`andrea.bettinelli@optit.net`

Abstract. Building routes that are optimal in terms of reducing costs, impact on the environment, and human workload plays a key role in logistics. However, producing efficient routes that satisfy operational constraints does not guarantee that human planners and drivers will actually accept and execute them. It is a long standing problem in optimization: personal knowledge as well as subjective preferences are difficult to fully formalize in a model. Smart city applications sharpen challenges, but open also opportunities, given by the larger data collection possibilities. In this paper, we develop a classification model to recognize routes that meet the preferences of human planners. It exploits data-driven techniques, in a pipeline whose core component is a Bidirectional Long-Short-Term Memory Neural Network architecture, which is able to capture the relationships between points visited in sequence in accepted routes.
We evaluate our methodology on a real-world case study. Our experiments show that our methods clearly outperform Markovian models from the literature in terms of recognition accuracy, demonstrating their ability to identify longer dependencies in the sequence of points. Integrating our recognition model into optimization algorithms can lead to the generation of efficient routes that also accommodate the implicit preferences of drivers and planners, thus enhancing the acceptance rate of suggested solutions.

Keywords: vehicle routing · LSTM · preference learning

1 Introduction

The Vehicle Routing Problem (VRP) is a critical optimization problem in distribution logistics, concerning the determination of cost-effective routes for a fleet of vehicles to serve a set of customers. The significance of VRPs is evidenced by the extensive body of research and the proliferation of specialized software solutions. Modern decision support systems are designed to produce optimized daily routing plans, balancing objectives such as vehicle utilization and emission reduction, while adhering to operational constraints like time windows and

M. Pavone et al. (Eds.): DSA ISC 2025, LNCS 16405, pp. 139–153, 2026.
https://doi.org/10.1007/978-3-032-21811-7_10

vehicle capacity. However, a persistent challenge is the observed discrepancy between system-generated solutions and actual implementation. Human planners frequently deviate from these optimized routes, resulting in constraint violations and suboptimal objective function values. This behavior is primarily attributed to two factors: (1) the presence of data quality issues, including inaccuracies and missing information, which limits the optimization model's effectiveness, and (2) the human planners' reliance on tacit, context-dependent knowledge, which is difficult to formalize and incorporate into algorithmic solutions. Therefore, a key objective is to develop methodologies that can learn and integrate human planner preferences into VRP solutions. This paper investigates the application of Long Short-Term Memory (LSTM) models as core components of a pipeline, which is able to extract and predict these implicit preferences from historical routing data, focusing on the sequential relationships between delivery stops. We present a case study within the Food & Beverage distribution industry, showcasing the potential of LSTM-based approaches to enhance the practical applicability of VRP solutions.

2   Related Work

The Vehicle Routing Problem was formalized for the first time in an article written by Dantzig, G. B. and Ramser, J. H. [5] in 1959 and has received a lot of attention ever since. Over the years more rich variants of the VRP have been studied trying to capture more details of the real-world problems [2] [7].

More recently it has been suggested that in some cases it is easier to learn human planners and drivers preferences from examples rather than trying to formalize them explicitly in constraints. In Canoy et al. [3] Markovian models are used to grasp subjective driver preferences by learning a probabilistic transition matrix. Then the authors use existing arc routing VRP software to create the actual routings, optimizing both over distances and preferences at the same time. The results reported shows that this method successfully generates routes similar to those of the human planner without the need to explicitly model preferences.

Other approaches tackles the problem of learning drivers or planners preferences as the task of learning an unknown cost function. Song et al. [8] propose an Inverse Reinforcement Learning approach in which the drivers' utility function is approximated by a neural network. Zattoni et al. [9] developed a methodology based on Inverse Optimization to learn the drivers' preferences, with a hypothesis function, loss function, and stochastic first-order algorithm tailored to routing problems. They obtained promising results on the instances of the Amazon Last Mile Routing Research Challenge [1].

With the advent of human-centric Industry 5.0 new methods have been applied to understand the operator. In the work of Dragone et al. [6], an LLM was applied to translate operator feedback into quantifiable objectives for LP problems. The use of LLMs in industry is still being integrated due to the lack of comprehensive architecture design. In the work of Chen et al. [4] a Large Language Model Systems Engineering (LLM-SE) is proposed for laying a foundation

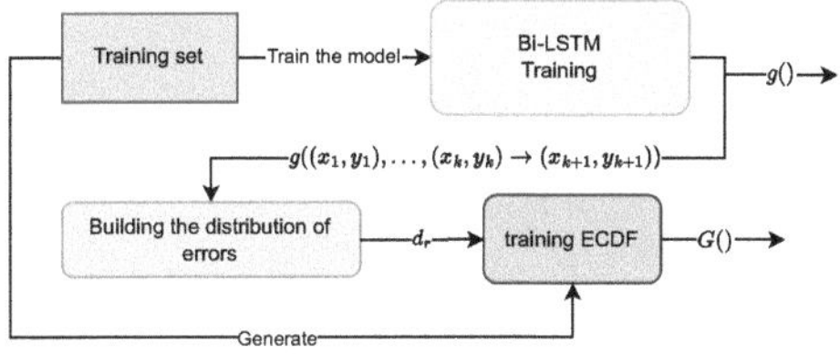

Fig. 1. Bi-LSTM training pipeline

for functional requirements and functionality requirements for LLM applications in industrial sectors.

3   Modeling

We formalize our setting as follows. Let I be the set of all customers. Each customer $i \in I$ is placed in a planar space, at coordinates (x_i, y_i). A *route* r is a sequence $(i_1, \ldots, i_{|r|})$ of $|r|$ customers serviced by the same vehicle during a daily duty. Each route is supposed to be the output of a decision support system, and might be accepted as it is by a customer, or not. Let R be the set of all possible routes.

Our **methodological task** is to find a function $f : R \to \{0, 1\}$ which classifies each route as either accepted (1) or not (0).

In order to find such a function, we assume that a set of routes $\bar{R} = \{r_1, \ldots, r_{|\bar{R}|}\}$ which *have been accepted by the decision maker* is known (e.g. it belongs to a database of routes implemented so far). Therefore, we employ a data-driven approach to rebuild $f()$ from $\bar{R}$. Our approach consists of a pipeline with the following structure. During training, a Long Short-Term Memory model (LSTM) is built, to predict the completion of partial routes, using data from $\bar{R}$ (Subsect. 3.1); the distribution of errors performed by the LSTM on training data is also built (Subsect. 3.2) in order to subsequently obtain the Empirical Cumulative Distribution Function (ECDF). The overall training pipeline is summarized in Fig. 1. During usage, a new route is observed, and the LSTM is used to predict the completion of each partial route; the overall error made w.r.t. the actual route is measured, and the probability of finding such an error is measured according to the distribution found in training. The probability is checked against a threshold, and the route is classified as either accepted (1) or not (0) depending on this check (Subsect. 3.3). The usage process is summarized in Fig. 2.

3.1   Predicting Customer Coordinates

The first (and main) ingredient of our approach is to find a function

$$g((x_1, y_1), \ldots, (x_k, y_k)) \to (x_{k+1}, y_{k+1})$$

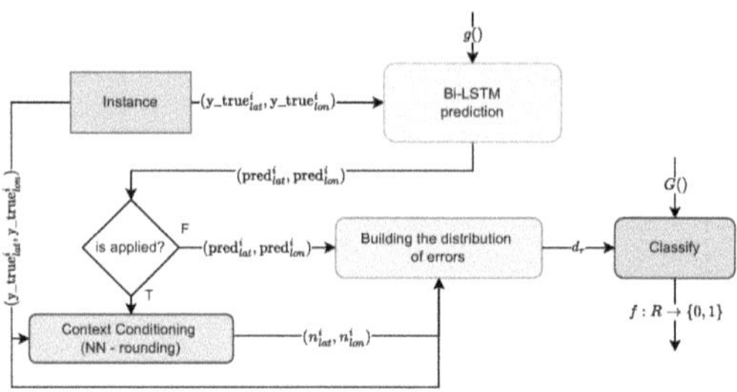

Fig. 2. Bi-LSTM usage pipeline

mapping a sequence of k pair of coordinates, which intuitively represent the coordinates of k consecutive customers in a route, to a new pair of coordinates, which intuitively approximate the position of the subsequent customer $k + 1$ in the same route. We find such a function by means of a Bidirectional Long Short-Term Memory (Bi-LSTM) model, as follows.

Bi-LSTM is a specialized form of RNN architecture. RNNs leverage previous data to enhance the neural network's ability to process and predict current and upcoming inputs. They have hidden state and recurrent connections, enabling the network to store and process information from sequences over time.

Simple RNNs struggle to capture and retain long-term dependencies effectively. To address this, LSTMs incorporate gates that regulate which information in the hidden state is retained, discarded, or passed on to the next state. This allows longer dependencies to be maintained and helps to make better predictions.

However, a simple architecture of LSTM, in our case, was not sufficient to obtain accurate predictions. It was instead necessary to design a custom architecture. Full details on our LSTM architecture are given in Sect. 4. Summarizing, a Bi-LSTM model, was able to achieve good results in terms of accuracy during coordinate predictions. Bi-LSTM learns dependencies in both directions between different time steps of sequences or time series. In this case our sequences are represented by the coordinates as described above. These dependencies are useful when we want the network to encompass the entire sequence at each time instant. BiLSTM networks allow advanced training because the input data is processed twice through the LSTM layer, thus improving the overall performance of the network (Fig. 3).

3.2   Building the Distribution of Errors

We stress that $g()$ cannot be used directly to our scope, since ours is not a predictive task, but a classification one.

Therefore we proceed as follows, for each route $r \in \bar{R}$, we measure the average distance between the predicted coordinates of a customer in the sequence, and the actual ones. That is, for each $r \in \bar{R}$, composed of customers with coordinates

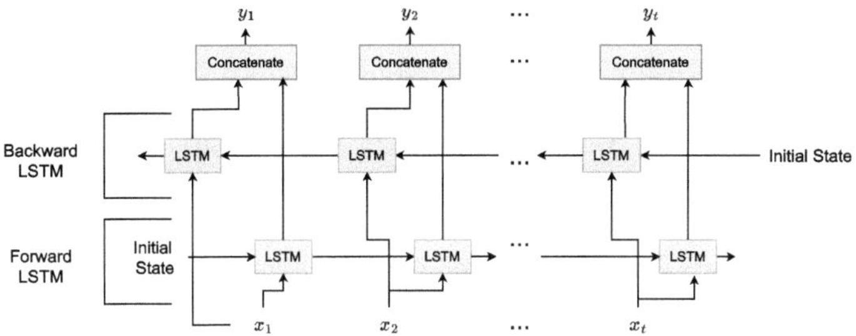

Fig. 3. Bi-LSTM architecture diagram with multiple t phases

$((x_1, y_1), \ldots, (x_{|r|}, y_{|r|}))$ we calculate

$$d_r = \frac{\sum_{i=k+1}^{|r|} \| g\left((x_{i-k}, y_{i-k}), \ldots, (x_{i-1}, y_{i-1})\right), (x_i, y_i) \|}{|r| - k}$$

Formally, the set of values $\{d_r, \forall r \in \bar{R}\}$ can be seen as $|\bar{R}|$ observations of a random variable D, describing the average error made by the Bi-LSTM in predicting the coordinates of customers in the routes of R. k is the size of the sliding window (number of customers to be considered for making the prediction). From that set of values we therefore rebuild an Empirical Cumulative Distribution Function (ECDF) $G(d) \to [0, 1]$, which approximates the CDF of D as described in the Algorithm 2.

3.3   Classification Model

Let us assume that both the function $g()$ and the corresponding function $G()$ have been found from set $\bar{R}$, as explained in the previous sections.

When a new, previously unseen, route s appears, we use $g()$ to predict the coordinates of its customers, and measure the average distance d_s between these predictions and the actual ones. The probability of finding such a value d_s (or lower) as a realization of D can be estimated by $G(d_s)$. We therefore fix a threshold τ and we classify s as *accepted* if $G(d_s) \leq \tau$, *rejected* otherwise, that is

$$f(s) = \begin{cases} 1 & \text{if } G(d_s) \leq \tau \\ 0 & \text{otherwise} \end{cases}$$

The constant τ becomes a parameter of our model: the lower the value of τ, the more selective our model in classifying new routes as *accepted*.

3.4   Overall Framework

As shown in Fig. 1, the Bi-LSTM function $g()$ and the ECDF $G()$ are obtained as output from training on the set of routes $\bar{R}$. As detailed in Fig. 2, they become the input of the predictive model: the Bi-LSTM function $g()$ is used to obtain the predictions of coordinates of customers in the new route. These coordinates are passed as input to Context Conditioning, when the framework is configured to do so, which corrects them to the nearest-neighbour among the customers which are known to belong to the new route. Otherwise we will directly use the coordinates generated by the model. In both cases the distance between the predicted coordinates and the actual ones are measured; an average distance value over the route is measured, and its probability of being a realization of the random variable D is obtained by the ECDF function $G()$. This probability is checked against the threshold τ, according to function $f()$, producing the final classification.

4   A Custom LSTM-Based Framework

The main idea of our model is to use the Bi-LSTM to predict, given a sliding window of coordinates of k consecutive customers in a route, the coordinates of the next customer. The intuition is that if these are accurate, predictions on training data and new, previously unseen, accepted routes should produce similar (small) errors, while predictions on other types of routes should produce larger errors.

Architecture. Our approach consists of two convolutional layers using bidirectional LSTMs and GRUs, followed by dense layers with batch normalization and ReLU activation. We also utilize a custom loss function, which has been proven effective in our experiments.

1. **Input Layer**: Our input layer takes sequences of length `seq_length` and shape (`seq_length, features`).
2. **First Convolutional Layer**: We apply a bidirectional LSTM with 64 units and L2 regularization of 0.0005 to the input sequence.
3. **Batch Normalization**: After the first convolutional layer, we apply batch normalization to stabilize the activations.
4. **Dropout**: We introduce dropout regularization at a rate of `dropout_rate` to reduce overfitting.
5. **Second Convolutional Layer**: The output from the first layer is passed through a 32-unit GRU with ReLU activation and L2 regularization of 0.0005.
6. **Batch Normalization**: Batch normalization is applied again after the second convolutional layer.
7. **Dropout**: We introduce dropout regularization at a rate of `dropout_rate` to reduce overfitting.
8. **Dense Layers**: Dense layers of 16 units with ReLU activation and L2 regularization is used to transform the output from the GRU layer into intermediate step for the last dense layer.
9. **Batch Normalization**: Batch normalization is applied again after the dense layer

10. **Dropout**: We apply dropout regularization at a rate of `dropout_rate/2` to further reduce overfitting.
11. **Output Layer**: The final output is obtained by applying a single dense layer of 2 units.

Bi-LSTM Training Phase. The training set is in turn divided into a *training-validation set* 80%, 20%. During the training phase of the model, techniques were used to prevent overfitting such as using the `ReduceLROnPlateau` with a factor `patience=5` that monitors the validation loss and if it does not improve within the factor then reduces the learning rate which is initially set to 0.001 with lower bound set to 0.00001. We refer to the documentation of the Keras library. In addition a `EarlyStopping` is set which also monitors the validation loss and if there is no improvement within the factor `patience=5` then it stops the training.

Custom Loss. Adam, a stochastic gradient descent method was used as the optimizer while a customized function between the *Mean Square Error* and the *Haversine's distance* between prediction and ground truth was applied for training loss. This is because built-in loss functions are used for general tasks but in this case we also wanted to weight it based on the distances between the predictions and the actual coordinates. In formula:

$$\frac{MSE + Avg\_Hav\_Distance}{\beta}$$

where MSE is calculated in this way:

$$MSE = \frac{1}{n} \sum_{i=1}^{n} (y_{\text{true}}, y_{\text{pred}})^2$$

and β for our test was set to 10000.
Haversine distance formula: $y\_true = (lat, lon), \quad y\_pred = (lat, lon),$

$$\text{haversine\_loss}(y_{\text{true}}, y_{\text{pred}}) = R \cdot 2 \cdot \text{atan2}(\sqrt{a}, \sqrt{1-a}) \tag{1}$$

$$\text{where:} \tag{2}$$

$$\text{lat}_1, \text{lon}_1 = y_{\text{true}}[:, 0] \cdot \frac{\pi}{180}, y_{\text{true}}[:, 1] \cdot \frac{\pi}{180} \tag{3}$$

$$\text{lat}_2, \text{lon}_2 = y_{\text{pred}}[:, 0] \cdot \frac{\pi}{180}, y_{\text{pred}}[:, 1] \cdot \frac{\pi}{180} \tag{4}$$

$$\Delta lat = \text{lat}_2 - \text{lat}_1 \tag{5}$$

$$\Delta lon = \text{lon}_2 - \text{lon}_1 \tag{6}$$

$$R = 6371 \text{ (Earth radius in Kilometers)} \tag{7}$$

$$a = \sin^2\left(\frac{\Delta lat}{2}\right) + \cos(\text{lat}_1) \cdot \cos(\text{lat}_2) \cdot \sin^2\left(\frac{\Delta lon}{2}\right) \tag{8}$$

And the average of Haversine Distance:

$$Avg\_Hav\_Distance = \frac{1}{n} \sum_{i=1}^{n} \text{haversine\_loss}(y_{\text{true}}^{(i)}, y_{\text{pred}}^{(i)})$$

Enhancement of the Features. To improve the predictions, an enhancement of the features was performed. In fact, the Euclidean distance was first calculated:

$$euclidean\_distance = \sqrt{(\Delta lat)^2 + (\Delta lon)^2}$$

then, the direction of motion (bearing) is calculating between two geographic points using the arctan2 function and deltas respectively for latitude and longitude. In formula:

$$bearing = arctan2((\Delta lon, \Delta lat))$$

Empirical Cumulative Distribution Function Extraction. As formalized in Sect. 3.2 we calculated the empirical cumulative distribution function for the training set at the end of each epoch. Then only the best ECDF obtained during the different epochs is printed graphically as shown in Fig. 4. In the Algorithm 2 is possible to understand how probabilities are extracted from the ECDF.

Context-Conditioning. During the testing phase of the model, preprocessing can be applied before extracting the likelihood from the ECDF. This is done simply through a function

$$\rho(pred_{lat}^i, pred_{lon}^i, r^i) \rightarrow (n_{lat}^i, n_{lon}^i) \quad \forall i \in pred$$

that takes as input for each prediction made, its corresponding latitude and longitude and extrapolates the nearest coordinate present in the specific test route. This means providing *context* to the predictions made by the model. Conversely, if this does not happen, it means that we are not providing context (*no-context*) and therefore we are directly using the model's predictions as-is.

5   Industrial Case Study and Experiments

As a case study, we consider real data of an Italian logistic service provider specialized in Food & Beverage distribution. In particular we focus on the deliveries from a single warehouse in the north of Italy. The daily planning problem is a rich single-depot vehicle routing with heterogeneous fleet.

The dataset used in our study is an extraction of 321 instances where each route is the concatenation of all trips in a planning day, in the period from April 2022 to January 2024, for a total of 14365 vehicle trips with an average route length of 535 stops.

There are about 16448 unique customers scattered over the territory.

Each route starts and ends within the depot.

However, not all of them result in ending at the depot, either due to failure of the operator to report the end of the route, or by a trivial removal of the route stop line during dataset extraction. These inconsistencies, which are common in real data, were corrected by a suitable set of automatic data processing procedures.

We denote as $\mathcal{L}$ the set of all available instances. For experiments, $\mathcal{L}$ was split in a training set $\mathcal{L}_{train}$ and a set $\mathcal{L}_{test}$.

An instance is described through tuples in this format: $\mathcal{L}_t = \{(t, x^t, coordinates)\}$ where t represents the timestamp in YYYY-MM-DD format, x^t is the concatenation of all trips belonging to that timestamp t. The concatenation is done by keeping the initial and final IDs as that of the depot.

$$id\_depot \rightarrow id_1 \rightarrow id_2 \rightarrow ... \rightarrow id_n \rightarrow id\_depot \rightarrow ... \rightarrow id_i \rightarrow ...id\_depot.$$

Instead, *coordinates* are expressed as a list of tuples

$$coords_t = [(lat_0, lon_0), (...), (lat_n, lon_n)]$$

Thus structuring the dataset

$$\mathcal{H} = \{l_0, ..., l_{320}\} \quad l_j \in \mathcal{L} \quad \forall j \in \{0, ..., 320\}$$

it was possible to apply a split training/test set 80–20%.

A detailed pseudo-code of our training procedure is reported in Algorithms 1 and 2. The ECDF obtained after training is shown in Fig. 4 and its evolution during epochs in reported in Fig. 5. As the error distance increases, the probability value decreases.

From a technical point of view, the training data is further partitioned to obtain the validation set. Once the model shown in the Algorithm 1 has been trained, at the end of each epoch The ECDF is calculated according to the Algorithm 2 and only the best one is saved. This is finally used to determine the probability of a test route.

Bi-LSTM Evaluation. To evaluate the model and the goodness of its geographic predictions, some metrics were calculated on both latitude and longitude. What is obtained is an average for the 5 different runs performed on the test set. In this way, by analyzing them, it is possible to understand how the Bi-LSTM architecture performs and where improvements can be made. For example, the *MSE for latitude* is in average 0,012 while *MSE for longitude* is 0,06. This measures the average of the squares of the errors between predicted and actual values. The MSE for longitude is higher than latitude. This is common in geospatial modeling since longitude spans can cover greater distances than latitude spans at most locations on Earth.

The *RMSE (Root Mean Squared Error) for latitude* is 0,109 and *RMSE for longitude* is 0,245. RMSE is the square root of MSE and represents the standard deviation of the prediction errors. It's in the same unit as the original data

---

Algorithm 1. Training Bi-LSTM algorithm

Input: $train\_data, seq\_length = 5$
Output: $ECDF\_complementary$

   ▷ Prepare sequence data

$X_{train\_dict}, y_{df\_training} \leftarrow prepare\_sequences(train\_data, seq\_length)$
$X_{train}, X_{val}, y_{train}, y_{val} \leftarrow$ **train_test_split** $(X_{train\_dict}['X\_sequences'],$
             $y_{df\_training},$ **test_size=0.2**)

   ▷ Features enhancement

$X_{train\_enhanced} \leftarrow add\_engineering\_features(X_{train})$
$X_{val\_enhanced} \leftarrow add\_engineering\_features(X_{val})$

   ▷ Monitoring validation loss for reduce overfitting

$early\_stopping \leftarrow EarlyStopping(monitor \leftarrow' val\_loss', patience \leftarrow 5)$
$reduce\_lr \leftarrow ReduceLROnPlateau(monitor \leftarrow' val\_loss', factor \leftarrow 0.5,$
           $patience \leftarrow 5, min\_lr \leftarrow 0.00001)$

   ▷ Training the model over 20 epochs and save history

$history = model.fit(X_{train}, y_{train},$
           $validation\_data \leftarrow (X_{val}, y_{val}),$
           $epochs \leftarrow 20,$
           $batch\_size \leftarrow 64,$
           $callbacks = [early\_stopping, reduce\_lr, ECDFCallback()]),$

   ▷ Note that ECDFCallback() refers to the Algorithm 2

return $ECDF\_complementary$

Algorithm 2. Empirical Cumulative Distribution Function extraction algorithm

---

    Input: $predictions, y\_true$
    Output: $ECDF\_complementary$

$distances \leftarrow batch\_haversine\_distance(y\_true, predictions)$
for $length$ in y_true **do**
   $length\_per\_routes \leftarrow [length]$
end for
$sequence\_lengths \leftarrow [length\_per\_routes, dim = len(distances)]$
$normalized\_distances \leftarrow distances/max(sequence\_lengths, 1)$
$sorted\_values = sort(normalized\_distances)$
$ECDF\_complementary = 1 - \frac{sorted\_values}{len(sorted\_values)}$
return $ECDF\_complementary$

(degrees). Again, the longitude error is higher. To give perspective to the reader, 0.1 degrees of latitude is approximately 11 km at the equator.

We also consider the R^2 coefficient of determination:

$$R^2 = 1 - \frac{\sum_{i=1}^{n}(y_{\text{true},i} - y_{\text{pred},i})^2}{\sum_{i=1}^{n}(y_{\text{true},i} - \bar{y}_{\text{true}})^2}$$

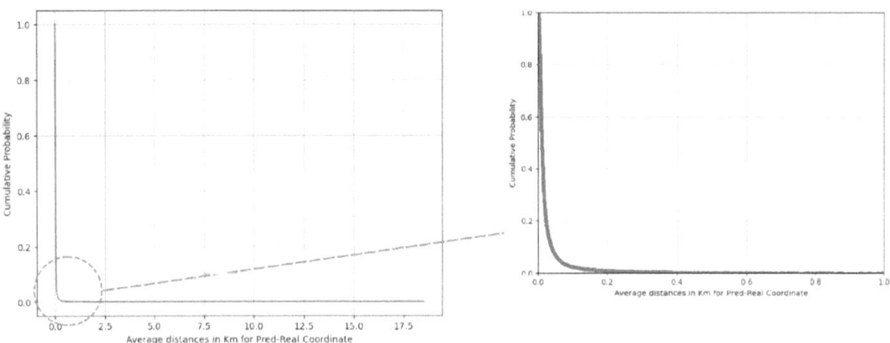

Fig. 4. Best train ECDF extracted during the training phase

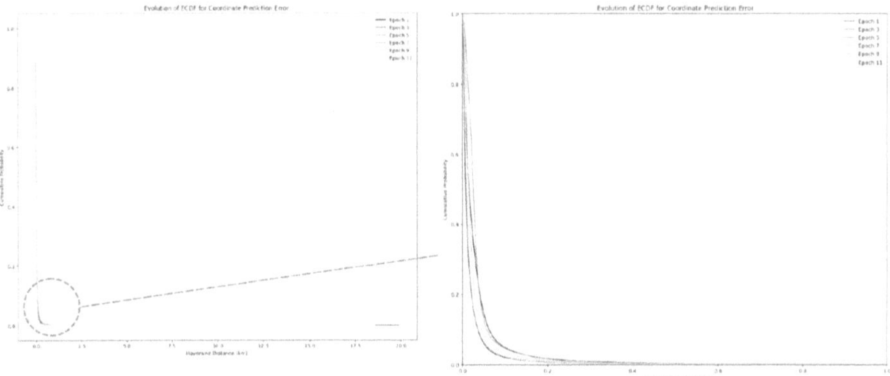

Fig. 5. Evolution of train EDCF in different epochs

Where:

- n is the number of samples
- $y_{\text{true},i}$ is the actual value for the i-th sample
- $y_{\text{pred},i}$ is the predicted value for the i-th sample
- $\bar{y}_{\text{true}}$ is the mean of the actual values: $\bar{y}_{\text{true}} = \frac{1}{n} \sum_{i=1}^{n} y_{\text{true},i}$

Its average value is 0,622 for latitude and 0,627 for longitude in our tests. This indicates that Bi-LSTM explains about 62% of the variance in both latitude and longitude predictions. We consider such a value to be suitable for our scopes, since accurate predictions of customer positions are not the main target of our framework.

The *MAPE (Mean Absolute Percentage Error) for latitude* is 0.113% while *MAPE for longitude* is 0.453%. This measures the average percentage difference between predicted and actual values. The latitude prediction is quite accurate

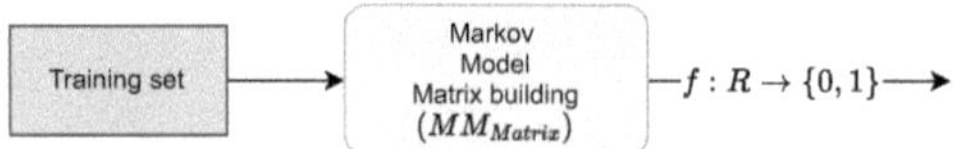

Fig. 6. Markov model training pipeline

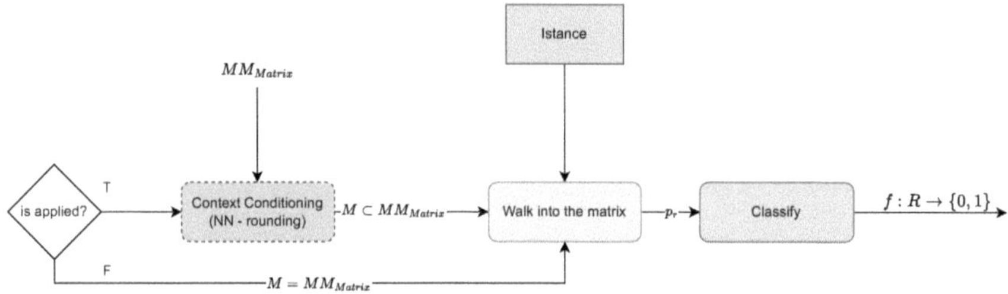

Fig. 7. Markov model usage pipeline

at only 0.113% error, while longitude has a higher error rate.

$$\text{MAPE} = 100 \times \frac{1}{n} \sum_{i=1}^{n} \left| \frac{y_{\text{true},i} - y_{\text{pred},i}}{|y_{\text{true},i}| + \epsilon} \right|$$

Where:

- n is the number of samples
- $y_{\text{true},i}$ is the actual value
- $y_{\text{pred},i}$ is the predicted value
- ϵ is a small constant (1e-10) added to prevent division by zero

The *Mean Distance (Haversine) is about 10.02 Km* This is perhaps the most intuitive metric for geographical predictions. It uses the Haversine formula to calculate the average error in kilometers between predicted and actual positions on Earth's surface. Bi-LSTM has an average error of about 10 km. Also this can be improved by adding more coordinates data.

5.1   Benchmarks

For the sake of comparison, we adapted the approach of [3] to our case. It is a Markovian approach. The corresponding training phase is summarized in Fig. 6, and the corresponding usage is reported in Fig. 7.

Markovian Model Without Context. From [3], we extracted Algorithm 1 for estimating a first order transition matrix from historical instances and repurposed it for our case. Briefly a summary of the steps applied:

- create the set $S^{all} = \cup_t S^t$ such that $\mu = |S^{all}|$ and S^{all} contains all the unique IDs of the available routes.

– construct the adjacency matrix $A^t_{\mu \times \mu}$ on the IDs of the training routes.
– construct the frequency matrix $F = \sum_t w_t A^t$ where in our case, we preferred not to weight the various adjacency matrices differently so $w_t = 1$
– Apply Laplacian smoothing on the frequency matrix to obtain

$$\hat{P}_{\mu \times \mu} = [p_{ij}]$$

that is, the final transition matrix, where

$$p_{ij} = Pr(X_{n+1} = s_j | X_n = s_i) \quad s_i, s_j \in S^{all}$$

From the transition matrix we want to extract the probability for each test route by performing a walk-through within it. The training process of the model is represented graphically in Fig. 6.

5.2   Markovian Model with Context

We want to perform the same procedure as seen previously for estimating the transition matrix but this time we want to add context as it was performed for the Bi-LSTM model in Sect. 4 and have a direct comparison. However, the method by which we provide context for the Markovian model is different. In fact, a renormalized submatrix is extracted $M \subset MM_{Matrix}$, starting with the first one containing all IDs belonging to $x^t \quad \forall l \in \mathcal{L}$. This simulates the fact that we already know which IDs to go to but do not know in which order. Running the walk for each submatrix extracts the probability of that specific route in the test case.

The usage part of the model is shown in Fig. 7

5.3   Performance Comparison Between Models

To generate the ground truth, half instances were extracted from the test instances by assigning the `label` = 1, while the other half are pseudorandom test instances with `label` = 0. A shuffle of the ground truth rows was performed. The following models were compared: Bi-LSTM without context (BiLSTM), Bi-LSTM with context (c-BiLSTM), Markov model without context (Markov), Markov model with context (c-Markov).

We performed 5 different runs in such a way as to obtain a more accurate estimate of performance. In Tab. 1 are given the average precision, recall, TPR, FPR metrics for each model in our experiments.

Average ROC Curve Plot. Figure 8 shows the average ROC Curve for the different models mentioned above. The Bi-LSTM with context or not are equivalent ($AUC = 0.55$ and $AUC = 0.56$). The markov model with context has reported as AUC score 0.15 compared to the markov model without context 0.09.

Table 1. Comparison of quality metrics (Precision, Recall, TPR, FPR) for different models

Model	Average Precision	Average Recall	Average TPR	Average FPR
Markov	26.236%	29.426%	29.46%	54.608%
c-Markov	31.794%	33.016%	20.578%	53.388%
BiLSTM	58.652%	52.876%	55.744%	50.214%
c-BiLSTM	58.56%	52.414%	54.262%	48.974%

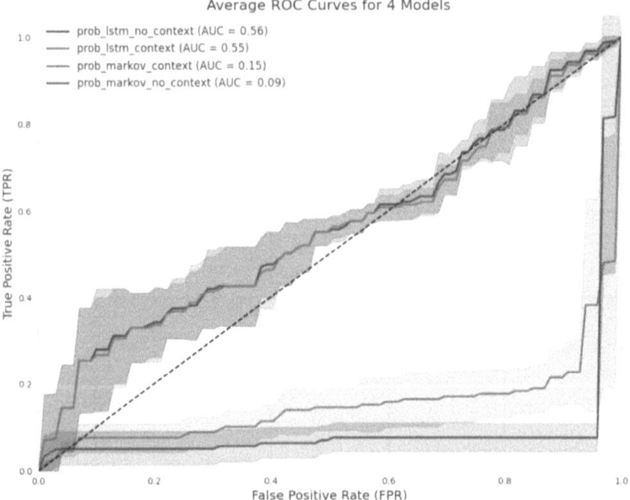

Fig. 8. Average ROC curve for different models

6   Conclusions

To achieve the goal of finding effective acceptance models, we designed a pipeline of methods. It is mainly composed by two core elements: a Bidirectional Long Short-Term Memory (BiLSTM) neural network, to predict the position of a customer after observing a sequence of k customers, and an empirical cumulative distribution function of errors in these predictions. When combined in a suitable classification pipeline, these two elements allow to clearly outperform an adaptation of the Markovian methods proposed in [3], as shown by the ROC Curve analysis (Fig. 8), as well as a set of quality metrics (Table 1).

From a methodological point of view, a key point has been to proper design a customized BiLSTM architecture, since a black-box use of LSTM was unsuccessful. Our custom LSTM is in fact able to capture longer dependencies and consequently have greater predictive ability. If we add a bidirectional layer to this, then the coordinate sequences will be examined more thoroughly in both directions capturing patterns that Markovian models cannot extract.

We mention that Markovian models benefit from context knowledge, that is knowing in advance the set of customers which are visited in a route (although their order of visit remains unknown). Our architecture does not show the same benefit. Indeed, adapting the concept of context in our architecture was not straightforward: other techniques might improve it, further boosting our methods.

Being able to understand drivers' preferences in more detail is just part of our research. Our final aim is to integrate our acceptance models within a mathematical route optimization mechanism, simultaneously taking into account both explicit factors such as minimization or maximization criteria and operational constraints, implicit preferences of the planner, or scoring of different routing options. Ultimately, our ambition is to deploy our methods in an industrial decision support tool.

References

1. Amazon.com, I.: Amazon last mile routing research challenge. https://routingchallenge.mit.edu/ (2021)
2. Caceres-Cruz, J., Arias, P., Guimarans, D., Riera, D., Juan, A.A.: Rich vehicle routing problem: survey. ACM Comput. Surv. (CSUR) **47**(2), 1–28 (2014)
3. Canoy, R., Bucarey, V., Mandi, J., Guns, T.: Learn-n-route: Learning implicit preferences for vehicle routing. CoRR abs/2101.03936 (2021). https://arxiv.org/abs/2101.03936
4. Chen, W., et al.: Systems engineering issues for industry applications of large language model. Appl. Soft Comput. **151**, 111165 (2024). https://doi.org/10.1016/j.asoc.2023.111165, https://www.sciencedirect.com/science/article/pii/S1568494623011833
5. Dantzig, G.B., Ramser, J.H.: The truck dispatching problem. Manage. Sci. **6**(1), 80–91 (1959). https://doi.org/10.1287/mnsc.6.1.80, https://doi.org/10.1287/mnsc.6.1.80
6. Dragone, R., Cerrone, C.: Integrating human-centric ai in fashion industry optimization: A step towards industry 5.0 (2025)
7. Drexl, M.: Rich vehicle routing in theory and practice. Logist. Res. **5**, 47–63 (2012)
8. Song, J., Shukla, A., Coad, J.: Inverse reinforcement learning for learning driver utility function for package delivery routing problems. Technical Proceedings of the Amazon Last Mile Routing Research Challenge (2021)
9. Zattoni Scroccaro, P., van Beek, P., Mohajerin Esfahani, P., Atasoy, B.: Inverse optimization for routing problems. Transportation Science (2024)

Monitoring Stress in Airline Pilots: Biomarker-Based Approaches to Operational Safety and Efficiency

Giacomo Belloni[1,2]([📧]) [iD], Francesca Antonella Aiello[2,3]([📧]) [iD], and Petru Lucian Curşeu[4,5] [iD]

[1] Faculty of management, Open Universiteit, Heerlen, The Netherlands
giacomo.belloni@ou.nl
[2] HUMAI Research, Abu Dhabi, United Arab Emirates
francesca.aiello@mathbiology.tech
[3] Math Biology srl, Rome, Italy
[4] Department of Organization, Open Universiteit, Heerlen, The Netherlands
petrucurseu@psychology.ro
[5] Department of Psychology, Babeş, Bolyai University, Cluj-Napoca, Romania

Abstract. Objective: This study investigates how stress affects airline pilots' performance and explores the potential of psychophysiological monitoring tools to assess and mitigate related risk.

Background: Modern commercial aviation is a hypercomplex socio-technical system in which safety margins are narrow and operational demands constantly evolve. While technological advancements have enhanced efficiency, they have simultaneously increased pilots' cognitive load by requiring the rapid integration of vast amounts of information. Stress, intensified by the complexities of the operational environment, undermines pilots' cognitive performance and constitutes a critical risk to the safety and efficiency of airline operations.

Methods: A structured and reasoned literature review was conducted using a snowballing technique to identify empirical and theoretical studies on psychophysiological indicators of stress and cognitive load in aviation.

Results: The review synthesises a range of biometric measures applied over time to capture pilots' cognitive and emotional states. These approaches demonstrate the feasibility of monitoring airline pilots' stress status in real time and highlight persistent challenges related to intrusiveness, ecological validity, and operational integration.

Conclusions: In an increasingly automated and data-intensive aviation environment, distress-related cognitive impairment constitutes a critical operational impediment for safety and efficiency. Addressing this challenge requires human-centred monitoring systems capable of detecting early physiological distress markers to enable adaptive interventions. Biomarker-based multimodal methods show promise, but their effectiveness relies on trustworthy data integration and accurate contextual interpretation. Emerging approaches, such as Deep Metabolic-processes Assessment (DMA®), may advance non-invasive detection of stress-induced metabolic changes. Recognising pilots as central agents of system resilience underscores the need to integrate cognitive state monitoring into future aviation system design to enhance safety and efficiency.

© The Author(s), under exclusive license to Springer Nature Switzerland AG 2026
M. Pavone et al. (Eds.): DSA ISC 2025, LNCS 16405, pp. 154–168, 2026.
https://doi.org/10.1007/978-3-032-21811-7_11

Keywords: Stress · mental workload · cognitive impairment · commercial aviation · airline pilots · biological stress biomarkers · biometric monitoring of workload · physiological sensors · Deep Metabolic-processes Assessment (DMA®) · neurophysiological response · cognitive state tracking

1 Monitoring Commercial Pilots' Stress and Cognitive Load

Modern commercial aviation operates within a highly complex socio-technical system, where safety and efficiency are paramount and tolerance for errors is minimal [1]. It is a hypercomplex domain [2] requiring the continuous integration of multiple inter-dependent agents to ensure safe and effective flight operations [3, 4]. This dynamic environment is characterised by continuous interactions between humans, technological systems, and organisational processes, demanding constant adaptation and synergy seeking [4].

While technological advancements have enhanced safety and efficiency, they have also introduced new challenges, increasing pilots' cognitive demands [5, 6]. The proliferation of digital interfaces necessitates greater mental effort to interpret and manage vast amounts of information, significantly elevating stress and cognitive load [7], impairing decision-making, delaying reactions, and leading to operational errors, particularly in high-pressure situations [8].

Despite being designed for efficiency [6], modern cockpits often require pilots to process overwhelming amounts of data in short timeframes, limiting their ability to act decisively under pressure [9, 10]. Such complexity reduces the likelihood of a synergetic human-technology interaction and ultimately creates room for errors with potentially devastating consequences.

Empirical evidence shows that human error, rather than mechanical failure, accounts for approximately 60–80% of aviation incidents [8, 11, 12]. These errors largely arise from operational complexities that generate stress and high cognitive load, making it difficult for pilots to make decisions under pressure or to sustain a comprehensive understanding of the operational context [13]. It underscores the need for strategies that enhance human performance and mitigate risks in high-stakes environments. Stress and the resulting cognitive overload impair pilots' situational awareness [14], problem-solving [15], and communication [16], thereby increasing the likelihood of procedural lapses and critical errors [11].

In commercial aviation, stress is a major contributor to cognitive overload, amplified by the unique challenges pilots face beyond those of ordinary workers [17]. Factors such as irregular schedules, sleep deprivation, fatigue, low oxygen levels [18], and unexpected stressors like adverse weather or technical malfunctions further compound the issue.

Stress is a biological, behavioural, and emotional multidimensional response that affects cognitive processes [17], a natural response to internal and external stimuli [18], enabling individuals to react to environmental demands [10].

Stress may be conceptualised as a process in which environmental demands tax or exceed an individual's coping resources [19].

Stress emerges from the dynamic interaction between individual and situational factors, mediated by the individual's appraisal of environmental demands (Lazarus &

Folkman, 1984). Through primary appraisal, stressors are evaluated as a potential harm, threat, or challenge [20]. A threat appraisal reflects the anticipation of loss or harm, whereas a challenge appraisal frames a demanding situation as an opportunity. Because stressors can be construed either negatively or positively, two distinct forms of stress are recognised: distress, denoting dysfunctional or maladaptive stress, and eustress, representing functional or adaptive stress [21].

Eustress is a constructive form of stress, characterised by positive emotional arousal that fosters activation and engagement under moderate demands. In contrast, distress is destructive, marked by negative emotional arousal that leads to dissatisfaction, disengagement, and impaired performance and typically arises under low or excessive demands [21]. For operators such as pilots in high-dynamic and complex contexts, eustress is expected to enhance performance, whereas distress undermines it and ultimately jeopardises safety.

Empirical evidence shows that individuals perform most effectively under eustress, where arousal is moderate and stress remains manageable. While eustress reflects an optimal and controllable stress level, distress occurs when individuals perceive little or no control over the situation [22]. More specifically, eustress arises when a stressor is appraised as a challenge, whereas distress emerges when it is perceived solely as a threat. In terms of human-technology fit, we expect that eustress increases the operational effectiveness, while distress reduces the human-technology synergy that is required for effective flight operations.

It is therefore evident that for individuals operating in compounding contexts, such as commercial aviation pilots, it is crucial to distinguish between these two, eustress and distress, to highlight when stress becomes dysfunctional so as not to impact the safety of operations.

Stress and cognitive functioning are closely interconnected. Lazarus and Folkman (1984) propose that individuals allocate mental and physiological resources when confronted with stressors. While this mechanism facilitates attentional focus, excessive demands can exceed coping capacity, impairing cognitive performance, attentional control, and working memory. Under stress, individuals become more vulnerable to distractions and attentional shifts, reducing goal-directed focus and increasing susceptibility to salient or threatening stimuli [23, 24]. A compromised cognitive control impacts decision-making, heightening the likelihood of errors and reducing safety margins [25].

Over the past decades, an increasing number of studies have made it possible to quantify stress by assessing the activation of the autonomic nervous system (ANS) in response to stressors. ANS is crucial in regulating physiological responses to work overload, primarily through activating the sympathetic and parasympathetic branches, triggering measurable physiological changes. These include glandular activation and the subsequent release of stress-related hormones such as cortisol and adrenaline, which influence cognitive and physical performance under stress and high workload conditions [26, 27].

1.1 Biometric Measures

Biometric measures, facilitated by sensor technology and computational miniaturisation advances, are increasingly employed for real-time monitoring in operational contexts

[28]. Applied alone or in combination, they enable detection of cognitive state fluctuations [13, 29, 30]. Favoured for their objectivity and quantifiability over self-reports or behavioural assessments [31], these measures permit passive, continuous, and multidimensional monitoring, supporting the creation of broadly applicable performance models.

Psychophysiological measures face notable limitations, including sensor intrusiveness, difficulty in discriminating overlapping mental states, vulnerability to noise and artefacts, need for extensive validation and baseline calibration, limited sensitivity to individual variability, and persistent concerns over privacy and operator acceptance [6].

Advances in technology have fostered research on quantifying mental impairment through autonomic nervous system (ANS) responses to stress.

Among the most widely used physiological parameters for assessing stress levels are heart rate (HR), heart rate variability (HRV), respiratory rate (RR), blood pressure (BP), body temperature, galvanic skin response (GSR), respiration rate (RR), eye activity, brain activity, muscle activation, and voice pattern.

Cardio-respiratory measures, used in aviation research since the 1960s, provide non-invasive indices of stress, fatigue and cognitive load [6]. Cardiac activity, regulated by ANS, is most reliably assessed via electrocardiography (ECG), which captures heart rate, RR intervals, and HRV—a sensitive marker of workload, stress, and fatigue [8, 33, 34]. Photoplethysmography (PPG), an optical alternative embedded in wearable devices, offers practical though less robust estimates of HR and HRV, with accuracy dependent on noise levels [35]. ECG and PPG together enable real-time monitoring of cardiovascular responses to operational demands.

(a) *Heart rate*, influenced by autonomic activity, is a widely used indicator of mental workload and stress in aviation. Values typically increase with task demands and sympathetic activation [30, 33, 36, 37]. Heart rate reflects heightened arousal and resource mobilisation, making it a standard parameter in ergonomics and applied cognitive research [38].

Studies show that HR rises during demanding aviation, including simulated flights with high cognitive load [39, 40]. In flight, brain blood flow peaks with HR during landing and is lowest post-flight [37].

However, HR lacks specificity, as it is also affected by physical activity, emotional arousal, circadian rhythms, and individual variability [28, 39]. Consequently, HR is best interpreted as a gross index of autonomic arousal and should be combined with measures such as HRV or electroencephalography (EEG) for more accurate workload assessment [32].

(b) *Heart rate variability* is a well-recognised non-invasive biomarker of autonomic nervous system activity, used to assess stress, workload, fatigue, and related states [6, 33, 34]. It reflects cardiac–neural interactions and supports classification of cognitive workload [27]. HRV emerges from complex, non-linear heart–brain–ANS dynamics, with variability underpinning adaptability and resilience [34].

In aviation, HRV is used to monitor pilots' autonomic balance and workload sensitivity [6, 12, 40, 41]. High-frequency power indicates parasympathetic activity, while low-frequency power reflects mixed influences and often rises under stress; the LF/HF ratio is common but controversial as a balance index [42].

Stress and overload typically reduce HRV, particularly HF power, whereas higher baseline HRV supports executive and emotional regulation [40, 43]. Phase-specific shifts occur during flight, with HF decreasing in landing, LF increasing in takeoff, and HF recovery post-flight [37].

HRV can be derived from ECG or PPG, though PPG is prone to motion artefacts [34, 35]. If heart rate often responds more rapidly to changing workload [12, 42], HRV is more reliable in controlled tasks, being sensitive to respiratory and other confounds [32]. Despite these limits, consistent reductions in high-frequency power under demand confirm HRV as a robust marker of cognitive workload and resilience [27].

(c) *Blood pressure*, particularly systolic blood pressure (SBP), is a less commonly used but promising marker of cognitive load, stress, and anxiety, reflecting sympathetic vasomotor activity [33, 39]. SBP typically rises during cognitively demanding tasks, such as mental arithmetic or dual-task performance, indicating cardiovascular adaptation to mental effort [44].

Today, wearable technologies facilitate continuous, non-invasive BP-monitoring in operational settings, though artefacts, calibration drift, lower temporal resolution, and individual variability remain challenges [40].

Consequently, SBP is best used as a complementary indicator alongside HR or HRV rather than as a standalone measure of mental workload [45].

Electrodermal activity (EDA), or galvanic skin response, reflects sympathetic nervous system activation and provides a non-invasive index of emotional arousal, stress, and cognitive workload [33, 46]. It consists of a tonic component, the skin conductance level (SCL) indicating baseline arousal and a phasic component, the skin conductance response (SCR) sensitive to transient stimuli, task events, or acute cognitive load [47, 48]. EDA is typically measured on high-sweat-density sites such as fingers, palms, or wrists, and modern sensors allow integration with other physiological streams.

While EDA provides real-time insight into sympathetic activation, it has limited specificity: changes may reflect emotional stress, physical exertion, or environmental influences [47]. Measurements are sensitive to motion artefacts, electrode placement, and environmental conditions, and long-term monitoring in operational settings requires careful calibration and context alignment. Consequently, EDA is most effective when combined with complementary indicators such as HRV, respiration, or EEG to assess cognitive and emotional states in dynamic operational environments like aviation [43].

Body temperature is one of the biosignals employed in psychophysiological monitoring, and it is one of aviation's most established and widely applied [36, 48]. Skin temperature reflects autonomic activity and sympathetic vasomotor responses to cognitive load and stress, mediated by changes in peripheral blood flow [33, 47]

Infrared thermography (IRT) allows non-contact, real-time monitoring of thermal fluctuations, with decreases in surface temperature typically indicating sympathetic activation [48]. Advantages include unobtrusive measurement and involuntary responsiveness, offering a reliable index of physiological state. Limitations arise from environmental factors, circadian and age-related variations, and the latency of thermal responses, which may constrain rapid detection in dynamic operational settings such as aviation [49].

Respiration monitoring is a non-invasive indicator of autonomic activity and psychophysiological states, including stress, cognitive load, fatigue, and startle responses [6, 13, 34, 46]. Respiratory rate, tidal volume, and variability respond rapidly to cognitive and emotional demands, with elevated rates and reduced variability reflecting sympathetic dominance. In aviation, stress or heavy workload similarly alters pilots' respiration, indicating sympathetic activation and reduced parasympathetic control [30].

Wearable sensors such as thoracic bands and capacitive plethysmography enable continuous in-flight monitoring and support real-time assessment of psychophysiological strain [50].

Interpretation is limited by confounders, including speech, movement, fatigue, altitude, and individual breathing patterns, which can produce similar respiratory patterns under different cognitive or emotional loads [28, 51, 52].

Despite these limitations, respiration can complement HR and HRV monitoring, and controlled breathing has been shown to modulate autonomic activity and support workload management in high-demand contexts (Veltman & Gaillard, 1996).

Eye tracking provides non-invasive metrics of cognitive and emotional states in aviation, including fixations, saccades, pupil diameter, blink rate, and PERCLOS [6, 29, 52]. Pupil dilation reflects controlled and in-depth cognitive processing, sympathetic activation, and emotional arousal [53], while blink and fixation metrics indicate attentional engagement, fatigue, or task-related cognitive load [28, 31, 41, 46]. Eye tracking captures involuntary and voluntary responses, making it a versatile tool for workload assessment [10, 13, 28, 31].

Wearable or head-mounted eye trackers enable natural movements and can be employed effectively even in dynamic settings such as airline cockpits. However, the quality of the recorded data remains highly sensitive to factors such as lighting conditions, head motion, occlusions, and calibration drift [34]. Blinks are regulated by visual task demands rather than cognitive load, limiting their interpretive specificity [39]. Multimodal integration with HR, HRV, EEG, or EDA improves validity and robustness in operational settings [52]. Eye tracking thus offers high-resolution access to cognitive-affective processes, but careful implementation is required in complex, high-stakes environments such as commercial aviation.

Brain activity provides a direct index of cognitive and affective processes, including workload, attention, emotion regulation, and situational awareness [9, 36]. In aviation, brain measures are increasingly valued as biomarkers of workload, stress, and startle responses [46], since changes in prefrontal and parietal activity reflect demands on working memory, decision-making, and attentional control [53]. Neural dynamics enable early detection of overload, fatigue, or emotional arousal that could undermine safety [53].

Recent advances in portable, low-cost, and minimally intrusive brain-sensing technologies have expanded applied cognitive research, particularly through EEG, functional near-infrared spectroscopy (fNIRS), Event-Related Potentials, and Transcranial Doppler Sonography (TCDS) [6, 29]. EEG and fNIRS have emerged as the most prominent tools for real-time monitoring of workload and alertness in operational aviation [13, 31, 38, 53].

EEG is widely adopted due to its high temporal resolution, portability, and wireless capability. It records scalp-level electrical activity, with frontal theta increases reflecting higher workload and parietal alpha decreases indicating cortical engagement [8, 38]. Though challenges remain, wearable EEG devices provide reliable signals in simulators and flight environments, particularly when integrated with other biosensors. Data collection requires baseline calibration and is prone to artefacts from motion, muscle activity, electromagnetic interference, ocular signals, and headsets that may cause discomfort [13, 33].

fNIRS offers complementary advantages, including better spatial resolution, minimal calibration, portability, and resistance to electromagnetic noise, making it suitable for simulators and in-flight use [54].

EEG and fNIRS provide robust workload assessments, especially when combined with behavioural or physiological data [55]. However, methodological, regulatory, and ethical challenges persist, including artefact management, temporal alignment with flight tasks, and concerns about continuous cognitive monitoring in safety-critical settings [53]. Future research increasingly points to multimodal approaches integrating EEG, fNIRS, and behavioural data as the most promising strategy for reliable operator-state assessment in aviation.

Voice patterns. Speech analysis has emerged as a non-intrusive method for assessing emotional state, stress, and cognitive workload in operational domains such as aviation [6, 10, 57]. Pilots' voice data can be captured through cockpit microphones and analysed via signal processing or machine learning to extract prosodic and spectral features, including pitch, amplitude, jitter, and speech rate, which are sensitive to stress-induced autonomic changes [57, 58]. Under stress, characteristic shifts occur—altered pitch, greater variability, reduced rate, and decreased articulation precision—making vocal parameters robust workload indicators [57]. Aviation incident analyses further confirm their relevance for detecting acute stress [58]. However, speech rate is considered less reliable: it could be affected by training, communication protocols, and ICAO/CAA guidelines promoting standardised delivery [57].

However, many challenges persist, such as noise contamination, microphone quality, and inter-individual variability linked to age, gender, language, and fatigue [59]. Moreover, speech inherently conveys personal information, raising ethical and privacy concerns in continuous monitoring [60]. Given these constraints, voice analysis remains an indirect measure and is best integrated with physiological and behavioural markers such as HR, RR, or EMG to strengthen reliability [58]. Overall, it offers a promising component within multimodal approaches to pilot state monitoring.

Muscle Activation. Electromyography (EMG) measures skeletal muscle activity and is increasingly recognised as an indicator of emotional arousal, stress, and mental load in operational settings [29, 44]. In the aviation setting, trapezius activation and neck pain are correlated with workload, stress, and fatigue [44].

Surface EMG (sEMG), the most common non-invasive method, allows real-time monitoring with minimal preparation [61]. Muscle choice is context-dependent: facial muscles signal emotional valence, though culturally modulated [62], while neck, shoulder, jaw, and limb muscles reflect tension or effort [31]. However, EMG is limited by interpretive ambiguity—since activity may stem from stress, posture, or discomfort [63]—and by artefacts from motion, vibration, and sensor noise in cockpits [45]. Its localised scope further underscores the need for integration with measures such as HRV, skin conductance, EEG, or respiration [64]. Thus, EMG is a complementary tool that strengthens workload and stress assessment when combined with multimodal data.

1.2 Biological Measurements

All these biometric measurements provide insight into autonomic regulation and have been increasingly integrated into wearable devices, making stress monitoring more accessible, even in everyday life [65, 66]. Smartwatches, for instance, can now measure some of these parameters, providing real-time feedback on physiological changes related to stress [67, 68]. Alongside these physiological indicators, a growing body of research focuses on biological biomarkers that study the body's internal response to cognitive and emotional stressors. Among the most validated is cortisol, the primary glucocorticoid released via HPA axis activation. Cortisol can be measured in sweat, blood, saliva, urine, or hair; notably, sweat cortisol offers a non-invasive, time-sensitive option that can be integrated into wearables [69]. Recent devices using molecularly imprinted polymers and flexible microfluidics now detect cortisol across physiologically relevant ranges and, in ambulatory settings, can even reproduce its expected diurnal pattern, enhancing their utility for mental-stress monitoring [70, 71].

In parallel, cumulative measures such as hair or nail cortisol continue to index chronic stress exposure and burnout risk. Comparisons with perceived stress often show limited concordance. For example, Schnall et al. (2024) [72] report high hair–nail agreement but no correlation with self-reported stress, supporting the view that subjective and biological indices capture complementary facets and should be jointly interpreted. Moreover, diurnal cortisol metrics (awakening response, daytime slope, area-under-the-curve) have been associated with specific burnout components in high-stress occupations, such as police work [73].

Catecholamines such as adrenaline (epinephrine) and noradrenaline (norepinephrine) are critical biomarkers of sympathetic nervous system activation, modulating cardiovascular tone, glucose metabolism, and overall stress reactivity. Clinically, they are typically assessed via plasma or urine sampling; on-body, real-time monitoring has long been hindered by their picomolar concentrations, chemical instability, and interference in sweat or other easily accessible fluids, which pose substantial technical barriers to continuous, non-invasive detection [74]. Recently, however, multiplexed microfluidic patches have demonstrated time-stamped, on-body quantification of cortisol and catecholamines (epinephrine, norepinephrine) in human sweat during physical, psychological, and pharmacological stress, marking a meaningful step toward integrating sympathetic–adrenal biomarkers into wearable platforms. These systems remain pre-commercial and require further validation for accuracy, specificity, and stability, but they revise the earlier view that catecholamines are wholly inaccessible to continuous, non-invasive sensing [75].

By contrast, wearable glucose monitoring has seen significant advancements, particularly through lab-on-a-chip technologies and electrochemical or optical biosensors. Many commercial continuous glucose monitoring (CGM) devices rely on minimally invasive or microinvasive approaches, most commonly interstitial-fluid sensing via subcutaneous filaments or microneedles, which are the methods with the strongest clinical validation to date [76–78]. Non-invasive solutions are actively being explored (e.g., optical, bioimpedance, ultrasound, millimetre-wave, reverse-iontophoresis), and watch-form prototypes have been demonstrated in research; however, these systems remain largely pre-commercial and not widely available [79].

From a metabolic perspective, blood glucose levels are highly sensitive to acute psychological stress, aligning with the organism's need to mobilise energy in anticipation of the "fight or flight" response. This stress-induced hyperglycemia, even in non-diabetic individuals, is primarily mediated by HPA-axis activation and catecholamine-driven hepatic gluconeogenesis [80]. Moreover, emerging lab-on-skin/ISF patch platforms facilitate naturalistic co-measurement of glucose and physiological signals, helping quantify these stress–metabolism interactions outside the lab [81].

Despite these advancements, research in the field of stress sensors remains highly active, driven by the need to enhance measurement accuracy and fully exploit the potential of monitoring operators when they are under pressure. Current limitations stem from the fact that analysing these parameters individually (e.g., HRV, EDA, or glucose alone) often fails to provide a sufficiently precise assessment of stress levels or cognitive load. Consequently, current efforts emphasise multimodal integration and improved algorithms; systematic reviews and recent deep-learning studies show that combining physiological streams, contextual data, and, where feasible, biochemical signals yield more robust and externally valid stress inference than any one signal alone [82–84].

2 New Perspective: DMA® and the Neurochemical Profile of Cognitive Stress a Subsection Sample

The Deep Metabolic-processes Assessment (DMA®) technology introduces a novel, non-invasive approach to understanding stress-related cognitive impairment by focusing on the metabolic demand of key neurotransmitters. Neurotransmitters such as dopamine, serotonin, acetylcholine, GABA, and glutamate govern core aspects of human behaviour (motivation, attention, emotional regulation) and decision-making. The brain's neurochemical environment, or "brain chemistry," reflects the relative dominance and interaction of these molecules, which in turn shape one's behavioural tendencies and coping strategies under stress.

Through the quantification of Bioelectrical Marker Response Indexes (BMRi) at specific anatomical points, DMA® allows for the assessment of neurotransmitter-related metabolic shifts in response to cognitive demands. By comparing DMA® scans taken before and after high-intensity cognitive stressors, such as those experienced during complex flight operations, subtle variations in metabolic profiles that correlate with behavioural rigidity, reduced adaptability, or maladaptive coping strategies can be detected.

This capability opens new avenues in predicting individual stress responses and tailoring interventions. For instance, identifying a dominance of glutamatergic stress-reactivity versus GABAergic inhibition may help anticipate maladaptive overactivation in certain pilots. DMA® could thus serve as both a diagnostic and training optimisation tool, helping to support long-term cognitive resilience and mental well-being in aviation professionals and similar high-load environments.

2.1 Additional Fields of Application

Understanding how to refine and use these measurements would help systems manage contexts characterised by operational complexity, even beyond the commercial aviation domain. In such environments, professionals—like airline pilots—must process vast amounts of information rapidly while managing limited resources and high-pressure conditions. Potential fields of application include (but are not limited to) the nuclear industry, where operators must continuously monitor critical systems; surgical and nursing care, where healthcare professionals make life-critical decisions under time-sensitive conditions; emergency response services, such as firefighting, where situational awareness and swift decision-making are paramount; and military operations, where personnel operate in dynamic and unpredictable environments [10, 85, 86].

We can improve system resilience by joining and harmonising efforts to produce exceptional, applicable knowledge beyond the single context. Only through a structured multidisciplinary approach capable of bringing together the wealth of experiences of individual domains and constant constructive and productive comparisons can we cope with the continuous increase in the complexity of the systems of the near future.

3 Conclusions

As the commercial aviation industry advances within an increasingly automated and data-intensive environment, the risk of cognitive impairment among pilots is becoming progressively more relevant. The considerable volume of information to be processed within constrained timeframes, in conjunction with an operational context that is ever more congested and complex, exacerbates the stress levels experienced by pilots. This study has illustrated that stress should not be regarded as a peripheral factor but rather as a central element with the potential to compromise operational safety, efficiency, and overall performance. Accordingly, stress must be addressed proactively by developing and integrating monitoring tools designed to evaluate pilots' mental states, with the specific capacity to differentiate between eustress and distress.

Biomarker-based systems offer promising avenues for non-invasively detecting early signs of stress. However, the effectiveness of such systems relies on the accuracy of data integration and the contextual interpretation of physiological responses. Future research and implementation efforts should focus on enhancing these capabilities within aviation and other high-reliability sectors. Recognising the pilot as a critical agent of system resilience rather than a liability demands a renewed emphasis on human-centred design and integrating cognitive state monitoring to support sustained performance in high-risk environments.

In this context, novel approaches such as Deep Metabolic-processes Assessment (DMA®) offer an innovative pathway for evaluating bioelectrical responses linked to stress-induced metabolic alterations. By capturing systemic changes without the need for invasive procedures, DMA® could enrich current monitoring strategies and facilitate a more holistic understanding of pilot resilience under pressure.

References

1. Belloni, G.: Beyond Evidence-Based Training. Analysis of the Reasons that Led to the Development of an Innovative Training Methodology Suited to the Aeronautical World's Ever-Increasing Complexity. City of London University (2020)
2. Dalla Quercia, C.: Complexity in the aviation sector. Bocconi Students for Digital Consulting (BSDC) (2017)
3. Salas, E., Wilson, K.A., Burke, C.S., Wightman, D.C.: Does crew resource management training work? An update, extension, and some critical needs. Hum. Factors. **48**(2), 392–412 (2010)
4. Adriaensen, A., Patriarca, R., Smoker, A., Bergström, J.: A socio-technical analysis of functional properties in a joint cognitive system: a case study in an aircraft cockpit. Ergonomics. **62**(12) (2019)
5. De Oliveira, C.G., De Oliveira, S.L.: Evaluation of workload in high complexity work place: an experiment during a real situation. In: International Symposium on Aviation Psychology, vol. 2005, pp. 177–182 (2005)
6. Lim, Y., et al.: Avionics human-machine interfaces and interactions for manned and unmanned aircraft. Prog. Aerosp. Sci. **102**, 1–46 (2018)
7. Vogl, J., et al.: A Literature Review of Applied Cognitive Workload Assessment in the Aviation Domain. U.S, Army Medical Research and Development Command Military Operational Medicine Research Program (2022)
8. Ji, L., Yi, L., Li, H., Han, W., Zhang, N.: Detection of pilots' psychological workload during turning phases using EEG characteristics. Sensors. **24**, 5176 (2024)
9. Çakır, M.P., Vural, M., Süleyman, Ö.K., & Toktaş, A. (2016). Real-Time Monitoring of the Cognitive Workload of Airline Pilots in a Flight Simulator with fNIR Optical Brain Imaging Technology.
10. Masi, G., Amprimo, G., Ferraris, C., Priano, L.: Stress and workload assessment in aviation—a narrative review. Sensors. **23**, 3556 (2023)
11. Casner, S.M., Gore, B.F.: Measuring and Evaluating Workload: a Primer. NASA (2010)
12. Wang, C., et al.: High-precision flexible sweat self-collection sensor for mental stress evaluation. Npj flexible. Electronics. **8**, 47 (2024)
13. Salvan, L., Paul, T.S., Marois, A.: Dry EEG-based mental workload prediction for aviation. In: IEEE/AIAA 42nd Digital Avionics Systems Conference (DASC), pp. 1–8 (2023)
14. Endsley, M.R., Robertson, M.M.: Training for situation awareness in individuals and teams. In: Endsley, M.R., Garland, D.J. (eds.) Situation Awareness Analysis and Measurement, pp. 349–365. Lawrence Erlbaum Associates (2000)
15. Sweller, J.: Cognitive load during problem-solving: effects on learning. Cogn. Sci. **12**(2), 257–285 (1988)
16. Helmreich, R.L., Foushee, H.C.: Why crew resource management? Empirical and theoretical bases of human factors training in aviation. In: Wiener, E.L., Kanki, B.G., Helmreich, R.L. (eds.) Cockpit resource management, pp. 3–45. Academic (1993)
17. Aubry, R., Carstens, D.S., Hight, M.: Perceived stress for pilots. In: Proceedings of the IEMS 2024 Conference (2024)

18. Giannakakis, G., Grigoriadis, D., Giannakaki, K., Simantiraki, O., Roniotis, A., Tsiknakis, M.: Review on psychological stress detection using biosignals. IEEE Trans. Affect. Comput. **13**, 440–460 (2019)
19. Lazarus, R.S., Folkman, S.: Stress, Appraisal, and Coping. Springer, New York (1984)
20. LePine, M.A.: The challenge-hindrance stressor framework: an integrative conceptual review and path forward. Group Org. Manag. **47**(2), 223–254 (2022)
21. Pluut, H., Curşeu, P.L., Fodor, O.C.: Development and validation of a short measure of emotional, physical, and behavioral markers of eustress and distress (MEDS). Health. **10**(2), 339 (2022)
22. Quick, J.C., Quick, J.D., Nelson, D.L., Hurrell, J.J.: Preventive Stress Management in Organizations. American Psychological Association (1997)
23. Eysenck, M.W., Derakshan, N., Santos, R., Calvo, M.G.: Anxiety and cognitive performance: attentional control theory. Emotion. **7**, 336–353 (2007)
24. ALPA: Pilot Peer Support Operations Manual. Air Line Pilots Association (2020)
25. Phillips-Wren, G., Adya, M.: Decision making under stress: the role of information overload, time pressure, complexity, and uncertainty. J. Decis. Syst. (2020)
26. Kim, H.G., Cheon, E.J., Bai, D.S., Lee, Y.H., Koo, B.H.: Stress and heart rate variability: a meta-analysis and review of the literature. Psychiatry Investig. **15**(3), 235–245 (2018)
27. Thayer, J.F., Åhs, F., Fredrikson, M., Sollers, J.J., Wager, T.D.: A meta-analysis of heart rate variability and neuroimaging studies: implications for heart rate variability as a marker of stress and health. Neurosci. Biobehav. Rev. **36**(2), 747–756 (2012)
28. Ayres, P., Lee, J.Y., Paas, F., van Merriënboer, J.J.G.: The validity of physiological measures to identify differences in intrinsic cognitive load. Front. Psychol. **12**, 702538 (2021)
29. Skinner, A., Sebrechts, M., Fidopiastis, C., Berka, C., Vice, J., Lathan, C.: Physiological measures of virtual environment training. In: Schmorrow, D.D., Fidopiastis, C.M. (eds.) Human Performance Enhancement in High Risk Environments, pp. 129–149. Springer (2010)
30. Charles, R.L., Nixon, J.: Measuring mental workload using physiological measures: a systematic review. Appl. Ergon. **74**, 221–232 (2019)
31. Tichon, J.: Affective Intensity and its Effects (AOARD 104011). Defense Technical Information Center (DTIC), Australia (2010)
32. Cain, B.: A review of the mental workload literature (Report No. RTO-TR-HFM-121-Part-II). Defence Research and Development Canada Toronto, Human Systems Integration Section (2007)
33. Rai, R.K., Singh, D.K.: Stress detection through wearable EEG technology: a signal-based approach. Comput. Electr. Eng. **126**, 110478 (2025)
34. Guo, D., et al.: Assessment of flight fatigue using heart rate variability and machine learning approaches. Front. Neurosci. **19**, 1621638 (2025)
35. Beh, W.-K., Wu, Y.-H., Wu, A.-Y.: Robust PPG-based mental workload assessment system using wearable devices. IEEE J. Biomed. Health Inform. **27**(5), 2323–2334 (2023)
36. Masi, G., Amprimo, G., Ferraris, C., Priano, L.: Stress and workload assessment in aviation—a narrative review. Sensors. **23**, 3556 (2023)
37. Kim, W.D., et al.: In-flight electrocardiography monitoring in a pilot during cross country flight. Korean J. Aerosp. Environ. Med. **34**(4), 101–107 (2024)
38. Wickens, C.D., Helton, W.S., Hollands, J.G., Banbury, S.: Engineering Psychology and Human Performance, 5th edn. Routledge (2022)
39. Veltman, J.A., Gaillard, A.W.: Physiological indices of workload in a simulated flight task. Biol. Psychol. **42**(3), 323–342 (1996)
40. Zhou, Z.-B., et al.: Wearable continuous blood pressure monitoring devices based on pulse wave transit time and pulse arrival time: a review. Materials. **16**(6), 2133 (2023)

41. Hidalgo-Muñoz, A.R., Mouratille, D., Matton, N., Causse, M., Rouillard, Y., El-Yagoubi, R.: Cardiovascular correlates of emotional state, cognitive workload and time-on-task effect during a realistic flight simulation. Int. J. Psychophysiol. **128**, 62–69 (2018)
42. Ribeiro, R.T., Cunha, J.P.: A regression approach based on separability maximization for modeling a continuous-valued stress index from electrocardiogram data. Biomed. Sign. Proc. Control. **46**, 33–45 (2018)
43. Hidalgo-Muñoz, A.R., Mouratille, D., El-Yagoubi, R., Rouillard, Y., Matton, N., Causse, M.: Conscientiousness in pilots correlates with Electrodermal stability: study on simulated flights under social stress. Safety. **7**, 49 (2021)
44. Lundberg, U., et al.: Psychophysiological stress and EMG activity of the trapezius muscle. Int. J. Behav. Med. **1**(4), 354–370 (1994)
45. Henelius, A.: A short review and primer on cardiovascular signals in human-computer interaction applications (2016)
46. Romano, F., Tritto, M., Di Cesare, M.G., Nocco, S., Cardone, D.: Evaluating startle in aviation: a focus on instrumentation and measurement techniques. In: IEEE International Workshop on Technologies for Defense and Security (2024) (TechDefense)
47. Boucsein, W.: Electrodermal Activity, 2nd edn. Springer (2012)
48. Luzzani, M., Di Flumeri, G., Sciaraffa, N., Borghini, G., Babiloni, F.: Advances in wearable biosensors for mental workload and stress assessment: a review. Sensors. **23**(4), 1984 (2023)
49. Liu, Q., et al.: Infrared thermography in clinical practice: a literature review. Eur. J. Med. Res. **30**, 33 (2025)
50. Cay, G., Zhang, Y., Yuce, M.R., Li, X.: SolunumWear: a smart textile system for dynamic respiration monitoring across various postures. Science. **27**(7), 110223 (2024)
51. Hussain, T., Ullah, S., Fernández-García, R., Gil, I.: Wearable sensors for respiration monitoring: a review. Sensors. **23**, 7518 (2023)
52. Cacioppo, J.T., Tassinary, L.G., Berntson, G.G.: Handbook of Psychophysiology, 3rd edn. Cambridge University Press (2007)
53. Dehais, F., Karwowski, W., Ayaz, H., Roy, R.N.: Monitoring pilot's cognitive fatigue with engagement features in simulated and actual flight conditions using a hybrid fNIRS-EEG passive BCI. IEEE Trans. Hum. Mach. Syst. **50**(4), 375–383 (2019)
54. Querino, E., et al.: Cognitive effort and pupil dilation in controlled and automatic processes. Transl. Neurosci. **6**(1), 168–173 (2015)
55. Gateau, T., Ayaz, H., Dehais, F.: In silico vs. over the clouds: on the fly mental state estimation of aircraft pilots, using a functional near infrared spectroscopy–based passive BCI. Front. Hum. Neurosci. **12**, 187 (2018)
56. Aricò, P., Borghini, G., Di Flumeri, G., Sciaraffa, N., Colosimo, A., Babiloni, F.: Passive BCI beyond the lab: current trends and future directions. Physiol. Behav. **176**, 139–148 (2016)
57. Bittner, R.M., Begault, D.R., Christopher, B.R.: Pilot workload and speech analysis: a preliminary investigation (paper 8985). In: 135th Audio Engineering Society Convention, New York, USA (2013)
58. Brenner, M.: Speech analysis. In: Harris, D., Li, W.C. (eds.) Engineering Psychology and Cognitive Ergonomics. HCII 2024. Lecture Notes in Computer Science, vol. 14692. Springer, Cham (2024)
59. Bottalico, P., Codino, J., Cantor Cutiva, L.C., & Rubin, A. (2020). Reproducibility of Voice Parameters: the Effect of Room Acoustics and Microphones.
60. Bäckström, T.: Privacy in speech technology. arXiv preprint arXiv:2305.05227 (2023)
61. Chowdhury, R.H., Reaz, M.B.I., Ali, M.A.M.: Surface electromyography signal processing and classification techniques. Sensors. **13**(9), 12431–12466 (2013)
62. Wingenbach, T.S.H.: Facial EMG – investigating the interplay of facial muscles and emotions. In: Social and Affective Neuroscience of Everyday Human Interaction: from Theory to Methodology. Springer (2023)

63. van Boxtel, A.: EMG as a tool in psychophysiology: applications in emotion, cognition, and human–computer interaction. In: Birbaumer, N., Öhman, A. (eds.) The Structure of Emotion: Psychophysiological, Cognitive, and Clinical Aspects, pp. 103–121. Hogrefe & Huber (2001)
64. Brouwer, A.-M., Hogervorst, M., Reuderink, B., van der Werf, Y., van Erp, J.: Physiological signals distinguish between reading emotional and non-emotional sections in a novel. Brain-Comput. Interfaces. **2**(2–3), 76–89 (2015)
65. Alugubelli, N., Abuissa, H., Roka, A.: Wearable devices for remote monitoring of heart rate and heart rate variability—what we know and what is coming. Sensors. **22**(22), 8903 (2022)
66. Nechyporenko, A., et al.: Galvanic skin response and Photoplethysmography for stress recognition using machine learning and wearable sensors. Appl. Sci. **14**(24), 11997 (2024)
67. Can, Y.S., Arnrich, B., Ersoy, C.: Stress detection in daily life scenarios using smartphones and wearable sensors: a survey. J. Biomed. Inform. **92**, 103139 (2019)
68. Jerath, R., Syam, M., Ahmed, S.: The future of stress management: integration of smart-watches and HRV technology. Sensors. **23**(17), 7314 (2023)
69. Gao, W., et al.: Fully integrated wearable sensor arrays for multiplexed in situ perspiration analysis. Nature. **529**(7587), 509–514 (2016)
70. Chen, Y., et al. (2025). A wearable molecularly imprinted electrochemical sensor for cortisol stable monitoring in sweat. Biosensors, 15(3), 194.
71. Wang, Z., Chen, M., Li, Y., Liu, H., Xu, J., Zhang, P.: Wearable multiplexed microfluidic patch for simultaneous monitoring of cortisol and catecholamines in human sweat. Nature. Biomed. Eng. (2024)
72. Schnall, R., et al.: Differences in self-reported stress versus. Hair and nail cortisol among adolescent and young adult males. Nurs. Res. **17**(73), 442–449 (2024)
73. McCanlies, E.C., et al.: Burnout and diurnal cortisol among police officers. Compr. Psychoneuroendocrinology. **4** (2020)
74. Fredj, Z., Sawan, M.: Advanced nanomaterials-based electrochemical biosensors for catecholamines detection: challenges and trends. Biosensors. **13**(2), 211 (2023)
75. Tu, J., , et al.: Stressomic: a wearable microfluidic biosensor for dynamic profiling of multiple stress hormones in sweat. Sci. Adv. 11 (2025)
76. Mansour, M., Darweesh, M.S., Soltan, A.: Wearable devices for glucose monitoring: a review of state-of-the-art technologies and emerging trends. Alex. Eng. J. **89**, 224–243 (2024)
77. Wu, Z., et al.: Interstitial fluid-based wearable biosensors for minimally invasive healthcare and biomedical applications. Commun. Mater. (2024)
78. Kim, G., Ahn, H., Chaj Ulloa, J., Gao, W.: Microneedle sensors for dermal interstitial fluid analysis. Med-X. **2** (2024)
79. Min, S., Geng, H., He, Y., Xu, T., Liu, Q., Zhang, X.: Minimally and non-invasive glucose monitoring: the road toward commercialization. Sens. Diagn. **4**, 370–396 (2025)
80. Vedantam, D., Poman, D.S., Motwani, L., Asif, N., Patel, A., Anne, K.K.: Stress Induced Hyperglycemia Consequences Manage. Cureus. **14**(7), e26714 (2022)
81. Vulpe, G., et al.: Lab on skin: real-time metabolite monitoring with polyphenol film-based subdermal wearable patches. Lab Chip. (2024)
82. Pinge, A., Gad, V., Jaisinghani, D., Ghosh, S., Sen, S.: Detection and monitoring of stress using wearables: a systematic review. Front. Comp. Sci. **6** (2024)
83. Kapogianni, N.-A., Sideraki, A., Anagnostopoulos, C.-N.: Using smartwatches in stress management, mental health, and Well-being: a systematic review. Algorithms. **18**(7), 419 (2025)
84. Xiang, J.Z., Wang, Q.Y., Fang, Z.B., Esquivel, J.A., Su, Z.X.: A multi-modal deep learning approach for stress detection using physiological signals: integrating time and frequency domain features. Front. Physiol. **16**, 1584299 (2025)

85. Helmreich, R.L., Merritt, A.C.: Culture at Work in Aviation and Medicine: National. Ashgate, Organizational and Professional Influences (1998)
86. Flin, R., O'Connor, P., Crichton, M.: Safety at the Sharp End: a Guide to Non-Technical Skills. CRC Press (2008)

Loss Capture Ratio: A Metric to Measure Basis Risk in Parametric Insurance Transactions

Guillermo Franco$^{(\boxtimes)}$[iD] and Roberto Guidotti[iD]

Guy Carpenter and Company, LLC, 1166 Avenue of the Americas,
New York, NY 10036, USA
{guillermo.e.franco,roberto.guidotti}@guycarp.com

Abstract. Parametric insurance is attracting interest as a device to expand coverage for catastrophic events. These policies pay based on a few measurable characteristics (parameters), rather than on the loss incurred. Since these characteristics are usually obtained quickly through a public and reputable technical agency, parametric mechanisms offer more transparency and speed than traditional insurance products. However, these positive features come at the cost of precision. Since there is no adjustment of the policy in the classical sense, the payout produced is commonly calculated using algorithms, models, or equations that aim to approximate the loss suffered by the insured. The insurance industry refers to this potential inaccuracy in recoveries as "basis risk," intuitively understood as the difference between the payout that one would expect from a traditional indemnity policy and the payout actually received from the parametric coverage. In this paper, we offer a formal description of basis risk using the Loss Capture Ratio (LCR), a metric that we have used successfully at Guy Carpenter to illustrate the performance of the parametric solutions we design for our clients.

Keywords: Parametric insurance · Basis risk · Catastrophe risk

1 Introduction

Parametric insurance, a type of insurance that was for many years restricted to the toolkits of large corporations and governments, is quickly becoming a presence in the retail commercial and homeowners' markets. This type of coverage bypasses the need to assess the loss of the insured, instead disbursing a payout that is determined with an algorithm and which depends on a set of factors, or parameters, that describe the event. Popular parametric methods for tropical cyclone risk transfer, for instance, consist of paying the insured a predetermined amount if a hurricane crosses a certain geography with a particular intensity [9]. Methods to design parametric solutions for earthquake risk transfer often involve measuring or estimating ground motions [11,15] or earthquake magnitudes and epicentral locations [3,12].

M. Pavone et al. (Eds.): DSA ISC 2025, LNCS 16405, pp. 169–182, 2026.
https://doi.org/10.1007/978-3-032-21811-7_12

Part of the allure of these solutions arises from new technological capabilities to remotely monitor, in near-real time, the fundamental characteristics of disasters with the usage of global sensor networks and satellite systems. Parametric policies that leverage these technologies tend to pay very fast, enabling insureds to limit their loss through quick action protocols and to access cash to fund the immediate emergency response.

Payouts are determined algorithmically with more transparency than loss adjustment protocols used in traditional indemnity insurance, which sometimes are slow and frustrating for the insured.

The problem that often plagues parametric insurance, though, is the "basis risk" that is carried by the purchaser of such products. Basis risk refers to the difference between a parametric payout and that which would be provided by an idealized indemnity policy. A simple way to understand basis risk is to picture the following scenario: A homeowner in Florida purchases a parametric policy that pays based on the category of the hurricane as it traverses a certain geography. A hurricane ends up affecting the coastline near the insured, but it does not come close enough to enter the geography defined in the contract. The payout of the policy is, therefore, zero. The hurricane did not produce any wind-related damages, but it did cause a significant storm surge that flooded the entire coast and produced major destruction at the insured's location. The homeowner is disappointed and left without the coverage that she assumed she had.

Basis risk can often be managed. In the previous example, for instance, one could argue that the policy was not well constructed. In addition to including a condition on the proximity of the hurricane, maybe it would have been helpful to add another payout condition that considered the level of flooding on the coast.

The design of parametric solutions in today's insurance market is often left to the particular company that commercializes the products. This causes great variability in the types of products that are available to clients, who often have difficulties understanding their differences. As parametric insurance becomes more readily available and the need grows for insureds to evaluate alternatives objectively, the development of standard tests of performance becomes a dire necessity.

In this paper, we present one such test that we, at Guy Carpenter, have been using as an internal, standard, and objective method to describe one important facet of the performance of parametric solutions, namely the degree to which the parametric solution is able, in average, to mimic a traditional indemnity transaction.

The following section, Sect. 2, addresses some preliminary considerations on the data and models that are common currency in the industry to design and evaluate these types of insurance products. We then describe the nomenclature of an insurance "layer" in Sect. 3, with the aim of defining the target against which the parametric solution will be evaluated. Section 4 formally introduces the metric we propose, and which we call the Loss Capture Ratio (LCR). Section 5 offers two illustrations of how we have used this test in the context of recent market transactions in China and India.

2 Data and Modeling Considerations

Actuaries at insurance companies use long historical records of loss data to determine the appropriate premiums of homeowners' policies. These data often include events such as fires, water leaks, or theft. Earthquakes or hurricanes that cause large losses, however, do not occur often enough to build a strong historical dataset from which to derive statistics. Instead, insurance companies use catastrophe risk models that leverage Monte Carlo simulation. This method consists of sampling the characteristics of catastrophes from probability distributions and combining them to produce synthetic but plausible scenarios [13]. Good quality models use up-to-date, sound scientific insights to define the probability distributions of hazards, building response, and their appropriate correlations and functional dependencies.

Catastrophes are complex phenomena. A particular feature such as the dominant period of vibration or the orientation of the ground motion in an earthquake may produce unremarkable consequences in most situations, but disastrous in some others. Therefore, it is important that the Monte Carlo implementation produces enough simulated scenarios that encompass a wide array of combinations of characteristics representative of a broad spectrum of outcomes.

The argument for large samples of synthetic events, or "catalogs," is strong. If a particular model contains the equivalent of fewer than 1,000 annual realizations of simulated seismic activity, the dataset might not be sufficient to properly represent the often-long tail of parameter distributions. It is common in the industry to find "10k year" catalogs that include 10,000 annual realizations of seismic or hurricane activity, but even those are quickly becoming outdated. As computational resources have improved, the demand has increased for larger catalog sizes of 100k, or even 1,000k years.

For our purposes, let us assume that a catalog of synthetic events is available and that it is sufficiently complete to accurately represent the risk faced by a portfolio of assets. We will denote the simulated events with the subscript $i = 1, 2, ..., N$, where N is the total number of events included in the dataset.

We will denote the simulated loss to the assets caused by each event i as L'_i. Note that we use the apostrophe to recall that these losses are a numerical approximation to the actual loss, L_i, that these events might cause in reality. If the catastrophe model is of good quality, then we can assume that $L'_i \approx L_i$.

3 Definition of a Risk Transfer Layer

Risk is most often transferred in "tranches" or layers, which precisely define the specific amounts of loss an insurance policy is responsible for covering. A layer is defined by two parameters, its "attachment" and its "exhaustion":

- Attachment (L_A): The minimum amount of loss necessary for the insurance coverage to respond. In a traditional homeowner's policy, this amount is often referred to as a "deductible." It is the responsibility of the insured to pay this portion of the claim before the insurer issues a payout.

– Exhaustion (L_E): The amount of loss, $L_E > L_A$, at which the responsibility of the insurer ceases. Losses in excess of this amount are born by the insured.

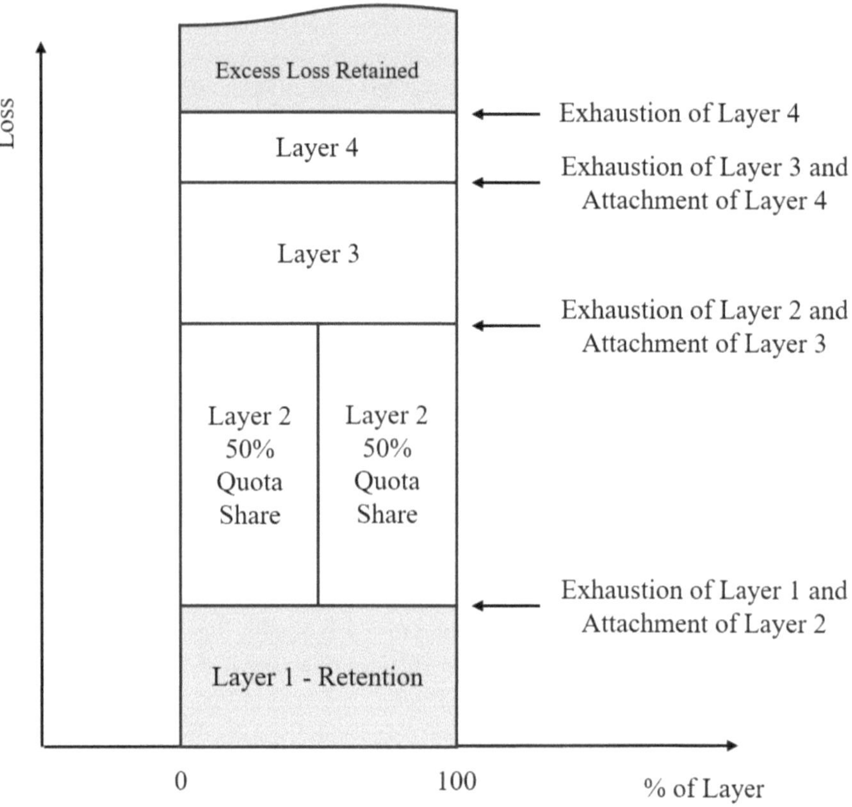

Fig. 1. This is an illustration of a reinsurance arrangement, in which the risk of the client is divided into portions. The first layer, from zero loss to the first exhaustion point represents the risk retained by the client, which is not insured. At that point, two reinsurers split the coverage of layer 2 in equal shares of 50%. At the exhaustion of layer 2, the coverage of layer 3 commences, which may be provided by a different reinsurer. When this coverage is exhausted, a fourth layer on top of the "tower" captures a portion of catastrophic (very high) losses. After all formal coverage is exhausted, the client is again responsible for retaining the remainder of the risk on top

Figure 1 illustrates the arrangement of several reinsurance layers for an insurer. This type of coverage structure allows the insured to spread the risk among participants with different risk appetites. Some markets might like to cover high probability, low loss layers that they can sell in exchange for a high annual premium. Others, more risk averse, might have appetite to only participate in the coverage of catastrophic layers, which have a very low probability of loss but also command a smaller premium.

The loss to an insured layer can be defined as:

$$\hat{L}_i = max(min(L_i, L_E) - L_A, 0) \tag{1}$$

As mentioned above, the actual loss of an event, L_i, is often replaced by the simulated loss, L'_i, because historical data is rarely available, and we need to resort to synthetic events. In that case, Eq. 1 can be reformulated as:

$$\hat{L}'_i = max(min(L'_i, L_E) - L_A, 0) \tag{2}$$

Although not strictly necessary for the discussion that follows, it is useful to be acquainted with these terms that often complement the previous definitions:

- Layer Limit: The maximum amount the insurer will pay for this particular coverage layer, or $L_E - L_A$.
- Retention: The total amount of loss that is born by the insured. Considering one single coverage layer, it includes any loss below the attachment point $min(L_i, L_A)$ plus any loss above the exhaustion point $max(L_i - L_E, 0)$.

4   Definition of the Loss Capture Ratio (LCR)

Parametric reinsurance is often designed with the purpose of "fitting" into an existing traditional indemnity insurance structure. Typically, this is expressed by designating a layer, with its attachment and exhaustion levels, in which the parametric solution needs to respond as if it were a traditional indemnity coverage. If we denote the loss of the parametric policy to the particular desired layer as $\hat{L}''_i$, then it is our objective to approximate $\hat{L}''_i \approx \hat{L}'_i$. Recall that the actual loss to the layer, $\hat{L}_i$, is impossible to know a priori. Therefore, our best compromise is to aim to approximate the simulated loss to the layer $\hat{L}'_i$.

Parametric models are not perfect, even if calibrated closely to modeled losses L'_i, and it is likely that the parametric loss will either under or overestimate the modeled loss. If the parametric loss to the layer, $\hat{L}''_i$, is smaller than the modeled loss, $\hat{L}'_i$, for a particular event i, we face an undesirable situation in which the insured recovers less than the amount the model indicates they would recover from a traditional indemnity policy.

As mentioned above, the discrepancy between the parametric payout and the actual loss is referred to as basis risk. Considering the losses to the layer, this error is defined as $\hat{L}_i - \hat{L}''_i$. Since the actual loss $\hat{L}_i$ is not known, we revert to approximating basis risk as $\hat{L}'_i - \hat{L}''_i$. If $\hat{L}'_i$ is greater than $\hat{L}''_i$, we refer to the ensuing basis risk as "negative," and, if large, it is considered highly detrimental to the transaction. If, on the other hand, the parametric loss to the layer, $\hat{L}''_i$, is greater than the modeled loss to the layer, $\hat{L}'_i$, we face a "positive" basis risk instance, in which the insured could potentially recover more than they would have under a traditional indemnity policy. Although this scenario may seem beneficial to the insured, in practice, it is not quite so. Over-recoveries are not permitted in insurance regulatory environments, as insurance products are not meant to generate a "net profit" for the insured, but rather to "restore" the insured to the situation prior to the loss. Therefore, even if from a numerical standpoint the parametric loss were greater than the actual loss, the insurer would not be able to pay the positive differential, and the insured would just recover the minimum of $\hat{L}''_i$ and $\hat{L}'_i$. Recovering more proceeds than necessary also potentially suggests that the insured overpaid their premium, which is undesirable for the insured. Therefore, parametric recoveries should be designed to approximate the modeled losses within reason.

We can now define the Loss Capture Ratio in a manner that reflects the above scenarios. The metric is based on a simple ratio, where the numerator is the minimum of the loss incurred by the insured or the expected parametric payout, and the denominator is the loss incurred. If the parametric payout is equal or larger than the loss incurred, then the parametric solution performs well, with a ratio of 1. Note that the parametric policy does not receive recognition for "positive" basis risk credits with this definition. Conversely, if the parametric payout is smaller than the incurred loss, the ratio decreases. Considering all events $i = 1, 2, ..., N$ in the synthetic catalog, the LCR takes the form of a ratio of sumproducts of all the individual recoveries and losses for all events in the set multiplied by their respective annual probability of occurrence, r_i.

$$LCR = \frac{\sum min(\hat{L}''_i, \hat{L}'_i)r_i}{\sum \hat{L}'_i r_i} \tag{3}$$

Using this well-defined, albeit simple, metric to calculate basis risk, allows the designer of parametric solutions to identify the realm in which the policy is expected to perform at its best, and even to guide the insured as to the optimal allocation of the protection within an existing insurance program.

5   Application Examples

We will illustrate the usage of the LCR metric in two cases, one for China and one for India, that involve a type of parametric solution known as "cat-in-a-grid" ("cat" stands for "catastrophe"). Prior to discussing the applications in detail, let us briefly describe these types of parametric triggers.

5.1   Cat-in-a-Grid Parametric Earthquake Transactions

This type of parametric construct involves the geographical discretization of the area of interest using a regular grid. A function or a table is assigned to each grid cell to describe the payout that corresponds to each level of hazard intensity.

In the particular examples below, we use the magnitude of the earthquake, measured in moment magnitude M_W, as the hazard intensity, and the payout is defined depending on the grid cell that contains the focus of the earthquake and on its magnitude. For example, a magnitude $M_W 6.0$ event in a particular grid cell may result in a 25% limit payout for the insured, whereas a $M_W 7.0$, much larger, earthquake within the same grid cell may cause a 100% limit payout. An earthquake located within a grid cell very far away from the exposures may require a much larger magnitude to produce a payout, because the energy of the earthquake dissipates as it travels long distances.

We have studied these types of solutions over the last fifteen years, documenting our findings and methods in several publications that may be useful to understand the transactions in more detail. The fundamental definitions are presented in, for instance, [3,4]. In addition to many applications in industry [5], several illustrations for applications in California, Greece, Romania, and Turkey, among others, have been documented in the public literature [6,8,10,14]. Methods to optimize these solutions so that they faithfully and efficiently represent the risk of the insured have been described in [1,2].

We use two features to describe these cat-in-a-grid parametric earthquake solutions: the "attachment return period" and the "level of smoothing."

Attachment Return Period. Insurers are especially interested in calculating the probability with which they have to disburse payouts to the insured. This probability is often referred to as the "attachment return period," because it is typically expressed in the form of a time span in which the insurer would have to pay, in average, one single time. For instance, an annual probability of payout of 1% would be expressed as a 100 year return period (the inverse of 1%), because it is expected to happen, in average, once every 100 years. A parametric transaction that requires a $M_W 7.0$ event to produce a payout, for example, will carry a smaller probability of attachment than another transaction that requires a $M_W 6.0$ event. Smaller events occur more frequently than larger events, their probability is higher, and their return period is, therefore, smaller. Low return periods are often associated with higher premiums, because the disbursement from the insurer's perspective is expected to happen more often. Transactions with lower attachment return periods, however, tend to offer more comprehensive coverage for the insured, because they are more likely to pay for smaller events. The choice of the attachment return period typically involves a compromise between obtaining as much coverage as possible while constraining the cost of the insurance to a given budget.

Level of Smoothing. Large swings in payment conditions across small distances are generally regarded as a sign of overfitting the parametric solution to the loss model. The insured and the insurer intuitively expect two events of similar magnitude that are not too distant from one other to produce similar payout responses. Sparse catalogs in catastrophe models, however, may lack enough samples to produce a nuanced variation of payout conditions. In those cases, some special techniques, such as using cross-validation or imposing curvature constraints during the calibration phase might be useful. The resulting designs are sometimes judged by the "level of smoothing," which we typically equate to the maximum or mean curvatures of the payout functions across space [7]. Flatter payout functions produce more uniform payouts across large distances, whereas high-curvature payout functions can produce highly volatile payout responses, even within short distances. The level of smoothing of a parametric solution should aim to balance the need to reproduce model results with the need to control volatility in its response.

5.2 Parametric Earthquake China Transaction

In order to illustrate some of the concepts above, let us consider the following parametric solution, designed to protect an insured's portfolio from earthquake losses in China. As part of this process, it is customary to build several prototypes that offer the insured different features. Figure 2 shows two such prototypes. The solution at the top of the figure features a return period of 10 years, and it is, therefore, designed to produce a payout to the insured, in average, every 10 years. This is a rather low threshold that may be of interest to insureds with very vulnerable assets at risk that expect to experience impacts with a relatively high recurrence. The solution at the bottom of the figure features a return period of 200 years, therefore offering the insured a payout with a much lower frequency. This is a rather high threshold that may be of interest to insureds with very resilient assets or with a low budget to pay for the coverage.

To illustrate how these solutions may fit optimally in a traditional insurance structure, we assume a set limit $(L_E - L_A)$, and we monitor the LCR as a function of the attachment level (L_A). The resulting LCR curves for three prototypes with return periods of 25, 40, and 50 years are shown in Fig. 3.

First, note that the maximum LCR value for the 25 year solution is attained at an attachment return period of 35 years. This discrepancy arises from the existence of basis risk. Were the parametric structure perfect in mimicking the modeled losses, the return period of the parametric solution would be equal to the attachment return period of the indemnity layer at a 100% LCR. In other words, the match would be perfect between the parametric and indemnity coverages. Since there is some basis risk (error), the 25 year parametric trigger completely substitutes the indemnity layer that attaches at a return period of about 35 years. This difference of 10 years, which carries an increase in premium, can be interpreted as the "additional cost" paid by the insured for using a parametric coverage that is imperfect. This effect is similar, although a bit more subdued, as the parametric and indemnity layers are positioned higher up in the tower, towards more catastrophic scenarios.

It tends to be true in most scenarios that parametric solutions have less error the more catastrophic the indemnity layer targeted is. At the top of the insurance tower, the truly catastrophic scenarios are few in number in the Monte Carlo sample contained in models, and they are usually easier to isolate with a parametric construct.

In this case, the discussion with the client would first revolve around budget available to determine what parametric return period they can afford. Then, assuming that their finances allow them to use the most expensive 25 year return

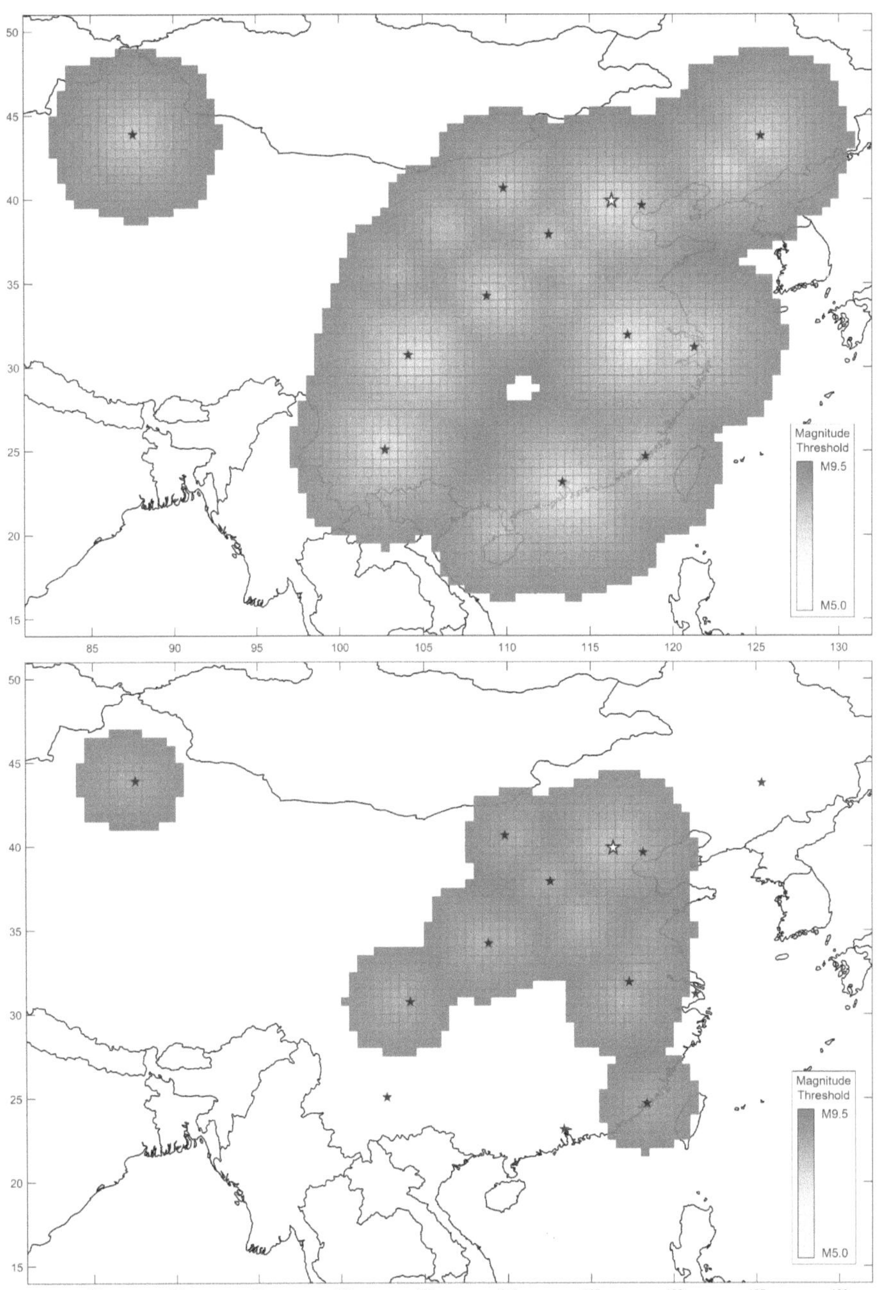

Fig. 2. Examples of cat-in-a-grid parametric solutions with different return periods designed to cover a vast portfolio of assets across China. Top: Expansive coverage which is very likely to produce a payout with a return period of 10 years; Bottom: Much more restricted coverage, which reduces the payout probability such that the return period increases to 200 years. Stars identify the location of the main cities where assets are concentrated

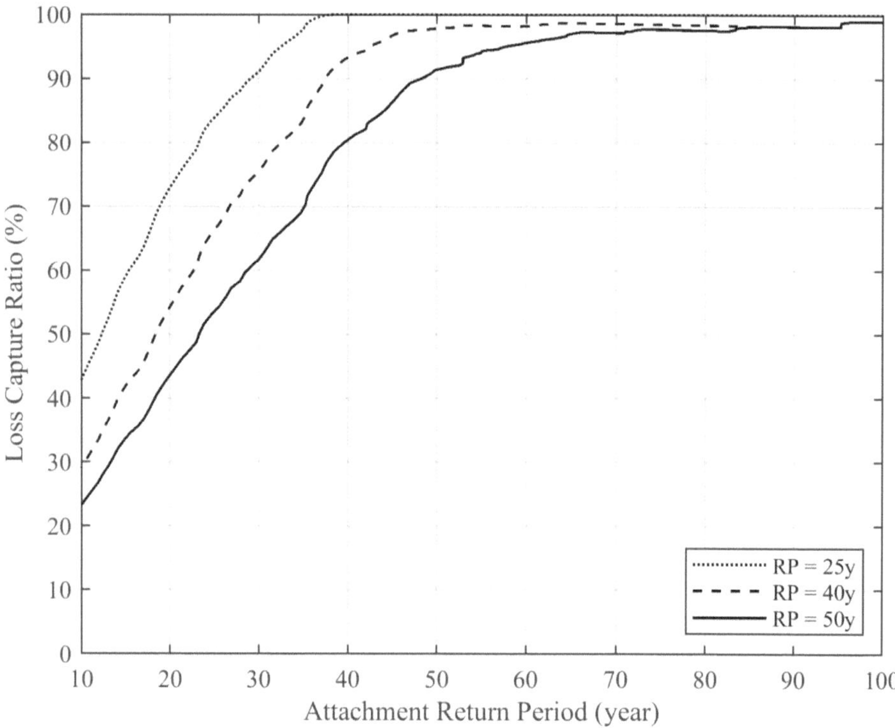

Fig. 3. Loss Capture Ratio (LCR) for different values of attachment of the indemnity layer targeted. The parametric solutions are designed to cover a vast portfolio of assets across China with different trigger return periods, which in turn correspond to different premium budgets

period trigger, we would recommend that this instrument is used to provide coverage somewhere above the 30 year attachment return period mark and below the 40 year return period mark. Assuming higher attachment return periods would be inefficient as they would be paying a high cost to offset far less likely losses. If their choice were to reduce cost and use the 50 year return period parametric, we would recommend to locate the target layer attaching at 50 years or higher, where the LCR is 90% or greater.

5.3 Parametric Earthquake India Transaction

In this second example, we consider a parametric solution that was designed to protect an insured's portfolio from earthquake losses in India. Several prototypes built during the design phase showed different levels of performance, including different approaches to the smoothing of the solutions. For instance, Fig. 4 shows two types of solutions, both calibrated to the same model results but one (top) imposing a very lax constraint on curvature, and another one (bottom) with a stricter constraint on curvature. The first solution is calibrated

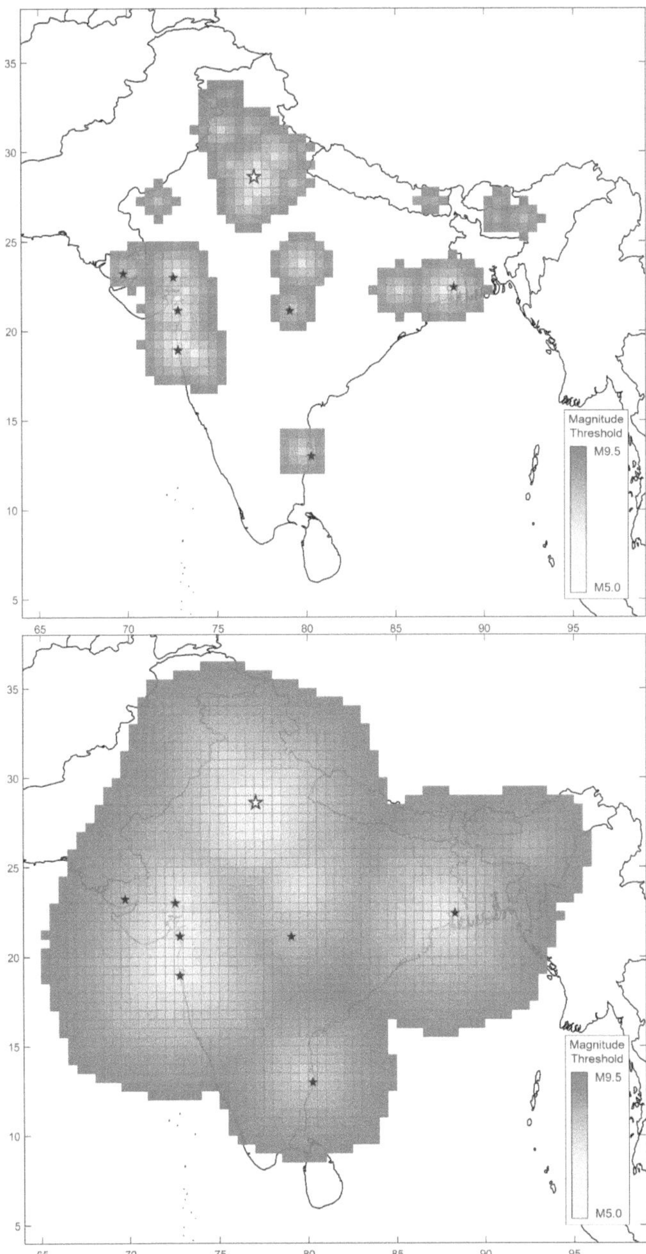

Fig. 4. Examples of cat-in-a-grid parametric solutions for India with the same return period of 10 years, calculated to display different smoothing levels, for India. Top: Low level smoothing or high curvature solution tightly fitted to the model results; Bottom: High level of smoothing with a decreased curvature and displaying a more nuanced variation of trigger conditions across the geography. Stars identify the location of the main cities where exposures are concentrated

very tightly to the model results and the variability of payment conditions across space, even in just a few grid cells, is very high. The second solution was calculated imposing a much lower maximum curvature, which reverts in much smoother transitions of payment conditions across space.

The second, smoother, solution is more resilient to variability in the parameters that drive the payouts, but imposing a higher curvature means the insured is purchasing coverage in regions where maybe the model contains events with very little loss. In other words, the insured might be over-purchasing coverage, based on the model, in order to decrease the potential variability of payout conditions. It is important for the insured to see these trade-offs reflected in a clear graph that guides them in the selection of the most appropriate coverage.

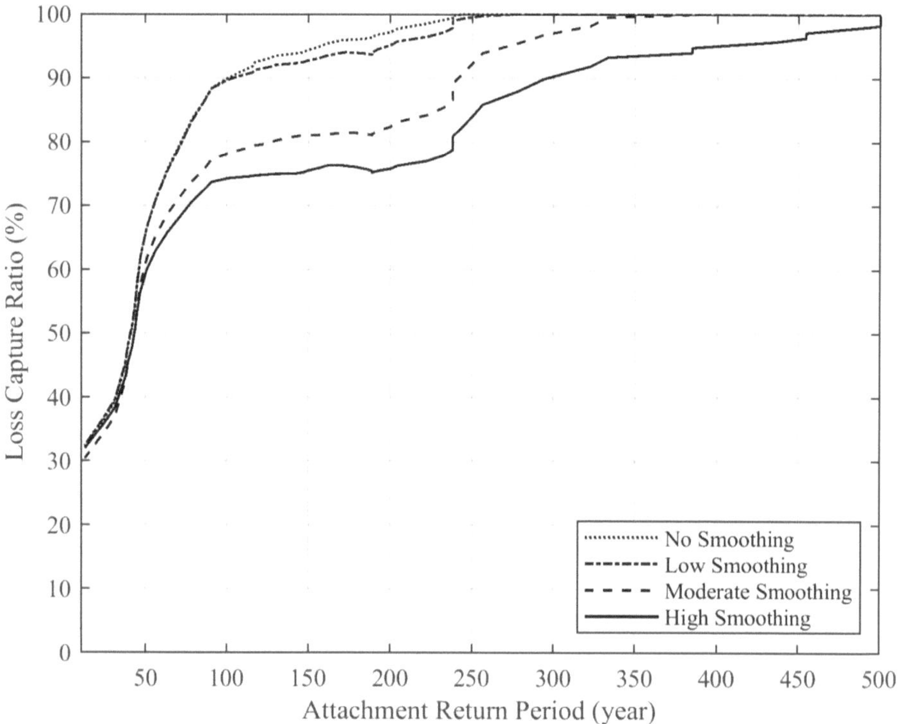

Fig. 5. Loss Capture Ratio for different values of attachment return period for India, considering four parametric solutions with the same return period, but with different levels of smoothing

The LCR plot of Fig. 5 includes four curves, computed for parametric cat-in-a-grid solutions with the same trigger return period but corresponding to different levels of smoothing. The choice for the client, in this case, is not about budget, since all solutions trigger with the same probability and, therefore, cost the same amount of premium. Rather, the choice consists of selecting a level of smoothing that avoids overfitting but that also provides an efficient risk transfer.

Consider, for instance, the curve with no smoothing, which represents a parametric solution that has been designed closely aligned to the model but without regard for the potential overfitting of the simulated losses. Its performance is good from an attachment return period perspective. At a 100 year attachment return period, the solution shows a relatively high LCR of about 90%. On the opposite side of the spectrum, the curve representing a high level of smoothing sits below all the others. Its associated LCR is lower for all attachment return periods considered. The curve attains an LCR value of 90% at 300 years of attachment of the indemnity layer. The efficiency of this solution from that perspective is low.

The LCR plot clearly informs us about one aspect of the performance of the solution, its alignment with the indemnity layer targeted. However, it is insufficient to portray the advantages that a high level of smoothing implies for the client. Indeed, we would need a complementary test that shows the reduced variability of the payout for highly smoothed solutions. In a situation like this, we would recommend caution choosing a solution that provides a moderate performance and a moderate level of smoothing, such as the second curve from the bottom. While the efficiency of the solution is not high, with an associated attachment return period of about 240 years, it is a less volatile solution than those with lower levels of smoothing. Other exhibits have to illustrate the consequences, outside of this plot, of choosing a parametric solution that is highly sensitive to the parameters of the event. Perhaps a cross-validation exercise, using only a portion of the event set to produce the LCR curves would be a solid methodology to improve the information provided by this type of analysis.

6   Conclusions

The Loss Capture Ratio (LCR) is a simple metric to quickly shed light upon the performance of a parametric solution that is designed with the purpose of responding to losses within a particular coverage layer. It is not a comprehensive evaluation of the parametric coverage but an inspection of one of its most important traits. The LCR curve, which summarizes the LCR values for different attachment points of a fixed-limit indemnity layer, helps the insured optimize the structuring of the parametric coverage, especially if this solution has to be fitted to complement other layers of indemnity coverage. In our experience, the use of this type of exhibit in client briefings in recent market transactions has been helpful to communicate the value and the limitations of these products, which in turn gives more confidence to the purchaser as they navigate the complex selection of their appropriate coverage.

Disclosure of Interests. The authors have no competing interests to declare that are relevant to the content of this article.

References

1. de Armas, J., Calvet, L., Franco, G., Lopeman, M., Juan, Á.A.: Minimizing trigger error in parametric earthquake catastrophe bonds via statistical approaches. In: Modeling and Simulation in Engineering, Economics and Management: International Conference, MS 2016, Teruel, Spain, July 4-5, 2016, Proceedings, pp. 167–175. Springer (2016)
2. Bayliss, C., Guidotti, R., Estrada-Moreno, A., Franco, G., Juan, A.A.: A biased-randomized algorithm for optimizing efficiency in parametric earthquake (re) insurance solutions. Comput. Oper. Res. **123**, 105033 (2020)
3. Franco, G.: Minimization of trigger error in cat-in-a-box parametric earthquake catastrophe bonds with an application to Costa Rica. Earthq. Spectra **26**, 983–998 (2010)
4. Franco, G.: Construction of customized payment tables for cat-in-a-box earthquake triggers as a basis risk reduction device. In: Proceedings of the International Conference on Structural Safety and Reliability (ICOSSAR), pp. 5455–5462 (2013)
5. Franco, G.: Earthquake mitigation strategies through insurance. Encyclopedia of Earthquake Engineering, pp. 1–18 (2014)
6. Franco, G., Guidotti, R.: Addressing focal depth in cat-in-a-box parametric earthquake transactions: The Vrancea seismic zone in Romania as a case study. In: 3rd European Conference on Earthquake Engineering and Seismology (2022)
7. Franco, G., Guidotti, R., Bayliss, C., Estrada, A., Juan, A., Pomonis, A.: Earthquake financial protection for Greece: a parametric insurance cover prototype. In: ICONHIC2019 2nd International Conference on Natural Hazards & Infrastructure (2019)
8. Franco, G., Guidotti, R., Field, E., Milner, K., Lee, Y., Stein, R.: An exploration of parametric earthquake risk transfer solutions that dynamically adapt to seismicity changes. In: Proceedings of the 17th World Conference on Earthquake Engineering (2020)
9. Franco, G., et al.: Typology and design of parametric cat-in-a-box and cat-in-a-grid triggers for tropical cyclone risk transfer. Mathematics **12**(11), 1768 (2024)
10. Franco, G., Tirabassi, G., Lopeman, M., Wald, D.J., Siembieda, W.: Increasing earthquake insurance coverage in California via parametric hedges. In: Eleventh National Conference on Earthquake Engineering (2018)
11. Goda, K.: Basis risk of earthquake catastrophe bond trigger using scenario-based versus station intensity-based approaches: a case study for southwestern British Columbia. Earthq. Spectra **29**(3), 757–775 (2013). https://doi.org/10.1193/1.4000164
12. Goda, K., Franco, G., Song, J., Radu, A.: Parametric catastrophe bonds for tsunamis: cat-in-a-box trigger and intensity-based index trigger methods. Earthq. Spectra **35**(1), 113–136 (2019). https://doi.org/10.1193/030918EQS052M
13. Grossi, P., Kunreuther, H., Windeler, D.: An introduction to catastrophe models and insurance. In: Catastrophe Modeling: A New Approach to Managing Risk, pp. 23–42. Springer (2005)
14. Guidotti, R., Franco, G., Tsatsis, A., Kourkoulis, R., Gelagoti, F.: Design and implementation of a parametric insurance cover to mitigate earthquake risk of infrastructure systems. In: ICONHIC2022 3rd International Conference on Natural Hazards & Infrastructure (2022)
15. Wald, D.J., Franco, G.: Financial decision-making based on near-real-time earthquake information. In: 16th World Conference on Earthquake Engineering, pp. 1–13. Santiago de Chile, Chile (2017)

Retrieval Augmented Generation with Iterative Critique: A Framework for Structured Task Solving

Alessio Mezzina[1,2]([📧]) [iD], Luca Naso[2] [iD], and Mario Pavone[1,3] [iD]

[1] Department of Mathematics and Computer Science, University of Catania, Viale Andrea Doria 6, 95125 Catania, Italy
alessio.mezzina@phd.unict.it, mpavone@dmi.unict.it, info@ANTs-lab.it
[2] Koexai S.R.L., Via Jose Maria Escriva 6, 95125 Catania, Italy
{alessio.mezzina,luca}@koexai.com
[3] ANTs Lab: Advanced New Technologies Research Laboratory, Via Rigolato 51, 95024 Acireale, (CT), Italy

Abstract. Large Language Models (LLMs) are increasingly employed in structured problem-solving, yet their unreliability in producing accurate, verifiable outputs limits their applicability in mathematically grounded domains such as decision science. In this work, we introduce a general framework inspired by self-refinement techniques, which combines Retrieval-Augmented Generation (RAG) with an iterative critique mechanism to enhance the correctness and consistency of generated solutions to decision-theoretic problems. Using four openly available LLMs, DeepSeek-R1, DeepSeek-V3, DeepSeek-R1T2 Chimera, and OpenRouter's Horizon-Beta, we evaluate our framework on core tasks in decision science. Our methodology involves three comparative settings: (*i*) a zero-shot baseline, (*ii*) a RAG-enhanced generator with task-specific retrieval, and (*iii*) the full pipeline loop, where the model iteratively evaluates and refines its own answers. Input–output formatting challenges using prompt engineering techniques, such as chain-of-thought reasoning, have also been addressed to guide the model through intermediate steps. Inspecting the outcomes, RAG–CRITIC reduces factual errors by up to 21 percentage points and increases precision by up to 13 times over the baseline. This work highlights the potential of loop-based reasoning architectures for trustworthy decision support and consolidates the foundations for future research in LLM-assisted judgment under constraints of rationality and optimality.

Keywords: Game Theory · Decision Science · Generative AI · Large Language Models (LLMs) · Retrieval Augmented Generation

1 Introduction

Large Language Models (LLMs) have demonstrated remarkable capabilities across a wide range of natural language processing tasks, including text generation, question answering, and reasoning. Their ability to generate fluent and

© The Author(s), under exclusive license to Springer Nature Switzerland AG 2026
M. Pavone et al. (Eds.): DSA ISC 2025, LNCS 16405, pp. 183–196, 2026.
https://doi.org/10.1007/978-3-032-21811-7_13

contextually relevant output has raised the question of how they behave in structured problem-solving domains such as decision science [3], game theory [2], and combinatorial optimization [21]. However, despite their impressive language fluency, LLMs often struggle to produce accurate, verifiable, and logically consistent solutions when addressed with mathematically grounded tasks or also when tackled with unreasonable math problems (UMP), such as those proposed in UMP benchmark [8]. This limitation becomes particularly critical in high-stakes domains, where correctness and interpretability are the main concern.

One of the key challenges underlying this issue is the tendency of LLMs to prioritize plausibility and fluency rather than factual accuracy, which can result in wrong conclusions [5]. This limitation is more pronunced in scenarios with sparse or ambiguous data, where the model lacks sufficient grounding to generate reliable output. To address these shortcomings, recent research has explored the addition of LLMs with external knowledge retrieval mechanisms, known as *Retrieval-Augmented Generation* (RAG) [22], first proposed in [7], which help provide a relevant and up-to-date context that can improve factual correctness. Additionally, iterative self-refinement techniques, in which models revise their own results, have shown promise in enhancing the consistency and reliability of generated content [9,14].

In this research work, a framework, termed *RAG-CRITIC* (RC), is proposed that combines retrieval-augmented generation with an iterative critique loop to improve LLM performance on structured decision-theoretic tasks. The proposed framework leverages task-specific retrieval to provide the model with pertinent domain knowledge, while an internal critic module evaluates the generated solutions and guides successive refinements. This iterative process aims to reduce factual errors and improve logical coherence without sacrificing the fluency of the responses. To demonstrate the backbone-agnostic nature of *RC*, we benchmarked it across four openly available LLMs served via OpenRouter on a suite of canonical decision-science tasks. The most pronounced gains were observed in the Nash-equilibrium (Sect. 4.1) and the Value of Perfect Information (VOPI) (also known as EVPI) (Sect. 4.2) benchmarks. Benchmark datasets have been designed to facilitate rigorous evaluation, with ground truth annotations and structured JSON outputs. Furthermore, three settings were compared: zero-shot generation, RAG-enhanced generation, and the full RC iterative pipeline.

The goals of this research work are twofold: (*i*) verify that an iterative self-refinement loop measurably improves performance; and (*ii*) demonstrate that LLMs can be effectively deployed to assist humans in scenarios where decisions must be made. The outcomes of the investigation presented show that RC significantly improves the accuracy and consistency of the solution, underscoring the promise of loop-based reasoning architectures for trusted AI-assisted decision support. Importantly, RC is not presented as a one-off codebase but as a reproducible, model-agnostic framework whose main strength is its modular design.

2   Related Work

Large Language Models store vast amounts of implicit knowledge in their parameters, but their factual accuracy makes less when the training distribution diverges from the query context. Retrieval-Augmented Generation (RAG) mitigates this weakness by coupling a parametric generator with a non-parametric memory that is queried on-the-fly.

The original RAG paper [7] jointly fine-tunes retriever and generator and evaluates on knowledge-intensive tasks, achieving state-of-the-art on open-domain QA (Natural Questions, WebQuestions, CuratedTrec) and strong results on TriviaQA; it also reports gains on knowledge-intensive generation (MS MARCO NLG, Jeopardy QG) and competitive fact verification on FEVER. Subsequent work has advanced scaling to longer and more structured contexts (e.g., hierarchical/tree retrieval) [12,23], explored latency-aware hybrid retrieval for real-time assistance [17], and produced survey-level syntheses that map common architectural patterns [4].

Several studies investigate RAG for mathematical and symbolic reasoning, showing that grounding the model with declarative facts (e.g., definitions, lemmas, or theorems) reduces hallucinations, but does not, by itself, enforce logical consistency [18,19]. Complementary to retrieval, iterative refinement treats the LLM as both *"solver"* and *"critic"*. Early work on "self-consistency" [15] proposed sampling multiple chain-of-thought (CoT) traces and selecting the modal answer.

Notable works are proposed in [20] that combines reasoning traces with tool invocation in ReAct, interleaving thoughts and actions to gather missing evidence; and in [9] that introduces *Self-Refine*, where the model writes, critiques, and rewrites its own output in a loop until a stopping criterion is met; later extensions, e.g., in [10], explore ensembles of critics to curb both factual and toxic errors. Compared with debate-style or tool-calling agents, critic loops are lightweight (single-model) and therefore attractive for plug-and-play integration into existing RAG stacks; however, their efficacy on *structured* tasks remains under-explored.

LLMs have recently been evaluated on domains that require formal correctness. In decision science, Eigner *et al.* studied how model explanations, uncertainty displays, and user expertise modulate human reliance on LLM suggestions [3] For game-theoretic reasoning, NeurIPS 2024 papers such as GTBench [2] and *"Emotional Decision-Making"* [11] assess whether models can identify Nash equilibria or adopt bounded-rational strategies in repeated play.

In combinatorial optimization, the research has explored prompt-as-program paradigms (e.g. Toolformer [13]) and program-of-thought approaches to synthesize verifiable solutions. Despite these advances, most evaluations are one-shot: the model produces an answer that is graded post-hoc. Few works test whether *iterative* LLM self-improvement, guided by external retrieval, can improve performance on algorithmic benchmarks where partial credit is rare.

The proposed RC framework sits at the intersection of the two strands above. It pairs task-specific retrieval with an internal critic that incrementally repairs reasoning chains and surface-forms. To the best of our knowledge, this is the first study to *jointly* quantify the impact of both self-reflection and retrieval on structured, verifiable outputs in decision science. We further release a reusable evaluation harness to support future studies on loop-based reasoning architectures and their trustworthiness.

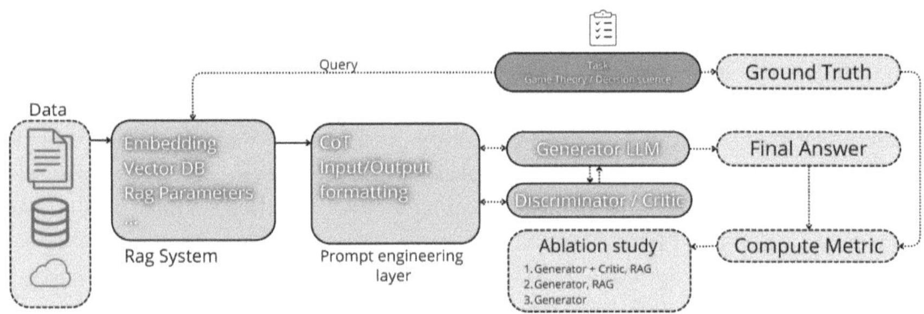

Fig. 1. Overview of the proposed RC framework. A query q is first embedded and used to retrieve a relevant context C_q from an external corpus $\mathcal{D}$. The generator G produces a candidate answer $\hat{a}_i$, which is critiqued by the discriminator D. If the quality score s_i exceeds a threshold τ, the loop terminates; otherwise the critique is appended to the context and the process repeats for up to k iterations

3   Methodology

Figure 1 illustrates the end-to-end data flow of RC, including retrieval, generation, critique, and the iteration/termination conditions referenced below. Given a space of formal queries $\mathcal{Q}$, the goal is to return a structured answer $\tilde{a}$ that matches the solution produced by a trusted solver. For a query $q \in \mathcal{Q}$, if we use a retrieval system, we first obtain an external context C_q using a similarity-based retriever:

$$C_q = \mathrm{Retrieve}(q, \mathcal{D}), \tag{1}$$

where $\mathcal{D}$ is a vector-indexed corpus embedded in $\mathbb{R}^d$, and $\mathrm{Retrieve}:\mathbb{R}^d \to 2^{\mathcal{D}}$ returns the top-k documents by cosine similarity.

Starting from $\langle q, C_q \rangle$, RC alternates between a generator G, with parameter θ_G, and a critic D, with parameter θ_D. Here θ denotes the learned weights of the model, while λ collects the fixed decoding hyper-parameters (e.g. temperature, top-p, max-tokens) used at inference time. In our experiments G and D share the same checkpoint, therefore $\theta_G = \theta_D$, but this distinction allows future variants to employ different models or sampling settings. The values of λ are summarized in Table 1.

Table 1. Inference hyper–parameters collected in the vector λ

Symbol	Description	Value (this work)
T	Sampling temperature	0.2
k_{RAG}	Documents retrieved per query	5
$I_{\max}$	Max generator–critic iterations	2 (unless stated)
τ	Critic acceptance threshold (1–10 scale)	6
$L_{\max}$	`max_tokens` returned by model	none (provider default)

At most k iterations have been considered for the loop displayed in Fig. 1 (cf. $I_{\max}$ in Table 1). In particular, at each iteration, the generator G proposes a candidate $\hat{a}_i$ and the critic D assigns a quality score $s_i \in \{1, \ldots, 10\}$ according to a rubric that balances the accordance with the format, the logical soundness, the numerical correctness, and the completeness. If $s_i \geq \tau$ (acceptance threshold in Table 1), the candidate is returned; otherwise, the critic provides a short structured feedback c_i (e.g., a JSON block with detected issues and suggested fixes) which is appended to the working context to guide the next refinement:

$$\mathcal{C}_q \leftarrow \mathcal{C}_q \cup \{c_i\}. \tag{2}$$

The loop terminates early upon acceptance,

$$
\begin{aligned}
& textfor\, i = 1, \ldots, k : \\
& \quad \hat{a}_i = G\big(q, \mathcal{C}_q; \theta_G, \lambda\big) \qquad \text{(candidate answer)} \tag{3} \\
& \quad s_i = D\big(q, \hat{a}_i, \mathcal{C}_q; \theta_D, \lambda\big) \qquad \text{(critic score on a 1–10 scale)} \\
& \quad \textbf{if } s_i \geq \tau \textbf{ then } \tilde{a} = \hat{a}_i, \textbf{ break} \tag{4}
\end{aligned}
$$

or if no candidate is accepted after k iterations, i.e. no candidate meets the threshold; in this case, the framework returns the highest-scoring response observed:

$$\tilde{a} = \arg\max_{i \leq k} s_i. \tag{5}$$

RC seeks to maximize *decision fidelity*, defined as the expected agreement between the model output and a task-specific exact solver (of course, neither the generator nor the critic ever see the task-specific answer provided by the exact solver):

$$\max_{\theta_G, \theta_D} \mathbb{E}_{q \sim \mathcal{Q}} \big[\mathrm{Match}\big(\tilde{a}, \mathrm{Solver}(q)\big)\big]. \tag{6}$$

where $\mathrm{Match}(\cdot, \cdot)$ is a binary indicator (or a graded similarity metric) that returns 1 when the structured fields of the two outputs coincide.

To facilitate verifiable reasoning, we employ a *chain-of-thought* (CoT) [16] template that explicitly requests intermediate steps and JSON-formatted outputs. The critic receives a complementary prompt containing an explicit scoring rubric that balances logical soundness, completeness, and strict format compliance when assigning the quality score s_i. Finally, a lightweight prompt-engineering layer injects identical input–output schemas into both roles, guaranteeing that every generator response and every critic report conforms to the JSON specification. Both the generator G and the critic D are instantiated from the same LLM checkpoint accessed through the OpenRouter API.

Decoding is performed at a temperature of $T = 0.2$ with the full context window enabled, and the retrieval module (when enabled) supplies the top $k = 5$ documents per query. The generator–critic loop ends as soon as the critic assigns a score of at least $\tau = 6$; this threshold is empirically selected as it is an hyper-parameter that balances answer quality against the number of iterations and is meaningful only with respect to the critic's scoring rubric. The critic returns an integer score $\in \{1, \ldots, 10\}$ that weights, in equal proportion, adherence to the prescribed format, logical soundness, factual or numerical accuracy, and the completeness and relevance of the response. Note that minor infractions reduce the score incrementally; serious violations, such as, for instance, an invalid schema, yield a minimum value of 1; and a flawless answer receives a score of 10.

As already stated, we evaluated different checkpoints to assess whether our proposed workflow maintains its benefits across different backbone models. This comparative setup allows us to understand whether the improvements brought by retrieval and critique are model-agnostic or sensitive to specific architectures. In particular, we have considered:

- **DeepSeek-R1** ("**R1**" in our tables): created 26 Dec 2024; 163840-token context window. A May-2025 refresh brings performance on a par with OpenAI o1 while retaining full openness: 671 B parameters overall, with only 37 B active per inference pass. Median latency on the OpenRouter endpoint is ≈ 3.1 s and throughput ≈ 22.9 tps[1]
- **DeepSeek-V3** ("**V3**" in our tables): the latest DeepSeek flagship, pre-trained on nearly 15 trillion tokens. Created 26 Dec 2024; retains the same 163840-token context window but delivers markedly faster serving: latency ≈ 1.3 s and throughput ≈ 61.1 tps. [2]
- **DeepSeek R1T2 Chimera** ("**TNG**" in our tables): a finetuned variant of DeepSeek-R1, provided by TNGtech. This checkpoint blends reasoning ability with compliance to structured output formats, making it a strong candidate for decision-science tasks. It is freely accessible on HuggingFace and OpenRouter.[3]
- **Horizon-Beta** ("**HZN**" in our tables): a model developed internally by OpenRouter ("in-house"). It provides a 256 k-token context window, median latency of ≈ 1.07 s, and throughput of ≈ 110.6 tps. During our experiments the model was freely available as a cloaked release intended to collect community feedback.[4]

---

[1] https://openrouter.ai/deepseek/deepseek-r1-0528:free.
[2] https://openrouter.ai/deepseek/deepseek-chat:free.
[3] https://huggingface.co/tngtech/DeepSeek-TNG-R1T2-Chimera.
[4] https://openrouter.ai/openrouter/horizon-beta.

4 Experimental Setup and Results

All experiments were run locally on two machines: (a) Windows 11 workstation – AMD Ryzen 9 7950X3D CPU, NVIDIA GeForce RTX 3060 Ti, 32 GB DDR5-7200 CL34 RAM and (b) MacBook Air (M3, 2024) – Apple M3 system-on-chip with integrated GPU and 16 GB unified memory. In both setups the only on-device workload is the sentence-embedding inference; all large-scale generation is handled remotely by the LLMs model accessed through the OpenRouter API.

It is worth noting that another decision science problem were also considered in our evaluation, that is a synthetic *expected value regression* benchmark. However, all tested configurations, including the baseline, achieved 100% precision in this task. This ceiling effect suggests that the benchmark was too simple to provide meaningful differentiation between methods, and, for this reason, these results are not introduced into the paper.

4.1 Pure Strategy Nash Equilibrium

For the pure–strategy task two synthetic sets have been considered, respectively, one of 67 distinct 2×2 normal–form games, and the other one of 100. Each game is labelled with one of five mutually exclusive tokens {*U/L*, *U/R*, *D/L*, *D/R*, *none*}, denoting the location (or absence) of a pure Nash equilibrium. To generate this dataset, we followed a templated procedure: five base payoff matrices, one per token, are first verified with `nashpy`[5] [6] to guarantee the intended equilibrium structure. Integer shifts drawn uniformly from $\{0, 1, 2\}$ are then added independently to every player's pay-offs to create surface diversity while preserving equilibrium positions. The resulting corpus is stored as a `jsonl` file, with each record containing an `id`, a human-readable question string, and the ground-truth token; afterward, we deterministically verify whether each prediction is correct to ensure to not have multiple equilibrium or wrong ground truth answer.

Table 2. Performance metrics on the 67-game Nash set (**R1**)

Method	Accuracy	Precision	Recall	F_1
Baseline	0.7015	0.7394	0.7055	0.7003
RAG	0.7761	0.8195	0.7791	0.7777
RC	**0.8507**	**0.8614**	**0.8538**	**0.8504**

Table 3. Agreement-pattern frequencies for the same experiment (**R1**)

Pattern	Count	Pattern	Count
TTT	38	TFT	3
FFT	8	FTF	2
FTT	8	FFF	2
TTF	4	TFF	2

Table 2 confirms the incremental benefit of retrieval (*rag*) and the additional lift brought by iterative critique (*rag+critic*). Also Table 3 shows the pattern analysis that explains *where* those gains originate. Roughly one-third of the

---

dataset (patterns **FTT** and **FFT**) are cases where either retrieval or critique salvages an error made by the baseline, while **TTF** highlights a small set (4 games) in which the critic introduces a new mistake. The 38 universal successes (**TTT**) indicate a pool of "easy" matrices whose equilibria are clear from pay-off dominance alone.

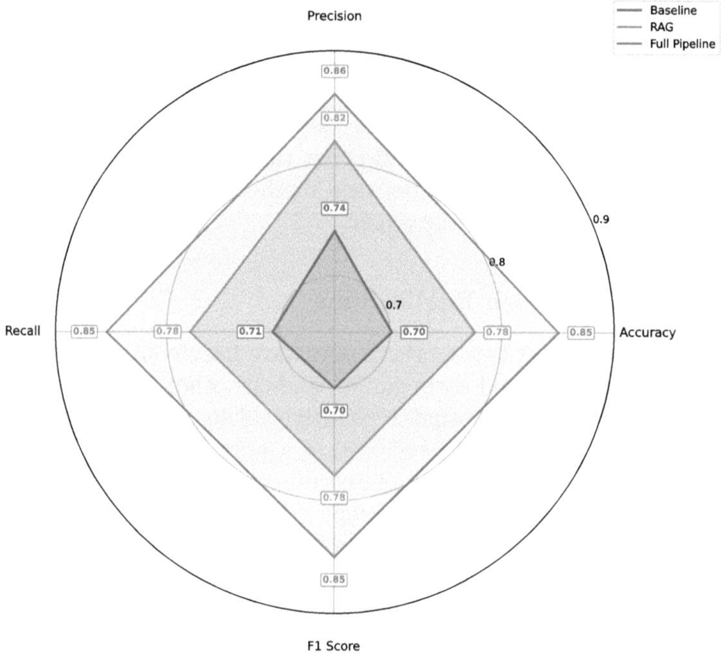

Fig. 2. Radar plot comparing the four metrics across the three evaluation modes

The radar plot in Fig. 2 visualises the same trends: every axis expands monotonically from *no_rag* to *rag* and *rag+critic*.

Table 4. Performance metrics for the 100-game Nash experiment (**V3**)

Method	Accuracy	Precision	Recall	F_1
Baseline	0.41	0.3947	0.41	0.3787
RAG	0.58	0.4880	0.58	0.4889
RC	**0.62**	**0.5524**	**0.62**	**0.5160**

Table 5. Agreement pattern frequencies in the Nash experiment (**V3**)

Pattern	Cnt	Pattern	Cnt
TTT	35	TFT	1
FFT	5	FTF	1
FTT	21	TTF	1
FFF	32	TFF	4

As noted above, the picture changes with the V3 model (Tables 4 and 5), considerably faster but less accurate. Since our aim is to assess the proposed framework, rather than the underlying model, metrics on the second Nash equilibrium dataset are also reported, which contains 100 instances. We can see how the accuracy increases from 41% for *no_rag* to 58% with plain retrieval and peaks at 62% once the critic is added. Even if absolute scores remain below those obtained with the stronger R1 backbone (Table 2), the *relative* improvement delivered by the RC framework is greater: the critic produces a +21 pp gain over the baseline, showing that the loop is most valuable when the underlying model is less robust. Table 5 shows that TTT accounts for one-third of the games, indicating a set of easy matrices even for V3. The FTT patterns (21 cases) and FFT (5 cases) reveal that retrieval fixes many baseline errors, while the critic still contributes unique corrections. Regressions caused by the critic are almost negligible (TTF+FTF=2), while the FFF block (32 cases) highlights really hard instances that neither retrieval nor critique can solve with this smaller model.

4.2   Value of Perfect Information (VOPI)

The Value of Perfect Information (VOPI) represents a fundamental concept in decision theory that quantifies the maximum amount a rational decision-maker should be willing to pay for perfect information about uncertain states of nature. Given a decision table with actions $A = \{a_1, a_2, ..., a_n\}$; states $S = \{s_1, s_2, ..., s_m\}$; probabilities $P = \{p_1, p_2, ..., p_m\}$ such that $\sum_{i=1}^{m} p_i = 1$; and payoffs $u(a_i, s_j)$; the VOPI is formally defined as:

$$\text{VOPI} = \text{EV}_{\text{perfect}} - \text{EV}_{\text{no info}} \tag{7}$$

where:

- $\text{EV}_{\text{perfect}} = \sum_{j=1}^{m} p_j \cdot \max_{a \in A} u(a, s_j)$ and represents the expected value when the decision-maker can observe the true state before choosing an action;
- $\text{EV}_{\text{no info}} = \max_{a \in A} \sum_{j=1}^{m} p_j \cdot u(a, s_j)$, and represents the expected value under the optimal action without additional information.

The proposed framework has been evaluated on a dataset of 101 VOPI problems, each of which presents a decision table with $2 - 3$ actions and $2 - 4$ states. The task requires computing three values with 2-decimal precision: the VOPI itself; the expected value without information ($\text{EV}_{\text{no info}}$); and the expected value with perfect information ($\text{EV}_{\text{perfect}}$). The experiments were conducted using the *OpenRouter Horizon-Beta model*[6] in our three operating modes. For the RAG-Critic mode, a critic score threshold of 6 was set and a maximum of 3 iterations. In addition, experiments with *DeepSeek R1T2 Chimera* have been conducted. Although this model achieved perfect accuracy (100%) in all test cases, it is unsuitable for comparative framework analysis, as it did not provide differentiation between enhancement techniques. Figure 3 presents the quantitative results in the three experimental modes.

The VOPI task proved particularly challenging for the evaluated framework, revealing several important insights:

---

[6] https://openrouter.ai/openrouter/horizon-beta.

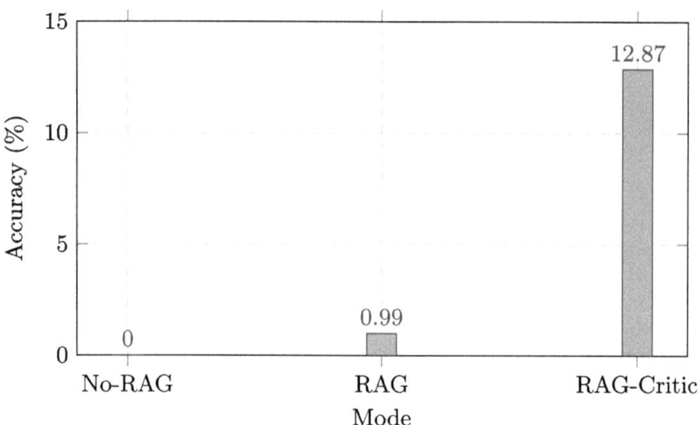

Fig. 3. Accuracy comparison across different operational modes for VOPI computation (**HZN**)

1. computational complexity, as the VOPI calculation requires precise multi-step reasoning: (i) computing expected values for each action under uncertainty; (ii) identifying the optimal action without information; (iii) calculating the expected value under perfect information by selecting optimal actions for each state; and (iv) computing the difference. The base model's 0% accuracy suggests fundamental difficulties with this sequential computation;
2. the marginal improvement from No-RAG (0%) to RAG (0.99%) indicates that access to our decision theory documentation provides negligible improvement. This suggests that the challenge lies not merely in knowledge access but in the application of multi-step mathematical reasoning;
3. the RC mode showed the most substantial improvement, achieving 12.87% accuracy with an average of 2.63 out of 3 iterations per problem. This result is particularly noteworthy when considering that the critic is the same model with identical reasoning capabilities as the generator, and crucially, *without access to ground truth answers*. The critic must evaluate the correctness of solutions based solely on its understanding of decision theory principles and the consistency of the mathematical reasoning presented.

Despite these constraints, the iterative refinement process achieved a remarkable 13-fold improvement over baseline performance. This suggests that the model has latent capabilities for error detection and correction that are not fully utilized in single-pass generation. The critic's ability to identify flaws in reasoning chains and propose corrections, even without knowing the correct answer, demonstrates that structured self-reflection can unlock significant performance gains. The average of 2.63 iterations indicates that the model typically required multiple refinement cycles to reach acceptable solutions (achieving the critic score threshold of 6). The investigation of the failure cases revealed, however, common error patterns: (i) *arithmetic errors*, that is incorrect multiplication or addition of probabilities and payoffs; (*ii*) *conceptual confusion*, that is mixing up $EV_{no\ info}$ and $EV_{perfect}$ calculations; (*iii*) *optimization errors*, that is failing to identify the correct maximum values; and (*iv*) *precision issues*, that is rounding errors propagating through multi-step calculations.

5   Conclusions

In this research work, RAG-CRITIC (RC) is proposed, a framework that combines Retrieval-Augmented Generation with iterative self-critique to enhance LLM performance on structured decision-theoretic tasks. Our evaluation across multiple benchmarks, including Nash equilibrium identification and Value of Perfect Information computation, demonstrates that the synergistic combination of task-specific retrieval and iterative refinement yields substantial improvements in solution accuracy and consistency. Our key findings reveal that: (*i*) the RC framework provides consistent performance gains across different model architectures, with improvements ranging from 12.87% to 21% over baseline approaches; (*ii*) the benefits are most pronounced when the underlying model tackles challenging tasks that require reasoning, as evidenced by the 13-fold improvement in VOPI computation accuracy; and (*iii*) remarkably, the critic module can identify and correct errors without access to ground truth, suggesting that LLMs possess latent self-evaluation capabilities that single-pass generation fails to exploit. The modular model-agnostic design of RC makes it readily adaptable to diverse decision science applications.

Several promising directions emerge from this research. While our investigation focuses on game theory and decision analysis, extending RC to broader operations research domains, such as stochastic optimization, queueing theory, and dynamic programming, would further validate its generalizability. These domains present indeed unique challenges that could reveal both the strengths and limitations of our iterative refinement approach. Although our current implementation uses fixed thresholds and iteration limits, the trade-off between computational cost and solution quality could be further optimized by developing adaptive strategies that dynamically adjust these parameters based on task complexity and intermediate performance. This adaptation challenge echoes similar problems in dynamic optimization contexts, where algorithms must respond to changing constraints in real-time [1].

Exploring ensemble critics with specialized expertise presents an intriguing avenue for improvement as well. For instance, one critic might focus on mathematical correctness while another ensures logical consistency, potentially capturing errors that a single critic misses, particularly for complex multi-step reasoning tasks. The integration of RC into human decision-making workflows represents another crucial research direction. Investigating how the iterative refinement process can be made transparent and interpretable to domain experts could enhance trust and adoption in high-stakes applications. This human-in-the-loop approach would allow experts to guide and validate the refinement process, combining human intuition with computational rigor.

From a theoretical perspective, developing formal guarantees on convergence properties and error bounds for the iterative refinement process would provide deeper insights into when and why the framework succeeds or fails. Such an analysis could inform principled design choices and help predict performance on novel task types. Furthermore, while our results demonstrate effectiveness, efficiency optimization remains an important practical consideration. Reducing API calls and latency through techniques such as early stopping criteria; caching intermediate results; or distilling the critic's capabilities into smaller models could make the framework more viable for real-time applications. As LLMs increasingly support decision-making in critical domains, frameworks like RC that enhance reliability through structured self-improvement represent a crucial step toward trustworthy AI systems. We hope that this work stimulates further research into loop-based reasoning architectures that can bridge the gap between impressive language fluency and the rigorous correctness demanded by mathematical and algorithmic tasks.

Acknowledgements. This research is supported by the project PIACERI - PIAno di inCEntivi per la Ricerca di Ateneo 2024/2026 – Linea di Intervento i "Progetti di ricerca collaborativa".

References

1. Cutello, V., Mezzina, A., Pavone, M., Zito, F.: A real-time adaptive tabu search for handling zoom in/out in map labeling problem. In: Festa, P., Ferone, D., Pastore, T., Pisacane, O. (eds.) Learning and Intelligent Optimization, pp. 108–122. Springer Nature Switzerland, Cham (2025)
2. Duan, J., et al.: GTBench: uncovering the strategic reasoning capabilities of LLMs via game-theoretic evaluations. Adv. Neural. Inf. Process. Syst. **37**, 28219–28253 (2024)
3. Eigner, E., Händler, T.: Determinants of LLM-assisted decision-making (2024). https://arxiv.org/abs/2402.17385
4. Fan, W., et al.: A survey on rag meeting LLMs: towards retrieval-augmented large language models. In: Proceedings of the 30th ACM SIGKDD Conference on Knowledge Discovery and Data Mining, pp. 6491–6501. KDD '24, Association for Computing Machinery, New York, NY, USA (2024). https://doi.org/10.1145/3637528.3671470
5. Huang, L., et al.: A survey on hallucination in large language models: principles, taxonomy, challenges, and open questions. ACM Trans. Inf. Syst. **43**(2) (2025). https://doi.org/10.1145/3703155
6. Knight, V., Campbell, J.: Nashpy: a python library for the computation of Nash equilibria. J. Open Source Softw. **3**(30), 904 (2018). https://doi.org/10.21105/joss.00904
7. Lewis, P., et al.: Retrieval-augmented generation for knowledge-intensive NLP tasks. In: Larochelle, H., Ranzato, M., Hadsell, R., Balcan, M., Lin, H. (eds.) Advances in Neural Information Processing Systems. vol. 33, pp. 9459–9474. Curran Associates, Inc. (2020). https://proceedings.neurips.cc/paper_files/paper/2020/file/6b493230205f780e1bc26945df7481e5-Paper.pdf

8. Ma, J., Dai, D., Sha, L., Sui, Z.: Large language models are unconscious of unreasonability in math problems. arXiv preprint arXiv:2403.19346 (2024)
9. Madaan, A., et al.: Self-refine: iterative refinement with self-feedback. In: Oh, A., Naumann, T., Globerson, A., Saenko, K., Hardt, M., Levine, S. (eds.) Advances in Neural Information Processing Systems. vol. 36, pp. 46534–46594. Curran Associates, Inc. (2023). https://proceedings.neurips.cc/paper_files/paper/2023/file/91edff07232fb1b55a505a9e9f6c0ff3-Paper-Conference.pdf
10. Mousavi, S., et al.: N-critics: Self-refinement of large language models with ensemble of critics. arXiv preprint arXiv:2310.18679 (2023)
11. Mozikov, M., et al.: EAI: emotional decision-making of LLMs in strategic games and ethical dilemmas. In: Globerson, A., et al. (eds.) Advances in Neural Information Processing Systems. vol. 37, pp. 53969–54002. Curran Associates, Inc. (2024). https://proceedings.neurips.cc/paper_files/paper/2024/file/611e84703eac7cc03f78339df8aae2ed-Paper-Conference.pdf
12. Sarthi, P., Abdullah, S., Tuli, A., Khanna, S., Goldie, A., Manning, C.D.: RAPTOR: recursive abstractive processing for tree-organized retrieval. In: The Twelfth International Conference on Learning Representations (2024)
13. Schick, T., et al.: Toolformer: language models can teach themselves to use tools. In: Oh, A., Naumann, T., Globerson, A., Saenko, K., Hardt, M., Levine, S. (eds.) Advances in Neural Information Processing Systems. vol. 36, pp. 68539–68551. Curran Associates, Inc. (2023). https://proceedings.neurips.cc/paper_files/paper/2023/file/d842425e4bf79ba039352da0f658a906-Paper-Conference.pdf
14. Shinn, N., Cassano, F., Gopinath, A., Narasimhan, K., Yao, S.: Reflexion: language agents with verbal reinforcement learning. In: Oh, A., Naumann, T., Globerson, A., Saenko, K., Hardt, M., Levine, S. (eds.) Advances in Neural Information Processing Systems. vol. 36, pp. 8634–8652. Curran Associates, Inc. (2023). https://proceedings.neurips.cc/paper_files/paper/2023/file/1b44b878bb782e6954cd888628510e90-Paper-Conference.pdf
15. Wang, X., et al.: Self-consistency improves chain of thought reasoning in language models (2023). https://arxiv.org/abs/2203.11171
16. Wei, J., Wang, X., et al.: Chain-of-thought prompting elicits reasoning in large language models. In: Koyejo, S., Mohamed, S., Agarwal, A., Belgrave, D., Cho, K., Oh, A. (eds.) Advances in Neural Information Processing Systems. vol. 35, pp. 24824–24837. Curran Associates, Inc. (2022). https://proceedings.neurips.cc/paper_files/paper/2022/file/9d5609613524ecf4f15af0f7b31abca4-Paper-Conference.pdf
17. Xia, M., Zhang, X., Couturier, C., Zheng, G., Rajmohan, S., Rühle, V.: Hybrid-RACA: hybrid retrieval-augmented composition assistance for real-time text prediction. In: Dernoncourt, F., Preoţiuc-Pietro, D., Shimorina, A. (eds.) Proceedings of the 2024 Conference on Empirical Methods in Natural Language Processing: Industry Track, pp. 120–131. Association for Computational Linguistics, Miami, Florida, US (2024). https://doi.org/10.18653/v1/2024.emnlp-industry.11, https://aclanthology.org/2024.emnlp-industry.11/
18. Yang, K., et al.: LeanDojo: theorem proving with retrieval-augmented language models. In: Oh, A., Naumann, T., Globerson, A., Saenko, K., Hardt, M., Levine, S. (eds.) Advances in Neural Information Processing Systems. vol. 36, pp. 21573–21612. Curran Associates, Inc. (2023). https://proceedings.neurips.cc/paper_files/paper/2023/file/4441469427094f8873d0fecb0c4e1cee-Paper-Datasets_and_Benchmarks.pdf
19. Yang, T., Yan, M., Zhao, H., Yang, T.: LemmaHead: RAG assisted proof generation using large language models (2025). https://arxiv.org/abs/2501.15797

20. Yao, S., et al.: ReAct: Synergizing reasoning and acting in language models. In: International Conference on Learning Representations (ICLR) (2023)
21. Ye, H., et al.: ReEvo: large language models as hyper-heuristics with reflective evolution. In: Globerson, A., Mackey, L., Belgrave, D., Fan, A., Paquet, U., Tomczak, J., Zhang, C. (eds.) Advances in Neural Information Processing Systems. vol. 37, pp. 43571–43608. Curran Associates, Inc. (2024). https://proceedings.neurips.cc/paper_files/paper/2024/file/4ced59d480e07d290b6f29fc8798f195-Paper-Conference.pdf
22. Yu, H., Gan, A., Zhang, K., Tong, S., Liu, Q., Liu, Z.: Evaluation of retrieval-augmented generation: a survey. In: Zhu, W., et al. (eds.) Big Data, pp. 102–120. Springer Nature Singapore, Singapore (2025)
23. Zhao, Q., et al.: LongRAG: A dual-perspective retrieval-augmented generation paradigm for long-context question answering (2024). https://arxiv.org/abs/2410.18050

A Centrality-Based Resource-Aware Microservice Orchestration in Cloud-to-Edge Continuum

Alberto Bertoncini$^{(\boxtimes)}$, Alberto Ceselli , and Christian Quadri

Computer Science Department, Università degli Studi di Milano, Milan, Italy
{alberto.bertoncini,alberto.ceselli,christian.quadri}@unimi.it

Abstract. Smart Cities and Industry 4.0 have opened new application scenarios posing new challenges to networking services. The continuously growing demand for high quality of service (QoS) and the intrinsic geographical distribution of this novel class of applications require novel and tailored deployment solutions. Microservice applications placement in edge computing requires balancing resource constraints, network efficiency, and end-to-end latency constraints. This work presents Centrality-based Resource-Orchestrator System (CeROs), a novel Mixed Integer Linear Programming (MILP) model that introduces an objective function based on the structural node importance based on its position within the network's topology, referred as node centrality, which promotes efficient resource usage across the network without directly optimizing latency or individual resource consumption. This approach ensures a generalizable and unbiased formulation that can be adjusted to various deployment scenarios. The experiments have been conducted over synthetic yet realistic cloud-to-edge network topologies and microservice applications, modeled to reflect real-world resource constraints, communication delays, and deployment challenges. The evaluations demonstrate that CeROs outperforms trivial placement strategies, achieving competitive latency even in strict scenarios and more balanced resource usage, resulting in a lower node costs. CeROs does not force optimization toward a single metric, making it more robust across different emerging environments as like smart cities with the Autonomous Vehicle Driving application.

Keywords: Service Orchestration · Cloud-to-Edge Continuum · Mathematical Optimization

This work has been funded by the Italian MUR in the framework of the PRIN 2022 project *CAVIA* (Grant 2022JAFATE, CUP: G53D23002850006) under MUR (Call PRIN 2022) and by project *SERICS* (PE00000014, CUP: G43C22002580001) under the MUR NRRP, funded by the European Union NextGenerationEU.

1 Introduction

Emerging smart city scenarios open several new perspectives and challenges. Just to mention a remarkable example, autonomous agents driving electric vehicles could allow safer, cleaner, and more efficient last-mile logistics. Augmented reality is another example of the access to services can be greatly enhanced by digital platforms.

Two common cross-domain features of these applications are the need to perform non-trivial computations and the sensitivity to latency. A well-established technique to cope with them at best, even on devices with limited computing power, is offloading: devices connect to the network that delegate all or part of their computation to remote servers. A current challenge in networking is to consider different network levels as a continuum in which applications can run. The cloud, as well as smaller urban edge servers, and the autonomous vehicles themselves can be used as computing elements of a distributed system, where microservices composing an overall application can be deployed and orchestrated to achieve high quality of service to end users. The heterogeneity of resource availability and computation isolation required for privacy and security pose challenges in service orchestration.

In this paper we face the problem of optimizing support for distributes systems that operate across vehicles and cloud infrastructure We cast the problem of optimally deploying microservice-based applications in such a continuum infrastructure as a Virtual Network Function Placing Problem (VNFPP) in which the aim is to place a sequence of virtual network functions. We assume that each microservice has a set of resource requests, and end-to-end dependencies exist among microservice chains. We propose a Mathematical Programming formulation, which encodes resource requests and end-to-end dependencies as constraints. As a novel feature, we employ an objective function which relies on the idea of *node centrality* to balance the load of the network also in the medium term. We adopt an exact solution method to rigorously evaluate the effectiveness and boundaries of the proposed model and to be the base for future more efficient heuristics

We empirically show the effectiveness of our methods by extensive experiments on realistic instances, designed around actual data of urban scenarios.

Background and Related Work. Our research focuses on the deployment of distributed microservice-oriented applications on an edge network; our interest arises mainly due to the relevance in the support of autonomous vehicles. These applications compose a Service Function Chain (SFC) as defined in [3]: multiple tasks that can be executed individually, each depending only on a subset of the preceding ones. In fact, each microservice may correspond to a component of a virtual network function, allowing us to model the decision problem of our application as a VNFPP, exploiting the vast literature on the topic. Various approaches have been explored for the general VNFPP, as highlighted by [11], such as heuristics, meta-heuristics, genetic algorithms, mathematical models

(Markov, BILP, ILP, MILP) and tree based algorithms. A variety of objectives and constraints have been considered in the literature, leading to many variants and refinements in the formulation of the problem: as highlighted in [1,2,5–8,12], the formulation of the problem strongly depends on the network type and on the target application. Key features commonly considered in objective functions include latency, delay, resource consumption, or a combination of them. However, research on metrics that do not go directly to optimizing one of the constraints has not been addressed

To the best of our knowledge, no specific formulation for the VNFPP with end-to-end constraint in edge networks has been proposed, focusing on *node centrality* as a usage metric to be optimized in the objective function. A key point in our modeling approach, in fact, is the following: single VNF resource requests, as well as end-to-end requirements, are encoded as constraints, while the objective function targets the relevance of network allocation nodes, to foster long-term network balancing.

2   System Modeling

In this section, we present the system model for orchestrating micro-services applications on top of a cloud-edge continuum physical network. The general overview of the system is depicted in Fig. 1. On the left, the application is modeled using the microservices architecture paradigm. On the right, we represent a generic multi-tier cloud-to-edge computing network comprising heterogeneous computing nodes and communication links. The *Orchestrator* receives application deployment requests and generates an *Application deployment plan* for the deployment phase, taking into account the physical network connectivity and the resource availability. In the following, we describe each component in detail.

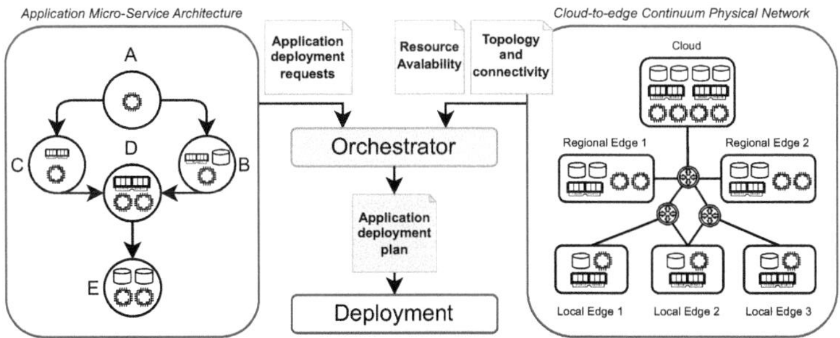

Fig. 1. System model schema

2.1   Microservices Application

The application is modeled as a directed acyclic graph (DAG), $M = (T, D)$, where nodes represent the microservices and the edges capture the dependencies among microservices. Each microservice, $u \in T$, is characterized by hardware resource requirements and execution time. Each link, $(u, v) \in D$, characterizes a functional dependency and the data flow among a pair of microservices in terms of required datarate. Finally, we consider applications characterized by multiple information flows each with specific end-to-end latency requirements.

The *application deployment request* (in Fig. 1) sent to the orchestrator includes the graph topology of the application, the microservices computing and data flow requirements, along with the limitations of the end-to-end constraint.

2.2   Cloud-to-Edge Continuum Physical Network Model

The *Cloud-to-Edge Computing Network* is modeled as a directed graph $G = (I, A)$, where I represents the network nodes and A the links connecting them. It reflects a cloud-to-edge continuum, organized in a hierarchical and multi-tier topology. Each computing node allows for the deployment of microservices, according to its available resources. The level of each node reflects its proximity to the edge, where the final devices are deployed. Nodes closer to the edge offer lower communication latency, but limited computational resources, while those in higher levels offer greater resources at the price of increased latency. We assume the existence of an idealized *Cloud* node with nearly infinite resources, having the incurring high latency cost for accessing it.

Physical Resources and Properties. Each node and link in the network has physical resources and properties. Node resources include standard computing capabilities, such as CPU power, memory, storage, I/O bandwidth, but also hardware accelerators like GPUs. The amount of physical resources available in each node varies according to the hierarchical level of the node.

Physical links are characterized by the available bandwidth and by the communication delay. This work assumes a simple time-invariant link delay model, disregarding queuing effects.

Computing Node Cost. In a continuum cloud-to-edge environment, computing nodes are different not only in terms of the amount of resources but also in terms of position in the network. Intuitively, a set of *central* nodes may exist, which are at higher risk of overloading than more peripheral ones, being highly connected or traversed by a large number of paths. To deploy several microservices in central nodes eases the enforcement of the application end-to-end flows requirements; nevertheless, exhausting the nodes capacity might prevent deployments of future applications. To drive the deployment algorithm (see Sect. 3), we assign a deployment cost to a node directly proportional to the centrality value of the node. In this work, we use the *betweenness centrality*[1] measure, as it captures both the microscopic and macroscopic role of the nodes.

2.3   Orchestration

The orchestrator, representative of our model of our model in Fig. 1, is responsible for producing an *application deployment plan* considering the application's resources and latency requirements and the status of the physical network (connectivity and physical resource availability). The deployment plan indicates the assignment of each application microservice to a physical node. In particular, the orchestrator has to *(i)* assign microservices to a suitable set of connected nodes, respecting their resource and dependencies requirements; *(ii)* meets the physical resources available on the selected nodes, and on the links belonging to the path connecting two dependent microservices location, preventing overloading; *(iii)* simultaneously enforces the latency constraints of all the information flows of the application.

Effective orchestration ensures a balance between resource allocation and the relative node usage cost, adhering to latency constraints and minimizing the usage of the network itself. In the following section, we present the optimization process performed by the orchestrator to produce the application deployment plan.

3   Mathematical Models and Optimization Algorithms

In this section, we present the mathematical formulation of the optimization process performed by the orchestrator. The *orchestrator* task described in Sect. 2 is formally modeled by the following entities:

- a set I of physical network nodes, available for microservices deployment;
- a set A of physical links (i, j) between pairs of network nodes $i \in I, j \in I$;
- a set R of consumable resources, in our case defined by:
  - CPU: the computational power in terms of cores;
  - RAM: the memory in terms of GB;
  - Hardware accelerator GPU: in terms of presence and cores;
  - Bandwidth: the amount of data manageable by the node or the data rate offered by a link, in terms of in Gb/s;
- a set T of application microservices to deploy;
- a set D of communication dependencies (u, v) between pairs of microservices $u \in T, v \in T$.

We assume that all microservices along with their communication dependencies are known at optimization time. We also assume that the application requirements and properties do not change over time. Moreover, the routing in the

---

[1] Betweenness centrality measures how often a node appears on the shortest paths between other nodes, indicating its role as a bridge in the network [4].

physical network is fixed and independent of the optimization process. Specifically, we assume that a path $\mathcal{P}_{ij}$, which connects nodes i and j, is a predefined sequence of links in A that establishes a physical route between them, and it remains unchanged throughout the optimization process. Formally, we introduce a set of binary coefficients a_{ij}^{lm}, taking the value 1 if the physical link $(l, m) \in A$ belongs to the path $\mathcal{P}_{ij}$, or conversely, takes the value 0.

We assume the following data to be given:

- for each node $i \in I$, w_i indicates the deployment cost of a microservice on node i, i.e., the value of the centrality measure for node i (see Sect. 2.2);
- for each node $i \in I$ and for each resource $k \in R$, Q_i^k indicates the availability of resource type k on node i;
- for each physical link $(i, j) \in A$, δ_{ij} and β_{ij} represent the communication delay and the available link bandwidth, respectively;
- for each microservice $u \in T$ and for each resource $k \in R$, the consumption q_u^k of resource type k yield by microservice u;
- for each microservice $u \in T$, δ_u is the processing delay for microservice u;
- for each dependency $(u, v) \in D$, q_{uv}^{β} is the bandwidth usage yield by (u, v).

Based on this data, we compute the total delay and the available bandwidth of the paths. The total communication delay Δ_{ij} of a path $\mathcal{P}_{ij}$ is defined as the sum of the communication delay of the physical links that form the route between the nodes i and j. Formally, it is defined as follows:

$$\Delta_{ij} = \sum_{(l,m)\in\mathcal{P}_{ij}} \delta_{lm} \tag{1}$$

As for the path bandwidth, it is determined by the physical bottleneck link along the path. Formally, the available bandwidth B_{ij} of the path $\mathcal{P}_{ij}$ is defined as follows:

$$B_{ij} = \min_{(l,m)\in\mathcal{P}_{ij}} \beta_{lm} \tag{2}$$

To ensure that the paths can accommodate the required bandwidth for microservices connections, we define a set of binary coefficients b_{uv}^{ij}. These coefficients are 0 if the bandwidth requirement q_{uv}^{β} for connection (u, v) exceeds the available path bandwidth B_{ij}, and 1 otherwise.

Finally, we model the multiple information flows of the application. Considering the DAG that describes the set of microservices and their dependencies, we define $\mathcal{C}$ as the set of microservice service chains. Each service chain $C \in \mathcal{C}$ represents a sequence of microservices that constitute a path in the application DAG. We define L_C as the end-to-end latency (communication and computation) requirement for each service chain $C \in \mathcal{C}$.

3.1   Optimization Problem

We formalize the orchestration task as a VNFPP problem using Mixed Integer Linear Programming (MILP) formulation, referred to as Centrality-based

Resource-Orchestrator System (CeROs). We introduce a set of binary decision variables $x_{ui} \in \{0, 1\}$ which are 1 if microservice $u \in T$ is assigned to node $i \in I$, 0 otherwise. The optimization problem is defined as follows:

$$\min \sum_{u \in T} \sum_{i \in I} w_i \cdot x_{ui} \tag{3}$$

$$\text{s.t.} \sum_{i \in I} x_{ui} = 1 \qquad\qquad \forall u \in T \tag{4}$$

$$x_{ui} + x_{vj} \leq 1 + b_{uv}^{ij} \qquad\qquad \forall i, j \in I; \forall u, v \in T \tag{5}$$

$$\sum_{u \in T} q_u^k \cdot x_{ui} \leq Q_i^k \qquad\qquad \forall k \in R, \forall i \in I \tag{6}$$

$$\sum_{i \in I} \sum_{j \in I} \sum_{u \in T} \sum_{v \in T} q_{uv}^{\beta} \cdot a_{ij}^{lm} \cdot x_{ui} \cdot x_{vj} \leq \beta_{lm} \qquad\qquad \forall (l, m) \in A \tag{7}$$

$$\sum_{u \in C} \sum_{i \in I} \delta_u \cdot x_{ui} + \sum_{(u,v) \in C} \sum_{i,j \in I} \Delta_{ij} \cdot x_{ui} \cdot x_{vj} \leq L_C \qquad\qquad \forall C \in \mathcal{C} \tag{8}$$

The objective function (3) aims to minimize the overall microservices' assignment costs. Constraints (4) ensure that each microservice is assigned to exactly one physical node. Constraints (5) guarantee that the paths between two physical nodes can accommodate the bandwidth required by microservices connections. Constraints (6) ensure that the total amount of resources consumed by microservices deployed on the physical nodes does not exceed the physical resources available on those nodes. Constraints (7) avoid exceeding the bandwidth available on the physical links. Finally, constraints (8) ensure the end-to-end latency requirement for each service chain, considering computation and communication delay components. We note that the set of constraints (5) and (7) ensure that, on the one hand, the paths are feasible by satisfying the requested bandwidth, and on the other hand, they prevent multiple paths sharing a common set of physical links from exceeding the available bandwidth on those links.

The resulting model is compact, having a polynomial number of variables and constraints. The last family of constraints is quadratic, but the quadratic part involves only the product of binary variables. These terms can be linearized, but given the size of instances we plan to solve, an ILP solver like GuRoBi[2] might also be able to manage them automatically.

4   Instances Generation Model

This section presents the experiment settings employed to generate a reproducible and realistic scenario regarding physical network topology and microservice applications.

---

[2] https://www.gurobi.com/.

4.1 Physical Network Topologies

To model realistic scenarios of cloud-edge continuum network topologies, we integrate real-world data from *LTE ITALY*[3] with synthetic transport network topologies. We selected the metropolitan area of Milan as a representative site, having high density and large variance of path latency.

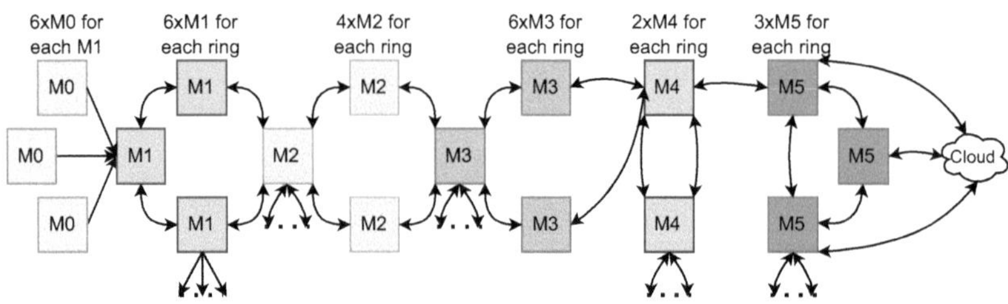

Fig. 2. The schema of the transport network ring topology, as described from [9]

As described in [9], the operator transport network, is organized in a hierarchical multiple rings architecture. In Fug. 2, we show an example of the rings architecture, where *M0* nodes represent the mobile base stations, while the other *M-nodes* indicate different aggregation levels from edge to cloud. To generate a realistic set of synthetic physical network topology, we exploit the LTE ITALY dataset to map the geographical location of the M0 nodes and we generate the remaining levels through the following steps.

Generation of Complete Ring Topology. The construction of the hierarchical structure is demanded to the *Algorithm 1*[4], which is a variant of the K-medoids algorithm. It selects the most central nodes for a given cluster size and connects the surrounding nodes to their centroid in a ring. These two steps are repeated on the elected medoids in the next iteration, up until the last level. In our case, we consider 6 levels, each having the following cluster sizes: (6, 6, 4, 6, 2, 3).

To represent a generic cloud datacenter we insert an additional node, named *cloud*, connected to the top of the node hierarchy through physical links having virtually infinite bandwidth and considerably long delays.

Extraction of Random Sub-network. To create the an instance of cloud-to-edge continuum network graph $G = (I, A)$ with a desired number of nodes n, we sample nodes from the complete ring topology obtained from the previous step.

---

[3] LTE ITALY is a collaborative collection of the Italian LTE antennas features: geographical position, coverage, radiated bands, and cells. https://lteitaly.it/it/.

[4] *Miller-Tucker-Zemlin* formulation on step 8 [10].

Algorithm 1. Hierarchical Structure Construction Algorithm

Require: Points dataset I, cluster sizes k_i, defined for n different levels
Ensure: Hierarchical structure with connected rings
 1: Initialize level $L \leftarrow 0$, active points $A \leftarrow I$
 2: **while** $|A| > 1$ **and and** cluster size $L \leq n$ **do**
 3:     **Apply K-medoid:** Cluster points A into groups of k elements
 4:     **for each** group $g \in \{1, \ldots, \lceil \frac{N}{k_L} \rceil\}$ **do**
 5:        **if** L=0 **then**
 6:           Directly connect the node of the cluster to the medoid
 7:        **else**
 8:           **Build Optimal Ring:** Connect g points in a ring structure via MTZ
 9:           **Connect Ring to Medoid:** Add edges from the ring to its medoid
10:        **end if**
11:     **end for**
12:     Update $A \leftarrow$ Medoids of the clusters at level L
13:     Increment level $L \leftarrow L + 1$
14: **end while**

We adopt a stratified node extraction with the number of nodes that exponentially decreases as the level increases. The links among the selected nodes are generated by computing the shortest path and collapsing the non-selected nodes into links by aggregating the delay properties and adding a processing delay for each traversed non-selected node.

Resource Availability Setting. The resources available on the nodes include hardware capacities, such as CPU, RAM, GPU, and I/O bandwidth, while for the links we only consider the bandwidth. We randomly equip nodes and links with resources drawn from normal distributions whose parameters are reported in Table 1. To model the differences in the availability of the resources among levels, we apply a linear scaling factor ($¿1$) for each level higher than level zero.

Table 1. Resource availability distribution parameters

Resource	Nodes								Links	
	CPU		RAM		GPU		Bw I/O		Bw	
	μ	σ	μ	σ	μ	σ	μ	σ	μ	σ
Availability	12	1	24	2.5	16	1.0	5.5	0.2	5.5	0.2

4.2   Microservices Application

We model a generic Autonomous Vehicle Driving Agent (AVDA) application that controls remote vehicles in the context of edge networks, particularly suited for smart cities, which provide broad network coverage and diverse types of available nodes. The application exploits multiple data sources to enhance context awareness. The microservice architecture of the application is depicted in Fig. 3. We consider multiple sources that transmit streams of raw data, these are placed in fixed specific locations, forcing CeROs to find an optimal solution starting by taking into account these constraints. The raw data flows are then processed and enhanced by a sequence of tasks and sent to a data fusion component where the multiple data streams are combined to produce augmented context information. Subsequently, the AVDA component utilizes this enriched information to generate control signals, which are then forwarded to the vehicle. In our synthetic microservice application (Fig. 3), *SVC* these sequence of steps, highlighting that the *chains* starts from different sources, they all converge on the Data Fusion, AVDA, and Vehicle microservices, making these components highly computationally demanding. As defined in the model, the specified resource requirements represent the minimum resources necessary for a microservice to operate within predefined end-to-end constraints. Microservice requirements are generated according to a normal distribution whose parameters are specified in Table 2, varying according to the role of the microservice.

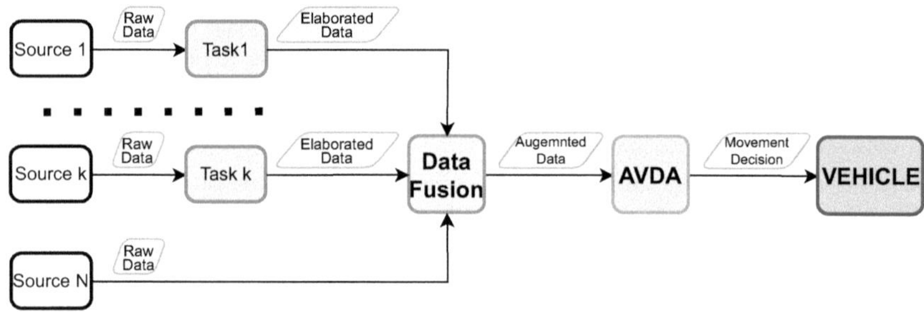

Fig. 3. Logical schema of a microservice graph

Table 2. Microservice Resource Requirements Parameters

Component	CPU		RAM		GPU		I/O Link		Proc. Time	
	μ	σ	μ	σ	μ	σ	μ	σ	μ	σ
Camera	4	0.0	8	0.5	4	0.0	2	0.1	3	0.5
Task	4	0.5	8	0.5	8	0.0	1	0.1	7	1
Fusion	8	0.5	16	0.5	4	0.0	1.5	0.1	3	0.5
AVDA	4	0.5	8	0.0	4	0.0	0.5	0.1	3	0.5

5   Results

This section presents the simulation results of the performance analysis.

5.1   Baseline Algorithms

We compare the performance of CeROs against two simple baselines that take into account only the node resource constraints, disregarding the end-to-end latency requirements in order to test the impact of this restrictive constraint on the feasibility.

Random Algorithm. The most basic baseline algorithm is the random assignment of microservices to nodes. Specifically, it generates a random permutation of the microservices and selects nodes randomly until it finds the first one with sufficient resources to accommodate the microservice. Once a node is selected, its available resources are updated accordingly.

Greedy Algorithm. This algorithm ensures the assignment of microservices to nodes by checking their resource availability. The microservice placement order is determined using a depth-first search (DFS) traversal of the microservice graph, starting from the most resource-intensive microservice, AVDA. The first sufficiently free node with lower centrality metric is selected for the deployment, updating its remaining resources.

As in the random algorithm, the end-to-end constraint is ignored; however, the choice of placement is thought to maximize the resource usage on low cost node, although not in global terms of the impact on the network as in the CeRoS case

5.2   Simulations Setup

We aim to evaluate the effectiveness of our model on the first complete placement. Referring to Fig. 1 we do not integrate the whole deployment loop for multiple, consecutive microservices applications, evaluating the goodness of feasible solutions. Based on the instance generation procedure described in Sect. 4, we generate different scenarios by varying the ratio between the number of microservices in the application graph and the number of nodes in the physical network. In particular, we select the network size among [5, 10, 15, 20] and the microservice-node ratio among [0.5, 0.75. 1.0, 1.25, 1.5]. For each combination, we generate 5 instances.

Regarding the end-to-end latency constraint, we choose three latency classes, reflecting three levels of real-time requirements. Specifically, we generate instances having end-to-end requirements extracted from the following normal distributions: 15 ms $\pm$ 0.5 ms, 30 ms $\pm$ 1 ms, and 45 ms $\pm$ 2 ms. The values have been selected to simulate different levels of control-loop real-time constraints.

The Gurobi solver was configured with the following parameters: *Node-FileStart* set to 0.5 GB to manage memory usage exploiting the disk; *PreSolve* and *PreSparsify* both set to 2 for aggressive model reduction; *MIPGap* set to enforce a 1% optimality gap; and a *TimeLimit* of 1200 seconds.

5.3 Performance Analysis

In the following, we present the results of the performance evaluation.

Feasibility. The first KPI we consider is the feasibility percentage of the instances. For each latency class, we compute the percentage of instances for which CeROs is able to find a feasible solution. We observe that CeROs finds an optimal solution for 100% of instances in the 45 ms class, the value decreases to 94% of instances in the 30 ms class. Regarding the 15 ms latency class, CeROs cannot find any feasible solutions, mainly due to physical topological constraints. While individual paths satisfy the latency constraint, as shown in Fig. 4, their combination easily leads to a violation of the end-to-end constraint, making feasible solutions not achievable with the fixed locations of source sensors and vehicle.

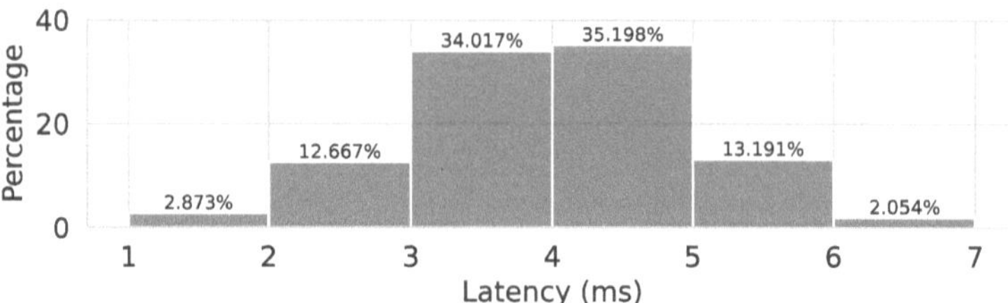

Fig. 4. Milan node-to-node path latency

End-to-End Latency. We compare CeROs algorithm against the two baselines, by instigating the end-to-end latency achieved by the three algorithms. In Fig. 5, we report the boxplot of the distribution of the percentage of end-to-end latency achieved, grouped by algorithm, latency class, and microservice-node ratio. The results show that CeROs is able to guarantee the end-to-end latency constraint, while the baselines fail to guarantee the latency requirements for every instance, as they only focus on meeting the node resource constraints disregarding the latency. The greedy algorithm exhibits competitive results only with the 45 ms latency class, by marginally exceeding the latency constraint (around 5–10% with 1.25 of microservice-node ratio). This is because in this latency class, CeROs has more deployment options that satisfy the latency constraint, preferring deployment plans that minimize the assignment costs. On the contrary, in the case of a more stringent latency requirement, the greedy algorithm fails to offer competitive latency, reaching values above 150%, jeopardizing the application's performance. Finally, the random baseline exhibits the worst performance under all the considered scenarios.

Focusing on the CeROs result, we can observe a common trend moving from the lowest to the highest microservice-node ratio, the latency percentage distribution progressively shifts toward 100% and becomes more compact. This behavior reflects the increasing challenges of the scenarios.

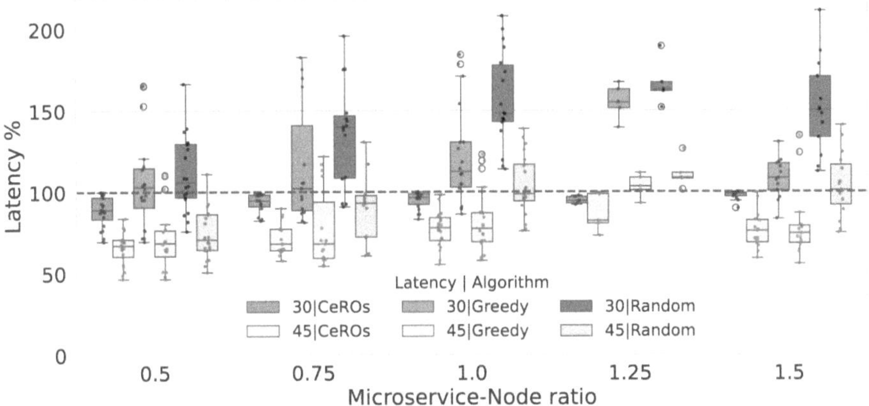

Fig. 5. Latency % by Microservice - Node ratio, Algorithm and Latency Class

Nodes and Links Resources Usage. Another KPI is resource usage at the single node and link level. CPU usage is taken as a representative resource consumption in nodes due to its strong correlation with overall resource requirements (see Table 2). As for the links we consider the used bandwidth. In Fig. 6, we report the boxplot of the distribution of the mean of the CPU (on the left) and link bandwidth (on the right) usage, grouped by algorithm, latency class, and microservice-node ratio. While not directly optimized, resources are carefully managed by CeROs, ensuring room for future deployments. In contrast, the *Greedy* algorithm prioritizes lower-cost nodes, leading to higher stacking, while the *Random* algorithm scatters solutions with very low stacking level.

As the microservice-node ratio increases, CeROs assigns more and more microservices to a limited set of nodes, minimizing the assignment costs and guaranteeing the end-to-end latency requirements. This behavior is common across all instances up to 1.25. This is confirmed by the distribution that becomes more and more compact. On the contrary, when the microservice-node ratio is 1.5, we observe a more diverse behavior of CeROs across instances due to the high number of microservices compared to the available physical nodes. In these scenarios, CeROs is forced to spread the application microservices across more nodes in some instances, leading to more dispersed distribution.

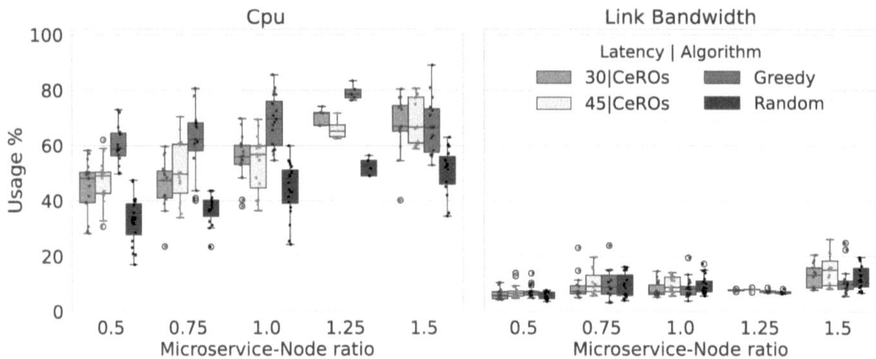

Fig. 6. Latency % by Microservice - Node ratio, Algorithm and Latency Class

5.4 CeROs vs Elastic-CeROs

The results have shown that the main cause of instance infeasibility is the violation of the end-to-end latency constraints. To quantify the extra latency required to make the instances feasible, we developed a comparative model in which we have relaxed the end-to-end latency constraints. Formally, we modify the model Sect. 3.1 by relaxing the constraints (10) allowing the latency limit to be exceeded and paying the excess in the objective function. We refer to this version of the model as Elastic-CeROs (E-CeROs) and is defined as follows:

$$\min \sum_{u \in T} \sum_{i \in I} w_i \cdot x_{ui} + \lambda \cdot l \tag{9}$$

$$\text{s.t. } (4), (5), (6), (7)$$

$$\sum_{u \in C} \sum_{i \in I} \delta_u \cdot x_{ui} + \sum_{(u,v) \in C} \sum_{i,j \in I} \Delta_{ij} \cdot x_{ui} \cdot x_{vj} \leq L_C + l \qquad \forall C \in \mathcal{C} \tag{10}$$

$$l \geq 0 \tag{11}$$

The weight of l is increased in (9) by λ to be heavily penalized compared to the centrality costs of the nodes.

In Fig. 7, we report the boxplot of the distribution of the percentage of end-to-end latency w.r.t. the constraint value, grouped by algorithm, latency class, and microservice-node ratio. As discussed in Sect. 5.3, most of the infeasible instances belong to the most stringent latency class (15 ms). In this setting, E-CeROs can resolve the instances by significantly exceeding the latency limit, with a median value between 140% and 170%, with a peak above 200%. In contrast, for the 30 ms latency class, E-CeROs successfully resolves infeasible instances with only a narrow margin exceeding the end-to-end limit. The results highlight that the physical topological constraints prohibit the E-CeROs solutions from offering sensible latency QoS.

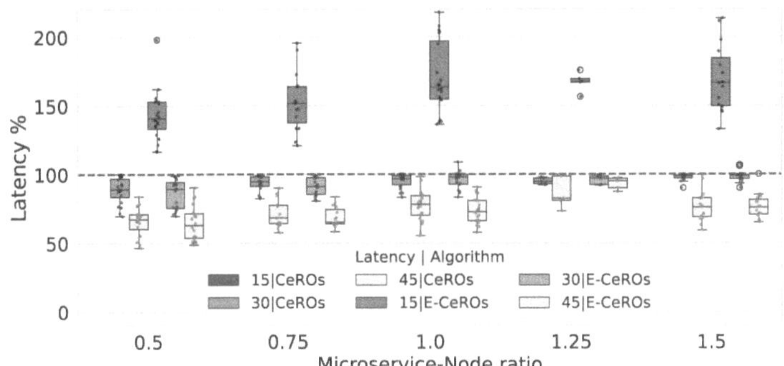

Fig. 7. Latency % comparison by Latency Class, Algorithm and Microservice-Node ratio

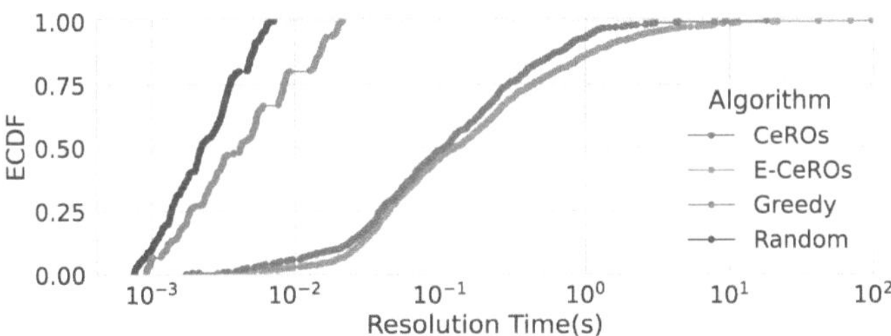

Fig. 8. Resolution time by algorithm

5.5 Resolution Time

Finally, we analyze the resolution time of the two versions of CeROs and the baseline algorithms. As we can observe in Fig. 8 they operate on different resolution time scales, with the baseline algorithms running an order of magnitude faster. CeROs exhibits a stable and smooth curve, whereas the baseline algorithms, especially the Greedy one, show distinct steps that indicate varying the hardness of the instance for the algorithm.

6 Conclusions and Perspectives

This paper presents CeROs, a novel MILP formulation to optimize microservice placement in cloud-to-edge continuum scenarios. The results have shown that CeROs achieves significantly lower latency and a more balanced resource distribution than the baseline algorithms, ensuring feasibility in a wide range of scenarios. The comparison with the elastic version of the model further highlights CeROs's effectiveness, showing that its solutions closely approach the best-relaxed configurations before hitting infeasibility. This confirms that CeROs ability to maximize performances while strictly adhering to constraints.

Future work could explore heuristic refinements and adaptive strategies to extend its applicability to even more complex deployments and scale to larger networks and microservices topologies.

References

1. Abdelaal, M.A., Ebrahim, G.A., Anis, W.R.: Efficient placement of service function chains in cloud computing environments. Electronics **10**(3) (2021). https://doi.org/10.3390/electronics10030323
2. Bari, M.F., Chowdhury, S.R., Ahmed, R., Boutaba, R.: On orchestrating virtual network functions. In: 2015 11th International Conference on Network and Service Management (CNSM), pp. 50–56 (2015). https://doi.org/10.1109/CNSM.2015.7367338
3. Bhamare, D., Jain, R., Samaka, M., Erbad, A.: A survey on service function chaining. J. Netw. Comput. Appl. **75**, 138–155 (2016). https://doi.org/10.1016/j.jnca.2016.09.001
4. Brandes, U.: On variants of shortest-path betweenness centrality and their generic computation. Soc. Netw. **30**(2), 136–145 (2008). https://doi.org/10.1016/j.socnet.2007.11.001
5. Cziva, R., Anagnostopoulos, C., Pezaros, D.P.: Dynamic, latency-optimal VNF placement at the network edge. In: IEEE INFOCOM 2018 - IEEE Conference on Computer Communications, pp. 693–701 (2018). https://doi.org/10.1109/INFOCOM.2018.8486021
6. Elias, J., Martignon, F., Paris, S., Wang, J.: Efficient orchestration mechanisms for congestion mitigation in NFV: models and algorithms. IEEE Trans. Serv. Comput. **10**(4), 534–546 (2017). https://doi.org/10.1109/TSC.2015.2498176
7. Gil Herrera, J., Botero, J.F.: Resource allocation in NFV: a comprehensive survey. IEEE Trans. Netw. Serv. Manage. **13**(3), 518–532 (2016). https://doi.org/10.1109/TNSM.2016.2598420
8. Leivadeas, A., Kesidis, G., Ibnkahla, M., Lambadaris, I.: VNF placement optimization at the edge and cloud. Future Internet (2019)
9. Martín-Pérez, J., Cominardi, L., Bernardos, C.J., de la Oliva, A., Azcorra, A.: Modeling mobile edge computing deployments for low latency multimedia services. IEEE Trans. Broadcast. **65**(2), 464–474 (2019). https://doi.org/10.1109/TBC.2019.2901406
10. Miller, C.E., Tucker, A.W., Zemlin, R.A.: Integer programming formulation of traveling salesman problems. J. ACM **7**(4), 326–329 (1960). https://doi.org/10.1145/321043.321046
11. Sun, J., Zhang, Y., Liu, F., Wang, H., Xu, X., Li, Y.: A survey on the placement of virtual network functions. J. Netw. Comput. Appl. **202**, 103361 (2022). https://doi.org/10.1016/j.jnca.2022.103361
12. Yusupov, J., Ksentini, A., Marchetto, G., Sisto, R.: Multi-objective function splitting and placement of network slices in 5G mobile networks. In: 2018 IEEE Conference on Standards for Communications and Networking (CSCN), pp. 1–6 (2018). https://doi.org/10.1109/CSCN.2018.8581714

Metabolic Pathway Models: An Overview of Existing Approaches and Emerging Perspectives

Aiello Francesca Antonella[1]([envelope]), Mario Lepore[2], Raffaele Maccioni[1], Elvira Plenzich[2], and Roberto Tufano[2]

[1] Math Biology, Diogeniano di Eraclea Street 89, 00124 Rome, RM, Italy
{francesca.aiello,raffaele.maccioni}@mathbiology.tech
[2] University of Salerno, Giovanni Paolo II Street 132, 84084 Fisciano, SA, Italy
{marlepore,eplenzich,rtufano}@unisa.it

Abstract. Metabolic pathways regulate essential biochemical processes in biological systems, and their modeling has provided fundamental tools for understanding cellular metabolism and its applications in fields such as precision medicine, biotechnology, and metabolic disease diagnostics. Over the years, numerous mathematical and computational models have been developed to describe these networks, employing approaches ranging from flux balance analysis to dynamic systems and data-driven methods. This work presents an overview of existing models for describing metabolic pathways, emphasizing their role in representing biochemical processes and their scientific and industrial applications. Building on this review, new perspectives are explored regarding the integration of metabolic models with innovative experimental data, particularly physiological signals that may contain implicit information yet to be validated. One specific case involves the use of surface electrical currents measured on human skin, which could provide insights into the metabolic demand of an organism and require validation through existing modeling frameworks. Furthermore, the potential for integrating metabolic models with advanced computational approaches is analyzed, with a focus on their contribution to the development of Human Digital Twins. These models enable "what-if" simulations to predict the effects of metabolic variations in a controlled virtual environment. This perspective opens new opportunities for research and applications in metabolic modeling, with implications for diagnostics, personalized therapies, and the development of new biomedical analysis tools.

Keywords: Metabolic Pathway Modeling · Systems Biology · Digital Twin · Machine Learning

1 Introduction and Rationale

Metabolism encompasses the set of biochemical reactions that sustain life by converting nutrients into energy and essential biomolecules [1]. These processes

© The Author(s), under exclusive license to Springer Nature Switzerland AG 2026
M. Pavone et al. (Eds.): DSA ISC 2025, LNCS 16405, pp. 213–227, 2026.
https://doi.org/10.1007/978-3-032-21811-7_15

are traditionally divided into two main categories: catabolism–the breakdown of complex molecules to release energy–and anabolism–the synthesis of complex molecules from simpler ones, which requires energy. For example, glycolysis and the tricarboxylic acid (TCA) cycle are key catabolic pathways that generate ATP and metabolic intermediates [2], while fatty acid and amino acid biosynthesis are anabolic routes involved in building cellular structures and signaling molecules.

These reactions are not isolated but are structured into metabolic pathways: chains of enzyme-catalyzed steps that transform specific substrates into intermediates and final products [3]. Metabolic pathways are highly interconnected and regulated, forming complex networks that enable cells to maintain homeostasis and respond to physiological and environmental changes.

The main components of these networks include metabolites (e.g., sugars, amino acids, lipids), enzymes that catalyze specific reactions, biochemical transformations linking substrates and products, and regulatory mechanisms such as allosteric modulation, covalent modification, and gene expression control [4]. As an illustrative example, the tricarboxylic acid (TCA) cycle is shown in Fig. 1, highlighting the core sequence of reactions involved in aerobic energy metabolism.

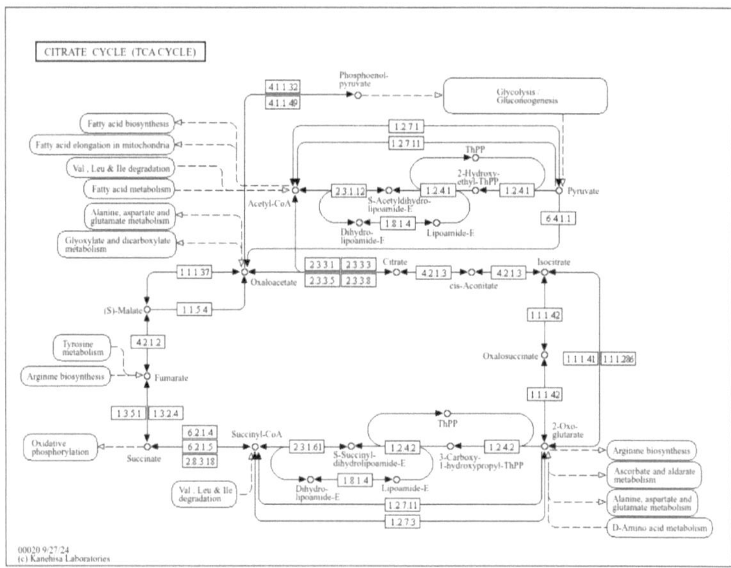

Fig. 1. The image represents the schematic visualization of the TCA cycle, a central metabolic pathway that oxidizes acetyl-CoA to CO_2, producing NADH and $FADH_2$ for ATP generation. It also provides key intermediates for biosynthetic processes and is tightly regulated to adapt to cellular energy demands. Image adapted from KEGG PATHWAY database [5]

Understanding these structures and their dynamic interactions is essential for studying both normal physiology and disease states. Alterations in metabolic pathways have been implicated in numerous conditions, including cancer, diabetes, neurodegenerative diseases, and inherited metabolic disorders [6,7].

Metabolism also plays a central role in applied fields. In pharmacology, for example, drugs such as statins target metabolic enzymes to modulate cholesterol biosynthesis [8]. In biotechnology, metabolic engineering is used to optimize the microbial production of biofuels, antibiotics, and other high-value compounds [9].

Given the complexity and scale of metabolic networks, especially in the era of high-throughput omics and multi-modal data, their investigation increasingly relies on computational and mathematical models capable of capturing their dynamic, multi-scale behavior.

The sheer complexity and interconnectivity of metabolic pathways, along with their regulation across multiple biological layers, make it difficult to rely on purely qualitative or descriptive approaches. Recent advances in high-throughput techniques–such as genomics, transcriptomics, proteomics, metabolomics, and fluxomics–have enabled the generation of massive datasets that offer a detailed, quantitative view of cellular metabolism. For instance, metabolomics provides snapshots of small-molecule activity [10], proteomics reveals enzyme levels and modifications [11], and fluxomics captures dynamic metabolic flows under varying conditions [12].

To support the interpretation of this wealth of data, several knowledge bases have been developed. Among the most prominent is the Kyoto Encyclopedia of Genes and Genomes (KEGG), which catalogs thousands of reactions and metabolites [5]. Other widely used resources include MetaCyc [13], Reactome [14], TRANSPATH [15], and BioModels [16].

However, these databases, while foundational, are not sufficient to describe the dynamic behavior of metabolic systems. Fully understanding such systems requires formal computational models capable of integrating heterogeneous data sources, representing interactions quantitatively, and enabling predictive simulations under different biological scenarios.

Over the past decades, a wide range of modeling approaches have been developed to represent and analyze metabolic pathways. These include stoichiometric and constraint-based models [17], kinetic models based on systems of differential equations [18], stochastic formulations that account for intrinsic biological variability [19], and more recent data-driven frameworks based on machine learning techniques [20]. Each approach offers a different lens through which to explore metabolism, from steady-state flux distributions to dynamic concentration profiles and predictive simulations.

These models serve as a bridge between experimental observations and mechanistic understanding, enabling in silico experimentation that would otherwise be costly, time-consuming, or even infeasible in vivo [21]. Despite the proliferation of modeling techniques, the literature still lacks a concise and up-to-date synthesis in light of current methodological and technological advancements.

This work aims to address this gap by reviewing the principal paradigms used for modeling metabolic pathways, highlighting their assumptions, data requirements, temporal dynamics, and application domains. In addition to presenting the main modeling approaches, this work also examines emerging directions in metabolic modeling research. Particular attention is given to the integration between explicit biological knowledge and implicit information derived from novel physiological signals. Moreover, the potential of embedding metabolic models within multi-scale digital twin frameworks is discussed, with the aim of enabling scenario-based simulations and personalized evaluations of metabolic function [22,23].

These emerging directions are key to enabling context-aware simulations, improving model explainability, and supporting personalized medicine through a systems-level understanding of metabolism.

The rest of this paper is structured as follows: Sect. 2 surveys the main classes of metabolic models proposed in the literature. Section 3 outlines new research perspectives and open challenges. Finally, Sect. 4 presents concluding remarks.

2   Modeling Approaches for Metabolic Pathways

This section presents a selection of modeling approaches that have been proposed in the literature to represent and analyze metabolic pathways. These models differ in terms of formalisms, assumptions, temporal dynamics, and data requirements. For each approach, representative works are briefly discussed to illustrate their conceptual framework and applications. The aim is not to provide an exhaustive comparison, but rather to highlight the diversity of methodologies currently available and their relevance to different research contexts. The selected contributions were identified based on their scientific relevance, influence in the field, and their ability to exemplify distinct modeling paradigms, including stoichiometric, kinetic, stochastic, and data-driven approaches.

2.1   Constraint-Based Models

Constraint-based modeling (CBM) provides a powerful framework for analyzing metabolic networks without requiring detailed kinetic parameters. It starts from a genome-scale reconstruction of the metabolic network, which includes genes, enzymes, metabolites, and reactions [24]. These reconstructions are refined by integrating biochemical and functional annotations [25], resulting in a stoichiometric representation suitable for computational analysis.

The behavior of a metabolic network is governed by mass balance equations for each metabolite:

$$\frac{dX_i}{dt} = \sum_j S_{ij} v_j, \tag{1}$$

where X_i is the concentration of metabolite i, v_j is the flux of reaction j, and S_{ij} is the stoichiometric coefficient.

Assuming a steady-state condition ($dX_i/dt = 0$), the system reduces to a set of linear constraints:

$$S \cdot v = 0, \tag{2}$$

subject to bounds $a_j \leq v_j \leq b_j$ that represent thermodynamic and biological constraints.

The space of feasible flux distributions forms a convex polyhedral cone:

$$C = \{v \mid v = \sum_{i=1}^{k} w_i p_i, \quad w_i \geq 0\}, \tag{3}$$

where p_i are the extreme pathways.

A widely used constraint-based approach is Flux Balance Analysis (FBA), which applies linear programming to identify an optimal flux distribution that maximizes (or minimizes) a biological objective, such as biomass production:

$$\max Z = \sum_{j} c_j v_j, \quad \text{subject to} \quad S \cdot v = 0, \quad a_j \leq v_j \leq b_j. \tag{4}$$

FBA has been extensively applied in metabolic engineering, gene knockout analysis, phenotype prediction, and drug target identification [26]. For instance, it has been used to simulate growth of *Escherichia coli* under varying oxygen conditions [26].

Despite its success, FBA has limitations. The results depend on the completeness of the metabolic reconstruction and on the choice of the objective function, which may not always reflect biological reality. Several extensions address these issues: MoMA (Minimization of Metabolic Adjustment) [27], ROOM (Regulatory On/Off Minimization) [28], and BOSS (Biological Objective Solution Search) [29] propose alternative optimization criteria. However, these often require high-resolution experimental data.

Another limitation is the inability of FBA to capture regulatory and signaling processes. Extensions such as Dynamic FBA (DFBA) [30] and Integrated FBA (IFBA) [31] attempt to incorporate temporal dynamics and regulatory constraints, but challenges remain due to the multi-scale nature of cellular systems.

Overall, constraint-based models, and in particular FBA, provide a scalable and interpretable framework for exploring metabolic phenotypes, especially in systems where kinetic parameters are unavailable or uncertain.

2.2 Kinetic Models

Kinetic models describe the dynamic evolution of metabolite concentrations over time by using ordinary differential equations (ODEs) derived from reaction kinetics [18]. Unlike constraint-based models, which rely on steady-state assumptions, kinetic models explicitly define reaction rates as functions of substrate concentrations and enzymatic activity. This allows the simulation of transient responses and regulatory effects, but requires detailed knowledge of kinetic parameters, which are often unavailable or context-dependent [32].

A widely used formulation is the Michaelis-Menten equation, which models the reaction rate v as a function of substrate concentration $[S]$:

$$v = \frac{d[P]}{dt} = \frac{V_{\max}[S]}{K_M + [S]},\tag{5}$$

where $[P]$ is the product concentration, $V_{\max}$ the maximum rate, and K_M the Michaelis constant [33].

Despite its utility, the Michaelis-Menten model has limitations, especially when applied to complex systems. Alternative approaches based on mass-action kinetics can offer greater mechanistic detail, but require even more parameters and assumptions [34].

Defining pathway boundaries and resolving inconsistencies across data sources is another major challenge. Standardized resources like KEGG, MPW, and BRENDA offer consensus pathway representations that help reduce variability in network structure [35].

Recent efforts in kinetic modeling focus on integrating multiple layers of regulation, including gene expression and signaling, to better reflect cellular complexity [36]. Multi-omics datasets are increasingly used to inform kinetic models, linking transcript, protein, and metabolite levels under different conditions [37]. These approaches support more robust, predictive simulations and pave the way for personalized applications in medicine and metabolic engineering [38].

2.3   Petri Net Models

Petri Nets (PNs) are an effective tool for modeling and analyzing metabolic pathways, offering a structured representation of complex biochemical systems. Originally developed in computer science to model concurrent processes, PNs have been applied to biological networks due to structural analogies between the two domains. Both describe systems composed of interacting entities–reactions that consume and produce resources–and share an intuitive graphical representation that supports comprehension and analysis [32].

Formally, a Petri Net is defined as a tuple $N = (P, T, W, M_0)$, where:

- P is the set of places;
- T is the set of transitions;
- $W : (P \times T) \cup (T \times P) \to \mathbb{N}$ is the weight function;
- M_0 is an n-dimensional vector of non-negative integers representing the initial marking of the net.

In the graphical representation, places (drawn as circles) correspond to metabolites or molecular species, while transitions (rectangles) represent biochemical reactions. Directed arcs connect places and transitions, indicating consumption and production of tokens. Tokens within places denote the quantity or availability of a metabolite, and a transition is enabled when all input places contain a sufficient number of tokens based on the arc weights.

This formalism allows the application of analytical techniques from PN theory to biochemical networks, supporting visualization, structural analysis, and dynamic simulation of metabolic processes.

When modeling metabolic pathways using PNs, there is a natural correspondence between biochemical network elements and PN components [32], as illustrated in Table 1:

Table 1. Correspondence between metabolic pathway elements and Petri Net components [32]

Pathway Element	PN Element	Description
Metabolites, proteins, enzymes	Places	Represent molecular species in the system.
Chemical reactions	Transitions	Represent events modifying the system state.
Substrates, reagents	Input places	Provide resources to a transition (reaction inputs).
Reaction products	Output places	Receive resources after a transition fires.
Stoichiometric matrix	Incidence matrix	Defines the quantitative relationship between reactants and products.
Stoichiometric coefficients	Arc weights	Indicate the quantity of resources consumed or produced.
Metabolites, enzymes, compounds quantities	Number of tokens on places	Represent the availability of a resource at a given time.
Reaction kinetic laws	Transition rates	Regulate reaction speed within the system.

The analysis of a Petri Net model enables the study of behavioral properties–such as reachability, boundedness, and liveness–as well as structural properties that depend solely on the network topology [39]. Several extensions have been developed to overcome the limitations of the basic formalism in representing real-world biological systems. Qualitative variants enhance expressiveness by introducing new arc types or colored tokens, while quantitative extensions incorporate temporal or probabilistic elements to simulate dynamic and uncertain behaviors [32]. Advanced models, such as colorized, timed, stochastic, continuous, and hybrid Petri nets, support more detailed simulations and can integrate kinetic laws, as discussed in the previous section. Although these extensions improve modeling capabilities, they also increase system complexity, limiting the availability of efficient analysis algorithms [32].

2.4 Stochastic Models

The adoption of stochastic models in metabolic pathway modeling allows for capturing the variability and unpredictability of these systems, particularly when molecule numbers are low, providing a more realistic representation compared to deterministic approaches [40, 41]. These models describe metabolic processes as probabilistic events, accurately reflecting intrinsic biochemical fluctuations. Among the most widely used techniques, Markov processes model the temporal evolution of metabolic networks, facilitating the study of transitions between different metabolic states [42]. In [43], metabolic network evolution is described as a discrete-space, continuous-time Markov process, where reaction addition or removal depends on neighboring reaction fractions. Another important approach is based on Fokker-Planck equations, which describe the probability distribution of metabolite concentrations over time, offering a more detailed view of biochemical dynamics [44].

A fundamental tool for simulating these models is the Gillespie algorithm, which determines reaction execution times and types based on event probabilities [45]. This method is particularly effective for systems with low metabolite concentrations, where deterministic assumptions fail.

Several software tools facilitate stochastic metabolic network modeling. COPASI supports both deterministic and stochastic simulations [46], StochPy provides a specialized Python package for biochemical system simulations [47], and libRoadRunner offers a high-performance library for analyzing and simulating System Biology Markup Language (SBML)-based metabolic models [48]. These tools enhance the exploration of metabolic networks, improving our understanding of cellular processes and their biomedical and biotechnological applications.

2.5 Machine Learning-Based Models

The integration of Machine Learning (ML) and Artificial Intelligence (AI) into metabolic network modeling has significantly enhanced our ability to understand and predict the behavior of complex biological systems. These approaches overcome the limitations of traditional models, providing more accurate predictions and facilitating the optimization of metabolic pathways.

For instance, a recent review highlighted how ML applications in genome-scale metabolic modeling and multi-step metabolic pathway optimization have improved metabolic engineering efforts [49]. Additionally, integrating ML with constraint-based models has enabled the combination of experimental and in silico data, enhancing knowledge mapping in biological systems [50]. Although to a lesser extent compared to other applications, these methodologies have also been used to exploit the vast availability of genomic and metabolomic data, contributing to the optimization and analysis of metabolic network models. Specifically, ML has supported the development of metabolic networks, the estimation of parameters for stoichiometric and kinetic models, and the identification of key features for optimal bioreactor applications [20].

In this context, software tools like COBRApy offer a Python interface for constraint-based modeling, supporting advanced analyses of metabolic networks and integrating ML techniques to improve the predictability of models [51]. These tools combine data-driven and constraint-based approaches, providing valuable resources for optimizing metabolic systems.

AI has also shown significant promise in designing dynamic metabolic pathways. A recent review discussed how AI is accelerating the pathway design cycle by identifying hidden patterns in data and rapidly evaluating vast collections of pathway designs [52]. Such advancements highlight the potential of AI to enhance pathway engineering and optimization.

Moreover, ML applications in metabolic engineering have demonstrated considerable progress, particularly with hybrid neural-mechanistic models that combine the strengths of mechanistic models with the flexibility of neural networks, improving phenotype predictions [53]. An example of this is DeepMetabolism, a deep learning system designed to predict cellular phenotypes from genomic sequencing data [54]. By integrating both unsupervised and supervised learning approaches, this algorithm improves the accuracy of phenotypic predictions.

Another AI-driven application is eselaGen, which aids in the discovery, characterization, and optimization of enzymes, pathways, and strains for high-value molecule production. This system represents a significant advancement in applying AI to predict cellular behavior, ultimately improving the design of metabolic networks and biotechnological applications

3 New Perspectives and Open Challenges

Despite the significant progress made in modeling metabolic pathways, several open challenges remain at the intersection of biology, data science, and computational modeling. Advances in sensing technologies, artificial intelligence, and physiological data acquisition offer unprecedented opportunities to expand the scope and applicability of metabolic models beyond traditional frameworks. This section explores emerging directions that could shape the next generation of metabolic modeling, including the integration of unconventional physiological signals. The conceptual interplay between explicit and implicit knowledge is also examined, along with the importance of explainable data-driven approaches and the potential role of simulation in the development of Human Digital Twins.

3.1 Novel Data Streams for Enhancing Metabolic Modeling

Among the emerging directions in metabolic modeling, a promising line of investigation involves the integration of novel physiological data sources acquired through non-invasive techniques. One such initiative is currently being pursued by the startup Math Biology, which has developed a proprietary method known as Deep Metabolic-processes Assessment (DMA) for real-time, non-invasive metabolic analysis. This approach is designed to detect variations in weak electrical signals measured on the surface of the human body in response to controlled exposure to specific molecular markers.

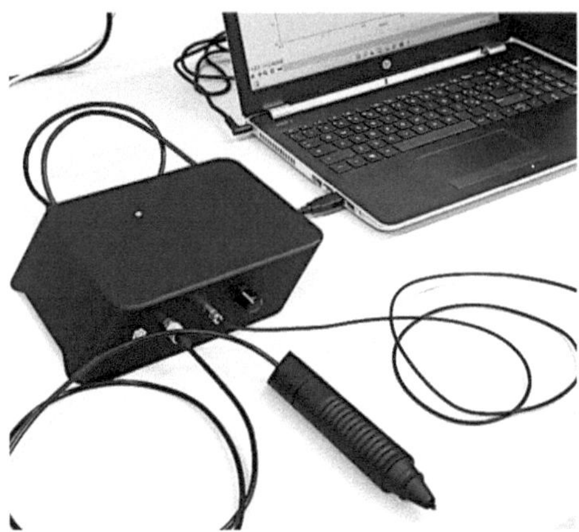

Fig. 2. The hardware device was developed in collaboration with the CNR-MISTER and used by Math Biology for the acquisition of bioelectrical signals

The DMA technology is supported by a custom-designed amperometric device, depicted in Fig. 2 and developed in collaboration with the CNR-MISTER[1], is used by Math Biology for the acquisition of ultra-low electrical currents. These signals, derive from the bioelectrical activity of the human body, which reflect underlying metabolic activity changes linked to physiological and pathological states. The analysis process typically involves an initial phase of baseline signal acquisition, followed by a phase of exposure to specific metabolic markers to elicit signal variation and a final AI-based data processing stage. The result is a Single Marker Response (SMR) index, which provides insight into metabolic imbalances, hormonal changes, stress-related signals, and early signs of disease. The potential of this technology is currently under evaluation in the context of the PREVEDO project, an initiative funded by the Italian Ministry of Economic Development (MISE) and carried out in collaboration with various departments of the University of Salerno. The project aims to assess the feasibility of using DMA-derived signals to support early diagnosis and therapy monitoring in cardiology and oncology, domains where metabolic dysregulation plays a central role.

While DMA and the PREVEDO project are still in an early experimental phase, their relevance to metabolic pathway modeling lies in the opportunity they offer to explore new forms of physiological data that may reflect underlying metabolic activity. From a modeling perspective, such data streams–if properly validated–could serve as inputs for enriching or calibrating existing models of metabolic pathways, particularly in the context of patient-specific monitoring or real-time prediction. In this regard, initiatives like PREVEDO illustrate how emerging technologies may pose new challenges and opportunities for extending

---

[1] https://www.cnr.it.

classical modeling frameworks and for designing novel approaches that integrate external physiological signals with internal biochemical representations.

3.2 Integrating Implicit and Explicit Knowledge in Metabolic Modeling

The increasing availability of physiological data obtained through novel sensor technologies introduces new opportunities, but also significant interpretative challenges. These data streams–often collected in real time and in non-invasive ways–may capture aspects of metabolic activity not directly addressed by traditional biochemical assays. However, such signals typically represent implicit knowledge: patterns emerging from empirical observation that are not yet grounded in a mechanistic framework. In order to interpret these signals and assess their biological relevance, one possible strategy is to relate them to explicit knowledge, such as that encoded in mathematical models of metabolic pathways. These models describe known reactions, fluxes, and regulatory mechanisms and can be used to infer what kinds of physiological responses are expected under specific conditions. If a correlation were observed between a given physiological signal–such as a surface electrical current variation in response to a biochemical marker–and a known metabolic reaction involving the same molecule, this alignment could suggest that the signal reflects a real underlying metabolic demand. While such correspondence would not constitute definitive proof, it could provide indirect support for the biological significance of the signal and the reliability of the acquisition method. This reasoning, currently being explored in relation to DMA technology, may be extended to other types of physiological signals obtained through emerging diagnostic tools. In all cases, explicit models may serve as a useful interpretative layer to anchor data-driven findings within a biologically coherent structure, thus supporting hypothesis generation, signal validation, and informed model refinement.

3.3 Metabolic Pathways and the Digital Twin Approach

Digital twin (DT) technology is increasingly recognized as a transformative framework for the future of metabolic modeling. Rather than serving merely as digital replicas of physiological processes, DTs offer a way to integrate mechanistic representations of metabolic pathways with patient-specific, time-dependent data streams–ranging from omics profiles to clinical biomarkers and physiological signals. In this perspective, DTs could act as dynamic containers for formal metabolic models, allowing simulations to be continuously updated as new data become available. This opens the door to predictive and context-aware simulations of metabolic alterations, with potential applications in early diagnosis, therapy monitoring, and personalized intervention planning. Recent advances suggest the feasibility of embedding flux-based models and ODE-driven metabolic networks within DT architectures, especially in chronic metabolic disorders like diabetes [22,55]. Nevertheless, several challenges remain. These include the standardization of metabolic model integration, the validation of predictions in clinical settings, and the need for robust data pipelines that preserve explainability and traceability across different data sources. In research and industrial biotechnology, similar approaches have been explored to track microbial metabolism using multi-omics data and to guide real-time optimization of bioproduction processes [56]. These developments reinforce the potential of DTs as adaptive infrastructures for metabolic knowledge, but also highlight the importance of modularity, interoperability, and the harmonization between explicit model-based reasoning and implicit data-driven learning. In the context of metabolic systems biology, the incorporation of digital twins represents not just a technical evolution, but a conceptual shift–from static model deployment to continuous, simulation-based decision support.

4   Conclusions

This paper has reviewed the main mathematical and computational approaches used to model metabolic pathways, including constraint-based, kinetic, Petri net, and machine learning-based models. Each of these paradigms offers a different lens for analyzing the structure, dynamics, and regulation of metabolism, with implications ranging from basic research to metabolic engineering and precision medicine. Among future directions, the integration of novel physiological signals–such as those acquired via bioelectrical measurement technologies–could play a central role in refining and personalizing simulations. Within this context, the PREVEDO project aims to investigate the potential of these signals to inform metabolic modeling and enhance predictive capabilities in clinical applications. This work is part of a broader research effort aiming to bridge theoretical modeling and clinical decision-making through explainable AI and service-oriented infrastructures. Initial steps in this direction include the design of architectural solutions for AI-based decision support and patient stratification systems, developed and tested in collaboration with MathBiology [57,58]. Future research will focus on validating the proposed hypotheses and further developing integrative frameworks that connect mechanistic models with data-driven insights, with the long-term goal of supporting personalized metabolic analysis and simulation.

Disclosure of Interests. The authors have no competing interests to declare that are relevant to the content of this article.

References

1. Salway, J.G.: Metabolism at a Glance. John Wiley & Sons (2016)
2. Krebs, H.A.: The tricarboxylic acid cycle. In: Chemical pathways of metabolism, pp. 109–171. Elsevier (1954)
3. Dagley, S., Nicholson, D.E.: An introduction to metabolic pathways. (1970)
4. Sharma, A., Anand, S.K., Singh, N., Dwivedi, U.N., Kakkar, P.: Amp-activated protein kinase: an energy sensor and survival mechanism in the reinstatement of metabolic homeostasis. Exp. Cell Res. **428**(1), 113614 (2023)

5. Kanehisa, M., Goto, S.: Kegg: kyoto encyclopedia of genes and genomes. Nucleic Acids Res. **28**(1), 27–30 (2000)
6. Stelling, J., Klamt, S., Bettenbrock, K., Schuster, S., Gilles, E.D.: Metabolic network structure determines key aspects of functionality and regulation. Nature **420**(6912), 190–193 (2002)
7. Shlomi, T., Benyamini, T., Gottlieb, E., Sharan, R., Ruppin, E.: Genome-scale metabolic modeling elucidates the role of proliferative adaptation in causing the warburg effect. PLoS Comput. Biol. **7**(3), e1002018 (2011)
8. Oesterle, A., Laufs, U., Liao, J.K.: Pleiotropic effects of statins on the cardiovascular system. Circ. Res. **120**(1), 229–243 (2017)
9. Lee, G., Lee, S.M., Kim, H.U.: A contribution of metabolic engineering to addressing medical problems: metabolic flux analysis. Metab. Eng. **77**, 283–293 (2023)
10. Saudubray, J.-M., García-Cazorla, A.: Clinical approach to inborn errors of metabolism in paediatrics. In: Inborn Metabolic Diseases: Diagnosis And Treatment, pp. 3–123. Springer (2022). https://doi.org/10.1007/978-3-642-15720-2_1
11. Aebersold, R., Mann, M.: Mass spectrometry-based proteomics. Nature **422**(6928), 198–207 (2003)
12. Sauer, U.: Metabolic networks in motion: 13c-based flux analysis. Mol. Syst. Biol. **2**(1), 62 (2006)
13. Caspi, R., et al.: The metacyc database of metabolic pathways and enzymes and the biocyc collection of pathway/genome databases. Nucleic Acids Res. **36**(suppl_1), D623–D631 (2007)
14. Jupe, S., Akkerman, J.W., Soranzo, N., Ouwehand, W.H.: Reactome-a curated knowledgebase of biological pathways: megakaryocytes and platelets. J. Thromb. Haemost. **10**(11), 2399–2402 (2012)
15. Krull, M., et al.: Transpath®: an information resource for storing and visualizing signaling pathways and their pathological aberrations. Nucleic Acids Res. **34**(suppl_1), D546–D551 (2006)
16. Novere, N.L., et al.: Biomodels database: a free, centralized database of curated, published, quantitative kinetic models of biochemical and cellular systems. Nucleic Acids Res. **34**(suppl_1), D689–D691 (2006)
17. Varner, J., Ramkrishna, D.: Mathematical models of metabolic pathways. Curr. Opin. Biotechnol. **10**(2), 146–150 (1999)
18. Rohwer, J.M.: Kinetic modelling of plant metabolic pathways. J. Exp. Bot. **63**(6), 2275–2292 (2012)
19. Ao, P.: Metabolic network modelling: including stochastic effects. Comput. Chem. Eng. **29**(11–12), 2297–2303 (2005)
20. Cuperlovic-Culf, M.: Machine learning methods for analysis of metabolic data and metabolic pathway modeling. Metabolites **8**(1), 4 (2018)
21. Carstensen, P.E., Bendsen, J., Reenberg, A.T., Ritschel, T.K.S., Jørgensen, J.B.: A whole-body multi-scale mathematical model for dynamic simulation of the metabolism in man. IFAC-PapersOnLine **55**(23), 58–63 (2022)
22. Mosquera-Lopez, C., Jacobs, P.G.: Digital twins and artificial intelligence in metabolic disease research. Trends in Endocrinology & Metabolism (2024)
23. Dudas, B., Miteva, M.A.: Computational and artificial intelligence-based approaches for drug metabolism and transport prediction. Trends Pharmacol. Sci. **45**(1), 39–55 (2024)
24. Pitkänen, E., Rousu, J., Ukkonen, E.: Computational methods for metabolic reconstruction. Curr. Opin. Biotechnol. **21**(1), 70–77 (2010)

25. Feist, A.M., Herrgård, M.J., Thiele, I., Reed, J.L., Palsson, B.Ø.: Reconstruction of biochemical networks in microorganisms. Nat. Rev. Microbiol. **7**(2), 129–143 (2009)
26. Orth, J.D., Thiele, I., Palsson, B.Ø.: What is flux balance analysis? Nat. Biotechnol. **28**(3), 245–248 (2010)
27. Segre, D., Vitkup, D., Church, G.M.: Analysis of optimality in natural and perturbed metabolic networks. Proc. Natl. Acad. Sci. **99**(23), 15112–15117 (2002)
28. Shlomi, T., Berkman, O., Ruppin, E.: Regulatory on/off minimization of metabolic flux changes after genetic perturbations. Proc. Natl. Acad. Sci. **102**(21), 7695–7700 (2005)
29. Gianchandani, E.P., Oberhardt, M.A., Burgard, A.P., Maranas, C.D., Papin, J.A.: Predicting biological system objectives de novo from internal state measurements. BMC Bioinform. **9**, 1–13 (2008)
30. Henson, M.A., Hanly, T.J.: Dynamic flux balance analysis for synthetic microbial communities. IET Syst. Biol. **8**(5), 214–229 (2014)
31. Covert, M.W., Xiao, N., Chen, T.J., Karr, J.R.: Integrating metabolic, transcriptional regulatory and signal transduction models in escherichia coli. Bioinformatics **24**(18), 2044–2050 (2008)
32. Baldan, P., Cocco, N., Marin, A., Simeoni, M.: Petri nets for modelling metabolic pathways: a survey. Nat. Comput. **9**, 955–989 (2010)
33. Johnson, K.A., Goody, R.S.: The original michaelis constant: translation of the 1913 michaelis-menten paper. Biochemistry **50**(39), 8264–8269 (2011)
34. Breitling, R., Gilbert, D., Heiner, M., Orton, R.: A structured approach for the engineering of biochemical network models, illustrated for signalling pathways. Brief. Bioinform. **9**(5), 404–421 (2008)
35. Kanehisa, M.: The kegg database. In: Silico'simulation Of Biological Processes: Novartis Foundation Symposium, vol. 247, pp. 91–103. Wiley Online Library (2002)
36. Klipp, E., Nordlander, B., Krüger, R., Gennemark, P., Hohmann, S.: Integrative model of the response of yeast to osmotic shock. Nat. Biotechnol. **23**(8), 975–982 (2005)
37. Hasin, Y., Seldin, M., Lusis, A.: Multi-omics approaches to disease. Genome Biol. **18**, 1–15 (2017)
38. Beale, D.J., Karpe, A.V., Ahmed, W.: Beyond metabolomics: a review of multi-omics-based approaches. In: Microbial metabolomics: Applications in Clinical, Environmental, and Industrial Microbiology, pp. 289–312 (2016)
39. Murata, T.: Petri nets: properties, analysis and applications. Proc. IEEE **77**(4), 541–580 (1989)
40. Wilkinson, D.J.: Stochastic modelling for systems biology. Chapman and Hall/CRC (2018)
41. Gillespie, D.T.: Stochastic simulation of chemical kinetics. Annu. Rev. Phys. Chem. **58**(1), 35–55 (2007)
42. Golightly, A., Gillespie, C.S.: Simulation of stochastic kinetic models. In Silico Syst. Biol., 169–187 (2013)
43. Mithani, A., Preston, G.M., Hein, J.: A stochastic model for the evolution of metabolic networks with neighbor dependence. Bioinformatics **25**(12), 1528–1535 (2009)
44. Wei-Hua, M., Ou-Yang, Z.-C., Li, X.-Q.: From chemical langevin equations to fokker-planck equation: application of hodge decomposition and klein-kramers equation. Commun. Theor. Phys. **55**(4), 602 (2011)
45. Gillespie, T.: The relevance of algorithms. Media Technol. Essays Commun. Materiality Soc. **167**(2014), 167 (2014)

46. Mendes, P., Hoops, S., Sahle, S., Gauges, R., Dada, J., Kummer, U.: Computational modeling of biochemical networks using copasi. Syst. Biol., 17–59 (2009)
47. Maarleveld, T.R., Olivier, B.G., Bruggeman, F.J.: Stochpy: a comprehensive, user-friendly tool for simulating stochastic biological processes. PLoS ONE $8(11)$, e79345 (2013)
48. Somogyi, E.T., et al.: Jlibroadrunner: a high performance sbml simulation and analysis library. Bioinformatics $31(20)$, 3315–3321 (2015)
49. Cheng, Y., et al.: Machine learning for metabolic pathway optimization: a review. Comput. Struct. Biotechnol. J. 21, 2381–2393 (2023)
50. Kundu, P., Beura, S., Mondal, S., Das, A.K., Ghosh, A.: Machine learning for the advancement of genome-scale metabolic modeling. Biotechnol. Adv., 108400 (2024)
51. Ebrahim, A., Lerman, J.A., Palsson, B.O., Hyduke, D.R.: Cobrapy: constraints-based reconstruction and analysis for python. BMC Syst. Biol. 7, 1–6 (2013)
52. Merzbacher, C., Oyarzún, D.A.: Applications of artificial intelligence and machine learning in dynamic pathway engineering. Biochem. Soc. Trans. $51(5)$, 1871–1879 (2023)
53. Faure, L., Mollet, B., Liebermeister, W., Faulon, J.-L.: A neural-mechanistic hybrid approach improving the predictive power of genome-scale metabolic models. Nat. Commun. $14(1)$, 4669 (2023)
54. Guo, W., Xu, Y., Feng, X.: Deepmetabolism: a deep learning system to predict phenotype from genome sequencing. arXiv preprint arXiv:1705.03094 (2017)
55. Surian, N.U., et al.: A digital twin model incorporating generalized metabolic fluxes to identify and predict chronic kidney disease in type 2 diabetes mellitus. NPJ Digital Med. $7(1)$, 140 (2024)
56. Helmy, M., Elhalis, H., Rashid, M.M., Selvarajoo, K.: Can digital twin efforts shape microorganism-based alternative food? Curr. Opin. Biotechnol. 87, 103115 (2024)
57. Lepore, M., Plenzich, E., Tufano, R., Cerulli, R., Maccioni, R.: Improving patient's medical history classification using a feature construction approach based on situation awareness and granular computing. Neural Comput. Appl. $36(35)$, 22461–22484 (2024)
58. Cerulli, R., Lepore, M., Maccioni, R., Plenzich, E., Tufano, R.: A Service oriented architecture for clinical decision support systems based on artificial intelligence. In: Juan, A.A., Faulin, J., Lopez-Lopez, D. (eds.) Decision Sciences. DSA ISC 2024. LNCS, vol. 14779. Springer, Cham (2025). https://doi.org/10.1007/978-3-031-78241-1_15

An Explainable Machine-Learning Framework for Detecting Fraudulent Bitcoin Addresses with Graph–Temporal Features

Mario Trerotola[1(✉)], Davide Calvaresi[2], and Mimmo Parente[3]

[1] Politecnico di Torino, Torino, Italy
`mario.trerotola@polito.it`
[2] University of Applied Sciences of Western Switzerland, Sierre, Switzerland
`davide.calvaresi@hevs.ch`
[3] Università degli Studi di Salerno, Fisciano, Italy
`parente@unisa.it`

Abstract. The pseudonymous nature of blockchain transactions poses a significant challenge for identifying fraudulent activity in decentralized financial systems. This study presents a comprehensive framework for classifying Bitcoin wallets as fraudulent or legitimate by integrating multi-source data, graph-based transaction modeling, and machine learning. Our methodology builds upon publicly available datasets–namely Elliptic++, Chainabuse, and a curated sample of recent transactions–and integrates structural, temporal, and monetary features extracted from the Bitcoin transaction graph. Through systematic experiments across three distinct labeling scenarios, we demonstrate that ensemble methods such as Random Forests offer strong performance even under label noise, achieving F1-scores up to 0.92. Moreover, an explainability framework grounded, in SHAP values, is used to systematically analyze feature contributions and elucidate behavioral patterns associated with financial fraud. Our approach bridges empirical robustness with forensic insight, contributing a scalable, transparent toolset for blockchain compliance and risk analysis.

Keywords: Bitcoin · Blockchain Analytics · Wallet Classification · Financial Fraud · Machine Learning · Explainable AI

1 Introduction

Cryptocurrencies like Bitcoin have redefined financial transactions, enabling decentralized, borderless value transfer. However, their adoption has been accompanied by growing concerns over fraudulent uses including scams, money laundering, and darknet commerce. Identifying fraudulent actors on-chain is a non-trivial task due to the pseudonymous nature of blockchain addresses, the high volume of transactions, and the lack of ground truth labels.

Existing approaches to wallet classification rely on supervised learning, requiring both engineered features and labeled data. Yet, the scarcity of high-quality labels, combined with behavioral variability among actors, complicates detection efforts. Moreover, heuristic-based clustering approaches have shown promising results in attributing ownership based on transaction patterns.

Our work tackles these challenges through:

- Multi-source data integration and pre-processing strategies for constructing high-quality labeled datasets, inspired by the address and transaction table extraction pipelines used in large-scale studies [7].
- A principled set of features capturing structural, temporal, and monetary characteristics of wallets, analogous to transaction patterns such as peel chains, sweep, relay, and self-spending, which have been successfully exploited to improve address clustering [7].
- Evaluation of machine learning models under varying data quality scenarios, as similarly conducted through Gini impurity measures in clustering validation [7].
- Application of SHAP for model explainability, allowing forensic interpretation of fraud indicators, complementing the heuristic interpretability present in traditional pattern-based techniques.

The structure of this study is organized as follows. Section 2 details our data collection and cleaning pipeline. Section 3 introduces our feature engineering framework. Section 4 describes the benchmark dataset assembly. Section 5 describes model training, evaluation results, and comparative analysis and it presents interpretability findings. Section 6 concludes the study by outlining future research directions.

2   Data Collection and Pre-processing Pipeline

This section outlines the data sources, the rationale behind our sampling strategy, and the theoretical foundations of the cleaning operations applied to obtain a high-quality dataset for wallet classification.

Data Sources and Sampling. Data acquisition proceeded in two main phases. In the first phase, wallet addresses were collected from three primary sources:

- **Elliptic++[1]**: an expanded version of the original Elliptic dataset, comprising 14,267 Bitcoin addresses labeled as fraud and 7,378 labeled as licit, an enriched version of the original Elliptic data set that adds richer entity labels and more recent blocks, aimed at benchmarking anti-money laundering (AML) models[2] [5].
- **Chainabuse[3]**: a collection of crowdsourced fraud reports from which 6,910 fraudulent bitcoin addresses were extracted. The entries were gathered

---

[1] Public release at https://github.com/git-disl/EllipticPlusPlus.

[2] Anti-money laundering (AML) models combine human expertise, automated software processes, artificial intelligence (AI) and machine learning (ML) to identify money laundering activities.

[3] Chainabuse is an open-source, crowdsourced reporting platform operated by TRM Labs, Binance, and Nansen. It collects addresses associated with scams, ransomware, phishing, and other fraudulent activities. Repository: https://chainabuse.com.

by scraping the Chainabuse platform and subsequently filtered to retain only Bitcoin addresses. Corresponding labels were obtained through the `blockchain.info` REST API.
- **Sample of Recent Transactions**: a uniform sample of 10,000 Bitcoin addresses extracted from the most recent 12,000 on-chain transactions, retrieved using https://blockchain.info REST API here again.

3   Graph–Temporal Feature Engineering

To systematically distinguish between fraudulent and legitimate Bitcoin wallets, we commence by extracting all on-chain transactions associated with each wallet of interest. Each wallet is represented as a subgraph of the global Bitcoin transaction network, modeled as a directed multigraph $\mathcal{G} = (\mathcal{V}, \mathcal{E})$, where:

- $\mathcal{V}$ denotes the set of unique addresses (nodes).
- $\mathcal{E}$ denotes the set of transaction edges, each edge defined as $e = (u, v, \tau, a)$, representing a transfer of amount a from address u to address v at timestamp τ. In the following the components of an edge e will be denoted using dot notation, e.g. the amount a will be denoted $e.a$

From this foundational representation, we derive three complementary categories of features:

Pruning of Low-Activity Addresses. For addresses $u, v \in \mathcal{E}$ let us define the timestamp multisets

$$T_u^{\text{out}} = \{\tau \mid (u, v, \tau, a) \in \mathcal{E}\}, \qquad T_v^{\text{in}} = \{\tau \mid (u, v, \tau, a) \in \mathcal{E}\},$$

with cardinalities n_i^{out} and n_i^{in}. We retain i only if $n_i^{\text{out}} \geq 2$ and $n_i^{\text{in}} \geq 2$, ensuring that

$$\Delta t_i^{\text{out}} = \frac{\max T_i^{\text{out}} - \min T_i^{\text{out}}}{n_i^{\text{out}} - 1}, \quad \Delta t_i^{\text{in}} = \frac{\max T_i^{\text{in}} - \min T_i^{\text{in}}}{n_i^{\text{in}} - 1}$$

are well-defined for both directions. Addresses failing either criterion are discarded from further analysis.

Label Merging and Deduplication. We merge the three strata and resolve conflicts by prioritizing the "fraud" label whenever an address appears in multiple sources. Let $L_i \in \{\text{licit}, \text{fraud}\}$ be the label for address i. The merged label is

$$\tilde{L}_i = \begin{cases} \text{fraud}, & \exists \text{ source } s : L_i^{(s)} = \text{fraud}, \\ \text{licit}, & \text{otherwise}. \end{cases}$$

Address-Level Metrics. For each address $v \in \mathcal{V}$, we denote the incoming transactions into node v and the outgoing transactions from node v as:

$$\mathcal{I}(v) = \{e \in \mathcal{E} \mid e = (u, v, \tau, a)\}, \quad \mathcal{O}(u) = \{e \in \mathcal{E} \mid e = (u, v, \tau, a)\}.$$

We compute the following address-specific metrics:

- **In-degree** $\text{in\_deg}(v) = |\mathcal{I}(v)|$ and **Out-degree** $\text{out\_deg}(v) = |\mathcal{O}(v)|$. The collections $\mathcal{I}(v)$ and $\mathcal{O}(v)$ are formally regarded as *multisets* of incoming and outgoing transactions, respectively, thereby preserving the full transaction multiplicities associated with address v.
- **Unique counterparties.** Let

$$C^{\text{in}}(v) = \{\, u \mid (u, v, \tau, a) \in E \,\}, \qquad C^{\text{out}}(v) = \{\, v \mid (u, v, \tau, a) \in E \,\},$$

where each element of $C^{\text{in}}(v)$ (resp. $C^{\text{out}}(v)$) appears exactly once, irrespective of transaction multiplicity. We define

$$\text{uniq\_in}(v) = |C^{\text{in}}(v)|, \qquad \text{uniq\_out}(v) = |C^{\text{out}}(v)|,$$

which quantify the number of *distinct* origin and destination addresses interacting with v, thus capturing the diversity of its transactional counterparties.
- **Average Transaction Amount** $\bar{a}_{\text{in}}(v) = \frac{1}{|\mathcal{I}(v)|} \sum_{e \in \mathcal{I}(v)} e.a$ and likewise for $\bar{a}_{\text{out}}(v)$, reflecting fragmentation or bulk transfer patterns.
- **Mean Inter-transaction Interval** $\overline{\Delta t}_{\text{in}}(v) = \frac{\tau_{\max}(v) - \tau_{\min}(v)}{|\mathcal{I}(v)| - 1}$ (and analogously for outbound edges), indicating temporal regularity or burstiness.
- **Net Balance** $\text{balance}(v) = \sum_{e \in \mathcal{I}(v)} e.a - \sum_{e \in \mathcal{O}(v)} e.a$: measures fund retention versus immediate pass-through.
- **Local Clustering Coefficient** $C(v) = \frac{2\,T(v)}{k(v)(k(v)-1)}$, where $T(v)$ is the count of cycles of length at least 3 containing node v and $k(v) = \text{in\_deg}(v) + \text{out\_deg}(v)$, capturing mixer-like subgraph density.

Wallet-Level Aggregates. Aggregating the address-level metrics across all addresses $v \in \mathcal{V}$ belonging to a given wallet $\mathcal{V}$, we define:

$$D_{\text{in}}(\mathcal{V}) = \sum_{v \in \mathcal{V}} \text{in\_deg}(v), \qquad D_{\text{out}}(\mathcal{V}) = \sum_{v \in \mathcal{V}} \text{out\_deg}(v),$$

$$R_{\text{recv}}(\mathcal{V}) = \sum_{v \in \mathcal{V}} \sum_{e \in \mathcal{I}(v)} e.a, \qquad R_{\text{sent}}(\mathcal{V}) = \sum_{v \in \mathcal{V}} \sum_{e \in \mathcal{O}(v)} e.a,$$

$$\text{net\_bal}(\mathcal{V}) = R_{\text{recv}}(\mathcal{V}) - R_{\text{sent}}(\mathcal{V}), \qquad \bar{d}(\mathcal{V}) = \frac{1}{|\mathcal{V}|} \sum_{v \in \mathcal{V}} (\tau_{\max}(v) - \tau_{\min}(v)).$$

These aggregates quantify total transaction counts, financial throughput, net fund retention, and mean address lifetime within each wallet.

Composite Ratios and Indices. To normalize across wallets of varying scale and to highlight disproportionate behaviors, we introduce:

$$\text{in\_out\_ratio} = \frac{D_{\text{out}} + \alpha}{D_{\text{in}} + \beta} \qquad \text{volume\_ratio} = \frac{R_{\text{sent}} + \alpha}{R_{\text{recv}} + \beta}$$

$$\text{activity\_idx} = \frac{D_{\text{in}} + D_{\text{out}}}{\bar{d}}$$

where $\alpha, \beta > 0$ are small smoothing constants. An elevated activity_idx signals condensed bursts of transactional activity, while extreme flow ratios flag unbalanced input-output dynamics.

This feature taxonomy not only synthesizes key structural, monetary, and temporal dimensions of on-chain behavior, but also integrates dispersion-based indicators which have emerged in more recent forensic analyses. To contextualize our contributions with respect to the state of the art, Table 1 compares our set of 20 engineered features against those adopted in leading studies on Bitcoin fraud detection and wallet profiling [3,4,6,10,11,13,14]. A checkmark ($\checkmark$) indicates that a given feature or an equivalent formulation is employed in the cited paper, while a dash denotes its absence. As shown, our framework comprehensively spans the major feature families and incorporates recent heuristics such as transaction dispersion ratios and micro-transaction indicators, which are underrepresented in earlier work.

4   Benchmark Dataset Assembly

This section first outlines three distinct scenarios, obtained using the datasets introduced in Sect. 2, either combined or standalone. Then it presents two complementary data-quality assessment methods: first, a Confident Learning–based technique for estimating and removing label noise in the Chainabuse dataset (Sect. 4.1); and second, an unsupervised anomaly-filtering approach driven by an Isolation Forest applied to the most recent sample (Sect. 4.2).

To assemble our benchmark suite, we first extract on-chain transaction histories for each wallet under investigation. Wallets are grouped via multi-input clustering heuristics applied to the global Bitcoin Unspent Transaction Output (UTXO) graph. From these raw transactions, the full set of address-level and wallet-level features described in Sect. 3 is computed. Based on this feature matrix, three distinct classification datasets are then constructed.

1. **Chainabuse and recent transactions addresses:** fraudulent labels from the Chainabuse community-reported dataset are paired with unlabeled addresses uniformly sampled from the most recent 10,000 on-chain transactions. This scenario tests robustness against noisy benign samples.
2. **Chainabuse and Elliptic++ licit addresses:** fraudulent addresses from Chainabuse are contrasted against high-quality licit labels drawn from the Elliptic++ academic dataset. This setting simulates a realistic deployment with reliable ground truth on both ends.
3. **Elliptic++:** both fraudulent and legitimate addresses originate from Elliptic++, controlling for source-domain bias by using identical provenance for each class.

4.1   ChainAbuse and Potential Label Bias

Chainabuse is a community-driven registry of cryptocurrency addresses reported for scams, ransomware, phishing and other illicit schemes. Since reports are

Table 1. Mapping of the features published in the literature with respect to the set of 20 variables adopted in this work

Feature/Family	Chang 18 [3]	Toyoda 18 [13]	Weber 19 [14]	Nerurkar 21 [11]	Chen 21 [4]	Monamo 16 [10]	İşcan 23 [6]	Our Paper
Count/Structural								
In-degree (addr)	✓	✓	✓	✓	✓	✓	–	✓
Out-degree (addr)	✓	✓	✓	✓	✓	✓	–	✓
Unique counterparties (in/out)	–	✓	✓	✓	–	–	–	✓
Local clustering coeff.	✓	–	–	–	–	–	–	✓
Monetary								
Avg. tx amount (in)	–	✓	✓	✓	✓	✓	✓	✓
Avg. tx amount (out)	–	✓	✓	✓	✓	✓	✓	✓
Net balance (address)	–	–	✓	✓	✓	–	–	✓
Temporal								
Mean inter-tx interval(in)	–	✓	✓	✓	✓	✓	–	✓
Mean inter-tx interval(out)	–	✓	✓	✓	✓	✓	–	✓
Address lifetime	–	✓	–	✓	✓	✓	–	✓
Wallet/Aggregate								
$\sum$ in-degree (D_{in})	–	–	–	✓	✓	–	–	✓
$\sum$ out-degree (D_{out})	–	–	–	✓	✓	–	–	✓
Total received (R_{recv})	–	–	–	✓	✓	–	✓	✓
Total sent (R_{sent})	–	–	–	✓	✓	–	✓	✓
Wallet net balance	–	–	–	✓	✓	–	✓	✓
Mean addr duration ($\bar{d}$)	–	–	–	✓	–	–	–	✓
Composite/Ratios								
In/Out ratio (tx ount)	–	✓	✓	✓	✓	✓	–	✓
Volume ratio (BTC sent/recv)	–	✓	✓	✓	✓	✓	✓	✓
Activity index (tx/ lifetime)	–	✓	–	–	✓	–	✓	✓
Dispersion/Pattern-specific								
Unique-out ratio	✓	–	–	–	–	–	–	✓
Time-interval ratio	–	–	–	–	✓	–	–	✓
Weighted avg tx (micro-tx)	–	–	–	–	✓	–	✓	✓

crowdsourced and not uniformly vetted, the dataset may suffer from two main biases: (i) *reporting bias*, whereby high-profile scams are overrepresented and low-value frauds remain unreported; and (ii) *confirmation bias*, arising if multiple users submit the same address without independent verification. Consequently, Chainabuse's "fraud" labels may include false positives or miss certain illicit addresses.

The reporting bias and confirmation bias inevitably introduce label noise into the ChainAbuse registry–in the form of *false positives* (benign addresses incorrectly flagged) and *false negatives* (illicit addresses that go unreported). To rigorously quantify and mitigate the noise generated by *false positives* and improve downstream model reliability, the Confident Learning framework [12], as implemented in the Python module *cleanlab*, is adopted. This approach systematically identifies examples whose observed labels conflict with a classifier's high-confidence predictions, providing a principled estimate of mislabeling rate.

A combined dataset of 6,910 ChainAbuse "fraud" addresses and 7,378 Elliptic "licit" addresses was used to train a balanced Random Forest (200 trees, max_depth=20) using five-fold, out-of-fold probability estimates. Leveraging the Confident Learning framework (via the `cleanlab` module[4]), we automatically

---

[4] https://github.com/cleanlab/cleanlab.

flagged 294 addresses (≈ 2.2 %) whose original labels conflicted with high-confidence predictions. After removing these suspected mislabels and retraining the model, the F1 score on the same held-out test split improved from 0.912 to 0.936 ($\Delta = +0.024$), demonstrating both the presence of label noise and the effectiveness of this methodology.

4.2   Anomaly Filtering in the Recent Sample

The *Recent Transactions* wallets, drawn uniformly from on-chain activity and lacking explicit labels, may still harbor latent fraudulent behavior. To mitigate the inclusion of potentially malicious wallets in our benign sample, we employ an unsupervised, two-stage anomaly filtering pipeline on the feature representations:

1. **2D PCA Projection.**
   - Compute the covariance matrix

$$\Sigma = \frac{1}{m-1} X^{\mathsf{T}} X,$$

   where $X \in \mathbb{R}^{m \times d}$ is the feature matrix of the m wallets across d attributes.
   - Solve the eigenproblem $\Sigma \, \mathbf{u}_k = \lambda_k \, \mathbf{u}_k$ and select the top two eigenvectors $\mathbf{u}_1, \mathbf{u}_2$ corresponding to the largest eigenvalues λ_1, λ_2.
   - Project each feature vector x_i into the 2D subspace:

$$z_i = \left[\mathbf{u}_1^{\mathsf{T}} x_i, \ \mathbf{u}_2^{\mathsf{T}} x_i \right]^{\mathsf{T}},$$

   preserving the cumulative variance fraction $(\lambda_1 + \lambda_2) / \sum_{k=1}^{d} \lambda_k$.
2. **Outlier Removal with Isolation Forest ($\alpha = 0.05$).**
   - Fit an Isolation Forest on the set of projections $\{z_i\}_{i=1}^{m}$, specifying a contamination level $\alpha = 0.05$ to excise the top 5% most extreme points.
   - The anomaly score for a point z is computed as

$$s(z) = 2^{-\mathbb{E}[h(z)]/c(n)},$$

   where $\mathbb{E}[h(z)]$ is the average path length for z in the forest and

$$c(n) = 2\,H(n-1) - \frac{2(n-1)}{n}, \quad H(k) \approx \ln k + \gamma$$

   is the normalizing constant for tree size n.
   - We adopt $\alpha = 0.05$ to perform robust trimming of heavy tails and extreme observations–not only true frauds, but any anomalous behavior that could distort mean and covariance estimates–in line with classical robust statistical practices [8,9].
   - Wallets for which $s(z_i)$ exceeds the threshold defined by α are removed, yielding a purified "Recent" sample free from potential outliers (Fig. 1).

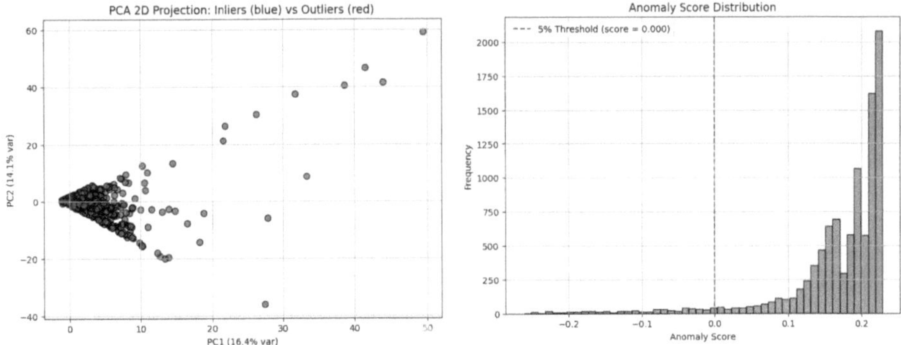

Fig. 1. Left: 2D PCA projection of the Recent Transactions wallets. Red points denote outliers identified by the Isolation Forest ($\alpha = 0.05$). **Right:** Distribution of anomaly scores $s(z_i)$; the dashed line marks the 95th percentile threshold

5   Experiments and Discussion

This section presents the experimental framework used to train and evaluate our classification models, followed by a discussion of the results. In Sect. 5.1, we describe the preprocessing pipeline, justify our choice of algorithms, outline the hyperparameter optimization strategy, and provide a comparative evaluation of three classification approaches: Random Forest, Multi-Layer Perceptron (MLP), and Support Vector Machine (SVM). In Sect. 5.2, we interpret the findings, highlight key performance trends, and discuss their implications.

5.1   Experimental Results

To assess the effectiveness of our classification framework, we designed a controlled experimental pipeline encompassing standardized preprocessing, model selection, hyperparameter optimization, and performance evaluation. This section outlines the methodological foundations underpinning our experiments, with an emphasis on reproducibility and comparability across different model families. All experiments were conducted on the three benchmark datasets introduced in Sect. 2, using the full suite of engineered features described in Sect. 3.

Preprocessing. To ensure consistency and reproducibility, all datasets underwent an identical preprocessing pipeline. First, the feature matrix was separated from the target class labels. A type-inference stage distinguished between numerical and categorical variables. Numerical attributes were subjected to median imputation followed by z-score normalization to reduce scale-related biases and the impact of outliers. Categorical features, where present, were imputed with a constant placeholder and transformed through one-hot encoding, allowing for compatibility with downstream models.

The resulting processed dataset was encoded entirely in numerical form and subsequently partitioned into training and test sets using a stratified 80:20 split to preserve class balance.

Random Forest Classifier. We trained a Random Forest classifier with class weights set to 'balanced' to compensate for minor class imbalances. A grid search was conducted over a range of hyperparameters using 5-fold cross-validation with accuracy as the optimization criterion. The optimal configuration consisted of 200 estimators, a maximum depth of 20, and the square root of the number of features as the maximum feature subset size.

The results are summarized in Table 2. Notably, in Scenario 2–where Chainabuse-provided fraudulent labels are paired with licit samples from Elliptic++–the Random Forest achieved an F1-score of 0.92 for both classes, indicating excellent discriminative performance.

Table 2. Random Forest Classification Results

Scenario	Class	Precision	Recall	F1-score
1	Fraud	0.78	0.79	0.79
	Licit	0.85	0.84	0.85
2	Fraud	0.93	0.91	0.92
	Licit	0.92	0.93	0.93
3	Fraud	0.92	0.92	0.92
	Licit	0.87	0.87	0.87

Feed-Forward Neural Network. We implemented a fully–connected *multi-layer perceptron* and performed *hyperparameter tuning* using a random search strategy. The search space and the best configuration discovered are reported in Table 3. Unless stated otherwise, all layers use the RELU activation.

Table 3. Hyper-parameter search space explored

Hyper-parameter	Search space	Best value
Number of hidden layers	$n_{\text{layers}} \in \{2, 3\}$	3
Units per layer	$u_i \in \{32, 64, 96, 128\}$	$[96, 96, 32]$
Dropout rate	$p_i \in \{0.2, 0.3, 0.4\}$	0.30
Learning rate (Adam)	$\alpha \in \{10^{-2}, 10^{-3}, 5 \times 10^{-4}\}$	10^{-3}
Batch size	$B \in \{32, 64, 128\}$	64
L2 weight decay	λ (fixed)	10^{-4}

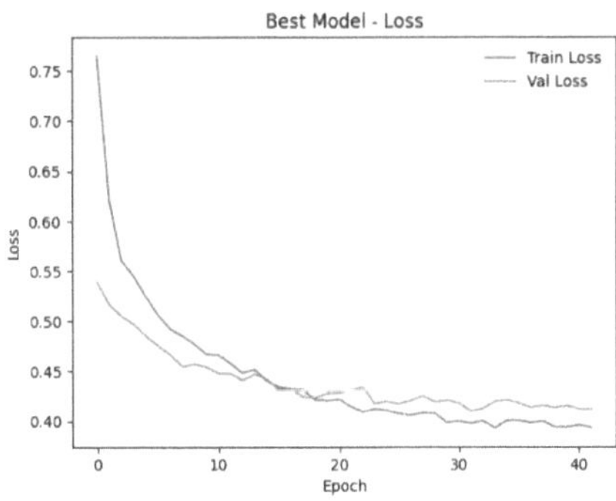

(a) Training and validation loss.

(b) Training and validation accuracy.

Fig. 2. Learning curves of the optimised MLP over 40 epochs

Table 4. MLP—Classification Results

Scenario	Class	Precision	Recall	F1-score
1	Fraud	0.66	0.78	0.71
	Licit	0.82	0.71	0.76
2	Fraud	0.89	0.78	0.83
	Licit	0.82	0.91	0.87
3	Fraud	0.81	0.85	0.83
	Licit	0.73	0.67	0.70

Figure 2 shows that the validation curves closely follow the training ones: the loss stabilises around 0.40, while accuracy converges near 0.85 without any noticeable divergence, indicating good generalisation.

Overall, the model attains an **accuracy of 0.85** on the test split, with balanced F1-scores across the fraud and licit classes (Table 4). These results confirm the effectiveness of the selected architecture and regularisation strategy for the task.

Support Vector Machine. To further validate the generality of our feature set, we trained a Support Vector Machine (SVM) classifier. The model was embedded in the same preprocessing pipeline adopted for the other learners, and class imbalance was mitigated via the `balanced` option. Hyperparameters were tuned through a 5-fold grid search over $C \in \{0.1, 1, 10\}$, `kernel`$\in$ {linear, rbf}, and `gamma`$\in$ {scale, auto}. The optimal configuration–shared across scenarios–was obtained with $C = 10$, an RBF kernel, and `gamma=scale` (Table 5).

The SVM attains performance comparable to the MLP, particularly in Scenario 2 where both precision and recall exceed 0.80 for each class. Nevertheless, its sensitivity to noisy or weakly-labeled data (Scenario 1) remains evident, reinforcing the advantages of ensemble models in heterogeneous operational settings.

Table 5. Support Vector Machine—Classification Results

Scenario	Class	Precision	Recall	F1-score
1	Fraud	0.65	0.76	0.70
	Licit	0.80	0.71	0.76
2	Fraud	0.83	0.80	0.81
	Licit	0.83	0.85	0.84
3	Fraud	0.82	0.87	0.85
	Licit	0.77	0.69	0.73

5.2 Discussion

Our experiments demonstrate that ensemble models, particularly Random Forests, offer a favorable performance. Scenario 2 yielded the best results, confirming the value of combining crowd-sourced and expert-labeled datasets. In contrast, the neural network showed reduced resilience to label noise and distributional shifts.

To benchmark our methodology, we compared classification accuracy with previous results in Table 6. Our Random Forest model matches or surpasses state-of-the-art approaches, validating the effectiveness of our feature set and classification pipeline.

Table 6. Accuracy reported in related works compared to our best-performing model (Random Forest, Scenario 2)

Authors (Year)	Model	Dataset	Accuracy
Chang et al. (2018) [3]	Random Forest/SVM	Bitcoin tx patterns	89%
Toyoda et al. (2018) [13]	k-NN/Decision Tree	Blockchain service wallets	85%
Nerurkar et al. (2021) [11]	CNN (Deep Learning)	Transaction flows	91%
Işcan et al. (2023) [6]	LightGBM	Wallet-based behaviors	93%
Our Results (2025)	Random Forest	Chainabuse + Elliptic++	**92%**[*]

[*]Inferred from F1-scores of 0.92 for both fraud and licit classes

Explainability via SHAP. To interpret the predictions of our Random Forest classifier, we compute SHAP values [2] on the binary fraud/licit model (fraud samples from Chainabuse; licit from Elliptic++). SHAP values provide both global feature importance and the direction in which each feature drives the model's output toward "fraud."

Figure 3 shows the SHAP summary plot: horizontal spread indicates each feature's contribution to output variance, and color denotes effect direction. Several key patterns emerge:

- **Dispersion-oriented features dominate.** Both `unique_out_ratio` and `total_unique_out` appear at the top of the ranking, underscoring that wallets which fan out funds to a large number of distinct counterparts are markedly more likely to be classified as fraudulent. This accords with classical forensic typologies—such as peel chains and mixer operations—where rapid splitting hinders traceability.
- **Temporal compactness signals illicit behaviour.** Features that capture *burstiness* (`time_interval_ratio`, `avg_out_time_interval`) and *ephemeral lifetimes* (`wallet_lifetime_sec`) exert a strong positive influence on the fraud score, reflecting the operational practice of using short-lived "burner" addresses for one-off laundering rounds.
- **Micro-transactions and rapid depletion are red flags.** Low-value transfers (`weighted_avg_tx`, `avg_out_transaction`) coupled with near-zero `net_balance` drive the model towards the fraud class, consistent with "smurfing" or "dusting" strategies designed to fragment illicit proceeds below exchange reporting thresholds.

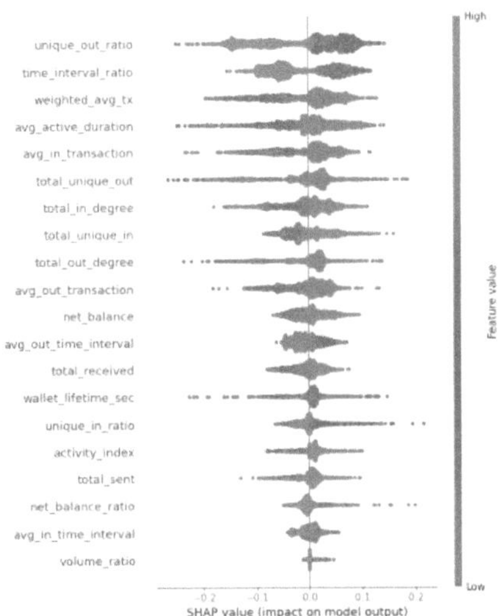

Fig. 3. SHAP summary plot: points show feature SHAP values; features are ordered by mean absolute impact

Table 7 distils these insights, reporting for each feature its tier ranking, the polarity of its dominant contribution, and a succinct forensic interpretation.

The SHAP analysis shows that rapid-dispersion (e.g. `unique_out_ratio`, `time_interval_ratio`) and micro-slicing (e.g. `weighted_avg_tx`, `avg_out_transaction`) features most strongly indicate fraud, offering concrete compliance and forensic insights.

Table 7. Global Feature Importance and Direction via SHAP

Rank	Feature	↑ Fraud	Rationale
1	unique_out_ratio	high (↑)	Many distinct recipients – typical of fund dispersion in scams or mixers.
2	time_interval_ratio	high (↑)	Denser outflows than inflows – rapid unloading of funds.
3	weighted_avg_tx	low (↓)	Numerous micro-transactions suggest smurfing or dusting.
4	avg_active_duration	low (↓)	Short-lived wallets often serve as "burner" addresses.
5	avg_in_transaction	low (↓)	Small inbound amounts frequently precede fraudulent aggregation.
6	total_unique_out	high (↑)	Large number of unique outgoing addresses reinforces dispersion.
7	total_in_degree	low (↓)	Few incoming transactions relative to outflows.
8	total_unique_in	low (↓)	Concentrated inbound sources indicate planned layering.
9	total_out_degree	high (↑)	High total count of outgoing transactions.
10	avg_out_transaction	low (↓)	Many small outgoing slices – indicative of splitting funds.
11	net_balance	low (↓)	Near-zero or negative balance from rapid depletion.
12	avg_out_time_interval	low (↓)	Tight timing between successive outflows.
13	total_received	low (↓)	Few receipts but many outgoing payments is anomalous.
14	wallet_lifetime_sec	low (↓)	Short operational lifetime increases suspicion.
15	unique_in_ratio	low (↓)	Low diversity of inbound sources – concentrated deposits.
16	activity_index	high (↑)	Very high daily transaction rate – hyperactivity.
17	total_sent	high (↑)	Large absolute outgoing volume.
18	net_balance_ratio	low (↓)	Re-sends nearly all received funds.
19	avg_in_time_interval	low (↓)	Closely spaced inbound deposits.
20	volume_ratio	high (↑)	Sends much more than receives (out/in elevated).

Ablation Study. Building upon the optimal Random Forest configuration described, we conducted a comprehensive ablation analysis by retraining the model omitting each individual feature. The baseline macro-averaged F1 score with the full feature set is **0.9255**. Table 8 reports the ten individual features whose removal produces the largest declines in macro-averaged F1; ΔF1 denotes the absolute drop relative to the baseline.

These results corroborate the SHAP-based interpretability analysis: omitting `weighted_avg_tx`, which captures micro-transaction patterns, causes the single largest drop (0.88 pp); elimination of `wallet_lifetime_sec` and `avg_out_transaction` similarly degrades performance by over 0.6 pp, underscoring the importance of bursty temporal behavior.

Scalability and Deployment. While the present study concentrates on the *offline* evaluation of our wallet-classification framework, operating at blockchain scale and low-latency processing introduces further engineering challenges. In an extended version of this work we plan to: (i) design a streaming architecture capable of ingesting live Bitcoin transactions; (ii) demonstrate incremental maintenance of the graph–temporal features introduced in Sect. 3; and (iii) empirically assess latency, throughput, and cost trade-offs under realistic workloads. Preliminary sketches suggest that combining lightweight stream processing for rapid event filtering with scheduled micro-batch jobs for feature aggregation could offer an effective balance between responsiveness and resource utilisation, but a rigorous validation is left for future work.

Table 8. Single-feature ablation – top 10 ΔF1 (top 3 in bold)

Feature removed	F1$_{\mathrm{macro}}$	ΔF1
weighted_avg_tx	0.9167	**−0.0088**
wallet_lifetime_sec	0.9186	**−0.0069**
avg_out_transaction	0.9192	**−0.0063**
total_unique_in	0.9193	−0.0062
avg_active_duration	0.9196	−0.0059
avg_in_time_interval	0.9196	−0.0059
total_sent	0.9203	−0.0052
total_in_degree	0.9204	−0.0051
unique_out_ratio	0.9204	−0.0051
total_out_degree	0.9211	−0.0044

6 Conclusions and Future Work

In this study, we have developed and evaluated a comprehensive framework for classifying Bitcoin wallets as illicit or legitimate. By integrating three disparate data sources–Elliptic++, ChainAbuse and a uniformly sampled set of recent on-chain wallets–we produced a balanced corpus of over 31,000 labeled entities, resolving conflicts in favor of the fraud label to maximize recall (cf. [11]). Our feature engineering pipeline extracted twenty metrics spanning structural counts, monetary aggregates, temporal patterns and composite ratios; these draw upon and extend prior heuristics [4,13] while incorporating dispersion indicators recently highlighted in live-forensics research [5]. In comparative experiments, Random Forest consistently achieved F1-scores above 0.92 under high-quality labeling, whereas LightGBM was notably effective at halving the false-alarm rate in an industrial setting [6]. Further, SHAP analyses elucidated that unique-out-ratio, time-interval-ratio and micro-transaction averages are the strongest predictors of fraud.

Nonetheless, crowdsourced labels and anomaly filtering may still misclassify borderline cases, and concentrating solely on-chain metrics can overlook crucial off-chain evidence (e.g., exchange KYC). The UTXO-centric feature set requires adaptation for account-based or privacy-focused blockchains; and fourth, our evaluation on static snapshots must give way to a streaming analytics architecture for real-time risk scoring.

Moving forward, we will leverage self-supervised and contrastive learning to exploit large volumes of unlabeled transactions for richer graph embeddings [1], integrate heterogeneous GNNs (e.g., GAT, HGT) to model complex laundering networks beyond pairwise links [14], and extend our pipeline to additional chains (e.g., Ethereum, BSC) while fusing on-chain analytics with off-chain data such as exchange records and darknet market reports.

References

1. Agarwal, U., Rishiwal, V., Tanwar, S., Yadav, M.: Blockchain and crypto forensics: investigating crypto frauds. Int. J. Netw. Manage **34**(2), e2255 (2024)
2. Van den Broeck, G., Lykov, A., Schleich, M., Suciu, D.: On the tractability of SHAP explanations. J. Artif. Intell. Res. **74**, 851–886 (2022)
3. Chang, T.H., Svetinovic, D.: Improving bitcoin ownership identification using transaction patterns analysis. IEEE Trans. Syst. Man Cybern. Syst. **50**(1), 9–20 (2018)
4. Chen, B., Wei, F., Gu, C.: Bitcoin theft detection based on supervised machine learning algorithms. Secur. Commun. Netw. **2021**(1), 6643763 (2021)
5. Elmougy, Y., Liu, L.: Demystifying fraudulent transactions and illicit nodes in the bitcoin network for financial forensics. In: Proceedings of the 29th ACM SIGKDD Conference on Knowledge Discovery and Data Mining, pp. 3979–3990 (2023)
6. Iscan, C., Kumas, O., Akbulut, F.P., Akbulut, A.: Wallet-based transaction fraud prevention through LightgBM with the focus on minimizing false alarms. IEEE Access **11**, 131465–131474 (2023)
7. Lin, C., Liao, H., Tsai, F.: A systematic review of detecting illicit bitcoin transactions. In: Proceedings of the 26th International Conference on Knowledge-Based and Intelligent Information & Engineering Systems (KES 2022). Procedia Comput. Sci. **207**, 3211–3219 (2022)
8. Liu, F.T., Ting, K.M., Zhou, Z.H.: Isolation forest. In: 2008 Eighth IEEE International Conference on Data Mining, pp. 413–422. IEEE (2008)
9. Liu, F.T., Ting, K.M., Zhou, Z.H.: Isolation-based anomaly detection. ACM Trans. Knowl. Discov. Data (TKDD) **6**(1), 1–39 (2012)
10. Monamo, P., Marivate, V., Twala, B.: Unsupervised learning for robust bitcoin fraud detection. In: 2016 Information Security for South Africa (ISSA), pp. 129–134. IEEE (2016)
11. Nerurkar, P., Bhirud, S., Patel, D., Ludinard, R., Busnel, Y., Kumari, S.: Supervised learning model for identifying illegal activities in bitcoin. Appl. Intell. **51**, 3824–3843 (2021)
12. Northcutt, C., Jiang, L., Chuang, I.: Confident learning: estimating uncertainty in dataset labels. J. Artif. Intell. Res. **70**, 1373–1411 (2021)
13. Toyoda, K., Ohtsuki, T., Mathiopoulos, P.T.: Multi-class bitcoin-enabled service identification based on transaction history summarization. In: 2018 IEEE International Conference Internet of Things (iThings), GreenCom, CPSCom & Smart-Data, pp. 1153–1160. IEEE (2018)
14. Weber, M., et al.: Anti-money laundering in bitcoin: experimenting with graph convolutional networks for financial forensics. arXiv preprint arXiv:1908.02591 (2019)

Addressing Prior Dependence in Hierarchical Bayesian Modeling for PTA Data Analysis I: Methodology and Implementation

L. D'Amico[1]([✉]) [iD], E. Villa[2] [iD], Fatima Modica Bittordo[1] [iD], A. Barca[1] [iD], F. Alì[1] [iD], M. Meneghetti[3] [iD], and L. Naso[1] [iD]

[1] Koexai Srl, Via Josemaria Escrivà 6, 95125 Catania, Italy
`luigi.damico@koexai.com`
[2] INAF Istituto di Astrofisica Spaziale e Fisica cosmica di Milano (IASF-MI), Via Alfonso Corti 12, 20133 Milano, Italy
[3] INAF Osservatorio di Astrofisica e Scienza dello Spazio di Bologna, Via Piero Gobetti 93/3, 40129 Bologna, Italy

Abstract. Complex inference tasks, such as those encountered in Pulsar Timing Array (PTA) data analysis, rely on Bayesian frameworks. The high-dimensional parameter space and the strong interdependencies among astrophysical, pulsar noise, and nuisance parameters introduce significant challenges for efficient learning and robust inference. These challenges are emblematic of broader issues in decision science, where model over-parameterization and prior sensitivity can compromise both computational tractability and the reliability of the results. We address these issues in the framework of hierarchical Bayesian modeling by introducing a reparameterization strategy. Our approach employs Normalizing Flows (NFs) to decorrelate the parameters governing hierarchical priors from those of astrophysical interest. The use of NF-based mappings provides both the flexibility to realize the reparametrization and the tractability to preserve proper probability densities. We further adopt `i-nessai`, a flow-guided nested sampler, to accelerate exploration of complex posteriors. This unified use of NFs improves statistical robustness and computational efficiency, providing a principled methodology for addressing hierarchical Bayesian inference in PTA analysis.

Keywords: Hierarchical Bayesian modeling · Normalizing Flows · Pulsar Timing Array · Decorrelation in the parameter space · Decision science · Machine learning

1 Introduction

The analysis of Pulsar Timing Array (PTA) data plays a central role in the effort to detect and characterize the Stochastic Gravitational Wave Background (SGWB) at nanohertz frequencies. Evidence of a SGWB has recently been reported by multiple international PTA collaborations [1]. PTA sensitivity depends critically on modeling both the SGWB and complex noise processes intrinsic to pulsars and the measurement system. Millisecond pulsars are extremely precise, stable rotators emitting radiation like cosmic lighthouses. A typical PTA model includes physical parameters describing the SGWB spectrum–modeled as a power law with amplitude and spectral index–alongside noise parameters accounting for pulsar-specific contributions, such as white and red noise amplitudes and spectral indices, clock and ephemeris errors, and timing-model parameters [21]. The effects of the SGWB perturbations are encoded in the differencies between the observed Time Of Arrival (TOA) with respect to the theoretical predictions. The time residuals are given by

$$\delta t = M\epsilon + Fa + n, \tag{1}$$

where ϵ are physical parameters, a Fourier coefficients with design matrix F, M the matrix of residual derivatives, and n white noise. The term Fa includes correlated and uncorrelated low-frequency processes such as red noise, SGWB, intrinsic spin-noise and dispersion measure.

Hierarchical Bayesian modeling provides a comprehensive framework for PTA data analysis by allowing the inclusion of priors on noise parameters and subsequent marginalization to estimate posteriors of physical parameters [8,9,11]. However, posterior inferences are sensitive to prior choices–a well-known problem in Bayesian analysis. Recent PTA studies have explored strategies to mitigate this sensitivity, including parametric uniform priors [12], Gaussian priors [8], and Jeffreys priors for red noise processes [14].

In this work, we address prior sensitivity in hierarchical PTA modeling through a reparameterization strategy based on parameter orthogonalization [2,3,23]. The orthogonalization technique proposed in the context of Effective Field Theory in cosmology in [19] makes use of Generalized Additive Models (GAMs) to decorrelate cosmological and nuisance parameters. As a result the posterior of cosmological parameters is less sensitive to the nuisance prior. We extend this approach by introducing a hierarchical layer for the noise model, placing hyperpriors on noise parameter distributions, and employing Normalizing Flows (NFs) [13,16,18,20] to decorrelate hyperparameters from physical parameters. Inference is performed using `i-nessai` [26–28], a flow-guided nested sampling algorithm that efficiently explores high-dimensional, highly correlated parameter spaces, accelerating full Bayesian inference compared to standard Parallel Tempered Markov Chain Monte Carlo (PTMCMC) approaches [25].

This paper is organized as follows. Section 2 presents our parameter decorrelation methodology and its implementation via NFs, with Subsect. 2.1 focusing on training. Section 3 discusses the application to PTA data, including Subsect. 3.1 on the hierarchical Bayesian implementation and Subsect. 3.2 on our validation test. Section 4 discusses results, and Sect. 5 summarizes our conclusions.

2 Parameter Decorrelation Methodology

We present here in full detail the construction of the orthogonal reparametrization in a general hierarchical Bayesian setting[1]. Our framework will be specialized to the PTA context in Sect. 3. We consider some physical parameters ϑ whose prior distribution $\pi(\vartheta|\Lambda)$ is parametrized by the hyperparameters Λ, which in turn are distributed according to their hyperprior $\pi'(\Lambda)$, in general different from the distribution π. The two-level parameters joint posterior is given by

$$\mathcal{P}(\vartheta, \Lambda|\delta t) = \frac{\mathcal{L}(\delta t|\vartheta)\pi(\vartheta|\Lambda)\pi'(\Lambda)}{\mathcal{Z}}. \tag{2}$$

In the above equation, $\mathcal{L}(\delta t|\vartheta)$ is the likelihood and depends on the physical parameters only, and $\mathcal{Z}$ is the Bayesian evidence. The hierarchical structure is fully encoded in the parametrized prior term $\pi(\vartheta|\Lambda)$, giving the distribution of ϑ depending on the hyperparameters Λ, which are in turn distributed according to $\pi'(\tilde{\Lambda})$.

The decorrelation procedure is based on projecting out the component of the hyperparameter vector Λ that lies in the subspace spanned by the physical parameters ϑ. The orthogonal complement to this projection yields the transformed hyperparameter vector $\tilde{\Lambda}$, which is, by construction, orthogonal to the physical parameters. The reparametrization is thus expressed by the following transformation

$$\tilde{\Lambda} = \Lambda - P_\vartheta \Lambda = (I - P_\vartheta)\Lambda, \tag{3}$$

where

$$P_\vartheta \equiv \vartheta(\vartheta^T \vartheta)^{-1}\vartheta^T \tag{4}$$

is the projector onto the subspace spanned by ϑ. Geometrically, this means removing from Λ the component lying along the direction of ϑ. As a result, the transformed hyperparameters $\tilde{\Lambda}$ satisfy by construction the orthogonality condition $\vartheta^T \tilde{\Lambda} = 0$. Our goal is to obtain an equivalent representation, where physical parameters depend on decorrelated hyperparameters $\tilde{\Lambda}$, which are orthogonal to the physical parameter directions, together with the corresponding distribution of these transformed hyperparameters. In formulas, we want to obtain the transformation

$$\pi(\vartheta|\Lambda)\pi'(\Lambda) \quad \longrightarrow \quad \pi(\vartheta|\tilde{\Lambda})\tilde{\pi}'(\tilde{\Lambda}). \tag{5}$$

However, the orthogonal projection in Eq. 3 presents a fundamental problem: it does not admit an inverse. Consequently, the straightforward variable change in Eq. 5 is ill-suited. Nevertheless, this difficulty can be circumvented by directly modeling both the distributions $\pi(\tilde{\Lambda})$ and $\pi(\vartheta|\tilde{\Lambda})$ with NFs [13,16,20]. NFs are a class of generative models that transform complex and potentially singular probability distributions into simpler and tractable ones through a sequence of

---

[1] We follow the notation of [8,22] for the hierarchical posterior.

invertible mappings. The "normalizing" attribute refers precisely to their ability to regularize problematic distributions by mapping them to standard distributions, enabling both efficient sampling and exact computations through a collection of subsequent invertible transformations.

To explain our procedure, we start by sampling from the prior distribution $\pi(\mathbf{\Lambda})$ and backward through the parameter hierarchy to get draws of $\boldsymbol{\vartheta}$ and we finally obtain draws of $\tilde{\mathbf{\Lambda}}$ via the projection. That is, we:

1. sample $\mathbf{\Lambda}_i \sim \pi(\mathbf{\Lambda})$ for $i = 1, ..., N_{\text{samples}}$;
2. sample $\boldsymbol{\vartheta}_i \sim \pi(\boldsymbol{\vartheta}|\mathbf{\Lambda})$ for $i = 1, ..., N_{\text{samples}}$;
3. get samples of $\tilde{\mathbf{\Lambda}}$ by transforming the pairs $(\boldsymbol{\vartheta}_i, \mathbf{\Lambda}_i)$ through the projection $\tilde{\mathbf{\Lambda}}_i = (I - P_{\boldsymbol{\vartheta}_i})\mathbf{\Lambda}_i$ for $i = 1...N_{\text{samples}}$
4. check the orthogonality condition $\boldsymbol{\vartheta}_i^T \tilde{\mathbf{\Lambda}}_i = 0$ for the samples.

Once we have N_{samples} of the triple $(\boldsymbol{\vartheta}_i, \mathbf{\Lambda}_i, \tilde{\mathbf{\Lambda}}_i)$, we employ two complementary NFs: the first, that we call *Push-forward*, learns from the draws $(\tilde{\mathbf{\Lambda}}_i)$ the distribution $\pi'(\tilde{\mathbf{\Lambda}})$ of the decorrelated hyperparameters and then the second, that we call *Pull-backward*, learns the conditional distribution $\pi(\boldsymbol{\vartheta}|\tilde{\mathbf{\Lambda}})$. This yields the required quantities for the transformation in Eq. 5: by taking advantage of NFs we approximate the projection in Eq. 3 and regularize its inverse, while remaining compatible with the orthogonalization and the hierarchical structure of priors and hyperpriors. In the following, we describe in more detail the two algorithms introduced above.

Push-Forward Normalizing Flow (PF-NF): the first component models the distribution $\pi'(\tilde{\mathbf{\Lambda}})$ of the orthogonalized hyperparameters. It is implemented as a Masked Autoregressive Flow (MAF) [17], with three transformation blocks, each consisting of masked affine autoregressive transforms with 32 hidden units. To avoid artifacts from a fixed variable ordering, random permutations are introduced between successive blocks. The base distribution can be flexibly chosen as either a standard Gaussian, $\mathcal{N}(0, \mathbf{I})$, or a uniform distribution, $\mathcal{U}([0, 1])$. The MAF architecture ensures exact invertibility with a tractable Jacobian computation in $\mathcal{O}(m)$ time, a feature that is essential for both efficient sampling and accurate density evaluation.

Pull-Backward Conditional Normalizing Flow (PB-CNF): the second component is a conditional NF [24], that learns the distribution of physical parameters given the hyperparameters, i.e. $\pi(\boldsymbol{\vartheta}|\tilde{\mathbf{\Lambda}})$. This is realized as a conditional MAF where the context, namely the decorrelated hyperparameters $\tilde{\mathbf{\Lambda}}$, is injected at each transformation layer. The architecture is composed of three conditional masked affine transforms with shared hyperparameters across layers. This conditional structure allows the flow to capture the hierarchical dependency between physical parameters and hyperparameters, while preserving by construction the orthogonality constraint between the two spaces.

A diagram of the full procedure is shown in Fig. 1.

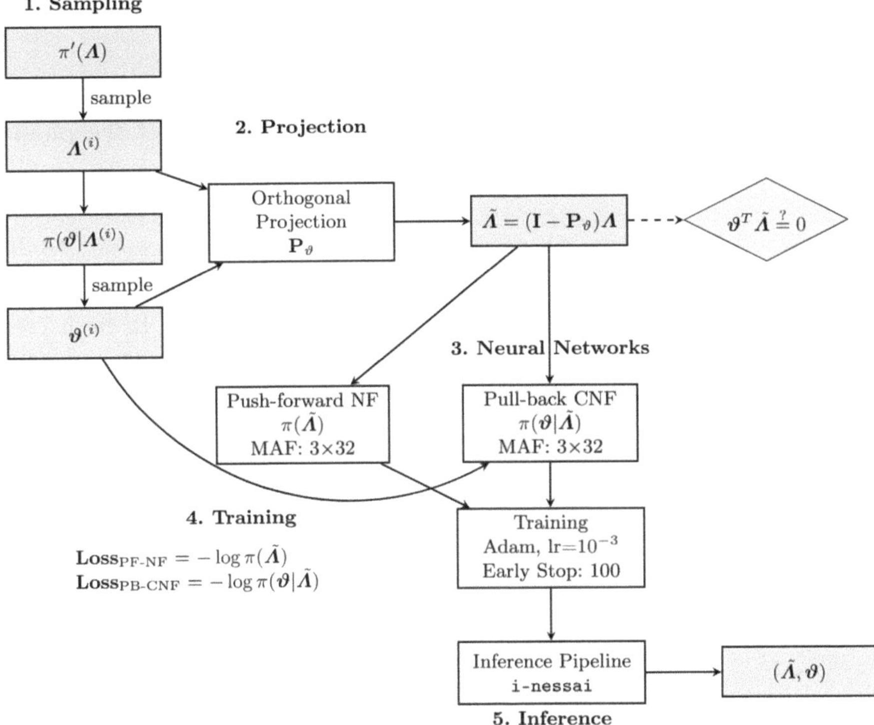

Fig. 1. Pipeline for hierarchical decorrelation: sample $(\boldsymbol{\theta}, \Lambda)$, project to $\tilde{\Lambda}$, learn $\pi(\Lambda)$ and $\pi(\boldsymbol{\theta}|\Lambda)$ with NFs, then infer with i-nessai

2.1 Push-Forward and Pull-Backward NFs Training

We trained the two NFs—the PF-NF and the PB-CNF—on a dataset consisting of $N_{\text{samples}} = 20000$ realizations generated from the priors of $\tilde{\Lambda}$ and θ. As a preliminary step, all previous samples were rescaled to the interval $[0,1]$, a transformation that generally improves the convergence of NFs. Although the available hardware would have allowed full-batch training, we adopted a mini-batch strategy with a batch size 256. This choice introduces stochasticity into the optimization process, helping the training escape poor local minima and improving the overall robustness of convergence.

The dataset was further split into training and validation subsets, with 10% of the samples reserved for validation. This separation serves two purposes: (i) it provides an unbiased evaluation of model performance on unseen data, and (ii) it enables the adoption of an early-stopping criterion, ensuring that the selected model corresponds to the minimum validation loss and reducing the risk of overfitting. The loss function used throughout training is the standard log-likelihood objective common to NFs optimization. In particular, the PF-NF is optimized by maximizing the log-likelihood of the decorrelated hyperparameters:

$$\mathbf{Loss}_{\text{PF-NF}} = -\frac{1}{N_{\text{batch}}} \sum_{i=1}^{N_{\text{batch}}} \log \pi_{\text{NF}}(\tilde{\mathbf{\Lambda}}^{(i)}) , \tag{6}$$

while the PB-CNF maximizes the conditional log-likelihood of the physical parameters given the hyperparameters:

$$\mathbf{Loss}_{\text{PB-CNF}} = -\frac{1}{N_{\text{batch}}} \sum_{i=1}^{N_{\text{batch}}} \log \pi_{\text{CNF}}(\boldsymbol{\vartheta}^{(i)} | \tilde{\mathbf{\Lambda}}^{(i)}) . \tag{7}$$

Figs. 2 display the training histories of PF-NF and PB-CNF. In both cases, the loss converges to low and stable values, demonstrating efficient and robust training. The training time takes approximately 9 min.

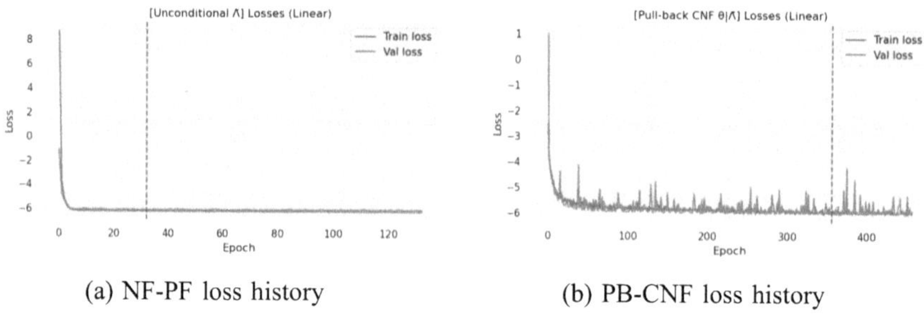

(a) NF-PF loss history   (b) PB-CNF loss history

Fig. 2. Training (blue) and validation (orange) losses for both unconditional Fig. (2a) and conditional networks Fig. (2b) on a linear scale. The vertical red dashed line indicates the selected early stopping epoch. Overall, the losses converge to low and stable values, confirming efficient and robust training (Color figure online)

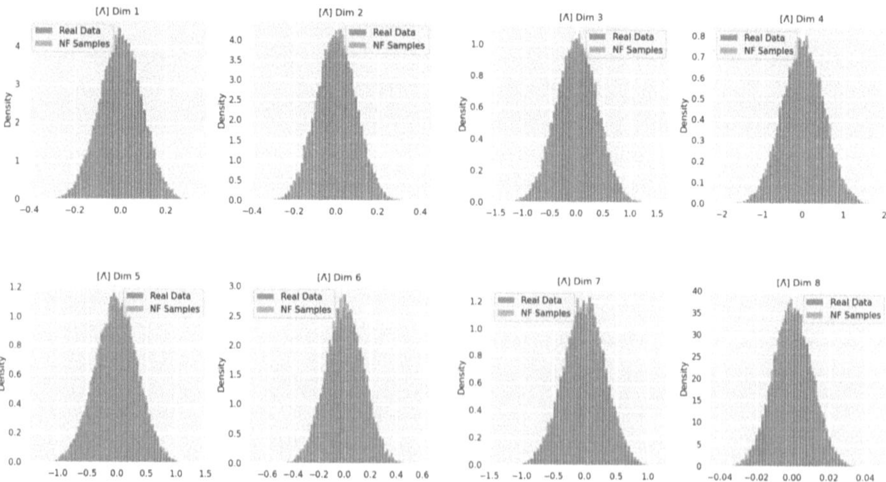

Fig. 3. Real data PDF (blue) vs PF-NF samples (orange) for eight $\tilde{\Lambda}$ dimensions; close agreement shows the model reproduces the target distribution (Color figure online)

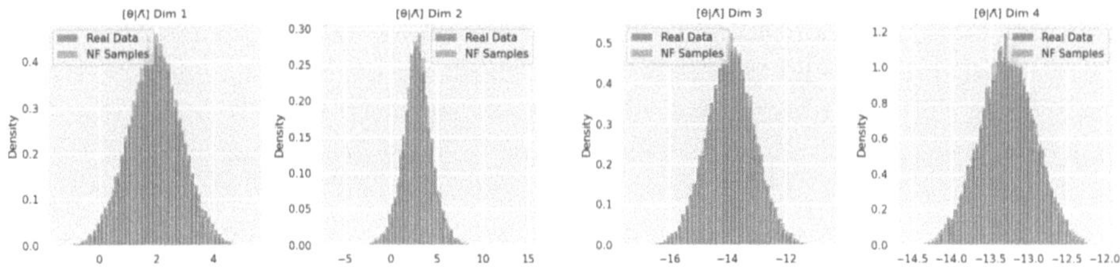

Fig. 4. Real data PDF (blue) vs PB-CNF samples (orange) for four ϑ dimensions; close agreement indicates the model reproduces the target distribution (Color figure online)

As an additional diagnostic, we compare samples generated from the trained flows with the validation data drawn from the priors. Figures 3 and 4 show the resulting distributions, where the generated samples visually reproduce the prior distributions with good fidelity. This agreement indicates that both flows have successfully captured the target probability structure.

3 Application to PTA Data Analysis

Our PTA inference pipeline relies on two complementary tools. The first is `Enterprise` (Enhanced Numerical Toolbox Enabling a Robust PulsaR Inference SuitE) [5], a Python-based software package that has become the de facto standard for PTA data analysis. It offers a modular architecture where pulsar noise and gravitational wave models are defined as components of a probabilistic model. This enables combining timing models, various stochastic noise processes, and common signals across the array, such as the SGWB. To perform the Bayesian inference we use `i-nessai` (Nested Sampling with Artificial Intelligence), a nested sampling algorithm that incorporates NFs. During the run, a NF is trained on the live points, allowing the sampler to capture complex posterior geometries and generate new samples following the likelihood contours. This greatly improves efficiency in high-dimensional correlated spaces, reduces the number of likelihood evaluations, and makes it particularly effective for PTA data analysis. In our reparameterized framework, the geometry of the posterior is simplified but remains non-trivial, and thus benefits directly from the flow-based sampling strategy.

To assess the validity of our reparameterization strategy, we apply our framework to the noise parameter inference of a single pulsar whose timing residuals are simulated from the DR2new release of the European Pulsar Timing Array dataset [7]. According to the standard prescription, we start by considering the PTA likelihood in the form that it assumes after the analytical marginalization over the timing model parameters, which is fully implemented in `Enterprise`. Moreover, in a good approximation, the white noise components are independent of other noise terms. Therefore, we fix the white noise parameters to their maximum-likelihood values and simply ignore them for the inference. Among the noise processes we account for the intrinsic Red Noise (RN) and the Dispersion

Measure (DM) noise. Both are modeled as power laws with two parameters each: the spectral index γ and the $\log_{10}$ of the amplitude A. Let us finally note that the SGWB signal is absent in a single-pulsar analysis, since it manifests itself as a correlated signal across a collection of pulsars. In the notation of Sect. 2 we thus have in total four components in the vector of physical parameters $\boldsymbol{\vartheta}$. Parametrized conditional priors $\pi(\boldsymbol{\vartheta}|\boldsymbol{\Lambda})$ with hyperpriors $\pi(\boldsymbol{\Lambda})$ are placed on the RN and DM parameters.

3.1   Implementation of the Reparametrized Hierachical Bayesian Framework in Enterprise and i-nessai

The implementation of our reparametrized hierachical Bayesian framework in Enterprise and i-nessai is the core of our work. It allows distinguishing between the parameters $\boldsymbol{\vartheta}$ and the hyperparameters $\boldsymbol{\Lambda}$ or $\tilde{\boldsymbol{\Lambda}}$, ensuring that the likelihood is correctly calculated only for the physical parameters $\boldsymbol{\vartheta}$, according to the hierarchical prescription for the joint posterior of Eq. 2. This structure is first implemented by setting the priors in Enterprise to be very wide uniform distributions, i.e. dummy priors, that do not play an effective role in the sampling process. Their only purpose is to register the physical parameters within Enterprise so that the likelihood can be computed through the standard get_lnlikelihood method, without building a specific extension of the PTA class. This approach ensures that the Bayesian estimate of the posterior is not biased by auxiliary parameters. All the specifications needed for priors $\pi(\boldsymbol{\vartheta}, \boldsymbol{\Lambda})$ and hyperpriors $\pi(\boldsymbol{\Lambda})$, together with their probability distributions— both before and after the reparametrization—are fully implemented and managed by i-nessai. This choice is motivated by practical considerations, as handling the interface on the i-nessai side proved to be much more efficient and flexible. The main characteristics of the implementation in i-nessai are:

Parameter separation for likelihood: only the $\boldsymbol{\vartheta}$ physical parameters directly affect likelihood, while all other parameters are considered only for posterior and log-prior calculation purposes. Thus, the function log_likelihood of i-nessai internally calls get_lnlikelihood of Enterprise only on the correct subset.

Sampling strategy: regardless of whether standard priors before reparameterisation or neural flows after reparameterisation are used, the sampling strategy is defined within the from_unit_hypercube method. This consists of computing the Inverse Cumulative Distribution Function (ICDF) for all parameters, thereby mapping unit-hypercube samples to the corresponding prior distributions. The nflows library [4] provides both the ICDF and the log-PDF evaluations, then i-nessai samples all parameters and calculates log-prior and log-likelihood separately, ensuring a precise estimate of the overall posterior.

The split implementation ensures that the Bayesian inference pipeline correctly evaluates the posterior while keeping Enterprise focused solely on likelihood evaluation. Thus, the integration of Enterprise with i-nessai via this user-defined class provides a useful framework for hierarchical posterior sampling in PTA analysis, while preserving the distinction between physical parameters and hyperparameters. Our implementation is straightforwardly adaptable to bigger or complex PTA datasets, as it deals mainly with the sampling functionalities in `i-nessai`, without modifying the internal architecture of `Enterprise`.

3.2   Validation Test

In the first place, as a validation test, we consider a single pulsar and we set a Gaussian distribution for the conditional parametrized prior $\pi(\vartheta|\Lambda)$. Here the vector ϑ has 4 components: γ_{RN}, γ_{DM}, $\log_{10} A_{RN}$, and $\log_{10} A_{DM}$. We want to show that the inference of these noise parameters ϑ is indeed affected by the choice of the hyperprior on Λ, $\pi(\Lambda)$. In order to quantify this effect we consider two widely-used different hyperprior classes: the Gaussian and the uniform distributions. Each of them adds two additional hyperparameters Λ: the mean and the standard deviation for the former, and the lower and upper bounds for the latter. As values for the eight hyperparameters in the hyperprior $\pi(\Lambda)$, we used the results of in [8]. In Fig. 5 we report the marginal posterior distributions for the RN and DM noise parameters with uniform Fig. (5a, Fig. 5b) and Gaussian Fig. (5c, Fig. 5d) hyperprior. The sampling is carried out with `i-nessai` and we vary the number of live points to optimize posterior accuracy. We find that with about 4000 live points, the posterior estimates are sufficiently precise to resolve the differences introduced by different hyperprior choices. Runs with fewer live points reproduce the main pattern, but exhibit

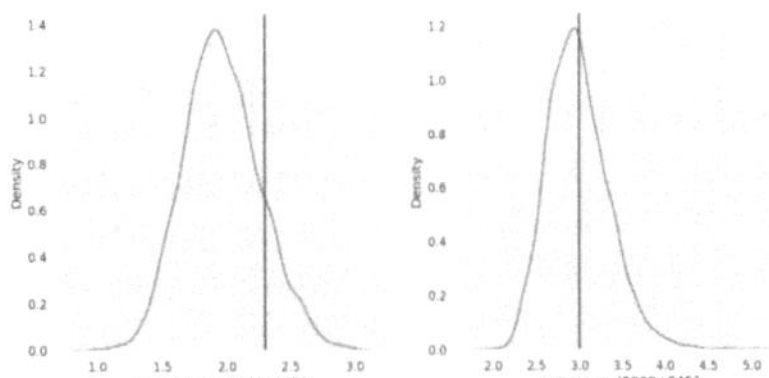

(a)   Gamma posteriors, uniform hyperprior.

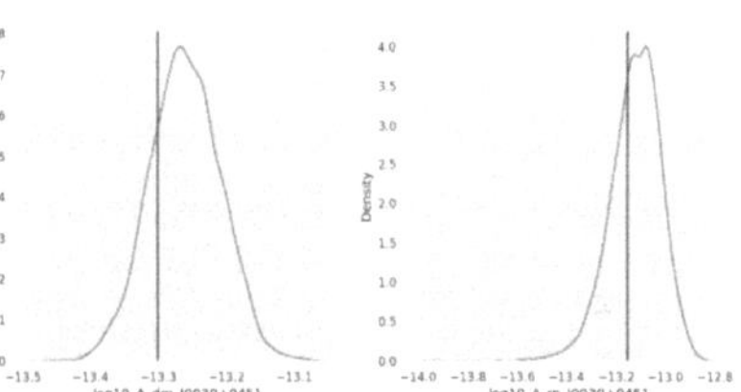

(b)   Amplitude posteriors, uniform hyperprior.

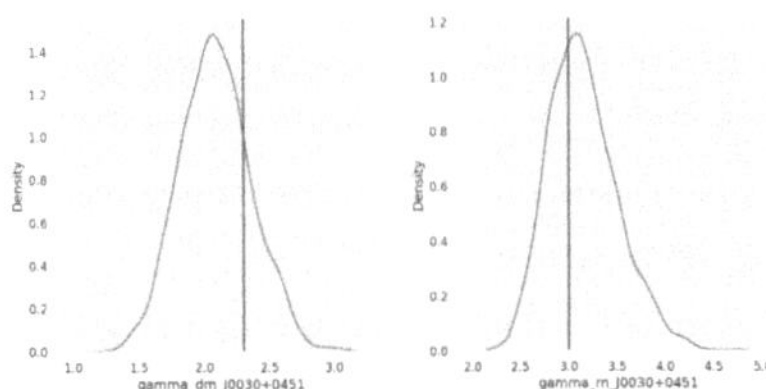

(c)   Gamma posteriors, Gaussian hyperprior.

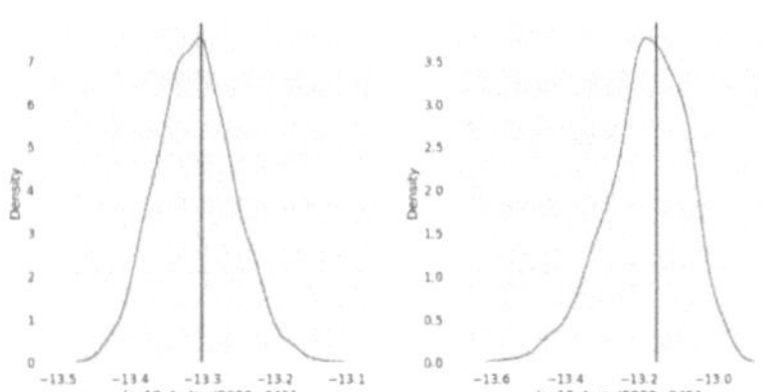

(d)   Amplitude posteriors, Gaussian hyperprior.

Fig. 5. Single-pulsar RN and DM posteriors under two hyperpriors: uniform Fig. (5a, Fig. 5b) vs Gaussian Fig. (5c, Fig. 5d); red lines mark injected values. The hyperprior choice materially affects the inferred parameters (Enterprise likelihood with `i-nessai`)

significant sampling noise in the tails of the distributions, which is completely consistent with expectations from nested sampling theory.

To summarize, we remark that the plots in Fig. 5 confirm that the detailed shapes of the posteriors depend on the specification of the hyperprior. Furthermore and most importantly, our validation test demonstrates that our implementation of the hierarchical Bayesian modeling of PTA in `Enterprise`, combined with the use of `i-nessai` for the sampling, provides a principle-based method for exploring the impact of hyperpriors on the PTA inference.

4   Discussion

To quantitatively assess the effectiveness of our reparametrization procedure, we employ two complementary metrics based on the variance decomposition principle. For any physical parameter ϑ_i and hyperparameter Λ_j (or transformed hyperparameter $\tilde{\Lambda}_j$), the law of total variance states that:

$$\mathrm{Var}(\vartheta_i) = \mathbb{E}[\mathrm{Var}(\vartheta_i|\Lambda_j)] + \mathrm{Var}(\mathbb{E}[\vartheta_i|\Lambda_j]) \tag{8}$$

where $\mathbb{E}[\mathrm{Var}(\vartheta_i|\Lambda_j)]$ represents the expected conditional variance (the variability in ϑ_i that remains after accounting for Λ_j), and $\mathrm{Var}(\mathbb{E}[\vartheta_i|\Lambda_j])$ quantifies the variance in ϑ_i explained by Λ_j. Based on this decomposition, we define two key metrics. We define the independence score $\mathcal{I}$ as:

$$(\vartheta_i, \Lambda_j) = \frac{\mathbb{E}[\mathrm{Var}(\vartheta_i|\Lambda_j)]}{\mathrm{Var}(\vartheta_i)} \tag{9}$$

which ranges from 0 to 1, with values approaching 1 indicating that ϑ_i is largely independent of Λ_j. This metric quantifies the fraction of variance in the physical parameter that is not explained by the hyperparameter, thus measuring the degree of statistical independence.

Conversely, we define the coefficient of determination R^2 as:

$$R^2(\vartheta_i, \Lambda_j) = \frac{\mathrm{Var}(\mathbb{E}[\vartheta_i|\Lambda_j])}{\mathrm{Var}(\vartheta_i)} = 1 - \mathcal{I}(\vartheta_i, \Lambda_j) \tag{10}$$

which represents the proportion of variance in ϑ_i that is predictable from Λ_j. Lower values of R^2 indicate better decorrelation. We computed two metrics above for each pair of parameters (ϑ_i, Λ_j) and $(\vartheta_i, \tilde{\Lambda}_j)$, using kernel ridge regression in the estimation of $\mathbb{E}[\vartheta|\Lambda]$ to capture non-linear dependencies.

Figure 6 shows the correlation structure in the original parameterization. While most parameter pairs exhibit high independence ($\mathcal{I} > 0.9$), notable exceptions include the coupling between $\log_{10} A_{\mathrm{dm}}$ and the hyperparameters $\mu_{\gamma_{\mathrm{rn}}}$ ($\mathcal{I} = 0.60$) and $\sigma_{\log_{10} A_{\mathrm{rn}}}$ ($\mathcal{I} = 0.61$). These correlations reflect a well-known characteristic of power-law noise processes in PTA data analysis: the amplitude and spectral index parameters exhibit strong anticorrelation [10,15]. This arises because, for a fixed dataset, a steeper spectrum (larger γ) can be partially compensated by a larger amplitude, creating a degeneracy in the likelihood surface.

This anticorrelation is particularly pronounced for red noise, where typical values yield $\rho \approx -0.7$ to -0.9 between $\log_{10} A$ and γ [6]. A similar but weaker anticorrelation exists for DM variation noise, as both processes share the same power-law spectral form. Crucially, these physical correlations persist even with hierarchical hyperprior structure, as evidenced by the moderate independence scores in Fig. 6 and the correlation in the 2D joint posterior distribution in Fig. 7. See also Figs. 1 and 2 in [8]. As expected, the hierarchical structure on the noise priors does not eliminate the inherent parameter degeneracies in the underlying modeling of the noise signal.

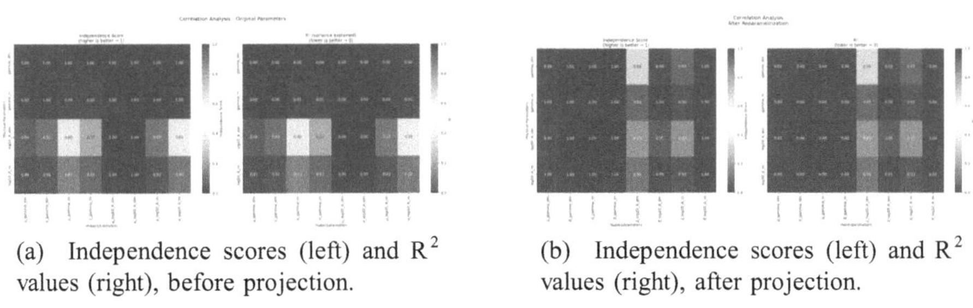

(a) Independence scores (left) and R^2 values (right), before projection.

(b) Independence scores (left) and R^2 values (right), after projection.

Fig. 6. Independence scores and R^2 before Fig. (6a) and after Fig. (6b) projection: decorrelation increases independence scores and reduces R^2

After applying the orthogonal projection to obtain $\tilde{\Lambda}$ Fig. 6., we observe a mixed but revealing pattern of decorrelation. The projection successfully eliminates several specific correlations present in the original parameterization: the couplings between $\log_{10} A_{\mathrm{dm}}$ and both $\mu_{\gamma_{\mathrm{rn}}}$ and $\sigma_{\gamma_{\mathrm{rn}}}$ improve from moderate correlation ($\mathcal{I} = 0.60$ and 0.77, respectively) to complete independence ($\mathcal{I} = 1.00$). Similarly, the correlation between $\log_{10} A_{\mathrm{dm}}$ and $\sigma_{\log_{10} A_{\mathrm{rn}}}$ is fully removed (from $\mathcal{I} = 0.61$ to 1.00). In contrast, the weak correlation between $\log_{10} A_{\mathrm{dm}}$ and $\mu_{\log_{10} A_{\mathrm{rn}}}$ remains largely unchanged (from $\mathcal{I} = 0.89$ to 0.83), suggesting that this particular coupling is not addressed by the orthogonal projection. Most notably, the amplitude parameter $\log_{10} A_{\mathrm{dm}}$ exhibits anomalous behavior with respect to its own transformed hyperparameters. This indicates that the orthogonal projection, rather than decorrelating these parameters, has concentrated approximately 81% of the variance in $\log_{10} A_{\mathrm{dm}}$ into its transformed mean hyperparameter. This selective failure — affecting primarily $\log_{10} A_{\mathrm{dm}}$ while leaving other parameters successfully decorrelated — suggests that the issue is not systemic but rather specific to how the projection interacts with the amplitude parameter under Gaussian priors. The concentration of residual correlations in the DM amplitude parameters may reflect the combined effect of the inherent amplitude-spectral index anticorrelation in power-law processes and the constraints imposed by the Gaussian hyperprior structure. These results indicate that there is room for improving the reparametrization procedure, particularly in the way it handles pre-existing anticorrelations peculiar of the PTA data and hierarchical modeling choices. Let us first comment on our prior and hyperprior choices. In the initial implementation presented in this work, we chose to adopt Gaussian priors for both physical parameters and hyperpriors, with hyperparameters means and standard deviations taken from Table 2 in [8], who performed inference on the hyperparameters directly. This choice makes our analysis methodologically robust, as it is grounded in estimates derived from the EPTA dataset. However, the effects after reparametrization may be partially attributable to our choice of Gaussian priors with relatively tight hyperpriors. As shown explicitly in Fig. 6, the mean of $\log_{10} A_{\rm dm}$ exhibits a particularly anomalous behavior: under Gaussian priors with means and variances from [8], the hyperparameter $\mu_{\log_{10} A_{\rm dm}}$ controls the mean of a relatively narrow distribution for $\log_{10} A_{\rm dm}$. The orthogonal projection, in removing the component of $\mu_{\log_{10} A_{\rm dm}}$ that lies in the subspace spanned by the physical parameters, may concentrate the remaining variation into a direction that is maximally aligned with $\log_{10} A_{\rm dm}$ itself. This effect is enhanced by the tight Gaussian hyperpriors, which limit the available parameter space and make the transformed parameter $\tilde{\mu}_{\log_{10} A_{\rm dm}}$ essentially a rescaled version of the physical parameter it was meant to be decorrelated from. This observation suggests that the combination of Gaussian priors and orthogonal projection may be particularly unsuitable for amplitude parameters in hierarchical PTA noise models. The failure could be specific rather than systemic: alternative prior specifications, particularly uniform priors on amplitudes or more flexible hyperprior distributions, may avoid this pathological behavior by providing more degrees of freedom that survive the projection.

Let us finally comment on the fundamental correlations inherent to the PTA noise models. Figure 7 shows that the characteristic anticorrelation between amplitude and spectral index parameters (Fig. 7a) also persists after the projection (Fig. 7b). This preservation is crucial, as these anticorrelations are not artifacts of the hierarchical structure but rather reflect the intrinsic degeneracies in the power-law noise modeling. Despite the complex procedure involved in the reparametrization of the hyperparameters, our method correctly captures and maintains the structure of the modeling of the lower hierarchical level. This robustness is essential for ensuring that any gains from reparametrization do not come at the cost of losing underlying parameter relationships. The ability of the NFs-based approach to maintain these intrinsic correlations while attempting to decorrelate hierarchical dependencies represents both a strength and a challenge. On one hand, it demonstrates that the method does not artificially destroy the essential structure of the noise model, which would compromise the physical interpretability of the results. On the other hand, it highlights the fundamental difficulty in distinguishing between correlations that arise from the hierarchical prior structure (which we aim to mitigate) and those that are inherent to the modeling of the process (which must be preserved).

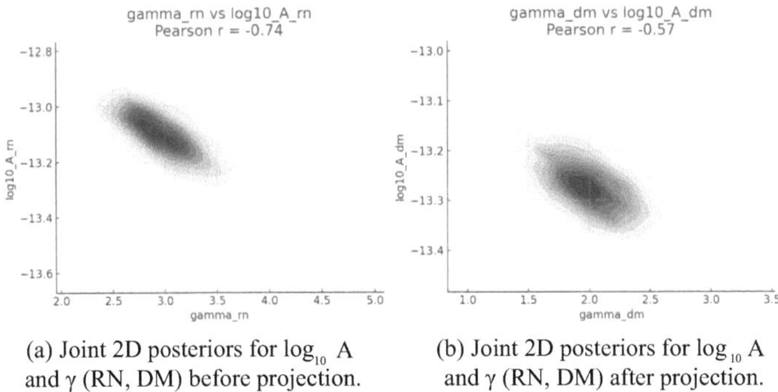

(a) Joint 2D posteriors for $\log_{10} A$
and γ (RN, DM) before projection.

(b) Joint 2D posteriors for $\log_{10} A$
and γ (RN, DM) after projection.

Fig. 7. Joint 2D posteriors for $\log_{10} A$ and γ (RN, DM) with Gaussian prior-hyperprior before and after reparameterization; Pearson coefficient shown

5   Concluding Remarks

In this paper, we have presented in full detail a hierarchical Bayesian framework for PTA noise analysis that systematically addresses prior dependence. The starting point is the introduction of hyperpriors on pulsar noise parameters: rather than using fixed priors on individual pulsar noise parameters, we introduced hyperpriors to describe the population-level distribution of these parameters, building up a hierarchical structure where hyperparameters govern the overall noise characteristics across the array of pulsars. We developed an orthogonal reparametrization strategy with the aim to address the correlations between physical parameters and hyperparameters. It is based on the employment of NFs, which provide flexible and tractable mappings between the two parameter spaces and model directly both the conditional distribution for the physical parameters and their hyperprior in the transformed parameter space.

The validation test in Sect. 3.2 confirms that combining hierarchical modeling in `Enterprise` with `i-nessai` sampling offers a consistent and statistically grounded framework to investigate how hyperpriors affect PTA inference. As a first application of our approach, we present the effects of the reparametrization on the RN and DM variation noise parameters for a single-pulsar with observed TOA simulated from the European Pulsar Timing Array dataset DR2new.

In summary, our work shows that orthogonal projection provides a principled first step toward reducing prior dependence in hierarchical PTA models: while preserving intrinsic parameter correlations of the underlying noise modeling, it does not fully disentangle them from those arising from the hierarchical structure considered here. The residual dependencies observed indicate that further refinements are required, in particular: (i) the use of more flexible prior specifications, such as a uniform prior on the physical parameters, that could potentially perform better, since the Gaussian assumption may impose a rigid hierarchical structure that, when combined with the orthogonal projection, may overly constrain the transformed parameter space and (ii) improvements of the NFs-guided reparametrization that can explicitly differentiate between power-law modeling and hierarchical correlations, potentially through physics-informed neural network architectures or by incorporating domain knowledge directly into the flow design.

The use of NFs is ubiquitous in our work: they are employed not only to realize orthogonal reparametrization, but also within the sampling algorithm. Specifically, we adopted `i-nessai`, a flow-guided nested sampler that leverages NFs to accelerate exploration of high-dimensional and computationally expensive posterior distributions. This combination ensures that both the statistical formulation and the computational implementation are consistently supported by flow-based methods.

Acknowledgments. We would like to thank Marco Bonici for his valuable insights on the details of the approach used in [19] and for stimulating discussions on our work. This paper is supported by the Fondazione ICSC, Spoke 3 Astrophysics and Cosmos Observations. National Recovery and Resilience Plan (Piano Nazionale di Ripresa e Resilienza, PNRR) Project ID CN-00000013 "Italian Research Center on High-Performance Computing, Big Data and Quantum Computing" funded by MUR Missione 4 Componente 2 Investimento 1.4: Potenziamento strutture di ricerca e creazione di "campioni nazionali di R&S (M4C2-19)" - Next Generation EU (NGEU). All HPC tests and benchmarks of this work have been carried out on the cluster at the Osservatorio di Astrofisica e Scienza dello Spazio (OAS), Bologna (Italy). We are grateful to Alessandro Tacchini for providing us access to the computing cluster.

References

1. Agazie, G., Anumarlapudi, A., Archibald, A.M., et al.: The nanograv 15 yr data set: evidence for a gravitational-wave background. Astrophys. J. Lett. **951**(1), L8 (2023). https://doi.org/10.3847/2041-8213/acdac6
2. Christensen, O.F., Roberts, G.O., Sköld, M.: Robust markov chain monte carlo methods for spatial generalized linear mixed models. J. Comput. Graph. Stat. **15**(1), 1–17 (2006)
3. Cox, D.R., Reid, N.: Parameter orthogonality and approximate conditional inference. J. Royal Statist. Soc. Ser. B (Methodol.) **49**(1), 1–39 (1987)
4. Durkan, C., Bekasov, A., Murray, I., Papamakarios, G.: nflows: normalizing flows in PyTorch (2020). https://doi.org/10.5281/zenodo.4296287
5. Ellis, J.A., Vallisneri, M., Taylor, S.R., Baker, P.T., Hazboun, J.S., Vigeland, S.J.: Enterprise: enhanced numerical toolbox enabling a robust pulsar inference suite (Sep 2020). https://doi.org/10.5281/zenodo.4059815, https://doi.org/10.5281/zenodo.4059815
6. EPTA Collaboration, InPTA Collaboration, Antoniadis, J., et al: The second data release from the european pulsar timing array. iii. search for gravitational wave signals. Astronomy & Astrophysics **678**, A50 (2023). https://doi.org/10.1051/0004-6361/202346844
7. EPTA Collaboration, InPTA Collaboration, Antoniadis, J., et al.: The second data release from the european pulsar timing array i. the dataset and timing analysis. Astronomy & Astrophys. **678**(A48), A48 (2023). https://doi.org/10.1051/0004-6361/202346841,https://arxiv.org/abs/2306.16224

8. Goncharov, B., Sardana, S.: Ensemble noise properties of the european pulsar timing array. Monthly Notices Royal Astronomical Soc. **537**(4), 3470–3479 (2025). https://doi.org/10.1093/mnras/staf190

9. Goncharov, B., et al.: Reading signatures of supermassive binary black holes in pulsar timing array observations (2025). https://arxiv.org/abs/2409.03627

10. van Haasteren, R., Vallisneri, M.: Gravitational wave detection using pulsar timing arrays. Mon. Not. R. Astron. Soc. **446**, 1170–1174 (2014)

11. van Haasteren, R.: Pulsar timing arrays require hierarchical models. Astrophys. J. Supplement Ser. **273**(2), 23 (2024). https://doi.org/10.3847/1538-4365/ad530f

12. van Haasteren, R.: Use model averaging instead of model selection in pulsar timing. Monthly Not. Royal Astronomical Soc. Lett. **537**(1), L1–L6 (2024). https://doi.org/10.1093/mnrasl/slae108

13. Kobyzev, I., Prince, S.J., Brubaker, M.A.: Normalizing flows: an introduction and review of current methods. IEEE Trans. Pattern Anal. Mach. Intell. **43**(11), 3964–3979 (2021). https://doi.org/10.1109/tpami.2020.2992934

14. Laal, N., Taylor, S.R., van Haasteren, R., Lamb, W.G., Siemens, X.: Solving the pta data analysis problem with a global gibbs scheme. Phys. Rev. D **111**(6) (2025). https://doi.org/10.1103/physrevd.111.063067

15. Lentati, L., et al.: Wide-band profile domain pulsar timing analysis. Mon. Not. R. Astron. Soc. **458**, 2161–2187 (2016)

16. Papamakarios, G., Nalisnick, E., Rezende, D.J., Mohamed, S., Lakshminarayanan, B.: Normalizing flows for probabilistic modeling and inference. J. Mach. Learn. Res. **22**(57), 1–64 (2021)

17. Papamakarios, G., Pavlakou, T., Murray, I.: Masked autoregressive flow for density estimation. In: Advances in Neural Information Processing Systems (2017)

18. Papaspiliopoulos, O., Roberts, G.O., Sköld, M.: A general framework for the parametrization of hierarchical models. Stat. Sci. **22**(1), 59–73 (2007)

19. Paradiso, S., Bonici, M., et al.: Reducing nuisance prior sensitivity via non-linear reparameterization, with application to eft analyses of large-scale structure. J. Cosmol. Astropart. Phys. **2025**(07), 005 (2025). https://doi.org/10.1088/1475-7516/2025/07/005

20. Rezende, D.J., Mohamed, S.: Variational inference with normalizing flows. In: Bach, F., Blei, D. (eds.) Proceedings of the 32nd International Conference on Machine Learning. Proceedings of Machine Learning Research, vol. 37, pp. 1530–1538. PMLR, Lille, France (2015)

21. Taylor, S.R.: The nanohertz gravitational wave astronomer (2021). https://arxiv.org/abs/2105.13270

22. Thrane, E., Talbot, C.: An introduction to Bayesian inference in gravitational-wave astronomy: parameter estimation, model selection, and hierarchical models. Publ. Astron. Soc. Austral. **36**, e010 (2019). https://doi.org/10.1017/pasa.2019.2

23. Tibshirani, R., Wasserman, L.: Some aspects of the reparametrization of statistical models. Can. J. Stat. **22**(1), 163–173 (1994)

24. Trippe, B.L., Turner, R.E.: Conditional density estimation with bayesian normalising flows. arXiv preprint arXiv:1802.04908 (2018)

25. Villa, E. and Shaifullah, G. and Possenti, A. and Carbone, C.: Improving Bayesian inference in PTA data analysis: importance nested sampling with Normalizing Flows. Astron. Comput. **55**, 101061 (2026). https://doi.org/10.1016/j.ascom.2026.101061

26. Williams, M.J.: nessai: Nested sampling with artificial intelligence (Feb 2021). https://doi.org/10.5281/zenodo.4550693

27. Williams, M.J., Veitch, J., Messenger, C.: Nested sampling with normalizing flows for gravitational-wave inference. Phys. Rev. D **103**(10) (2021). https://doi.org/10.1103/physrevd.103.103006
28. Williams, M.J., Veitch, J., Messenger, C.: Importance nested sampling with normalising flows. Mach. Learn. Sci. Technol. **4**(3), 035011 (2023). https://doi.org/10.1088/2632-2153/acd5aa

Sensitivity of Ground Motion Index Parametric Solutions to Polynomial Order and Data Splitting

Roberto Guidotti[1]($\boxtimes$) , Xabier A. Martin[2] , Elnaz Ghorbani[2,3] ,
and Guillermo Franco[1]

[1] Guy Carpenter and Company, LLC, 1166 Avenue of the Americas, New York,
NY 10036, USA
{roberto.guidotti,guillermo.e.franco}@guycarp.com
[2] Research Center on Production Management and Engineering,
Universitat Politécnica de Valéncia, Alcoy, Spain
xamarsol@upv.es
[3] Research Center on Production Management and Engineering, Universitat Oberta
de Catalunya, Barcelona, Spain
eghorbanioskalaei@uoc.edu

Abstract. Ground motion index-based parametric mechanisms are commonly used in the risk transfer industry. They depend on actual or estimated ground motion parameters, such as peak ground acceleration or pseudo-spectral acceleration values at certain periods of interest. Our research focuses on improving these mechanisms, particularly in reducing basis risk, i.e., the discrepancy between expected payouts and actual payouts triggered by parametric policies. In this paper we assess the sensitivity of a proposed ground motion index-based parametric formulation for Morocco to: i) the polynomial order used to represent the relationship between ground motion values and losses; and ii) to the data partitioning, calibrating distinct functions based on earthquake characteristics such as magnitude, depth, and location. To prevent overfitting, we introduce a data split between training and testing data, and a 5-fold cross-validation process. Results show that when 100% of data are used for training and no cross-validation is considered, parametric models with larger number of partitions relying on high-order polynomial functions produce the lowest errors, whereas the baseline approach consisting of a polynomial function of order 3 without data partition proves to be an excellent and robust model when data overfitting is minimized. The methodology is illustrated for Morocco, but it is general and applicable to any region with an available catastrophe model.

Keywords: Parametric Earthquake Risk Transfer · Index-based Parametric Solutions · Sensitivity Analysis

1   Introduction

Parametric earthquake insurance has been used since the 1990s as a critical tool for enhancing community resilience. Rather than depending on a lengthy claims assessment process, these products enable rapid and transparent payments to the insured for a pre-agreed sum if the physical, measurable attributes of the seismic event attain specific criteria (see e.g., [2,13,14]). Two typology of parametric earthquake mechanisms are frequently utilized in the industry: 1) Cat-in-a-grid mechanisms, which trigger a payout based on the primary physical characteristics of a seismic event (i.e., hypocenter location and magnitude); and 2) Ground motion indices (GMI) mechanisms, which respond based on actual or estimated ground motions at the exposure site, such as Peak Ground Acceleration (PGA), Peak Ground Velocity (PGV), Pseudo-Spectral Acceleration (PSA) at specified periods, and Modified Mercalli Intensity (MMI). Each approach has its own advantages and disadvantages, which may vary in relevance depending on the specific objectives of the stakeholders (see e.g. [1,3–6,11]). Independently on the typology, parametric solutions are prone to basis risk, i.e., the discrepancy between expected payouts and actual payouts triggered by parametric policies. To quantify the (model) basis risk we assess the performance of the parametric model comparing its estimated losses with those from a reference catastrophe model. In this paper we consider the R^2 value as error metric: highest the R^2 value, lowest the model basis risk, and highest is the performance of the tested parametric model.

In this paper we focus on the ground motion index parametric typology, and we build on the formulation proposed in [8], exploring the sensitivity of the parametric model to: i) high-order polynomial functions, and ii) data partitions. Namely, the two hypothesis we test are: i) whether using high-order polynomial functions may allow for a more accurate representation of the relationship between ground motion values and potential losses, and ii) whether partitioning the input data to estimate distinct functions based on earthquake characteristics such as magnitude, depth, and location may allow for tailored estimations that better reflect the different conditions of different seismic events.

In the next Section we recall the Ground Motion Index parametric formulation presented in [8], introducing the error metrics adopted in this paper. Section 3 discusses the index parametric solution for Morocco, the reference catastrophe model and the assumptions in the baseline solution. The methodology is illustrated for Morocco, but it is general and applicable to any region with an available catastrophe model. Section 4 presents the results of the sensitivity analysis, introducing the proposed data partitioning and the two different model calibrations analyzed in this paper. Results in Sect. 4 show that when the entire data set is used for calibration and no cross-validation is considered, parametric models with larger number of partitions relying on high-order polynomial functions produce the lowest errors. However, high values of R^2 may be the result of data overfitting, i.e. the parametric model is mimicking the underlying catastrophe model to a level that even small variations from the model may produce unpredictable results. To prevent the overfitting phenomenon, in

Sect. 4 we discuss an alternative model calibration consisting of a 80%–20% split between training and testing data, and a 5-fold cross-validation process. In this case, the results suggest that the baseline approach consisting of a polynomial function of order 3 without data partition proves to be an excellent and robust model. Section 5 draws the main conclusions.

2   Ground Motion Index Parametric Solution

The design goal for a ground motion index parametric solution is to find an optimal relationship between the ground motion at given locations and the loss it causes to the insured's assets. Parametric design usually relies on the outcomes of catastrophe models (see e.g. [7]), which incorporate various components to estimate assets' losses within a specific area. The hazard component generally includes numerous simulated stochastic earthquakes, providing both primary parameters (such as hypocenter locations and magnitudes) and the predicted ground motion values at exposure sites, typically derived from a collection of ground motion prediction equations (GMPEs) integrated into the model. The vulnerability component evaluates the estimated damage to a specific exposure based on the simulated intensity measure values, whereas a financial module converts the physical damage into monetary loss.

2.1   Parametric Model Formulation

Following the approach discussed in [8], the parametric design problem in an index-based formulation could be formulated as an optimization problem, namely as a minimization problem, where the target function to minimize is an error function g that captures the discrepancy between parametric losses (L_P) and modeled losses (L_M):

$$\min \sum_{e \in E} g(L_P^e - L_M^e) \tag{1}$$

where e represents a stochastic event in the catastrophe model's event set E, and L_P^e and L_M^e represent the losses estimated from the parametric model and from the stochastic model for the stochastic event e, respectively. The area of interest is divided into a evenly spaced grid of S virtual stations, and to each station s is assigned a weight α_s proportional to the exposure, such that:

$$\sum_{s \in S} \alpha_s = 1 \tag{2}$$

Each stochastic event e generates a footprint of the intensity measure (IM) of interest (i.e., PGA, PGV, PSA, MMI). For each the stochastic event e we extract the value of the IM of interest at each one of the virtual stations within the IM footprint. We also keep track of the model loss L_M^e caused by the stochastic event e to the insured's exposure. For each station s we can define an index L_s, which is function f of the IM of interest, as follows:

$$L_s = f(IM_s) \tag{3}$$

We assume the same functional form of L_s for each virtual station s. The total parametric loss L_P^e for event e is obtained by summing for all the S virtual stations the product of the index L_s by its corresponding weight α_s, as shown in Eq. 4:

$$L_P^e = \sum_{s \in S} \alpha_s \cdot L_s \tag{4}$$

The optimization problem consists in finding the parameters of the functional form L_s that minimize discrepancies between the parametric loss L_P^e and the modeled loss L_M^e for each event e in set E. The problem can be formulated as follows:

$$\min \sum_{e \in E} g \left(\sum_{s \in S} \alpha_s \cdot L_s - L_M^e \right) \tag{5}$$

Different error functions g can be used, such as squared error (which emphasizes large discrepancies) or absolute error (which is more robust but less differentiable).

2.2   Accuracy Metrics

In this paper, we consider a metric typically used to capture the model accuracy, (i.e., the differences between model outputs and observations), namely the coefficient of determination (R^2). The R^2, between 0 and 1, measures the proportion of the variance in the dependent variable that is explained by the adopted model. In general, the higher the R^2 value, the higher the model's accuracy; higher R^2 values indicate that the independent variables in the model explain well the variability in the dependent variable and are thus more desirable. The following equation illustrates how this evaluation metric is computed: y_i represents the true value of the target variable (for element i of a sample of n elements), $\hat{y}_i$ is the predicted value, and $\bar{y}_i$ is the average of value of the target variable:

$$R^2 = 1 - \frac{\sum_{i=1}^{n} (y_i - \hat{y}_i)^2}{\sum_{i=1}^{n} (y_i - \bar{y})^2} \tag{6}$$

In the context of catastrophe models, where the dependent variable is the loss to a particular asset or portfolio induced by a natural event, it is of interest to assess the amount of loss that would be exceeded with a certain return period. This information is provided by exceedance probability (EP) curves, used in the catastrophe (re)insurance industry to predict the monetary impact of extreme events, such as earthquakes and hurricanes, on a considered portfolio [9]. Therefore, in addition to the accuracy evaluation metric introduced above, in this paper, we compare the EP curves obtained with the different models against the EP curve obtained with the reference catastrophe model.

3   Index Parametric Solution for Morocco

3.1   Reference Catastrophe Model

Morocco is located at the intersection of the African and Eurasian Plates, and therefore faces significant seismic risk, particularly in the northern regions where the movement of these plates has caused the most destructive earthquake recorded in the country's history. The calibration of the parametric model is based on a catastrophe model. In this paper, we utilize the results from Guy Carpenter's Middle East and North Africa (MENA) Earthquake model (GCAT Model). This model features a seismic hazard module based on a modified implementation by [10]. It incorporates an event set spanning over 10,000 years, generated using GEM's OpenQuake Engine [12]. Ground motion prediction equations were selected to effectively capture both between-event and within-event variability in ground motions. The seismic hazard catalog is created on a grid with variable resolution, achieving a maximum resolution of about 850 m. The GCAT Model features a highly detailed built environment module that encompasses a wide array of risk types and secondary modifiers, covering residential, commercial, and industrial portfolios. Regarding the vulnerability module, the GCAT Model includes an extensive suite of analytical and empirical fragility and vulnerability functions that have been calibrated to align with historical loss data, based on the GEM-UCL database [15].

3.2   Baseline Solution

We could assume as baseline solution a cubic polynomial function for the index L_S, without any data partition in either magnitude, depth or geography. Under such assumptions, Eq. 3 becomes:

$$L_s = a \cdot IM_s^3 + b \cdot IM_s^2 + c \cdot IM_s \tag{7}$$

in which the intercept term is set to zero, based on the evidence that no IM would result in no loss for the insured. As a result, the optimization problem outlined in Eq. 5 assumes the following form:

$$\min_{a,b,c} \sum_{e \in E} g \left(\sum_{s \in S} \alpha_s (a \cdot IM_s^3 + b \cdot IM_s^2 + c \cdot IM_s) - L_M^e \right) \tag{8}$$

The optimization problem consists in finding the parameters (a, b and c) of the functional form used for the index L_s, such to minimize the discrepancies between the parametric loss L_P^e and modeled loss L_M^e for each event e in set E. Solving Eq. 8 for the baseline solution, considering as IM the Pseudo Spectral Acceleration (PSA) at period $T = 0.3s$, the optimal parameters assume the following values: $a = -4.288 \times 10^10$, $b = 2.044 \times 10^11$ and $c = -1.159 \times 10^10$. Such solution attains the following value for the accuracy metric: $R^2 = 0.938$.

4 Sensitivity Analysis Results

In this Section, we discuss the sensitivity of a Morocco-wide index parametric cover, varying: i) the order of the polynomial function used for the index L_S, and ii) the number of intervals in which the original data set is partitioned. Table 1 lists the intervals considered in the sensitivity analysis, for a maximum number of 6 intervals in Magnitude values, 7 intervals in Depth and 16 Area intervals (based on Longitude and Latitude values). When partitioning the original dataset in intervals, we are going to solve Eq. 5 for a subset of the dataset, considering only the portion of data (i.e. stochastic events) that has the Magnitude, Depth and Location values within the considered interval. Necessarily, considering a higher order of the polynomial function, and higher number of intervals, would result in an increase of the computational burden and in a progressively more complex and less transparent parametric model, increasing the risk of data overfitting. All these considerations have to be taken into account in selecting the optimal value of the polynomial order and number of intervals considered.

Table 1. Intervals and the number of partitions for each model parameter.

Parameter	Intervals	Number of Intervals
Magnitude	$[0, \infty)$	1
	$[0, 6), [6, \infty)$	2
	$[0, 5), [5, 6), [6, 7), [7, \infty)$	4
	$[0, 5), [5, 5.5), [5.5, 6), [6, 6.5), [6.5, 7), [7, \infty)$	6
Depth [km]	$[0, \infty)$	1
	$[0, 20), [20, \infty)$	2
	$[0, 10), [10, 20), [20, 30), [30, \infty)$	4
	$[0, 5), [5, 10), [10, 15), [15, 20), [20, 25), [25, 30), [30, \infty)$	7
Area	Lon: $(-\infty, \infty)$ Lat: $(-\infty, \infty)$	1
	Lon: $(-\infty, -7.5°), [-7.5°, \infty)$ Lat: $(-\infty, 30°), [30°, \infty)$	4
	Lon: $(-\infty, -13.75°), [-13.75°, -7.5°), [-7.5°, -1.25°), [-1.25°, \infty)$ Lat: $(-\infty, 25°), [25°, 30°), [30°, 35°), [35°, \infty)$	16

4.1 Model Calibration: 100% Train, No Cross-Validation

In this section we perform the sensitivity analysis considering polynomial functions from order 2 up to order 20, and considering up to 4 data partitions in the Magnitude parameter (up to 6 intervals), up to 4 data partitions in the Depth parameter (up to 7 intervals), and up to 3 data partitions in the Location parameter (up to 16 intervals), see Table 1. This results in a total of 912 different

Table 2. Top 10 Ground Motion index models ranked by decreasing R^2 value considering different polynomial orders and number of data intervals.

Rank	M	D	A	Number of Intervals	Order of Polynomials	R^2
1	6	7	16	672	17	0.986
2	6	7	16	672	18	0.986
3	6	7	16	672	19	0.986
4	6	7	16	672	20	0.986
5	6	7	16	672	16	0.986
6	6	7	16	672	15	0.985
7	6	7	16	672	14	0.985
8	6	7	16	672	13	0.984
9	6	7	16	672	12	0.983
10	6	4	16	384	17	0.982
854	**1**	**1**	**1**	**1**	**3**	**0.938**

ground motion index models, each with a different combination of polynomial order, and number of magnitude, depth and area intervals.

Figure 1 illustrates the variation of R^2 with respect to order of the polynomial function used for the index L_S (Eq. 3), and with the respect to the number of intervals used to partition the original dataset. Increasing the order of the polynomial results in an increasing value of R^2 up to polynomial order 17, with a steep decrease from polynomial order 19 to polynomial order 20; similarly, considering a larger number of intervals would result in higher values of R^2. We can observe that varying the functional form of the index L_S from a quadratic to a cubic function results in a considerable increase in the R^2 value, while further increasing the order of the polynomial reflects moderate increases in the R^2 value. Similarly, we observe that the effect of a larger number of intervals in the Magnitude and Depth parameter has a larger effect on the R^2 value, than a corresponding increase in the number of interval considering the Area parameter.

Table 2 ranks the 912 combinations based on decreasing value of R^2 and provides the top 10 ranked ground motion index parametric models. As expected, parametric models relying on higher polynomial order and larger number of intervals figure at the top of the table, with the highest ranked model having $R^2 = 0.986$, polynomial order 17 and a total number of intervals of 672, combining magnitude, depth and area partitions. The baseline model consisting of a cubic polynomial function for L_S (Eq. 7), and no data partitions, ranks 854th, toward the end of the list, notably with a handful of models having higher polynomial order and higher number of intervals, but lower value of R^2. Figure 2 visualizes the results listed in Table 2, illustrating the R^2 values as a function of the number of intervals considered. The size of the circles represents the different order of the polynomials, with larger circles representing higher order of poly-

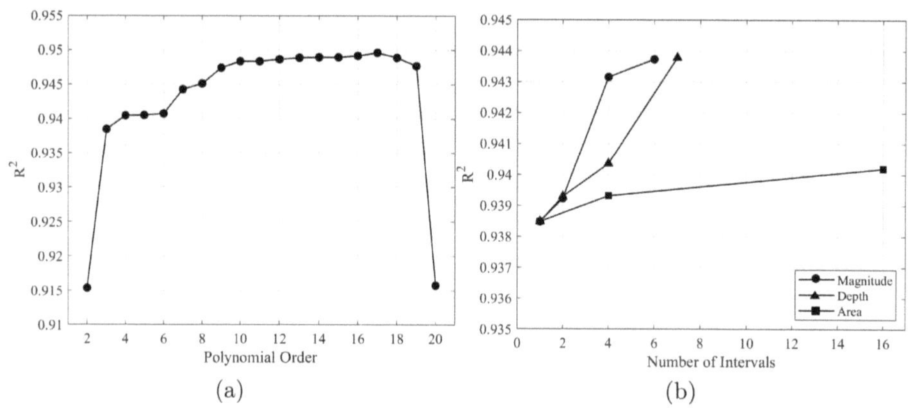

Fig. 1. R^2 variation with polynomials order (a) and number of data intervals (b).

nomials. The baseline model (Order = 3, Number of Intervals = 1) is illustrated with a white marker.

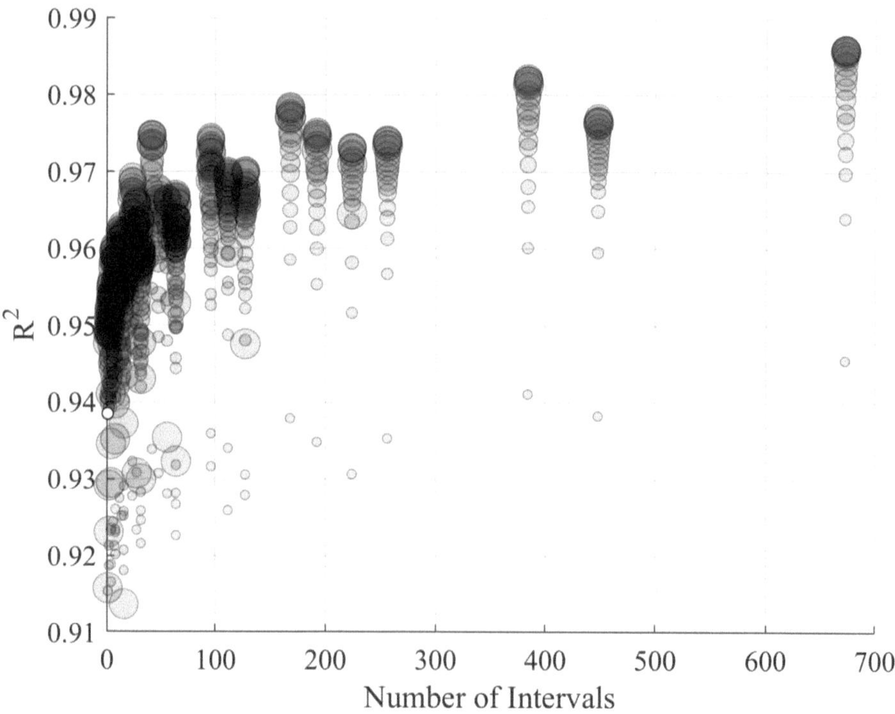

Fig. 2. Relationship between the number of intervals and R^2 values across different configurations for ground motion index parametric models.

Table 2 and Fig. 2 suggest that ground motion index parametric models with high order of polynomials for the index function L_S and calibrated on a high number of data partitions may increase the value of R^2 of more than 5%, up to $R^2 = 0.986$, a significant increase considering the already relatively high value of R^2 in the baseline model.

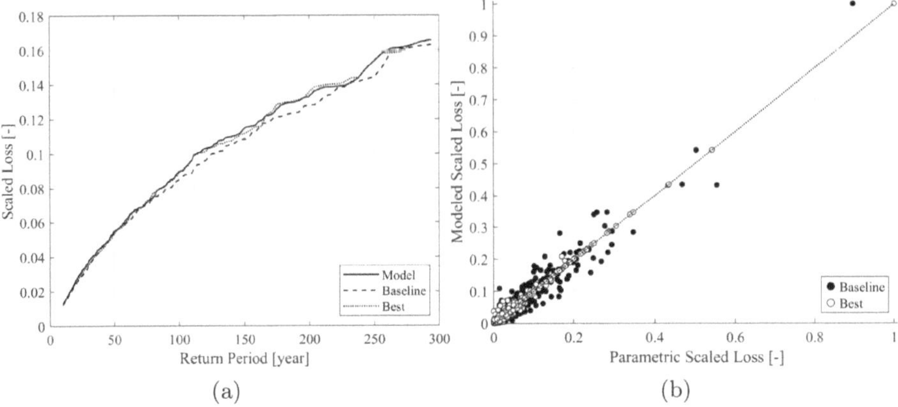

(a)   (b)

Fig. 3. EP Curve (a) and loss scatter plot (b) of the best and baseline ground motion index parametric model.

Figure 3 provides the exceedance probability (EP) curves obtained with the top ranked model (best model, Order = 17, Number of Intervals = 672) and with the baseline model, against the EP curve obtained with the reference catastrophe model. A scatter plot of parametric loss (computed with best and baseline model) vs. modeled loss is also provided. We can observe that both considered parametric models have EP curve close to the reference one, with the best model matching very closely the model EP curve. Similarly, the best model has a scatter plot with less dispersion around the identity line, suggesting that this parametric model is able to provide loss estimates very close to the loss estimated from the reference model. Figure 3 also suggest that in order to achieve such results the best model considered here may be heavily over-fitted to the reference model. The next section tackles the overfitting issue.

4.2   Model Calibration: 80% Train-20% Test, 5-Fold Cross-Validation

In this section, we tackle the overfitting issue observed in Sect. 4.1, considering a data split between training (80%) and testing data (20%). A 5-fold cross-validation process is implemented to increase the robustness of the results.

Similarly to Sect. 4.1, we perform the sensitivity analysis considering polynomial functions from order 2 up to order 20, and considering up to 4 data partitions in the Magnitude parameter (up to 6 intervals), and up to 4 data partitions in the Depth parameter (up to 7 intervals), see Table 3. We do not include the Area partitions in this sensitivity analysis for the following reasons: i) splitting the data in training and testing data set may not allow enough samples when higher number of Area intervals are considered to draw significant results; ii) the effect of Area partitions on R^2 values is less relevant with respect to the effect of Magnitude and Depth partitions (see Fig. 1b). This results in a total of 304 different ground motion index models, each with a different combination of polynomial order, and number of magnitude and depth intervals.

Table 3. Top 10 GMI models based on binned parameter configurations and different polynomial orders after cross-validation.

Rank	M	D	Number of	Order of	R^2	R^2
1	4	1	4	3	0.943	0.928
2	4	2	8	3	0.944	0.928
3	2	1	2	4	0.942	0.927
4	1	1	1	4	0.940	0.926
5	2	2	4	4	0.943	0.926
6	1	2	2	4	0.941	0.925
7	1	2	2	3	0.939	0.925
8	**1**	**1**	**1**	**3**	**0.938**	**0.925**
9	6	1	6	3	0.944	0.925
10	2	2	4	3	0.940	0.924

Figure 4 illustrates the variation of both the R^2 value for the training set and the R^2 value for the testing set with respect to order of the polynomial function used for the index L_S (Eq. 3), and with the respect to the number of intervals used to partition the original dataset. It is possible to observe that while increasing the order of the polynomial results in an increasing value of R^2 in the training set with a trend similar to the one observed in Fig. 1, the R^2 value in the testing set reaches higher values for polynomial order $= 3$ and polynomial order $= 4$, suggesting that further increasing the polynomial order does not necessarily result in an increase of the error metric, with large fluctuations and instability of the R^2 value for higher polynomial orders; similarly, considering a larger number of intervals would result in higher values of R^2 for the training set, but not necessarily in the testing set, with fluctuations around the baseline value of no data partitions.

Table 3 ranks the 304 combinations based on decreasing value of R^2 for the testing set, and provides the top 10 ranked ground motion index parametric models. Notably the top ranked models have either 3 or 4 as order of polynomials, with a generally small number of intervals. Even more notably we can observe that the baseline solution (Order $= 3$, Number of Intervals $= 1$) ranks in top 10, 8th out of 304 considered models.

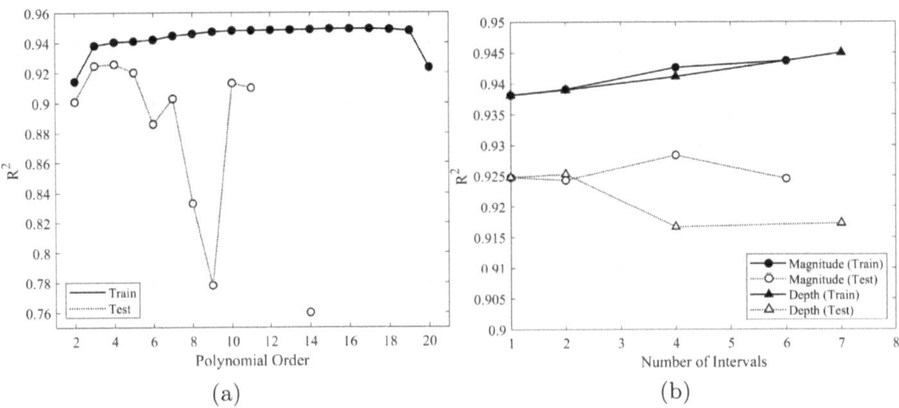

Fig. 4. R^2 variation for the training and testing set with polynomials order (a) and number of data intervals (b).

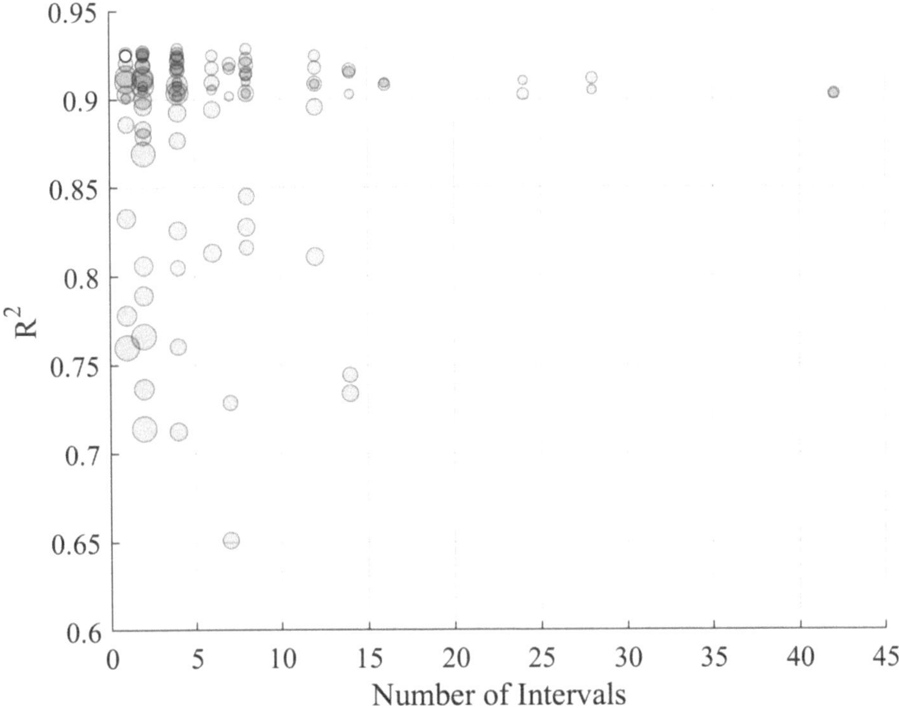

Fig. 5. Relationship between the number of intervals and average R^2 values across different configurations after cross-validation for ground motion index parametric models.

Figure 5 visualizes the results listed in Table 3, illustrating the R^2 values as a function of the number of intervals considered. Similarly to Fig. 2, the

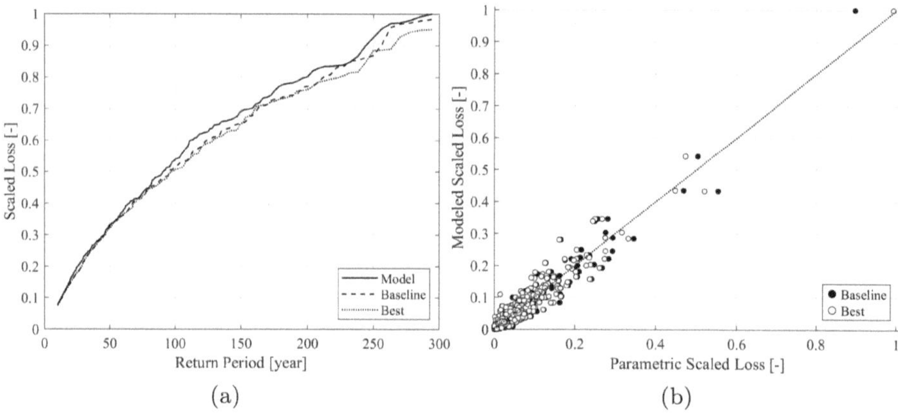

Fig. 6. EP Curve (a) and loss scatter plot (b) of the best and baseline ground motion index parametric model, when data splitting and cross-validation are considered.

size of the circles represents the different order of the polynomials, with larger circles representing higher order of polynomials. The baseline model (Order = 3, Number of Intervals = 1) is illustrated with a white marker.

Table 3 and Fig. 5 suggest that when overfitting is minimized through a data split between training and testing datasets, and cross-validation is considered, the best ground motion index parametric model (Order = 3, Number of Intervals = 4) shows only a relatively small increase in the value of R^2 of about 0.3% with respect to the baseline model.

The similarity between best and baseline model is further illustrated in Fig. 6, which provides the exceedance probability (EP) curves obtained with the top ranked model (best model, Order = 3, Number of Intervals = 4) and with the baseline model, against the EP curve obtained with the reference catastrophe model, and a scatter plot of parametric loss (computed with best and baseline model) vs. modeled loss. We can observe that the two considered parametric models have similar EP curves, close to the reference one. Similarly the two models have similar dispersion around the identity line in the scatter plot, suggesting that the baseline model has an overall performance similar to the best option, and therefore can be adopted confidently in risk management operations.

5   Conclusion

In this paper we assess the sensitivity of a proposed ground motion index parametric formulation for Morocco to the polynomial order used to represent the relationship between ground motion values and losses; and to the data partitioning, calibrating distinct functions based on earthquake characteristics (i.e., magnitude, depth, and location). The methodology is illustrated for Morocco, but it is general and applicable to any region in which a catastrophe model is available. Results show that when 100% of data are used for training and no cross-validation is considered, parametric models with high-order polynomial functions and a larger number of partitions produce the highest R^2 values, at the risk of overfitting the model data. The baseline model consisting of a polynomial function of order 3 without data partition proves to be an excellent and robust ground motion index parametric model when data overfitting is minimized, introducing a data split between training and testing data, and a 5-fold cross-validation process.

Ground motion index parametric models discussed in this paper offer a direct link between hazard intensity measures (namely Spectral Acceleration at period T=0.3 s) at the exposure locations and actual losses, potentially reducing basis risk compared to simpler cat-in-a-box triggers. However, their implementation requires dense station networks or reliable ground motion prediction equations (GMPEs), which may not be available in all regions. The possible gain in accuracy needs to be weighed against the trade-off in simplicity and transparency. Because the calculations require downloading extensive datasets and executing relatively complex computations, stakeholders often find it challenging, if not impossible, to calculate the index on their own, leading to a dependence on third-party calculation agents. In contrast, cat-in-a-grid parametric solutions are typically regarded as simpler and more transparent than ground motion indices, as they depend on a small set of parameters that are generally readily and quickly available to all parties involved, enabling stakeholders to determine potential payouts soon after an earthquake occurs. Future developments in the design of earthquake parametric solutions would consist in the combination of ground-motion indices and cat-in-a-grid parametric solutions with the goal to design more effective, reliable and robust (re)insurance solutions, which would -in turn- improve community resilience, ensuring that stakeholders receive fair and timely compensation in the aftermath of earthquakes.

Acknowledgments. The authors are grateful to Prof. A. Juan for his valuable comments and insights, functional to the development of the paper.

Disclosure of Interests. The authors have no competing interests to declare that are relevant to the content of this article.

References

1. Bayliss, C., Guidotti, R., Estrada-Moreno, A., Franco, G., Juan, A.A.: A biased-randomized algorithm for optimizing efficiency in parametric earthquake (re) insurance solutions. Comput. Oper. Res. **123**, 105033 (2020)
2. Franco, G.: Minimization of trigger error in cat-in-a-box parametric earthquake catastrophe bonds with an application to Costa Rica. Earthq. Spectra **26**, 983–998 (2010)
3. Franco, G.: Construction of customized payment tables for cat-in-a-box earthquake triggers as a basis risk reduction device. In: Proceedings of the International Conference on Structural Safety and Reliability (ICOSSAR), pp. 5455–5462 (2013)
4. Franco, G.: Earthquake mitigation strategies through insurance. Encycl. Earthquake Eng., 1–18 (2014)

5. Goda, K.: Seismic risk management of insurance portfolio using catastrophe bonds. Comput. Aided Civ. Infrastruct. Eng. **30**(7), 570–582 (2015)
6. Goda, K., Franco, G., Song, J., Radu, A.: Parametric catastrophe bonds for tsunamis: cat-in-a-box trigger and intensity-based index trigger methods. Earthq. Spectra **35**(1), 113–136 (2019). https://doi.org/10.1193/030918EQS052M
7. Grossi, P., Kunreuther, H., Windeler, D.: An introduction to catastrophe models and insurance. In: Catastrophe modeling: a new approach to managing risk, pp. 23–42. Springer (2005). https://doi.org/10.1007/0-387-23129-3_2
8. Guidotti, R., et al.: Design of a hybrid ground motion index and cat-in-a-grid parametric earthquake trigger for Morocco. In: Proceedings of the 18th World Conference on Earthquake Engineering (2024)
9. Kunreuther, H.: Risk analysis and risk management in an uncertain world. Risk Anal. Int. J. **22**(4), 655–664 (2002)
10. Poggi, V., Garcia-Peláez, J., Styron, R., Pagani, M., Gee, R.: A probabilistic seismic hazard model for North Africa. Bull. Earthq. Eng. **18**(7), 2917–2951 (2020)
11. Pucciano, S., Franco, G., Bazzurro, P.: Loss predictive power of strong motion networks for usage in parametric risk transfer: Istanbul as a case study. Earthq. Spectra **33**(4), 1513–1531 (2017)
12. Silva, V., Crowley, H., Pagani, M., Monelli, D., Pinho, R.: Development of the openquake engine, the global earthquake model's open-source software for seismic risk assessment. Nat. Hazards **72**, 1409–1427 (2014)
13. Wald, D.J., Franco, G.: Financial decision-making based on near-real-time earthquake information. In: 16th World Conference on Earthquake Engineering, pp. 1–13. Santiago de Chile, Chile (2017)
14. Wald, D.J., Franco, G.: Money matters: rapid post-earthquake financial decision-making. Nat. Hazards Observ. **40**(7), 24–27 (2016)
15. Yepes-Estrada, C., et al.: The global earthquake model physical vulnerability database. Earthq. Spectra **32**(4), 2567–2585 (2016)

From Hype to Judgment: Reassessing the Gartner Cycle for Decision-Making in the Age of Generative AI

Juan Pablo Mora-López[1,3]([✉]), David Lopez-Lopez[2], and Olga Rivera-Hernaez[1]

[1] Deusto Business School, Universidad de Deusto, San Sebastián, Spain
{j.mora,olga.rivera}@deusto.es
[2] Marketing Department, ESADE Business School, Barcelona, Spain
david.lopez1@esade.edu
[3] Pontificia Universidad Javeriana, Bogotá, Colombia
j.mora@javeriana.edu.co

Abstract. The Gartner Hype Cycle (GHC) remains a cornerstone of technology strategy; however, the rapid acceleration of Generative AI (GAI) challenges its predictive validity. This study evaluates the GHC's utility through a systematic review of 41 Web of Science-indexed articles, classifying academic discourse into supportive, critical, and adaptive positions. By contrasting GHC stage assignments with empirical adoption signals, including user engagement, media sentiment, and investment trends, we identify a significant divergence, particularly regarding the *"Trough of Disillusionment."* Our analysis demonstrates that this misalignment undermines the model's strategic value in high-velocity contexts. Consequently, we propose a replicable, data-driven framework utilizing publicly available indicators to complement traditional GHC analysis, ensuring that decision science remains commensurate with the rapid evolution of intelligent systems.

Keywords: Technology Adoption Models · Generative Artificial Intelligence (GAI) · Decision-Making under Uncertainty · Strategic Technology Evaluation · Gartner Hype Cycle Critique

1 Generative AI and Measurements of Diffusion and Adoption and Potential Overhype

Generative Artificial Intelligence (GAI) synthesizes human-like multimedia, including text, images, and video, from existing datasets. This technology integrates reinforcement (RL), machine (ML), and deep learning (DL), architectures to create novel content from established data sources (García-Peñalvo & Vázquez-Ingelmo, 2023).

While traditional AI optimizes organizational efficiency, GAI has radically disrupted the digital landscape. Freemium models for platforms such as ChatGPT, Gemini, and Midjourney have resulted in exponential adoption across both individual and organizational sectors (Ritala et al., 2024).

© The Author(s), under exclusive license to Springer Nature Switzerland AG 2026
M. Pavone et al. (Eds.): DSA ISC 2025, LNCS 16405, pp. 273–291, 2026.
https://doi.org/10.1007/978-3-032-21811-7_19

The rapid adoption of multimodal Large Language Models (LLMs) has triggered concerns that vendor overhyping may inflate expectations and extend unwarranted investments with sub-optimal outcomes. This pattern could mirror previous cycles observed in big data (Lorica, 2024) and augmented reality (Zenith Media, 2020), where technical capabilities lagged behind media attention.

Consequently, business leaders require strategic tools to balance agile competitive positioning against hype-driven risks. Effective information and communication technologies (ICT) investment depends less on acquisition and more on strategic diffusion and utilization, critical for bridging the digital divide between developing and developed economies (Opesade, 2011).

Traditional frameworks, including Diffusion of Innovations (Rogers, 1995), the Technology Acceptance Model (Davis, 1986), and the Theory of Planned Behavior (Fishbein & Ajzen, 1975), often fail to capture the unique adoption dynamics of GAI. Consequently, the Gartner Hype Cycle (GHC) has emerged as a primary tool for identifying emerging technologies and informing managerial decisions on disruptive ICT investments (Chen & Han, 2019).

Despite its industry popularity for categorizing technology maturity and penetration, GHC faces significant academic scrutiny. Critics highlight persistent empirical inconsistencies and methodological flaws that undermine its reliability as a theoretical framework (Dedehayir & Steinert, 2016), (Steinert et al., 2010), (O'Leary, 2008a).

Meanwhile, major IT consulting and development firms have acknowledged the significant current and potential impacts of GAI due to its rapid and iterative advancements. At the same time, GHC reports from 2020 to 2023 consistently identified GAI as an *"Innovation Trigger."* During this period, initial industry projections estimated technological maturity within one to three years.

Gartner's 2024 report revised GAI's maturity timeline to 2–5 years, citing excessive media coverage and widespread misconceptions regarding actual capabilities. Notably, some AI technologies, such as smart robots have bypassed standard GHC progression, exhibiting fluctuating trajectories rather than a linear path toward maturity (Fig. 1).

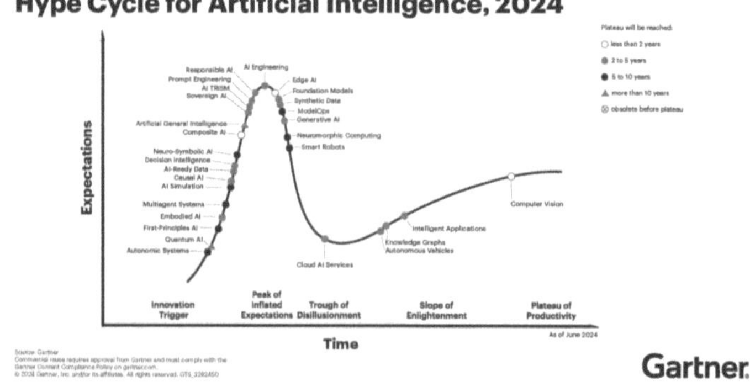

Fig. 1. HGC for AI technologies, 2024 (Jaffri, 2024). Source: Gartner Inc.

As a result, this study proposes a complementary methodology that cross-references Gartner's GAI assessments with open-source data to enhance transparency. This framework enables managers to prioritize strategic projects, optimizing returns on investment while mitigating unexpected technical and functional risks.a

2   Literature Review and Emerging Research Question

To examine the GHC as a predictive model, we conducted a systematic literature review (SLR) via Web of Science (WoS). WoS was selected for its rigorous curation and prestige in innovation management, ensuring data traceability and quality.

The SLR identified 41 articles across disciplines—including information systems, healthcare, and engineering, utilizing the GHC as a predictive tool. Inclusion prioritized studies employing the GHC as a core methodological or conceptual framework rather than a peripheral citation in technological diffusion processes.

Semantic positioning analysis classified articles as supportive, critical, or neutral. From this corpus, we selected a representative subset for in-depth comparative analysis based on methodological transparency and the clarity of the underlying arguments. Simultaneously, expert elicitation from academics and industry professionals validated hypotheses regarding GHC limitations and reformulations. This dual feedback ensured that the resulting analytical framework maintained both theoretical rigor and practical relevance.

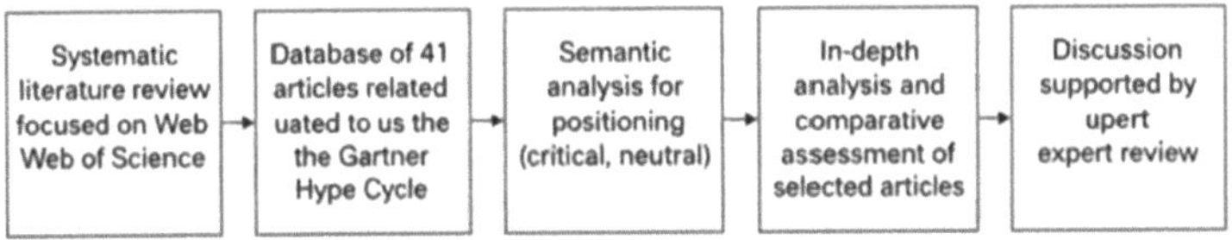

Fig. 2. Methodological Flowchart of the Literature Review and Expert-Based Evaluation Process. Source: Own elaboration

This mixed-methods framework (Fig. 2) evaluates the GHC's scientific legitimacy, its adaptability to GAI, and its dual utility for academic research and strategic decision-making.

2.1   From Industry Metaphor to Academic Framework: The Enduring Influence of the Gartner Hype Cycle

Introduced by Jackie Fenn in 1995, the GHC models technology adoption through five distinct phases: Innovation Trigger, Peak of Inflated Expectations, Trough of Disillusionment, Slope of Enlightenment, and Plateau of Productivity. It remains a global benchmark for lifecycle positioning and strategic investment.

Industry cases illustrate this trajectory: Cloud computing peaked in 2010 before maturing into foundational infrastructure, while augmented reality transitioned from 2015 over-expectation to its current pragmatic application in industrial training, healthcare, and field services.

Leading technology firms, including Microsoft, Google, and IBM, utilize GHC to align product development with market readiness. Similarly, consultancies like Accenture and Deloitte integrate the framework into client advisory services, while organizations in regulated sectors, such as Pfizer and Boeing, employ it to map digital transformation trajectories.

The GHC has evolved from a descriptive management tool into an explanatory and predictive instrument in academic research. Table 1 summarizes a curated selection of peer-reviewed studies that utilize the Hype Cycle as a scientific framework across diverse domains.

Table 1. Overview of Academic Applications of the Gartner Hype Cycle in Peer-Reviewed Research. Source: Own elaboration.

Author(s)	Context / Sector	Research Focus	Methodology	Use of the Hype Cycle
(Prinsloo & Deventer, 2017)	Education / Gamification	Evaluation of gamification adoption in learning environments	Qualitative: stakeholder interviews and policy analysis	Interpretive lens for institutional readiness
(Jayasundara, 2021)	Higher Education (Sri Lanka)	Adoption of electronic books in academic settings	Mixed methods: surveys and contextual analysis	Temporal framework to understand adoption timing
(Dargha, 2013)	Cloud Computing / Industry	Cloud computing's path from hype to implementation	Conceptual, supported by market data	Narrative scaffold to assess technological maturity
(Lima et al., 2014)	Augmented Reality / Google Glass	Rendering techniques for AR devices (Google Glass)	Design-based research (DBR)	Justifies study topic based on Hype Cycle placement
(Samuel, 2011)	Technology Adoption Theory	Proposal of a J-Curve model as an extension of the Hype Cycle	Theoretical / Conceptual	Expands the model to include socio-behavioral dynamics

In education, the GHC serves as a conceptual lens to interpret institutional readiness. (Prinsloo & Deventer, 2017) used stakeholder interviews and policy analysis to evaluate gamification adoption, while (Jayasundara, 2021) employed a mixed-methods approach to contextualize the timing and sequencing of e-book adoption in Sri Lanka.

(Dargha, 2013) utilized the GHC as a conceptual scaffold to map the maturity of cloud computing. By analyzing technology market data, the study framed the transition from inflated expectations to practical implementation.

In augmented and mixed reality, (Lima et al., 2014) used the GHC to evaluate the novelty of Google Glass rendering techniques. The framework provided the rationale for evaluating emerging applicability within a design-based research approach.

Theoretically, (Samuel, 2011) extended the GHC into a J-curve framework to incorporate socio-economic dimensions. This model accounts for institutional friction and adoption lags, providing a more nuanced understanding of contextual adoption factors beyond the standard cycle.

Synthesis reveals the GHC's role in structuring temporal diffusion dynamics and legitimizing readiness studies. Future research priorities include longitudinal empirical testing (Jayasundara, 2021), the integration of user-centric performance indicators (Lima et al., 2014), and sector-specific comparative analyses (Samuel, 2011). As such, the GHC remains a vital conceptual element, bridging industry forecasting with academic theory. It provides a standardized vocabulary for navigating the nonlinear complexities of innovation and technological adoption.

2.2 Challenging the Narrative: Limitations and Critiques of the Hype Cycle as a Predictive Model

Despite its prevalence, the GHC faces increasing scrutiny regarding its predictive validity, theoretical foundations, and transparency. As synthesized in Table 2, these critiques suggest the methodology may reflect media and market dynamics rather than actual technological maturation processes.

Critiques center on the tension between the GHC's communicative utility and its lack of empirical testability. (Badenhorst et al., 2023) found that misalignments between projected stages and actual adoption curves in South African firms undermine the GHC's forecasting reliability. They suggest the model functions as a symbolic device to justify investment narratives rather than a predictor of innovation.

(Steinert et al., 2010) argue that the GHC's popularity erodes its epistemic credibility, creating a *"self-fulfilling prophecy."* This performative effect causes organizations to align decisions with GHC labels rather than underlying data, potentially distorting technological trajectories. In fact, they argue that the lack of a transparent classification methodology introduces significant subjectivity and inconsistency.

Taking a broader perspective (Dedehayir & Steinert, 2016) identify a critical lack of theoretical grounding in the GHC. Unlike established models such as the Diffusion of Innovations, the GHC lacks causal mechanisms explaining phase transitions and fails to account for sectoral dynamics, applying a uniform curve across disparate domains.

In high-stakes sectors like healthcare, the GHC's inflation of expectations lacks clinical validation, risking misinformed investment by clinicians and administrators (Oosterhoff & Doornberg, 2020). The authors propose a more evidence-based, outcome-driven approach to evaluating technological readiness in medical contexts. Similarly, in engineering, publication data reveals that the GHC fails to reflect the rhythm of academic knowledge production, which relies on cumulative peer-reviewed evidence rather than commercial cycles (Cha et al., 2023).

The GHC integrates technological potential, social expectation, and commercial viability into a single trajectory, obscuring the non-linear impacts of regulatory hurdles,

Table 2. Overview of Scholarly Critiques Questioning the Predictive Validity of the Gartner Hype Cycle. Source: Own elaboration.

Author(s)	Context / Sector	Critique Focus	Methodology	Main Argument Against the Hype Cycle
(Badenhorst et al., 2023)	Organizational adoption (South Africa)	Misalignment between theoretical stages and actual adoption behavior	Empirical: comparative case study	Model lacks predictive alignment with firm-level reality
(Steinert et al., 2010)	Conceptual / General	Risk of self-fulfilling prophecy and lack of classification methodology	Theoretical critique	The model is performative and epistemologically fragile
(Dedehayir & Steinert, 2016)	Cross-sector / Literature review	Lack of grounding in innovation theory and absence of mechanisms	Systematic literature review	Incompatible with established innovation diffusion models
(Oosterhoff & Doornberg, 2020)	Healthcare / Orthopedic AI	Promotes premature adoption in clinical contexts lacking validation	Applied case analysis	Model fosters unrealistic expectations in sensitive medical settings
(Cha et al., 2023)	Engineering / Antennas Research	Disconnection from scientific research dynamics and publication cycles	Bibliometric analysis of publication data	Scientific advancement does not follow the Hype Cycle's market-driven assumptions

infrastructural constraints, and user adaptation. This simplified curvature fails to capture the multi-dimensional complexity of actual maturity.

To address these gaps, recent academic research advocates for sector-specific adaptations, longitudinal empirical validation, and integration with established innovation theories. Rather than a universal predictive model, the GHC should be reframed as a heuristic tool requiring contextual recalibration. Its evolution from a dominant practical element into a testable, pluralistic framework remains essential for rigorous academic and strategic application.

2.3 Between Validation and Recalibration: Sectoral Adaptations and Conceptual Extensions of the Hype Cycle

The tension between endorsement and critique has fostered a middle-ground body of research that refines and extends the GHC. As synthesized in Table 3, these studies define the model's sectoral constraints and offer pathways for its conceptual evolution.

Table 3. Sector-Specific Adaptations and Theoretical Extensions of the GHC in Academic Research. Source: Own elaboration.

Author(s)	Context / Sector	Main Contribution	Methodology	Perspective on the Hype Cycle
(Turnsek et al., 2020)	Sustainability / Aquaponics	Highlights system-level constraints and proposes hybrid modeling	Empirical / Case-based	Model lacks depth for complex, systemic innovations
(Kassarjian, 2019)	Healthcare / Musculoskeletal Radiology	Uses Hype Cycle alongside Amara's Law to analyze medical hype	Conceptual / Applied	Suggests adaptation to evidence-based clinical validation processes
(Shi & Herniman, 2023)	Innovation theory	Integrates Hype Cycle with Amara's Law to frame expectation cycles	Theoretical / Conceptual	Reframes model as part of a broader system of innovation dynamics
(Meihami & Alexander, 2024)	Education / ELT technologies	Applies model to analyze digital tech trends in language teaching	Interpretive / Longitudinal	Use Hype Cycle as pedagogical and discursive structuring tool
(Koltai, 2010)	Public Sector / E-Government (East Asia)	Shows divergence between policy innovation and Gartner trajectories	Comparative policy analysis	Advocates for selective, context-aware application of the model

It may seem that the GHC lacks the depth to account for systemic variables in complex technologies. (Turnsek et al., 2020) argue that sustainability-rooted innovations require hybrid models that integrate hype tracking with systems thinking to address regulatory, ecological, and supply-chain interdependencies absent from the standard cycle.

Sector-specific critiques emphasize that hype cycles must be filtered through independent validation processes. (Kassarjian, 2019) argues that in evidence-based domains like healthcare, the GHC requires discipline-specific adaptation to decouple clinical outcomes from media narratives and align with established principles like Amara's Law.

Integrative frameworks reconcile the GHC with Amara's Law, the tendency to overestimate short-term and underestimate long-term technological impacts. (Shi & Herniman, 2023) reframe the GHC as an iterative phase within a broader ecosystem of learning and reinvention, transforming it from a communicative device into a robust component of comprehensive innovation models.

Beyond high-technology sectors, (Meihami & Alexander, 2024) used the GHC as an interpretive framework to examine digital technology evolution in English Language Teaching. Here, the model functions as a structuring device to analyze how technology discourses shape teacher perceptions and institutional policy adoption.

Finally, in public policy, (Koltai, 2010) demonstrated that bureaucratic inertia and legacy systems create nonlinear trajectories that diverge from GHC projections. These findings challenge the model's universality, suggesting it be applied as a critical heuristic rather than a predictive rule for institutional innovation.

These perspectives invite a transition from treating the GHC as a predictive tool to applying it as a calibrated, theory-enriched framework. Scholars now favor a spectrum of contextual adaptations and empirical triangulations, analyzing how expectations, technologies, and institutions co-evolve across diverse domains.

3 Proposing Alternative Open Metrics for Emerging ITC Hype Assessment

To address the GHC's methodological limitations, this study defines open, replicable quantitative metrics for assessing GAI trajectories. Building on frameworks by (Järvenpää & Mäkinen, 2008), (Özmen, 2013) or (Bosch-Sijtsema et al., 2021), we integrate expectations, lifecycle analysis, bibliometrics, and software-specific indicators.

Assessing emerging technological trajectories is constrained by fragmented data. Following (Kampker et al., 2023), this study aggregates academic publications, conference proceedings, patent filings, and web-domain growth as quantitative proxies for the GHC phases, ensuring a transparent and replicable assessment of GAI maturity.

For this, we follow existing examples: In quantum technology, (Roberson et al., 2023) demonstrated the efficacy of analyzing national strategy documents and expert panel assessments to validate hype trajectories, providing a precedent for utilizing formal institutional discourse as a metric.

(Van Lente et al., 2013) expanded hype assessment by integrating longitudinal media tracking with academic bibliometrics. By quantifying article volume and coding narrative patterns, they identified shifts in expectations, ranging from inflation to recovery, demonstrating how media discourse serves as a proxy for social technological maturation.

As such, this study proposes an open-source data framework to augment proprietary GHC insights. By utilizing publicly accessible metrics, organizational leaders can mitigate the inherent biases of the GHC when evaluating GAI investments and maturation.

3.1 Proposed Alternative Open Metrics for GAI Assessment

Technical enhancements in GAI between 2023 and 2025 have surpassed historical forecasts, creating a decoupling between rapid model evolution and even the most optimistic established maturity assessments (Kanbach et al., 2024).

Consequently, organizations face significant pressure to integrate Large Language Models (LLMs) despite a lack of rigorous evaluation tools and a highly fragmented media landscape (Kronblad et al., 2024). As such, this study proposes four open-source metrics to empirically validate GAI's trajectory against GHC projections: 1. Venture capital (VC) investment levels, 2. Projected market size and/or website traffic Fig. 3. Global search (interest) figures and, 4. Global mediatic coverage and tonality figures.

3.2 Risk Investments as a Proxy for Corporate Interest in Emerging Technologies Such as GAI

Venture Capital (VC) serves as a critical leading indicator of technological trends, reflecting investor consensus on high-growth ICT sectors (Zider, 1998). By facilitating the translation of novel concepts into market-ready services, VC flow signals technical maturation (Hyun & Kim, 2024), though it remains a measure of market expectation rather than a definitive predictor of long-term success.

Beyond capital provision, VC involvement accelerates innovation through strategic guidance and management expertise. This resource concentration facilitates compressed cycles of design, development, and commercialization, effectively driving high-tech entrepreneurship and the rapid scaling of disruptive technologies (Timmons & Bygrave, 1986).

As a result, venture capital concentration serves as a lead signal for emerging technological trends, particularly in R&D-intensive sectors (Colombo et al., 2010). Recent investment surges in AI and GAI reflect a strategic consensus on their disruptive potential, reinforcing VC flow as a primary metric for tracking technology-intensive growth trajectories.

Corporate VC is driven by path-dependent technological capabilities. Firms strategically invest in areas of moderate or complementary expertise to leverage their existing knowledge base, while exploring disruptive frontiers (Ryu et al., 2024) using investment levels as a sensor for adjacent market growth and future integration

Investment levels in AI are directly correlated with *"knowledge coupling"*, the interconnectivity between technological capabilities and their operational environments. High coupling attracts significant capital, which subsequently accelerates patent filing volumes (Santos & Qin, 2019), signaling an investor preference for AI innovations with high integrative potential and broad applicability.

VC syndication and knowledge-sharing networks serve as stabilizing mechanisms in emerging sectors. (Hyun & Kim, 2024) demonstrate that firms integrated into experienced AI investment networks show higher independent entry rates. Such collaborative investment structures foster professional confidence, potentially flattening the GHC's *"inflated expectations"* through collective risk assessment and expertise-driven capital allocation.

This study formalizes four empirical proxies to quantify investment trends in GA. By triangulating venture capital and risk-investment flows, these metrics provide a transparent, data-driven alternative to the proprietary GHC methodology: 1. **US Venture Funding:** Percentage allocated to emerging technologies (GAI) startups and total funding in MUSD, for a defined period, 2. **Global VC Investment:** Segmented by region for emerging technologies (GAI), for a defined period, 3. **Median Round Size:** For emerging technologies (GAI) and rest of tech startups, for a defined period, and 4. **AI Unicorns:** Number of new and existing unicorns (>$1 billion valuation) developing emerging technologies (GAI) for a defined period.

While VC indicates capital allocation, it is not a deterministic predictor of technical success given the influence of volatile market conditions. To ensure a robust assessment, this methodology integrates additional publicly available metrics to quantify market demand and exogenous variables.

3.3 Website Traffic and Market Size as Proxy for Individual Interest in Emerging Technologies Such as GAI

Traditional academic reliance on self-assessed surveys often fails to capture real-time public interest in high-velocity technologies. Conversely, third-party web traffic analytics provide granular, behavioral datasets, enabling precise geographical segmentation and engagement clustering that offer a more objective metric for evaluating technological adoption and dissemination.

Web traffic serves as a vital proxy for public expectations and technological forecasting. Beyond general search indices, (Kämpf et al., 2015) demonstrated that Wikipedia traffic, integrated with context networks, offers superior granularity for identifying emerging sectors. By analyzing the Hadoop ecosystem, they established Wikipedia's knowledge-based metrics as a distinct, often more predictive, alternative to Google Trends.

(Carbonell et al., 2018) validated the use of open-access digital traces to monitor GHC-listed technologies through a multi-domain framework. By synthesizing metrics for *"overall activity"* and *"community interest"* across public (web and anchor hits), academic (citations), and industrial (patents) spheres, they demonstrated that internet-based data provides a robust, empirical proxy for tracking cross-sectoral diffusion patterns and maturity.

(Demers & Lev, 2001) established that digital engagement metrics, specifically *"reach"* and *"stickiness"*, are significant predictors of equity valuation. Their findings demonstrate that web traffic serves as an empirical anchor for financial performance, providing a methodological basis for identifying asset bubbles and overhyped technological trajectories.

To account for business-to-business (B2B) models where web traffic is an insufficient proxy, this study adopts growth-velocity metrics. Following (Jiang et al., 2024), we propose rapid scaling of this metric as a primary indicator of technological maturation.

In their study of VoIP, (Van Lente et al., 2013) demonstrated that hype peaks are driven by speculative projections of disruptive growth. These expectations eventually stabilize as technological capabilities and market share reach a plateau, providing an empirical marker for hype maturation.

(Floridi, 2024) synthesized five historical tech bubbles to isolate *"abnormal investor excitement"* as a lead indicator. In this framework, overrated market-share projections serve as the primary catalyst for the artificial inflation observed in AI trajectories.

As such, we propose two sets of measures to assess public or corporate interest in emerging technologies: 1. Time-series website traffic data for commercial emerging technologies from third-party sources (open or paid), and 2. Market share and/or size projections for emerging technologies based on investors and consulting firms.

3.4 Web Search Traffic as a Proxy for Public Interest in Emerging Technologies

The proliferation of the World Wide Web was catalyzed by search engine architectures. These systems revolutionized information retrieval by utilizing indexing and customized algorithms to transform unstructured online data into organized, prioritized query results (Schwartz, 1998).

The consolidation of the search market resulted in Google's sustained dominance. As of early 2025, Google maintains a 90% global market share, establishing a massive incumbent barrier against competitors like Microsoft Bing (Statista, 2025).

(Jun et al., 2014) demonstrate that search traffic granularity determines its predictive utility. While generic queries are weak indicators, brand- and attribute-specific searches, such as pricing, correlate strongly with sales. This refined analysis provides a high-fidelity tool for demand forecasting and identifying overhyped technological phases.

Google Trends (GT) provides real-time visibility to shifting public interest by aggregating search behavior across Google Search, News, and YouTube (Google, 2025). The platform generates a Search Volume Index (SVI), a normalized metric of search frequency relative to total volume, enabling comparative longitudinal and regional analysis. Despite critiques regarding sampling bias and algorithmic opacity quality (Franzén, 2023), (Cebrián & Domenech, 2023), GT remains a primary instrument for quantifying public interest and information-seeking behavior in high-velocity technology contexts.

(Jun, 2012) identifies a fundamental limitation in GHC methodology: its reliance on institutional data (equities and news) neglects *"visibility"* from the user perspective. By utilizing GT, search traffic serves as an empirical proxy for consumer expectations, mapping the *"information search"* stage of behavioral models.

In terms of AI technologies, (Omar et al., 2017) conceptualize the *"Google Sphere"*, comprising Trends, Scholar, and N-gram, as a multifaceted ecosystem for analyzing AI discourse. Their findings indicate that GT enables the identification of global interest patterns, revealing that AI engagement transcends traditional technological geographies and institutional hubs.

(Ahuja et al., 2023) validated the transition of AI into clinical ophthalmology by correlating GT results with capital investment and peer-reviewed output. The significant positive correlation among these three indices suggests that converged interest signals a shift from theoretical speculation to applied practice.

Given these factors, and to assess public and media engagement—key indicators of overhyping according to Gartner—we propose the use of GT to analyze: 1. **Weekly Global Searches:** Public search volume data for emerging technologies such as GAI, and 2. **News SVI:** Relative news coverage for emerging technologies such as GAI.

3.5 News Coverage and Tonality to Determine Global Interest and Perception of Emerging Technologies Such as GAI

News coverage has served as a primary proxy for media salience and thematic framing. However, empirical evidence suggests that while sentiment analysis identifies public perception trends, media volume alone does not confirm overhyping. Consequently, media metrics must be integrated with behavioral indicators to achieve higher predictive certainty.

(Kregel et al., 2021) utilized sentiment analysis and topic modeling on 95,000 news articles to evaluate the *"hype and fear"* cycle of Robotic Process Automation (RPA). Their findings indicate that despite non-linear correlations between volume, polarity and objectivity, the discursive evolution reflected a maturation of public perception, shifting from polarized narratives toward a realistic assessment of technological utility.

(Dudo et al., 2011) longitudinal analysis of nanotechnology coverage identified a *"rise and dip"* cycle characterized by journalist concentration and thematic bias. By prioritizing business and research over risk and regulation, media framing disproportionately influences initial public perception. This selective occurrence serves as a critical proxy for artificial hype inflation among consumers during the early stages of technological diffusion.

(Ahuja et al., 2023) utilized mixed-methods content analysis across four decades of elite English-language press to map the AI discourse. Their findings reveal a dramatic surge in media intensity since 2014, with approximately 16% of the coverage exhibiting significant capability overestimation. This pattern of *"marketing bluffs"* suggests systemic media-driven inflation of AI's functional maturity, signaling a peak in the hype cycle.

(Garvey & Maskal, 2020) challenged the *"Terminator Syndrome"* hypothesis: The assumption that dystopian media framing drives AI apprehension. Their sentiment analysis of high-volume news corpora revealed a paradox: despite the community's perception of negative bias, coverage is predominantly positive, suggesting that hype, rather than fear, characterizes the media landscape.

(Christner et al., 2022) highlight the *"integration crisis"* in media analysis, where the proliferation of disparate sources, from print to digital logging, impedes a holistic view of global trends. Due to high costs and limited replicability, researchers must navigate trade-offs between transparent proxies and invasive tracking (web-scrapping), necessitating a synthesized approach to mitigate source-specific technical limitations.

The *"Global Database of Events, Language, and Tone"* (GDELT 2.0) project provides a globally inclusive infrastructure for analyzing news-driven discourse across 100 languages. By synthesizing over 250 million records into a near real-time repository, it enables high-resolution event-count analysis and time-series aggregation. Its capacity to bridge Western and non-Western sources provides a robust empirical foundation for social science research on a planetary scale (Saz-Carranza et al., 2018).

As a result, and to facilitate data gathering, analysis and replicability of new coverage and tonality analysis surrounding emerging technologies in relation to potential overhype, we propose to include two key metrics: 1. **Volume Intensity:** Percentage of articles mentioning specific terms for emerging technologies (GAI) relative to overall global news coverage monitored by GDELT (0–100%), and 2.**Tone:** Sentiment analysis of articles based on the same topics, with values above 5 and below −5 indicating positive and negative sentiment, respectively.

4 Discussion-Reassessing Technology Adoption Models in the Age of Generative AI: a Data-Driven Imperative

Exiting literature reveals a fragmented consensus on the GHC, fluctuating between predictive validation and empirical skepticism. This conceptual ambiguity necessitates a methodological pivot, as the unprecedented velocity of GAI and LLMs challenges traditional adoption models and requires renewed analytical reflection.

GAI exhibits adoption dynamics that deviate from traditional Gartner projections. Despite its 2024 classification within the *"Trough of Disillusionment"* (Jaffri, 2024), empirical indicators signal sustained expansion. By late 2024, ChatGPT's 3 billion monthly visits and a surge in sectoral investment, from $500 million in 2016 to $40 billion, demonstrate a decoupling of public utility and capital intensity from speculative disillusionment.

From an academic standpoint, GAI serves as a critical case for refining technological emergence models. Its natively digital, platform-based architecture bypasses the linear, infrastructure-dependent paths of hardware-centric cycles. By leveraging cloud-based iteration and open APIs, GAI achieves frictionless integration across consumer and enterprise ecosystems, accelerating diffusion beyond traditional benchmarks (Kim & Gopal, 2021).

The availability of high-frequency datasets enables the systematic triangulation of public interest, capital flows, and media discourse. Integrating platforms like Google Trends and GDELT with web traffic analysis allows for the identification of thematic inflection points, validating a data-driven framework for real-time technology assessment (Lopez-Lopez et al., 2024).

These tools make it possible to measure hype not as an abstract narrative, but as a dynamic phenomenon unfolding across multiple observable dimensions ((Choi & Varian, 2009); (Jansen et al., 2022)). This transparency is particularly relevant given long-standing concerns about the lack of methodological clarity in the Gartner model ((Dedehayir & Steinert, 2016); (O'Leary, 2008b)).

The universal applicability of the *"Trough of Disillusionment"* is challenged by GAI's sustained user growth and capital backing. This divergence suggests that traditional frameworks may misclassify high-utility technologies, creating strategic risks for firms and investors who rely on skewed expectations of diminished engagement during sectoral maturation.

The complexity of GAI ecosystems necessitates analytical approaches that transcend static, linear models. (Rodriguez-Garcia et al., 2024) demonstrate that optimization-based frameworks, balancing financial risk, feasibility, and desirability, provide a robust multidimensional alternative. Such computational support is essential for designing resilient strategies under the multifaceted constraints of high-velocity technological innovation.

The divergence between GAI's empirical trajectory and existing theoretical heuristics necessitates a methodological shift toward transparent, replicable, and data-driven frameworks. Table 4 synthesizes these rationales, outlining the core justifications for re-evaluating the Gartner Hype Cycle through the lens of high-velocity generative technologies.

Table 4. Main justifications to reassess the results proposed by the GHC in Academic Research. Source: Own elaboration.

Justification Theme	Supporting Evidence or Argument	Source(s)
Empirical misalignment with GAI adoption	GAI remains in rapid growth while Gartner places it in the "trough of disillusionment"; traffic and VC investment continue rising	Jaffri (2024); CB Insights; Similarweb
Methodological transparency	Gartner model lacks empirical replicability and is perceived as a closed interpretive framework	Dedehayir & Steinert (2016); O'Leary (2008)
Digital-native innovation dynamics	GAI evolves via cloud platforms, APIs, and user-led feedback—differing from hardware-centric cycles	Kim & Gopal (2021)
Availability of open-source data	Tools like Google Trends, GDELT, Crunchbase allow for real-time, longitudinal observation of hype and adoption	Choi & Varian (2012); Jansen et al. (2022)
Relevance for multiple stakeholders	A new framework could inform scholars, strategists, investors, and digital commerce professionals	*This study's conceptual proposition, supported by expert review*

Based on these imperatives, we propose the Data-Driven Adoption Assessment Framework (DAAF). This tripartite structure integrates the interpretive logic of the Hype Cycle with empirical GAI adoption patterns and high-frequency, open-source datasets. The resulting framework provides a transparent, replicable, and context-aware methodology for evaluating technological diffusion in the digital-native era. A symbolic representation of this framework is presented in Fig. 3.

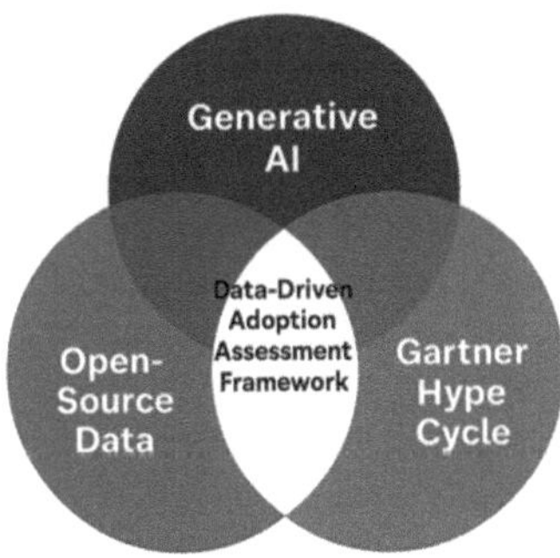

Fig. 3. Conceptual Structure of the Data-Driven Adoption Assessment Framework. Source. Own Elaboration.

Generative AI represents a dual technological and methodological inflection point, necessitating the evolution of legacy models like GHC. To maintain theoretical relevance, the academic community must shift from universalizing heuristics toward evidence-grounded frameworks. Recalibrating these models is essential for capturing the unique diffusion dynamics and practical utility of high-velocity innovations in a data-driven era.

5   Conclusions and Future Research Directions

This study has examined how the Gartner Hype Cycle is used, interpreted, and challenged within academic research. Through a systematic literature review and expert-supported analysis, we identified a lack of consensus around its predictive validity and conceptual robustness. While some scholars continue to find value in the model, others highlight its theoretical limitations, methodological opacity, and poor alignment with real-world innovation patterns.

The trajectory of Generative AI illustrates this challenge clearly. Despite its classification in the so-called "trough of disillusionment," GAI shows continued growth in user adoption, capital investment, and integration into core business processes. This raises questions about whether current models are still adequate for guiding technology-related decisions in fast-moving environments.

In response, this paper puts forward the need to build a new evaluative perspective grounded in evidence and adaptable to different sectors. We define the conceptual foundation for a Data-Driven Adoption Assessment Framework, located at the intersection of three elements: the Gartner model as a point of departure, the case of Generative AI as a contemporary stress test, and the increasing availability of open-source data as a methodological opportunity.

Future research should focus on transforming this conceptual space into a working framework. This includes identifying robust indicators of hype and adoption, designing sector-sensitive metrics, and validating the approach through longitudinal data and expert input. Doing so may help decision-makers better understand not only where technology stands, but also how it evolves and why it matters. In a time of accelerating change, more transparent and empirical models will be essential for making informed, forward-looking decisions.

References

Ahuja, A.S., Rahimy, E., Sridhar, J.: Tracking online interest in artificial intelligence in ophthalmology using Google trends. Semin. Ophthalmol. **38**(7), 644–647 (2023). https://doi.org/10.1080/08820538.2023.2204919

Badenhorst, D.P., Mashabane, S.T., Morele, J.X., Mutizwa, T., Vundla, P.Q.M., Grobbelaar, S.: Does Gartner's hype cycle theory match practice? In: Zimmermann, R., Rodriguez, J.C., Simoes, A., Dalmarco, G. (eds.) Springer Proceedings in Business and Economics, pp. 333–341. Springer, Cham (2023)

Bosch-Sijtsema, P., Claeson-Jonsson, C., Johansson, M., Roupe, M.: The hype factor of digital technologies in AEC. Constr. Innov. **21**(4), 899–916 (2021). https://doi.org/10.1108/CI-01-2020-0002

Carbonell, J., Sánchez-Esguevillas, A., Carro, B.: Easing the assessment of emerging technologies in technology observatories. Findings about patterns of dissemination of emerging technologies on the internet. Tech. Anal. Strat. Manag. **30**(1), 113–129 (2018). https://doi.org/10.1080/09537325.2017.1337886

Cebrián, E., Domenech, J.: Is Google trends a quality data source? Appl. Econ. Lett. **30**(6), 811–815 (2023). https://doi.org/10.1080/13504851.2021.2023088

Cha, Y.O., Ihalage, A.A., Hao, Y.: Antennas and propagation research from large-scale unstructured data with machine learning: a review and predictions. IEEE Antennas Propag. Mag. **65**(5), 10–24 (2023). https://doi.org/10.1109/MAP.2023.3290385

Chen, X., Han, T.: Disruptive technology forecasting based on Gartner hype cycle. In: IEEE Technology & Engineering Management Conference (TEMSCON), vol. 2019, pp. 1–6 (2019). https://doi.org/10.1109/TEMSCON.2019.8813649

Choi, H., Varian, H.: Predicting the Present with Google Trends (2009). http://CRAN.R-project.org

Christner, C., Urman, A., Adam, S., Maier, M.: Automated tracking approaches for studying online media use: a critical review and recommendations. Commun. Methods Meas. **16**(2), 79–95 (2022). https://doi.org/10.1080/19312458.2021.1907841

Colombo, M.G., Luukkonen, T., Mustar, P., Wright, M.: Venture capital and high-tech start-ups. Ventur. Cap. **12**(4), 261–266 (2010). https://doi.org/10.1080/13691066.2010.486153

Dargha, R.: Cloud computing: from hype to reality. Fast tracking cloud adoption. In: Proceedings of the International Conference on Advances in Computing, Communications and Informatics (2013). https://doi.org/10.1145/2345396.2345469

Davis, F.D.: A Technology Acceptance Model for Empirically Testing New End-User Information Systems [Massachusetts Institute of Technology] (1986). https://www.researchgate.net/publication/35465050

Dedehayir, O., Steinert, M.: The hype cycle model: a review and future directions. Technol. Forecast. Soc. Chang. **108**, 28–41 (2016). https://doi.org/10.1016/j.techfore.2016.04.005

Demers, E., Lev, B.: A Rude Awakening: Internet Shakeout in 2000. In: Review of Accounting Studies, vol. 6. Kluwer Academic Publishers (2001)

Dudo, A., Dunwoody, S., Scheufele, D.A.: The emergence of Nano news: tracking thematics trends and changes in U.S. newspaper coverage of nanotechnology. J&MC Q. **88**(1) (2011)

Fishbein, M., Ajzen, I.: Chapter 1. Belief, attitude, intention, and behavior: an introduction to theory and research. In: Reading, MA: Addison-Wesley (1975)

Floridi, L. (2024). Why the AI hype is another tech bubble. Philos. Technol. (Vol. 37, Issue 4). Springer Science and Business Media B.V. doi:https://doi.org/10.1007/s13347-024-00817-w

Franzén, A.: Big data, big problems: why scientists should refrain from using Google trends. Acta Sociologica (United Kingdom) (2023). doi:https://doi.org/10.1177/00016993221151118

García-Peñalvo, F., Vázquez-Ingelmo, A.: What do we mean by GenAI? A systematic mapping of the evolution, trends, and techniques involved in generative AI. Int. J. Interact. Multimedia Artif. Intell.., In Press (In Press), 1. (2023). https://doi.org/10.9781/ijimai.2023.07.006

Garvey, C., Maskal, C.: Sentiment analysis of the news media on artificial intelligence does not support claims of negative bias against artificial intelligence. OMICS A J. Integr. Biol. **24**(5), 286–299 (2020). https://doi.org/10.1089/omi.2019.0078

Google: Google Trends: See what's trending across Google Search, Google News and YouTube (2025). https://newsinitiative.withgoogle.com/es-mx/resources/trainings/google-trends-lesson/

Hyun, E.J., Kim, B.T.S.: Overcoming uncertainty in novel technologies: the role of venture capital syndication networks in artificial intelligence (AI) startup Investments in Korea and Japan. System. **12**(3) (2024). https://doi.org/10.3390/systems12030072

Jaffri, A: Explore Beyond GenAI on the 2024 Hype Cycle for Artificial Intelligence (Nov 2024). https://www.gartner.com/en/articles/hype-cycle-for-artificial-intelligence

Jansen, B.J., Jung, S.G., Salminen, J.: Measuring user interactions with websites: a comparison of two industry standard analytics approaches using data of 86 websites. PLoS One. **17**(5 May) (2022). https://doi.org/10.1371/journal.pone.0268212

Järvenpää, H.M., Mäkinen, S.J.: An empirical study of the existence of the hype cycle: a case of DVD technology. In: 2008 IEEE International Engineering Management Conference (2008). https://doi.org/10.1109/IEMCE.2008.4617999

Jayasundara, C.C.: Adoption of electronic books in a higher education setting: an exploratory case study based on diffusion of innovation and Garner's hype cycle paradigms. Ann. Libr. Inf. Stud. **68**, 258–267 (2021) https://www.researchgate.net/publication/355444046

Jiang, M., Yang, S., Gao, Q.: Multidimensional indicators to identify emerging technologies: perspective of technological knowledge flow. J. Informet. **18**(1) (2024). https://doi.org/10.1016/j.joi.2023.101483

Jun, S.P.: An empirical study of users' hype cycle based on search traffic: the case study on hybrid cars. Scientometrics. **91**(1), 81–99 (2012). https://doi.org/10.1007/s11192-011-0550-3

Jun, S.P., Yeom, J., Son, J.K.: A study of the method using search traffic to analyze new technology adoption. Technol. Forecast. Soc. Chang. **81**(1), 82–95 (2014). https://doi.org/10.1016/j.techfore.2013.02.007

Kämpf, M., Tessenow, E., Kenett, D.Y., Kantelhardt, J.W.: The detection of emerging trends using Wikipedia traffic data and context networks. PLoS One. **10**(12) (2015). https://doi.org/10.1371/journal.pone.0141892

Kampker, A., Heimes, H., Dorn, B., Neb, D., Clever, H., Machura, J.: Hype cycle assessment of emerging technologies for battery production. In: 5th Conference on Production Systems and Logistics (2023). https://doi.org/10.15488/15220

Kanbach, D.K., Heiduk, L., Blueher, G., Schreiter, M., Lahmann, A.: The GenAI is out of the bottle: generative artificial intelligence from a business model innovation perspective. Rev. Manag. Sci. **18**(4), 1189–1220 (2024). https://doi.org/10.1007/s11846-023-00696-z

Kassarjian, A.: Hip hype: FAI syndrome, Amara's law, and the hype cycle. Semin. Musculoskelet. Radiol. **23**(3), 252–256 (2019). https://doi.org/10.1055/s-0039-1677695

Kim, K., Gopal, A.: Soft but strong: software-based innovation and product differentiation in the IT hardware industry. Forthcoming MIS Quart. (2021). https://doi.org/10.2139/ssrn.3308277

Koltai, A.: Little tigers' big surprise - the evolution of telecommunication based on Gartner's hype cycle theory. Informacios Tarsadalom. **10**(3–4), 5–26 (2010)

Kregel, I., Koch, J., Plattfaut, R.: Beyond the hype: robotic process automation's public perception over time. J. Organ. Comput. Electron. Commer. **31**(2), 130–150 (2021). https://doi.org/10.1080/10919392.2021.1911586

Kronblad, C., Jonsson, A., Pemer, F.: Generative AI beyond the hype—new Technologies in the Face of organizing and organizations. J. Appl. Behav. Sci. (2024). https://doi.org/10.1177/002 18863241285454

Lima, J.P., Roberto, R., Texeira, J.M., Teichrieb, V.: [poster] device vs. user perspective rendering in Google glass ARApplications. In: 2014 IEEE International Symposium on Mixed and Augmented Reality (ISMAR) (2014). https://doi.org/10.1109/ISMAR.2014.6948449

Lopez-Lopez, D., Plaza-Navas, M.A., Torres-Pruñonosa, J., Martinez, L.F.: Navigating the landscape of e-commerce: thematic clusters, intellectual turning points, and burst patterns in online reputation management. Electron. Commer. Res. (2024). https://doi.org/10.1007/s10660-024-09893-8

Lorica, B.: Learning from the Past: Comparing the Hype Cycles of Big Data and GenAI. Gradient Flow (2 May 2024). https://gradientflow.substack.com/p/learning-from-the-past-comparing

Meihami, H., Alexander, C.: Adopting Gartner's hype cycles to investigate the lifespan of computer assisted language learning from EFL learners' perspectives. Asia-Pac. J. Second Foreign Lang. Educ. **9**(1) (2024). https://doi.org/10.1186/s40862-024-00256-2

O'Leary, D.E.: Gartner's hype cycle and information system research issues. Int. J. Account. Inf. Syst. **9**(4), 240–252 (2008a). https://doi.org/10.1016/j.accinf.2008.09.001

O'Leary, D.E.: Gartner's hype cycle and information system research issues. Int. J. Account. Inf. Syst. **9**(4), 240–252 (2008b). https://doi.org/10.1016/j.accinf.2008.09.001

Omar, M., Mehmood, A., Choi, G.S., Park, H.W.: Global mapping of artificial intelligence in Google and Google scholar. Scientometrics. **113**(3), 1269–1305 (2017). https://doi.org/10.1007/s11192-017-2534-4

Oosterhoff, J.H.F., Doornberg, J.N.: Artificial intelligence in orthopaedics: false hope or not? A narrative review along the line of Gartner's hype cycle. EFORT Open Rev. **5**(10), 593–603 (2020). https://doi.org/10.1302/2058-5241.5.190092

Opesade, A.O.: Strategic, value-based ICT investment as a key factor in bridging the digital divide. Inf. Dev. **27**(2), 100–108 (2011). https://doi.org/10.1177/0266666911401707

Özmen, F.: Virtual communities of practices (VCoPs) for ensuring innovation at universities-Fırat university sample. Eurasian J. Educ. Res. **53**, 131–150 (2013)

Prinsloo, T., Van Deventer, J.P.: Using the Gartner hype cycle to evaluate the adoption of emerging technology trends in higher education. In: Huang, T.-C., Rynson, L., Huang, Y.-M., Spaniol, M., Yuen, C.-H. (eds.) Emerging Technologies for Education. SETE 2017. Springer, Cham (2017). https://doi.org/10.1007/978-3-319-71084-6_7

Ritala, P., Ruokonen, M., Ramaul, L.: Transforming boundaries: how does ChatGPT change knowledge work? J. Bus. Strateg. **45**(3), 214–220 (2024). https://doi.org/10.1108/JBS-05-2023-0094

Roberson, T., Raman, S., Leach, J., Vilkins, S.: Assessing the journey of technology hype in the field of quantum technology. Zeitschrift Fur Technikfolgenabschatzung in Theorie Und Praxis. J. Technol. Assess. Theor. Pract. **32**(3), 17–21 (2023). https://doi.org/10.14512/tatup.32.3.17

Rogers, E.M.: Diffusion of Innovations, Fourth edn. The free press, New York (1995)

Ryu, W., Bae, J., Brush, T.H.: Where to invest? Effects of technological capabilities on corporate venture capital investments. Strateg. Entrep. J. **19**(1), 1–221 (2024). https://doi.org/10.1002/sej.1513

Samuel, J.C.: A theoretical technology adoption rate model. In: E-Adoption and Socio-Economic Impacts, pp. 321–331. IGI Global (2011). https://doi.org/10.4018/978-1-60960-597-1.ch016

Santos, R.S., Qin, L.: Risk capital and emerging technologies: innovation and investment patterns based on artificial intelligence patent data analysis. J. Risk Financ. Manag. **12**(4) (2019). https://doi.org/10.3390/jrfm12040189

Saz-Carranza, A., Maturana, P., & Quer, X. (2018). The Empirical Use of GDELT Big Data in Academic Research.

Schwartz, C.: Web search engines. J. Am. Soc. Inf. Sci. **49**(11), 973–982 (1998). https://doi.org/10.1002/(SICI)1097-4571(1998)49:11<973::AID-ASI3>3.0.CO;2-Z

Shi, Y., Herniman, J.: The role of expectation in innovation evolution: exploring hype cycles. Technovation. **119** (2023). https://doi.org/10.1016/j.technovation.2022.102459

Statista: Market share of leading search engines worldwide from January 2015 to March 2025 (2025). https://www.statista.com/statistics/1381664/worldwide-all-devices-market-share-of-search-engines/#:~:text=Global%20market%20share%20of%20leading%20search%20engines%202015%2D2025&text=In%20January%202025%2C%20the%20online,represented%20around%201.34%20percent.

Steinert, M., Leifer, L. J., Leifer, L.: Scrutinizing Gartner's hype cycle approach. PICMET 2010 Proceedings (18 July 2010). https://www.researchgate.net/publication/224182916

Timmons, J.A., Bygrave, W.D.: Venture capital's role in financing innovation for economic growth. J. Bus. Ventur. **1**(2), 161–176 (1986). https://doi.org/10.1016/0883-9026(86)90012-1

Turnsek, M., Joly, A., Thorarinsdottir, R., Junge, R.: Challenges of commercial aquaponics in Europe: beyond the hype. Water (Switz.). **12**(1) (2020). https://doi.org/10.3390/w12010306

Van Lente, H., Spitters, C., Peine, A.: Comparing technological hype cycles: towards a theory. Technol. Forecast. Soc. Chang. **80**(8), 1615–1628 (2013). https://doi.org/10.1016/j.techfore.2012.12.004

Zenith Media: Overhyped or Underplayed: Augmented Reality (2020). https://www.zenithmedia.co.uk/wp-content/uploads/2020/09/Overhyped-or-Underplayed-Augmented-Reality.pdf

Zider, B.: How venture capital works. Harvard Business Review (1998 Dec). https://hbr.org/1998/11/how-venture-capital-works

Can ChatGPT Take a Patient Medical History? Exploring the Role of Generative AI in Clinical Decision Support Systems

Mario Lepore[1] , Raffaele Maccioni[2], Elvira Plenzich[1] ,
and Roberto Tufano[1]([envelope])

[1] University of Salerno, Giovanni Paolo II Street 132, 84084 Fisciano, SA, Italy
`rtufano@unisa.it`
[2] Math Biology, Diogeniano di Eraclea Street 89, 00124 Rome, RM, Italy

Abstract. The integration of generative artificial intelligence into Clinical Decision Support Systems (CDSS) represents a promising direction for improving patient interaction and the quality of input data. This study explores the potential role of ChatGPT in the collection and structuring of patient medical histories through two applications: the direct classification of free-text medical histories, and the use of ChatGPT as an interactive agent to assist patients in providing medical information via natural language dialogue. Experimental results show that the direct classification of unstructured medical histories by ChatGPT yields suboptimal performance, highlighting the limitations of general-purpose language models in specialized clinical prediction tasks. Conversely, the interaction with ChatGPT to guide patients in narrating their medical history, followed by the structuring of this information, produces outputs that can be more effectively integrated into downstream classification modules. These findings suggest that, rather than serving as an autonomous decision-making system, ChatGPT could play a supportive role in facilitating early-stage data acquisition for clinical workflows.

Keywords: Clinical Decision Support System · Generative Artificial Intelligence · ChatGPT · Machine Learning

1 Introduction and Rationale

In recent years, artificial intelligence has achieved remarkable progress, largely driven by the development of advanced language models such as ChatGPT, created by OpenAI [1]. Based on transformer architecture, ChatGPT has demonstrated a high capacity for understanding and generating natural language, enabling its application across a wide range of domains, from educational and academic assistance to creative writing, from code generation and debugging to language translation, sentiment analysis, and text classification [2].
While ChatGPT's versatility makes it suitable for numerous applications, one of the most promising, and at the same time most cautiously approached, fields

M. Pavone et al. (Eds.): DSA ISC 2025, LNCS 16405, pp. 292–306, 2026.
https://doi.org/10.1007/978-3-032-21811-7_20

is healthcare [3,4]. In this context, the use of ChatGPT is gradually shifting from a general-purpose conversational assistant to a more targeted integration into clinical workflows [5,6]. One notable example is the automated generation of clinical reports, where ChatGPT has produced promising results in terms of both linguistic accuracy and content quality [7]. In [8], the model was also used as a comparative tool to evaluate the diagnostic safety and effectiveness of an AI-powered virtual assistant designed for neurological disease triage.

In the area of clinical decision-making, ChatGPT has been evaluated for its potential in the automated synthesis of therapeutic guidelines. For instance, Maniaci et al. [9] examined its application in the management of chronic rhinosinusitis. Similarly, Lee and Choi [10] conducted a systematic review exploring the capabilities and limitations of ChatGPT in clinical research within anesthesiology, highlighting its role in medical documentation, clinical decision support, and specialist training.

With regard to diagnostics, Fukuzawa et al. [11] emphasized the critical role of medical history-taking in enhancing AI-assisted diagnoses, identifying ChatGPT as a complementary tool that can reach a clinical diagnosis based on the medical history provided. In the context of medical education, Scherr et al. [12] demonstrated the model's effectiveness in clinical scenario simulations, promoting its use as a didactic tool to support interactive learning through case-based exercises.

The pharmaceutical sector is also exploring generative AI applications. According to Gangwal et al. [13], ChatGPT and similar models can be used in virtual screening for drug discovery, significantly reducing the time and cost associated with preclinical testing.

From the patient's perspective, ChatGPT may serve as a valuable tool in understanding information about symptoms, therapies, and medications, and in supporting the daily management of chronic diseases [14].

Altogether, these applications reflect the growing interest in integrating generative models into clinical, educational, and administrative aspects of medicine. However, alongside their potential, several critical challenges must be acknowledged. These include limitations in semantic understanding, data biases, risks related to safety and privacy, and concerns that unregulated use may undermine human creativity and independent reasoning [1]. In healthcare, where stakes are high and ethical implications significant, such issues are especially pronounced, reinforcing the need for precise and up-to-date medical data and a cautious and regulated approach to the deployment of these technologies [15,16]. In light of these challenges, a critical research question emerges regarding how generative AI can be responsibly integrated into clinical workflows. Specifically, understanding how such models can support – rather than replace – human decision-making becomes essential to harness their benefits while mitigating risks. Within this perspective, our ongoing research focuses on designing hybrid Clinical Decision Support Systems (CDSS) that combine structured information extraction and situation awareness to enhance clinical decision-making processes. Recent contributions have emphasized the importance of merging data-driven and knowledge-driven approaches to improve the quality and usability of clinical data [17,18]. Whereas, in this context, the present study addresses the following research question:

What role could ChatGPT play in a modern Clinical Decision Support System? Specifically, we explore whether generative language models can support the early phases of clinical workflows by assisting in the collection and structuring of patient medical histories.

To answer this question, two experimental applications were designed: the first evaluates ChatGPT's ability to directly classify free-text medical histories, while the second investigates its use as an interactive agent to guide patients in narrating their medical history through natural language dialogue. The results of these experiments aim to shed light on the potential and limitations of ChatGPT in the context of clinical data acquisition.

The remainder of the paper is organized as follows. Section 2 reviews the relevant background on Clinical Decision Support Systems and introduces the system we are currently developing. Section 3 presents the methodology and the experimental setup. Section 4 discusses the results. Finally, Sect. 5 provides the conclusions and outlines directions for future research.

2 CDSS Systems Overview

The widespread adoption of digital health technologies has led to the generation of vast and heterogeneous clinical data that clinicians must rapidly interpret to support decision-making [19]. This combination of high data volume and heterogeneity contributes to information overload, which is increasingly linked to physician burnout [20]. At the same time, interest is growing in the use of Artificial Intelligence (AI), particularly Machine Learning (ML) and Deep Learning, to improve the quality and efficiency of clinical decisions [21]. To build trust in these AI-driven systems, it is important to adopt strategies that enhance transparency, reliability, and user engagement, including, among others, human-in-the-loop approaches and participatory design frameworks [22–24].

Clinical Decision Support Systems (CDSS) have become a central component in the effort to address the complexity of modern healthcare [25]. These electronic tools assist clinicians by analyzing patient-specific data and providing tailored recommendations or assessments. By integrating computable biomedical knowledge with individual-level information, including demographics, medical history, and even genomic profiles, CDSS support clinical decision-making through advanced inference mechanisms [26]. In addition, CDSS help mitigate cognitive overload by streamlining access to relevant information and reducing the manual effort required to process complex datasets [27].

Typical functionalities of CDSS include real-time alerts for potential health risks, evidence-based clinical guidelines, condition-specific order sets, diagnostic assistance, and context-sensitive reference materials [28]. Recent studies have demonstrated the effectiveness of CDSS in improving healthcare outcomes. For instance, a review [29] analyzed 85 studies on the use of CDSS in diabetes care.

The findings highlighted that CDSS are not only effective and safe in improving disease management, but also serve as valuable support tools for physicians, particularly those with limited experience, and for patients with limited access to medical resources. Other reviews have reported that CDSSs have significantly impacted health professionals' adherence to preventive guidelines, appropriate medication prescribing, and the delivery of specific services [30–32]. However, these studies also point out that comparative research in this area tends to be of low quality, due to the wide range of heterogeneous findings and limited evidence supporting the widespread adoption of CDSS.

Indeed, a significant barrier to the effectiveness of CDSS is interoperability, which involves the difficulty of integrating data from various sources such as EHRs, wearable devices, and laboratory systems [33]. Standards such as Fast Healthcare Interoperability Resources (FHIR) and Health Level Seven (HL7) have been developed to address these challenges, but their adoption remains slow and complex, requiring alignment across institutions, healthcare systems, and countries [34]. The lack of standardized data formats and communication protocols makes it difficult to aggregate and use these data efficiently, and as a result, health care providers may have difficulty accessing complete and up-to-date patient information when they need to make critical decisions.

Taking into account these challenges, several innovative solutions have emerged in recent years. In [35], the authors propose CarePre, an intelligent clinical decision support system that uses deep learning techniques to predict future diagnostic events based on a patient's medical history, offering interactive visualizations to support clinical decisions. Another example is the NAIF (NAFLD-AI-Fibrosis) tool, which uses routine blood biochemistry and demographic parameters to diagnose advanced liver fibrosis [36]. In comparison to existing scoring systems like APRI and Fib4, NAIF has shown improved performance in sensitivity, precision, and specificity, highlighting its potential for clinical use. Another example is VisualDx, a system designed to enhance the accuracy and efficiency of dermatologic diagnosis through a comprehensive image database [37]. However, recent studies have shown that these tools still tend to underperform on richly pigmented skin due to limited representation in training datasets, highlighting the need for more diverse and inclusive data [38]. Similarly, the melanoma diagnostic tool presented in [39] uses epiluminescence dermoscopy and deep learning image analysis to improve early diagnosis. This tool accurately predicts the risk of metastasis and disease-free intervals in melanoma patients by analyzing simple serologic and histopathologic biomarkers.

However, while many recent advances focus on optimizing the analysis of traditional clinical data, the healthcare sector is also witnessing the emergence of novel data acquisition methods that could significantly enrich clinical workflows. Effectively integrating these new types of biomedical data into decision support systems poses additional challenges in terms of interoperability, standardization, and predictive modeling. Addressing these challenges requires flexible and adaptive system architectures capable of evolving alongside technological innovation.

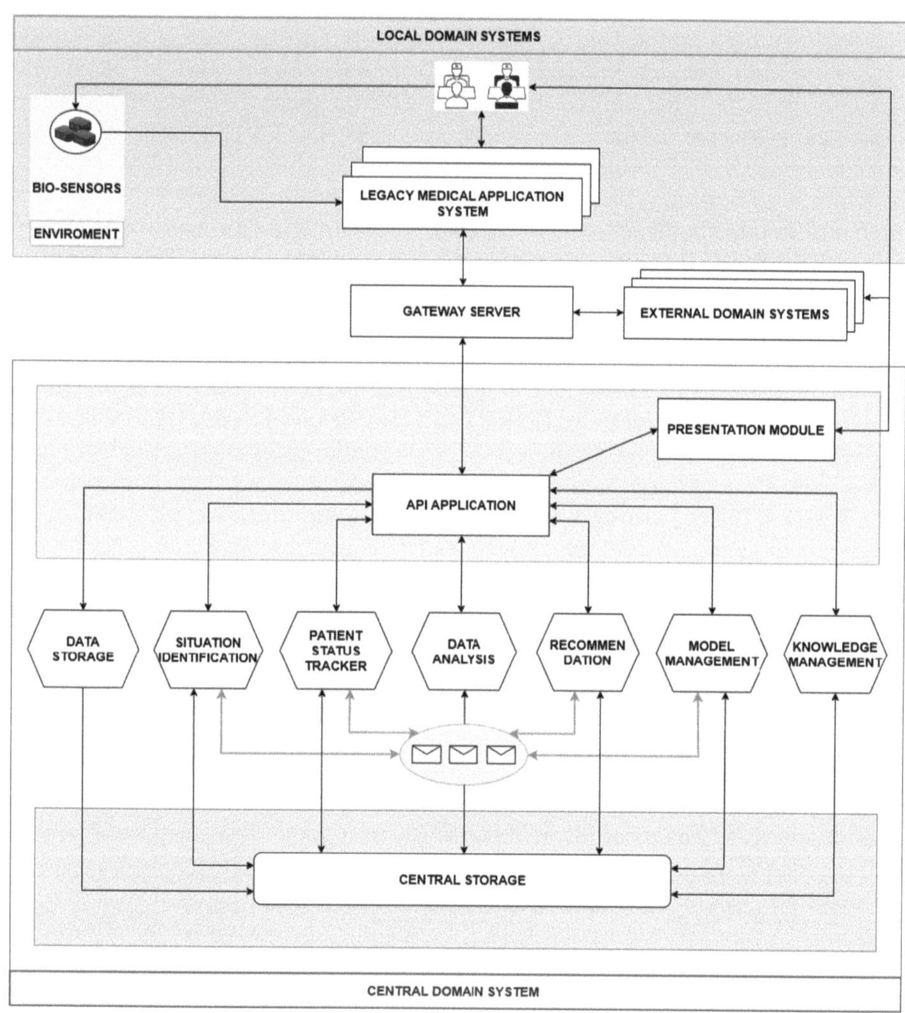

Fig. 1. The image shows a Service Oriented Architecture for Clinical Decision Support Systems based on Artificial Intelligence [17].

This is where start-ups like MathBiology play a crucial role [40]. Specializing in non-invasive early diagnosis and personalized treatments, MathBiology is developing advanced technologies that integrate novel data collection methods with AI-powered analysis. The company's biosensor technology detects nanocurrents at various points on the human body, enabling the collection of biomedical data that, when combined with traditional medical information, enhances clinical decision-making. These data are acquired within the framework of a protocol known as DMA (Deep Metabolism Assessment), a structured visit designed to analyze the patient's metabolic status through an integrated approach. The methodology is currently undergoing validation as part of the PREVEDO research project, funded by the Italian Ministry of Economic Development (MISE), involving multiple departments of the University of Salerno. The system currently used by Math Biology is capable of performing the core functionalities; however, it presents significant technological limitations. Specifically, it does not allow for the full exploitation of the predictive potential of the collected data nor easy integration with other healthcare systems. In this context, a new conceptual architecture has been developed to address these issues, providing a flexible, interoperable, and scalable infrastructure that effectively supports the deployment of AI in clinical settings.

The proposed architecture, shown in Fig. 1 and described in detail in [17], is based on a Service-Oriented Architecture (SOA) approach and enables the enhancement of data collected during the DMA visit, also through the integration of externally sourced information.

The key advantages derive from the modular nature of the approach: the software components are designed as independent services, interconnected via communication standards, thus promoting flexibility, reusability, and interoperability with other clinical systems. Furthermore, the system's deployment across geographically distributed domains, such as hospitals, diagnostic centers, and pharmacies, is made possible without compromising data consistency and integrity.

A central element of the infrastructure is the Gateway Server, which manages communication between local DMA system instances and the central data acquisition and processing domain. This module also handles the transformation of

Table 1. Description of proposed Clinical Decision Support System Services.

Service's name	Description
Data Storage	Stores data in a global database, Central Storage, standardized through FHIR for consistency and synchronization across domains.
Situation Identification	Describes the patient's current state using relevant data from various sources, leveraging Machine Learning and Soft Computing algorithms
Patient Status Tracker	Projects and evaluates the evolution of the patient's status over time using Machine Learning and AI
Recommendation	Provides feedback to assist clinicians in decision-making based on the identified situations and patient status evolution
Data Analysis	Offers statistical insights at both the individual and aggregated levels across application domains
Model Management	Defines and updates Machine Learning and AI models used by the system services, with asynchronous updates via MQTT protocol
Knowledge Management	Processes large-scale data to generate new insights useful for clinical decision-making

data into the FHIR standard format, enabling seamless integration with external systems such as EHRs, Laboratory Information Systems, and other clinical databases. Data security is ensured through the use of advanced encryption protocols and strict access control policies.

The core of the solution is the Central Domain System, which offers, via REST interfaces, a set of intelligent services, summarized in Table 1, to support clinicians in the decision-making process, providing AI-based tools capable of improving both accuracy and effectiveness.

3   Methodology

This section describes the experimental framework used to explore the potential applications of ChatGPT 4o within the Clinical Decision Support System (CDSS) we are currently developing, focusing on its ability to process and interpret patient medical histories. Two scenarios were investigated: the first assessed ChatGPT's performance in directly classifying free-text medical histories, while the second evaluated its role as an interactive agent to support the structured collection of patient information.

To contextualize the comparative evaluations, we briefly summarize the characteristics of the dataset employed in the study, originally developed and used in previous research [18]. The dataset consists of 1213 anonymized medical histories, extracted from a proprietary clinical database compiled by Math Biology [40], which includes over 20,000 records collected across three years. For the experiments, the dataset was partitioned into a training set (968 records) and a test set (245 records), with balanced representation across three major disease categories: cardiac, gastrointestinal, and thyroid conditions.

All preprocessing steps–such as duplicate removal, correction of formatting errors, and handling of missing values–were performed as described in the original study [18]. Each medical history is structured into 53 features, organized into four high-level categories:

- *Physiological medical history*, describing the patient's general health status and lifestyle factors [41];
- *Upcoming pathological medical history*, covering recent symptoms or conditions reported shortly before the clinical visit [42];
- *Remote pathological medical history*, relating to medical events occurring from birth to a few months prior to the visit [43];
- *Family medical history*, detailing hereditary and familial risk factors [44,45].

In the original study, multiple machine learning classifiers were trained on this dataset, including Support Vector Machines (SVM), Decision Trees, Random Forests, k-Nearest Neighbors (k-NN), Naïve Bayes, and Artificial Neural Networks (ANN). In addition to baseline models, advanced feature construction techniques based on Situation Awareness and Granular Computing were explored to enhance classification performance.

Against this background, the two experimental scenarios described in the following subsections aim to assess how ChatGPT can contribute to or enhance the clinical information processing pipeline within a CDSS.

3.1   Scenario 1: Direct Classification of Free-Text Anamnesis

The first scenario explores ChatGPT's ability to directly classify free-text anamneses. For this task, the test set from [18] was used, where ChatGPT was asked to read and understand each anamnesis and assign it to one of three diagnostic categories *cardiac*, *gastrointestinal*, or *thyroid* as a physician would during a real clinical visit. This scenario was inspired by the recent work of Fukuzawa et al. [11], which represents one of the few studies adopting advanced techniques for evaluating patients' medical history for diagnostic purposes. Their study tested ChatGPT's ability to produce a diagnosis using only patient history, achieving a 76.6% accuracy rate across 30 BMJ clinical vignettes. When physical examination and laboratory data were added, accuracy rose up to 93.3%, demonstrating that medical history can provide strong diagnostic performances. When the study by [18] was conducted, no such advanced artificial intelligence approaches were available in the literature beyond traditional machine learning models. The contribution by Fukuzawa therefore opened the way for a possible comparison with the methodology proposed in [18], through the use of ChatGPT for anamnesis classification. However, this comparison was limited to the results obtained using out-of-the-box classifiers, without taking into account the more advanced classification strategies proposed in [18].

In the chat sessions, it was clearly stated that the interaction was part of a scientific research study. ChatGPT was explicitly instructed to rely solely on its generative capabilities as a language model, rather than constructing implicit classification models using previously seen examples. Moreover, even in case of uncertainty, it was required to provide exactly one class evaluation, choosing the most likely among the three options based on its understanding.

Each anamnesis was prompted to ChatGPT individually to prevent cross-contamination of context or mixing of patient information.

Additionally, a second experiment was conducted where ChatGPT was first shown a few example-classified anamneses to observe whether performance would improve with prior exposure to correctly labeled cases.

3.2   Scenario 2: Guided Anamnesis via ChatGPT Dialogue

Given the suboptimal performance of ChatGPT in directly classifying free-text medical history, a second scenario was explored to observe its potential in a different role. In this scenario, the same 245 anamneses as before were used but rather than classifying the anamneses directly, ChatGPT was tasked with structuring and organizing the free-text medical history, providing a well-structured form that could then be used in a CDSS architecture.

At the start of the session, ChatGPT was introduced to the task and instructed to extract and schematize relevant information from each anamnesis. It was explicitly requested to convert qualitative descriptions into numerical representations (e.g., assigning severity scores to symptom descriptions based on intensity-related expressions) and to autonomously choose an appropriate format for the first example as in Fig. 2.

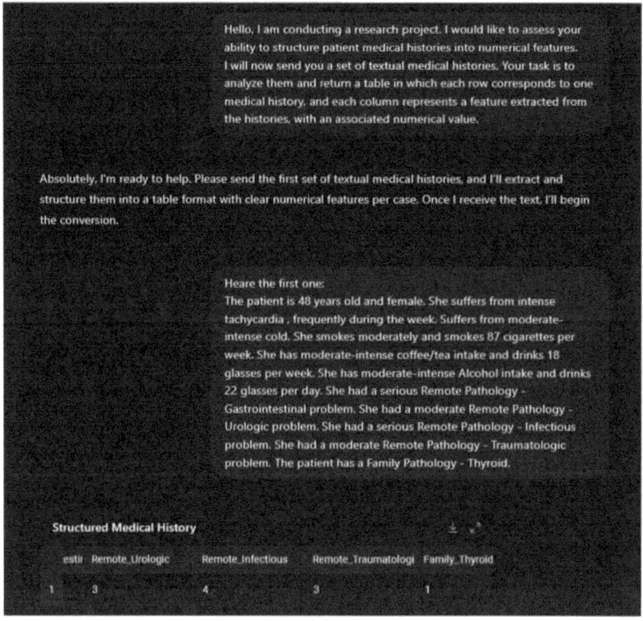

Fig. 2. This figure illustrates the first interaction with ChatGPT, in which the model was prompted to structure a medical history without receiving any contextual information beforehand.

Once the structuring logic was considered acceptable, ChatGPT was subsequently provided with the exact column names and order adopted in the structured dataset in [18]. This alignment was important to ensure compatibility during the following comparison.

Before beginning the full batch of 245 anamneses, ChatGPT was also given several "training" examples, and this helped the model to adjust and normalize the numerical scales it had initially used. Following this preparation phase, the 245 textual anamneses were submitted one at a time, and ChatGPT returned their structured versions accordingly to the previously instructions.

4   Experimental Evaluation and Disucssion

The effectiveness of ChatGPT in the two scenarios was assessed through quantitative evaluation metrics and comparative analysis.

In the first scenario, the accuracy of ChatGPT was assessed by comparing its predicted class for each anamnesis with the correct label. When ChatGPT was asked to classify each anamnesis without any prior examples, acting purely as a generative model, based on its knowledge, the accuracy reached 53%. In a second phase, a few correctly labeled example medical histories were provided to the model in advance, in an attempt to simulate a light training process. In this case, accuracy slightly improved to 63%.

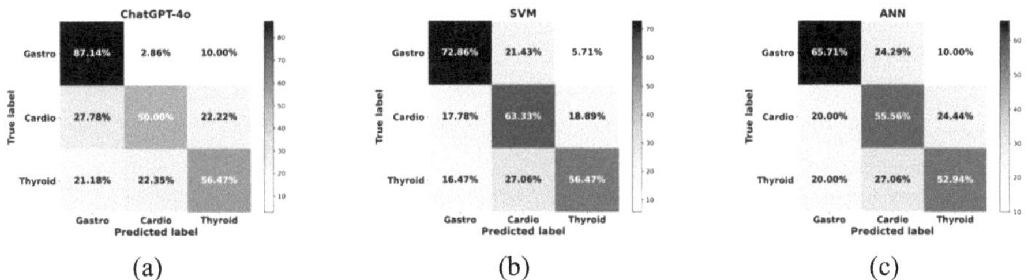

Fig. 3. Confusion matrices for ChatGPT-4o, SVM, and ANN classifiers.

In order to contextualize the performance of ChatGPT-4o, a comparative analysis with the results reported in [18] was conducted. In Fig. 3 the confusion matrices of ChatGPT-4o, Support Vector Machine (SVM), and Artificial Neural Network (ANN). SVM and ANN have been selected as representative classifiers from the original study–SVM as a conventional machine learning model, and ANN as the only neural architecture used in [18]. The comparison highlights the not encouraging classification performance of ChatGPT-4o, which exhibits many misclassifications across the three diagnostic categories, in particular for cardio and thyroid disease. These outcomes are supported by a comparison focused on the balanced accuracy metric, which was the primary evaluation criterion adopted in [18]. The bar plot in Fig. 4 summarizes the balanced accuracy scores of all classifiers tested in the original study, highlighting that ChatGPT-4o yields the lowest performance–even below the least effective out-of-the-box classifier examined in that work.

In light of a previous work by Fukuzawa et al. [11] that reported high classification performances using ChatGPT, we anticipated stronger performances but the obtained results were below expectations. However, key methodological differences could explain the discrepancy. In [11], the input to the model included additional diagnostic information, such as laboratory results, which are critical to accurately understanding and contextualizing a patient's condition. In our setting, the classification was performed based only on unstructured anamnesis text, without access to any supplementary data.

Moreover, the three disease categories used in our study (cardiac, gastrointestinal, and thyroid) share many overlapping symptoms [46–51]. This symptom overlap probably contributed to the difficulty of assigning a single clear diagnostic category in the absence of more discriminative features.

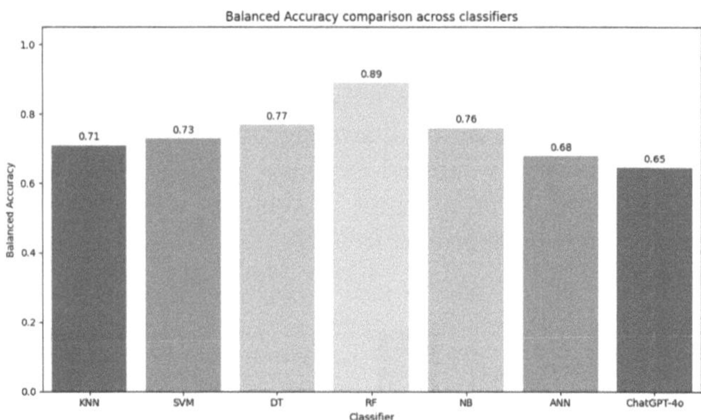

Fig. 4. Barplot with the balanced accuracy obtained by the different classifiers in [18].

The just mentioned limitations emphasize the need for caution when applying general-purpose LLMs like ChatGPT directly to clinical classification tasks.

Therefore, while ChatGPT is capable of good free-text comprehension, it struggles in providing consistent and reliable medical classification in the absence of explicit clinical training or contextual information that strongly characterizes a pathology.

In the second scenario, ChatGPT was used to convert unstructured medical histories into structured numerical data. The output produced by ChatGPT was compared with the structured ones in [18]. In order to measure how closely ChatGPT's output aligned with the reference structured data, the following statistical error metrics were computed:

- Mean Squared Error (MSE): 0.2248
- Root Mean Squared Error (RMSE): 0.4742
- Mean Absolute Error (MAE): 0.0697

These indicators were selected because they are widely used in the literature to assess the closeness between predicted and true values [52–54].

These findings suggest that ChatGPT may serve effectively as a semantic interpreter to structure a free-text medical history, bridging the gap between raw patient medical history and the structured data required by AI-powered diagnostic tools. Within a Service-Oriented Architecture for Clinical Decision Support Systems such as the one proposed by [17], ChatGPT could be integrated as a pre-processing assistant tasked with converting input anamneses into standardized ones. This would reduce the cognitive and operational burden on healthcare professionals where quick and scalable anamnesis interpretation could be critical and ensure consistency without compromising the quality of data processed by CDSS.

5   Conclusion and Future Work

This study investigated the potential role of ChatGPT within Clinical Decision Support Systems (CDSS), focusing on two distinct applications: direct classification of free-text medical histories and interactive assistance in structuring clinical information.

Experimental evidence showed that ChatGPT, despite its strong natural language processing capabilities, struggles to achieve consistent performance when directly tasked with clinical classification based solely on unstructured anamnesis data. These findings highlight the current limitations of general-purpose generative models in specialized medical predictive tasks without tailored clinical training or integration of additional diagnostic information.

Conversely, when used as an interactive agent to guide and structure patient-provided information, ChatGPT demonstrated promising capabilities. Its ability to organize medical narratives into structured formats suitable for downstream processing suggests a valuable supporting role in clinical workflows, particularly during the early stages of patient data acquisition.

Building on these insights, future research directions include: (i) the integration of domain-specific fine-tuning to improve ChatGPT's diagnostic reasoning in medical contexts; (ii) the exploration of hybrid approaches combining ChatGPT-assisted data structuring with traditional machine learning classifiers; and (iii) the evaluation of user acceptance and clinical usability of such systems through real-world pilot studies involving healthcare professionals and patients.

Ultimately, positioning ChatGPT not as an autonomous decision-maker, but as a cognitive assistant capable of enhancing data quality and clinical workflow efficiency, appears to be a promising path for responsible and effective integration of generative AI into healthcare.

Acknowledgements. The authors have no competing interests to declare that are relevant to the content of this article.

References

1. Partha Pratim Ray: ChatGPT: a comprehensive review on background, applications, key challenges, bias, ethics, limitations and future scope. Internet Things Cyber-Phys. Syst. **3**, 121–154 (2023)
2. Singh, S.K., Kumar, S., Mehra, P.S.: Chat GPT & Google bard AI: a review. In: 2023 International Conference on IoT, Communication and Automation Technology (ICICAT), pp. 1–6. IEEE (2023)
3. Li, J., Dada, A., Puladi, B., Kleesiek, J., Egger, J.: ChatGPT in healthcare: a taxonomy and systematic review. Comput. Methods Programs Biomed. **245**, 108013 (2024)
4. Vaishya, R., Misra, A., Vaish, A.: ChatGPT: is this version good for healthcare and research? Diab. Metab. Syndr.: Clin. Res. Rev. **17**(4), 102744 (2023)
5. Biswas, S.S.: Role of Chat GPT in public health. Ann. Biomed. Eng. **51**(5), 868–869 (2023)

6. Javaid, M., Haleem, A., Singh, R.P.: Chatgpt for healthcare services: an emerging stage for an innovative perspective. BenchCouncil Trans. Benchmarks, Stand. Eval. **3**(1), 100105 (2023)
7. Zhou, Z.: Evaluation of ChatGPT's capabilities in medical report generation. Cureus **15**(4) (2023)
8. Alessandro, L., et al.: Validation of an artificial intelligence-powered virtual assistant for emergency triage in neurology. Neurol., 10–1097 (2025)
9. Maniaci, A., et al.: Is generative pre-trained transformer artificial intelligence (Chat-GPT) a reliable tool for guidelines synthesis? A preliminary evaluation for biologic CRSwNP therapy. Eur. Arch. Oto-Rhino-Laryngol. **281**(4), 2167–2173 (2024)
10. Lee, S.-W., Choi, W.-J.: Utilizing ChatGPT in clinical research related to anesthesiology: a comprehensive review of opportunities and limitations. Anesth. Pain Med. **18**(3), 244–251 (2023)
11. Fukuzawa, F., et al.: Importance of patient history in artificial intelligence-assisted medical diagnosis: comparison study. JMIR Med. Educ. **10**, e52674 (2024)
12. Scherr, R., Halaseh, F.F., Spina, A., Andalib, S., Rivera, R.: ChatGPT interactive medical simulations for early clinical education: case study. JMIR Med. Educ. **9**, e49877 (2023)
13. Gangwal, A., Ansari, A., Ahmad, I., Azad, A.K., Kumarasamy, V., Subramaniyan, V., et al.: Generative artificial intelligence in drug discovery: basic framework, recent advances, challenges, and opportunities. Front. Pharmacol. **2024**, 15 (2024)
14. Momenaei, B., et al.: ChatGPT enters the room: what it means for patient counseling, physician education, academics, and disease management. Curr. Opin. Ophthalmol. **35**(3), 205–209 (2024)
15. Wang, C., Liu, S., Yang, H., Guo, J., Yuxuan, W., Liu, J.: Ethical considerations of using ChatGPT in health care. J. Med. Internet Res. **25**, e48009 (2023)
16. Sallam, M.: ChatGPT utility in healthcare education, research, and practice: systematic review on the promising perspectives and valid concerns. In: Healthcare, vol. 11, pp. 887. MDPI (2023)
17. Cerulli, R., Lepore, M., Maccioni, R., Plenzich, E., Tufano, R.: A service oriented architecture for clinical decision support systems based on artificial intelligence. In: Decision Science Alliance International Summer Conference, pp. 161–171. Springer (2024)
18. Lepore, M., Plenzich, E., Tufano, R., Cerulli, R., Maccioni, R.: Improving patient's medical history classification using a feature construction approach based on situation awareness and granular computing. Neural Comput. Appl. **36**(35), 22461–22484 (2024)
19. Adeghe, E.P., Okolo, C.A., Ojeyinka, O.T.: A review of wearable technology in healthcare: monitoring patient health and enhancing outcomes. OARJ Multi. Stud. **7**(01), 142–148 (2024)
20. Arnold, M., Goldschmitt, M., Rigotti, T.: Dealing with information overload: a comprehensive review. Front. Psychol. **14**, 1122200 (2023)
21. Aung, Y.Y., Wong, D.C., Ting, D.S.: The promise of artificial intelligence: a review of the opportunities and challenges of artificial intelligence in healthcare. Br. Med. Bull. **139**(1), 4–15 (2021)
22. Antoniadi, A.M., et al.: Current challenges and future opportunities for XAI in machine learning-based clinical decision support systems: a systematic review. Appl. Sci. **11**(11), 5088 (2021)

23. Budd, S., Robinson, E.C., Kainz, B.: A survey on active learning and human-in-the-loop deep learning for medical image analysis. Med. Image Anal. **71**, 102062 (2021)
24. Olufisayo Olusegun Olakotan and Maryati Mohd Yusof: The appropriateness of clinical decision support systems alerts in supporting clinical workflows: a systematic review. Health Inform. J. **27**(2), 14604582211007536 (2021)
25. Chen, Z., et al.: Harnessing the power of clinical decision support systems: challenges and opportunities. Open Heart **10**(2), e002432 (2023)
26. Garg, A. X., et al.: Effects of computerized clinical decision support systems on practitioner performance and patient outcomes: a systematic review. Jama **293**(10), 1223–1238 (2005)
27. Bezemer, T., et al.: A human (e) factor in clinical decision support systems. J. Med. Internet Res. **21**(3), e11732 (2019)
28. Chien, S.C., et al.: Alerts in clinical decision support systems (CDSS): a bibliometric review and content analysis. In: Healthcare, vol. 10, pp. 601. MDPI (2022)
29. Huang, S., Liang, Y., Li, J., Li, X.: Applications of clinical decision support systems in diabetes care: scoping review. J. Med. Internet Res. **25**, e51024 (2023)
30. Bright, T.J., et al.: Effect of clinical decision-support systems: a systematic review. Ann. Intern. Med. **157**(1), 29–43 (2012)
31. Sariköse, S., Şenol Çelik, S.: The effect of clinical decision support systems on patients, nurses, and work environment in ICUs: a systematic review. CIN: Comput., Inform., Nurs. **42**(4), 298–304 (2024)
32. Mebrahtu, T.F., et al.: Effects of computerised clinical decision support systems (CDSS) on nursing and allied health professional performance and patient outcomes: a systematic review of experimental and observational studies. BMJ Open **11**(12), e053886 (2021)
33. Canova-Barrios, C., Machuca-Contreras, F.: Interoperability standards in health information systems: systematic review. In: Seminars in Medical Writing and Education, vol. 1, pp. 7–7 (2022)
34. HL7. HL7 FHIR Release 5, 2024. https://hl7.org/fhir/. Accessed 04 Mar 2024
35. Jin, Z., Cui, S., Guo, S., Gotz, D., Sun, J.: CarePre: an intelligent clinical decision assistance system. ACM Trans. Comput. Healthc. **1**, 1–20 (2020)
36. Hassoun, S., et al.: NAIF: a novel artificial intelligence-based tool for accurate diagnosis of stage F3/F4 liver fibrosis in the general adult population, validated with three external datasets. Int. J. Med. Informatics **185**, 105373 (2024)
37. Vardell, E., Bou-Crick, C.: VisualDx: a visual diagnostic decision support tool. Med. Ref. Serv. Q. **31**(4), 414–424 (2012)
38. Cirone, K., Akrout, M., Simpson, R., Lovegrove, F., et al.: Investigating the performance of VisualDx on common dermatologic conditions in skin of color. SKIN J. Cutan. Med. **8**(5), 1788–1796 (2024)
39. Diaz-Ramón, J.L., et al.: Melanoma clinical decision support system: an artificial intelligence-based tool to diagnose and predict disease outcome in early-stage melanoma patients. Cancers **15**(7), 2174 (2023)
40. Math Biology. Math Biology Homepage (2024). https://mathbiology.tech. Accessed 29 Apr 2024
41. Porter, R.: The patient's view: doing medical history from below. Theory Soc. **14**, 175–198 (1985)
42. Haidet, P., Paterniti, D.A.: Building a history rather than taking one: a perspective on information sharing during the medical interview. Arch. Intern. Med. **163**(10), 1134–1140 (2003)

43. Ramsey, P.G., Curtis, J.R., Paauw, D.S., Wenrich, M.D.: History-taking and preventive medicine skills among primary care physicians: an assessment using standardized patients. Am. J. Med. **104**(2), 152–158 (1998)
44. Rich, E.C., et al.: Reconsidering the family history in primary care. J. Gen. Intern. Med. **19**(3), 273–280 (2004)
45. Yoon, P.W., et al.: Awareness of family health history as a risk factor for disease–united states, 2004. MMWR: Morb. Mortal. Weekly Rep. **53**(44) (2004)
46. Zhou, X.Z., Menche, J., Barabási, A.-L., Sharma, A.: Human symptoms-disease network. Nat. Commun. **5**(1), 4212 (2014)
47. Frissora, C.L., Koch, K.L.: Symptom overlap and comorbidity of irritable bowel syndrome with other conditions. Curr. Gastroenterol. Rep. **7**(4), 264–271 (2005)
48. Guang-Meng, X., Ming-Xin, H., Li, S.-Y., Ran, X., Zhang, H., Ding, X.-F.: Thyroid disorders and gastrointestinal dysmotility: an old association. Front. Physiol. **15**, 1389113 (2024)
49. Daher, R., Yazbeck, T., Jaoude, J.B., Abboud, B.: Consequences of dysthyroidism on the digestive tract and viscera. World J Gastroenterol: WJG **15**(23), 2834 (2009)
50. Ball, J.R., Miller, B.T., Balogh, E.P.: Improving Diagnosis in Health Care (2015)
51. Grais, I.M., Sowers, J.R.: Thyroid and the heart. Am. J. Med. **127**(8), 691–698 (2014)
52. Géron, A.: Hands-On Machine Learning with Scikit-learn, Keras, and TensorFlow: Concepts, Tools, and Techniques to Build Intelligent Systems. O'Reilly Media, Inc. (2022)
53. Bruce, P., Bruce, A., Gedeck, P.: Practical Statistics for Data Scientists: 50+ Essential Concepts Using R and Python. O'Reilly Media (2020)
54. Downey, A.: Think Stats: Exploratory Data Analysis. O'Reilly Media, Inc. (2014)

Human-AI Collaboration in High-Volume Project Risk Management: A Cross-Domain Framework Transfer Study

Marc Bara Iniesta[1]([envelope]) [ORCID] and Marisa Lostumbo[2] [ORCID]

[1] ESADE Business School, Barcelona, Spain
marcoantonio.bara@esade.edu
[2] Universitat Oberta de Catalunya (UOC), Barcelona, Spain
mlostumbo@uoc.edu

Abstract. As artificial intelligence tools generate increasingly complex risk scenarios, project managers face a critical challenge: hundreds of potential risks per project that far exceed human capacity to evaluate manually. While other high-risk domains have developed frameworks for human-AI collaboration in high-volume contexts, project management lacks systematic approaches to this emerging problem. Without proper frameworks, AI adoption risks creating an "intelligence paradox"—where AI's analytical power generates so many scenarios that managers become more overwhelmed than before. This study presents a cross-domain framework transfer analysis through a systematic review of 344 papers. We identify four proven human-AI collaboration frameworks from cybersecurity, robotics, security screening, and financial systems that successfully manage high-volume risk data. Our analysis reveals that while 17 studies document information overload in project management, none provide systematic solutions adapted from other domains. To address this gap, we develop a comprehensive framework consisting of: (a) a transfer matrix that maps each source framework to specific project management challenges, (b) an integrated system architecture preventing AI-induced overload, and (c) three critical principles—context preservation, threshold calibration, and integration requirements—that guide successful implementation. The framework provides practitioners with concrete pathways for adaptation: from cybersecurity's HAT framework for risk prioritization to weak-signal detection for early warning systems. This study offers the first systematic approach to transferring proven human-AI collaboration models to project management, providing both immediate solutions and a methodology that other fields facing AI-integration challenges could adapt.

Keywords: human-AI collaboration · project risk management · framework transfer · information overload · high-volume risk analysis · systematic literature review · AI-human teaming · cross-domain adaptation · intelligence paradox

© The Author(s), under exclusive license to Springer Nature Switzerland AG 2026
M. Pavone et al. (Eds.): DSA ISC 2025, LNCS 16405, pp. 307–320, 2026.
https://doi.org/10.1007/978-3-032-21811-7_21

1   Introduction

Project management is approaching a critical juncture in AI-driven transformation. As artificial intelligence tools become increasingly sophisticated in risk analysis, they promise to generate hundreds of potential risk scenarios per project, far exceeding human capacity to evaluate manually. This emerging challenge of high-volume risk scenario management represents a significant gap in current project management practice (Fig. 1).

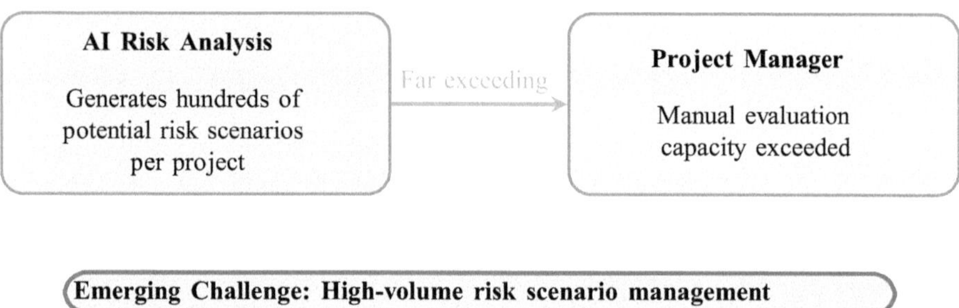

Fig. 1. The emerging challenge in AI-driven project risk management.

Other domains have successfully developed frameworks for human-AI collaboration in high-volume decision contexts. Financial trading systems process thousands of alerts daily through human-AI teaming (HAT) approaches. Healthcare manages diagnostic alert fatigue through strategic filtering mechanisms. Cybersecurity operations centers have evolved sophisticated prioritization frameworks. However, project management has not systematically explored how these proven frameworks could be adapted to its specific context.

Our preliminary investigation revealed that while several authors suggest the potential value of transferring financial and healthcare AI-human frameworks to project management [1–4], no study has systematically explored this adaptation. This represents a clear research gap that becomes increasingly critical as AI adoption in project management accelerates.

This study addresses the research question: **"What human-AI collaboration models from other domains could be adapted for high-volume risk scenario management in projects?"** We present the first systematic analysis of transferable frameworks, providing a foundation for project managers to prepare for the impending wave of AI-generated risk data.

Our approach not only addresses current challenges but anticipates a critical future risk: the "intelligence paradox" where AI's analytical capabilities could generate so many scenarios that managers become more overwhelmed than before AI adoption. By developing this framework now, we enable project management to harness AI's benefits while avoiding the information overload trap that naive implementation would create. Moreover, our cross-domain transfer methodology provides a model that other fields could adapt as they face similar AI-integration challenges.

2 Methodology

To develop a comprehensive understanding of transferable human-AI collaboration frameworks from high-risk domains to project management, we conducted a systematic literature review (SLR). SLRs provide the most rigorous approach for methodically reviewing and synthesizing diverse literature streams, offering critical insights for identifying research gaps and theoretical foundations [15]. They enable a detailed, organized synthesis of knowledge across domains, making them particularly suitable for our cross-disciplinary investigation [16].

Our research data was compiled from Scopus, which provides extensive coverage of both technical and management literature with over 27,000 peer-reviewed journals. This choice was guided by three considerations: first, Scopus offers superior coverage of engineering and computer science publications where human-AI collaboration research predominantly appears [17]; second, it indexes major project management journals comprehensively; and third, focusing on a single database ensures consistency and prevents duplication issues that can arise from multi-database searches [18].

2.1 Search Strategy

We employed a two-pronged search strategy to bridge two distinct literature streams: established human-AI collaboration frameworks in other domains (Search 1) and emerging challenges in project management (Search 2). This approach enabled us to identify both proven solutions and contextualized needs, essential for meaningful framework transfer. We limited our search to publications from 2020–2025 to capture the rapid evolution of AI capabilities and human-AI collaboration frameworks that emerged following the widespread digital transformation accelerated by COVID-19. This timeframe ensures currency while providing sufficient literature for meaningful synthesis. Following established practice in international management research, we focused on English-language publications to ensure conceptual consistency and enable detailed framework analysis [19].

Two complementary searches were performed in the Scopus database:

Search 1 - Human-AI Frameworks in High-Volume Risk Contexts:

```
("high volume" OR "alert fatigue" OR "information overload")
AND ("risk management" OR "risk assessment")
AND ("human AI" OR "human in the loop" OR "human oversight")
AND (framework OR model)
```

This search aimed to identify established frameworks for human-AI collaboration in domains experiencing high-volume risk data challenges.

Search 2 - Evidence of Complexity Challenges in Project Management:

```
("project management" OR "project risk")
AND ("scenario analysis" OR "multiple scenarios"
OR "information overload")
AND (AI OR "artificial intelligence")
AND (challenge OR limitation OR complexity)
```

This search sought evidence of information overload and scenario complexity challenges emerging in AI-enhanced project management contexts.

2.2 Inclusion and Exclusion Criteria

Studies were included based on the following criteria: (1) empirical or conceptual papers describing human-AI collaboration frameworks for risk management, (2) explicit consideration of high-volume data or information overload challenges, (3) implementation or validation in real-world contexts, and (4) publication in peer-reviewed venues. For the second search, papers were additionally required to provide specific evidence of complexity challenges in project management contexts. Purely theoretical proposals without empirical grounding were excluded.

2.3 Data Analysis

The analysis followed a three-stage process. First, framework characteristics were extracted from each study, including domain of application, human and AI role delineation, collaboration mechanisms, data volume specifications, and effectiveness metrics. Second, thematic analysis identified common patterns across domains [20]. Third, a synthesis matrix was developed to map identified frameworks to project management needs, considering both direct applicability and necessary adaptations. This systematic approach enabled identification of transferable elements while acknowledging domain-specific constraints.

3 Results of the Systematic Literature Review

The systematic search yielded 48 papers from Search 1 (human-AI frameworks) and 296 papers from Search 2 (PM challenges), providing comprehensive coverage of both framework availability and domain needs.

3.1 Human-AI Collaboration Frameworks in Other Domains

Analysis of the first search revealed four primary frameworks addressing high-volume risk management across different domains. Table 1 summarizes these frameworks and their key characteristics.

Table 1. Human-AI Collaboration Frameworks for High-Volume Risk Management

Framework	Domain	Volume	Human Role	AI Role
Human-AI Teaming (HAT) [7]	Cyber-security	Thousands daily alerts	Pattern recognition, context, final decisions	Process volumes, identify anomalies, prioritize
Alert Generation Framework [8]	Robotics	Multiple robot updates	Assess states, update assignments	Risk assessment, suggestions, simulation
Dual-Strategy Decision [9]	Security Screening	High-volume data	Learning, reasoning, verification	Normative processing, initial screening
Human-in-Loop Verification [10]	Early Warning	Continuous signals	Verify signals, tune thresholds	Detect weak signals, monitor

Common patterns emerged across frameworks: (1) AI systems consistently handled volume and speed challenges, (2) humans provided contextual understanding and final judgment, (3) clear delineation of responsibilities prevented both gaps and overlaps, and (4) feedback mechanisms allowed continuous system improvement.

A closer inspection of the *metadata* behind these 48 papers enriches our argument: (i) 76% were published between 2023 and 2025, underscoring how recent this topic is; (ii) research articles dominate (50%), followed by surveys or reviews (21%), with conference papers and books still marginal; (iii) only 31% appear in open-access venues, potentially slowing down replication; and (iv) six studies already operationalize explainable-AI (XAI) modules. These figures add empirical texture to the qualitative patterns just noted and serve as a baseline for the comparative synthesis with Sect. 3.2.

3.2 Evidence of Information Overload in Project Management

The second search revealed substantial evidence of emerging complexity challenges in project management. 45% of the papers were published in 2024–2025, underscoring a rapidly growing interest. Title–abstract keyword mining shows that 24 papers (8.1%) discuss some form of *volume- or complexity-related bottleneck*, and within that subset, 17 papers (5.7%) explicitly use the term "information overload." The phenomenon is therefore acknowledged but still peripheral in the literature.

Key findings include:

- The oil and gas sector reported that "information overload, where the sheer volume of documents makes it challenging to find specific information promptly" [11].
- Industry 4.0 contexts face "the burden of information overload, and the rapid obsolescence of knowledge" [12].
- Multiple papers highlighted cognitive load issues when processing AI-generated scenarios [13,14].

Although the challenge is repeatedly flagged, none of the 17 "overload" papers propose systematic mitigation strategies adapted from other high-volume domains. This diagnostic gap motivates the cross-domain transfer matrix developed in Sect. 4.1.

4 Framework Transfer and Adaptation

Building on the systematic review findings, this section presents our theoretical contribution: a framework for adapting human-AI collaboration models to project management contexts. We first develop a transfer matrix that maps specific frameworks to PM challenges, then synthesize these into an integrated system architecture, and finally identify three critical principles that guide successful adaptation.

4.1 Synthesis: Transferable Elements to Project Management

The analysis revealed a clear opportunity: while other domains have developed sophisticated frameworks for managing high-volume risk data (Sect. 3.1), project management faces similar challenges without adapted solutions (Sect. 3.2). To bridge this gap, we synthesized the findings to develop a transfer matrix that maps each identified framework to specific project management scenarios.

Our synthesis process considered three key factors: (1) the core mechanism of each source framework, (2) the specific PM challenges that mirror the original domain's problems, and (3) the necessary modifications for PM contexts. Table 2 presents the resulting transfer matrix, which provides practitioners with concrete adaptation pathways for each framework.

These frameworks address specific project management scenarios through distinct mechanisms:

Cybersecurity HAT → *Risk Alert Prioritization:* In cybersecurity, this framework [7] processes thousands of daily security alerts, using AI to identify patterns that might indicate coordinated attacks. In PM adaptation, imagine a large construction project with 500+ risks in the register. The AI continuously analyzes these risks, identifying dangerous combinations (e.g., "weather delay" + "supplier shortage" + "critical path activity") that might not be obvious when reviewing risks individually. The PM then focuses their limited time on these high-priority combinations rather than reviewing all 500 risks weekly.

Dual-Strategy → *Portfolio Risk Triage:* Advanced security screening systems [9] use AI to analyze passenger behavior patterns, item density in X-rays, and historical threat data to dynamically classify items as "clear" or "requires inspection." The AI learns from millions of scans to identify subtle anomalies. In PM adaptation, the AI would analyze historical project data to learn risk patterns—which combinations of factors typically escalate, which risks tend to resolve themselves, seasonal patterns, team-specific tendencies. For a portfolio of 20 projects, the AI dynamically adjusts its classification based on learned

Table 2. Framework Transfer Matrix for Project Management

Framework Name	What it Does (Original Domain)	PM Challenge it Solves	How to Adapt it
Cybersecurity HAT [7]	Processes 1000s of security alerts daily, identifies attack patterns among noise	PMs can't review 500+ risks weekly to spot dangerous combinations	AI identifies risk clusters and dependencies, PM focuses on high-threat combinations
Dual-Strategy (Security) [9]	Screens millions of items using "clear safe" vs "flag suspicious" logic	Portfolio managers waste time on trivial risks while missing critical ones	AI auto-clears risks below thresholds, escalates only those needing human judgment
Weak-Signal Detection [10]	Monitors continuous data streams for early warning signs of system failures	Critical risks often start as minor issues buried in emails/reports	AI scans all project communications for early warning patterns, alerts PM to investigate
Multi-Robot Coordination [8]	Coordinates multiple robots by aggregating status data and suggesting task reallocation	Multi-vendor projects have fragmented risk views, miss interdependencies	AI consolidates all team risks, identifies conflicts, suggests resource shifts

patterns: risks that match "benign" historical patterns are auto-classified as "monitor only," while those matching escalation patterns trigger human review. Unlike static threshold rules, the AI continuously learns which risk characteristics actually matter in your specific organizational context.

Weak-Signal Detection → *Early Risk Identification:* Financial systems [5] detect weak signals of market changes by monitoring vast data streams. For PM, the AI would continuously scan project artifacts—emails mentioning "concerns," status reports showing subtle trend changes, team communication patterns indicating stress. When the AI detects signal patterns similar to past project failures, it alerts the PM to investigate before the risk fully materializes.

Multi-robot Coordination → *Multi-Team Risk Synthesis:* In robotics, this framework [8] coordinates multiple autonomous units operating in uncertain environments. For large programs with multiple vendors and subteams, each team maintains their own risk register. The AI aggregates these, identifies where Team A's risk might impact Team B's deliverables, suggests resource reallocation, and flags conflicts for PM resolution.

4.2   Towards an AI-Augmented Project Risk Management System

The framework adaptations in Table 2 collectively define the functional architecture of an AI-augmented project risk management system that operates continuously across all project activities, not just during planning or review phases. Inspired by cross-domain analysis from other sectors, this approach is essential because naive AI adoption—simply using AI to generate more risks for traditional risk registers—would exacerbate rather than solve the information overload problem. Without intelligent filtering and synthesis, AI's ability to identify risks across vast datasets could overwhelm project managers with hundreds of additional items to review weekly.

The system emerges from the synthesis of four complementary capabilities:

- **Risk Pattern Recognition**: AI continuously identifies risk clusters and dependencies across project data flows, surfacing dangerous combinations that would be invisible to human reviewers processing risks individually during weekly reviews.
- **Intelligent Risk Filtering**: The system autonomously processes incoming risks, automatically clearing those below defined thresholds while escalating only those needing human judgment, preventing cognitive overload from trivial issues.
- **Proactive Signal Scanning**: AI scans all project communications—emails, reports, meeting transcripts—for early warning patterns, alerting PMs to investigate before minor issues escalate into critical risks.
- **Cross-Team Risk Synthesis**: The system consolidates risk data from all teams and vendors, identifies conflicts and interdependencies, and suggests resource shifts as project conditions evolve.

These capabilities represent a paradigm shift from reactive risk registers to proactive, intelligence-driven risk management. Without this integrated approach—informed by successful implementations in cybersecurity, finance, and infrastructure monitoring—AI adoption in project management risks creating an "intelligence paradox" where AI's analytical power generates so many risks and scenarios that human managers become even more overwhelmed than before. The system addresses this by transforming how project managers interact with risk data—from manual processing of ever-growing risk lists to strategic review of AI-surfaced insights.

4.3   Adaptation Principles

Successful implementation of cross-domain AI solutions in project management requires more than technical integration. The transfer process revealed three critical principles that differentiate project management adaptations from their source domains:

1. **Context Preservation:** While AI excels at pattern recognition across large datasets, project-specific context remains crucial for effective risk management. This principle differs significantly from domains like cybersecurity,

where threats often have universal characteristics (e.g., a malware signature is dangerous regardless of organization). In contrast, project risks are highly context-dependent. For example, a two-week delay might be catastrophic for a product launch but acceptable for an internal system upgrade. The human project manager must therefore retain responsibility for interpreting risks within their specific stakeholder dynamics, organizational culture, and strategic implications.

This context preservation requires designing AI systems that present risk information alongside contextual metadata rather than making autonomous decisions. The cybersecurity HAT framework's approach [7] of having AI flag anomalies while humans make final decisions provides a useful template, but PM adaptations must ensure richer contextual information is preserved and presented.

2. **Threshold Calibration:** Each source framework employs different intervention thresholds based on domain-specific trade-offs. Financial trading systems [5], for instance, may tolerate higher false-positive rates because automated trades can be quickly reversed with minimal cost. Healthcare systems prioritize high sensitivity to avoid missing critical diagnoses, accepting more false alarms as a necessary trade-off.

   Project management faces unique threshold challenges because decisions often involve irreversible resource commitments. Assigning a team to investigate a potential risk consumes time and budget that cannot be recovered. Therefore, PM adaptations must incorporate more conservative thresholds with explicit escalation paths. We propose a three-tier threshold system: (1) AI-managed monitoring for low-severity risks with minimal project impact, (2) AI-flagged human review for moderate-severity risks or those with uncertain impact, and (3) human intervention for high-severity risks or any risk that could significantly impact project objectives. These thresholds should be dynamically adjustable based on project context. During critical path activities, thresholds should be lowered (made more sensitive) to escalate more risks for human review.

3. **Integration Requirements:** Successful implementation requires seamless integration with existing PM tools and workflows. This presents a unique challenge compared to other domains. Cybersecurity operations typically use centralized Security Information and Event Management (SIEM) platforms, while healthcare has standardized Electronic Health Records (EHR). Project management, however, relies on diverse toolsets including scheduling software (MS Project, Primavera), risk registers (often Excel-based), communication platforms (Slack, Teams), and specialized tools for different industries (Fig. 2).

The integration challenge extends beyond technical compatibility to workflow disruption. Project managers already struggle with tool proliferation and context switching. Adding another AI-powered system risks exacerbating rather than solving the information overload problem [6]. Therefore, PM adaptations must prioritize integration approaches that enhance rather than replace existing

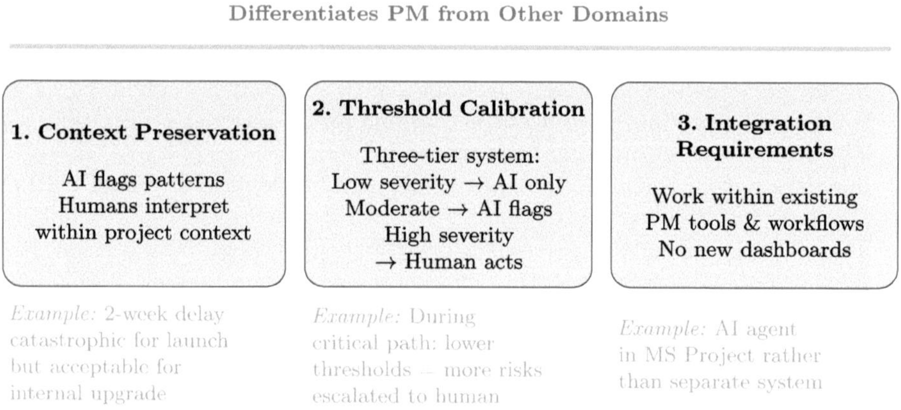

Differentiates PM from Other Domains

1. Context Preservation

AI flags patterns
Humans interpret
within project context

Example: 2-week delay
catastrophic for launch
but acceptable for
internal upgrade

2. Threshold Calibration

Three-tier system:
Low severity → AI only
Moderate → AI flags
High severity
→ Human acts

Example: During
critical path: lower
thresholds – more risks
escalated to human

3. Integration Requirements

Work within existing
PM tools & workflows
No new dashboards

Example: AI agent
in MS Project rather
than separate system

Fig. 2. Three critical principles for adapting human-AI collaboration frameworks to the unique context of project management.

workflows. This might involve AI agents that work within existing platforms, providing insights through familiar interfaces rather than requiring new dashboards.

These three adaptation principles—context preservation, threshold calibration, and integration requirements—form the foundation for successful transfer of human-AI collaboration frameworks to project management. They address the information overload challenge identified in our second search while respecting the unique characteristics of project-based work.

5   Discussion

This study provides the first systematic mapping of human-AI collaboration frameworks from high-risk domains to project management. Through our systematic review, we identified existing solutions in other domains and developed a concrete framework showing how to adapt these solutions to PM contexts. Our findings reveal both the promise of cross-domain transfer and specific pathways for implementation.

5.1   Theoretical Implications

Our systematic review confirms that successful human-AI collaboration models exist across finance, healthcare, cybersecurity, and robotics for managing high-volume risk data. Our primary contribution lies in demonstrating how these frameworks can be systematically transferred to project management contexts.

The framework transfer matrix (Table 2) provides the first structured approach to cross-domain adaptation in this field. By mapping specific source frameworks to analogous PM challenges and detailing the necessary modifications, we move beyond abstract suggestions of "learning from other domains" to provide concrete transfer mechanisms. For instance, our analysis shows how cybersecurity's HAT framework for processing thousands of alerts translates directly to managing 500+ project risks by focusing on dangerous risk combinations rather than individual assessments.

This systematic mapping approach contributes to technology transfer theory by demonstrating that successful adaptation requires understanding not just what a framework does, but how its core mechanisms can address analogous problems in different contexts. The adaptation principles we identified—context preservation, threshold calibration, and integration requirements—emerged from this mapping process as necessary considerations, though the transfer matrix itself represents our core theoretical contribution.

5.2   Practical Implications

Our framework transfer matrix (Table 2) provides immediate actionable guidance for practitioners. Rather than wondering whether solutions from other domains might help, project managers now have a clear menu of options:

- For organizations asking how to handle risk overload, the cybersecurity HAT framework provides a proven approach.
- For those questioning how to efficiently triage portfolio-level risks, the Dual-Strategy approach from security screening offers a tested solution.
- For teams seeking early warning capabilities, weak-signal detection from financial systems presents an established method.
- For programs struggling with multi-vendor risk coordination, the multi-robot framework demonstrates effective aggregation and synthesis techniques.

We recommend beginning with pilot implementations of the Dual-Strategy framework for portfolio risk triage, as it offers the lowest disruption while demonstrating clear value. The three-tier threshold system we propose allows organizations to maintain human oversight where PM expertise is crucial while leveraging AI for volume management. Success metrics should encompass not only risk detection accuracy but also PM time savings, decision quality, and user acceptance.

The urgency of implementation is underscored by our finding that 17 papers already document information overload challenges in project management. Without frameworks like those we propose, AI adoption risks exacerbating rather than alleviating the burden on project managers.

5.3   Limitations

Two main limitations bound the interpretation of our findings. First, our framework transfer matrix and adaptation principles, while grounded in systematic analysis, remain theoretical constructs awaiting empirical validation. Organizations implementing these frameworks may discover additional adaptation requirements we have not anticipated.

Second, we limited our search to English-language publications. While this ensured conceptual consistency in our analysis, frameworks published in other languages may offer additional insights or alternative approaches to human-AI collaboration in high-volume risk contexts.

5.4 Future Research Directions

Our work opens several promising research avenues:

- **Empirical Validation of the Framework:** Field studies should test our transfer matrix and adaptation principles in real project environments. Particularly valuable would be comparative studies examining different frameworks' effectiveness across project types.
- **Industry-Specific Refinements:** While our transfer matrix provides specific adaptation pathways, construction, software development, and pharmaceutical projects may require specialized modifications. Research should explore how these frameworks perform in different industrial contexts.
- **Maturity Model Development:** Organizations need guidance on progression from basic AI-assisted risk management to sophisticated human-AI collaboration. Future work should develop maturity models building on our framework.
- **Tool Integration Architectures:** Our integration requirements principle identifies the challenge but not the solution. Technical research should develop reference architectures for embedding AI capabilities within existing PM tool ecosystems.
- **Longitudinal Impact Studies:** As organizations implement these frameworks, longitudinal research should assess their impact on project success rates, risk management effectiveness, and PM job satisfaction.

6 Conclusions

This study makes two key contributions to addressing AI-generated risk scenario overflow in project management. First, through a systematic review of 344 papers, we provide the first comprehensive mapping of human-AI collaboration frameworks across high-risk domains, while documenting the emerging information overload crisis in project management. Second, we developed a comprehensive framework for adapting proven human-AI collaboration models to PM contexts, consisting of: (a) a transfer matrix mapping successful frameworks from other domains to PM challenges, (b) an integrated system architecture preventing AI-induced information overload, and (c) three critical adaptation principles ensuring successful implementation.

Our systematic review revealed a critical gap: while 17 papers documented information overload in PM, none explored solutions from domains that successfully manage similar challenges. Our framework addresses this gap by providing concrete adaptation pathways from cybersecurity, robotics, security screening, and financial systems to project risk management.

The urgency is clear: AI tools promise to generate hundreds of risk scenarios per project, far exceeding human evaluation capacity. Without proper frameworks, this technological advancement risks creating more problems than it solves. Our framework enables organizations to harness AI's analytical power for volume processing while preserving essential human judgment for context-dependent decisions.

As project management enters the AI era, success lies not in human-AI competition but in collaboration. Organizations that implement these frameworks now will be positioned to transform information overload from a threat into a competitive advantage. This study provides both the knowledge foundation and practical tools for that transformation.

Beyond project management, this cross-domain transfer methodology could inspire similar approaches in other fields facing AI-induced information overload. As AI capabilities expand across disciplines, the systematic framework transfer approach we demonstrate here offers a replicable model for preventing the intelligence paradox in any domain where human expertise must be preserved while leveraging AI's analytical power.

Disclosure of Interests. The authors have no competing interests to declare that are relevant to the content of this article.

References

1. Castellanos, N.G.: Augmenting human judgment in AI-powered project management: a framework for collaborative decision-making. Proj. Manag. J. **53**(4), 412–428 (2022)
2. Yazdi, M., Zarei, E., Adumene, S.: Navigating the power of artificial intelligence in risk management: a comparative analysis. Saf. Sci. **164**, 106–124 (2024). https://doi.org/10.1016/j.ssci.2024.106124
3. Fragiadakis, G., Diou, C., Kousiouris, G.: Evaluating human-AI collaboration: a review and methodological framework. arXiv preprint: arXiv:2407.12345 (2024)
4. Ahmad, S.S.W., Bibi, K., Talani, R.A., Raja, R.: Investigating the potential of collaborative human-AI systems to revolutionize decision-making in areas like medical diagnosis, financial forecasting and strategic planning. Soc. Sci. Rev. Arch. (2025)
5. Zetzsche, D.A., Arner, D.W., Buckley, R.P.: Artificial intelligence in finance: putting the human in the loop. CFTE Academic Paper Series, Centre for Finance, Technology and Entrepreneurship (2020)
6. Acebes, F., González-Varona, J. M., López-Paredes, A., Pajares, J.: Beyond probability-impact matrices in project risk management: a quantitative methodology for risk prioritisation. arXiv preprint: arXiv:2405.20679 (2024)
7. Jalalvand, F., Baruwal Chhetri, M., Nepal, S., Paris, C.: Alert prioritisation in security operations centres: a systematic survey on criteria and methods. ACM Comput. Surv. **57**(2) (2024). https://doi.org/10.1145/3695462
8. Al-Hussaini, S., Guan, Y., Gregory, J.M., Pollard, K., Khooshabeh, P., Gupta, S.K.: Assessing the impact of alerts on the human supervisor's decision-making

performance in multi-robot missions. ACM Trans. Hum.-Robot Inter. **14**(1) (2024). https://doi.org/10.1145/3689828

9. Huang, Y., Wang, X., Zhang, Y., Chen, L., Zhang, H.: Application of human-in-the-loop hybrid augmented intelligence approach in security inspection system. Front. Artif. Intell. **8** (2025). https://doi.org/10.3389/frai.2025.1518850

10. Černý, J., Potančok, M., Castro Hernandez, E.: Toward a typology of weak-signal early alert systems: functional early warning systems in the post-COVID age. Online Inf. Rev. **46**(5), 904–919 (2022). https://doi.org/10.1108/OIR-11-2020-0513

11. Mohamed, A.A., Mohammed, A., Saud, S., Veedu, A.K., Zubair, S.N., Al Sayari, A.S.: AI-enabled annexure parsing for oil & gas project exhibits. In: Society of Petroleum Engineers - ADIPEC 2024 (2024). https://doi.org/10.2118/222187-MS

12. Sherif, A., Salloum, S. A., Shaalan, K.: Systematic review for knowledge management in industry 4.0 and ChatGPT applicability as a tool. In: Studies in Big Data (2024). https://doi.org/10.1007/978-3-031-52280-2-19

13. Almalki, S.S.: AI-driven decision support systems in agile software project management: enhancing risk mitigation and resource allocation. Systems **13**(3) (2025). https://doi.org/10.3390/systems13030208

14. Metwally, A.B.M., Ali, S.A.M., Mohamed, A.T.I.: Thinking responsibly about responsible AI in risk management: the Darkside of AI in RM. In: 2024 ASU International Conference in Emerging Technologies for Sustainability and Intelligent Systems (ICETSIS 2024), pp. 1–5 (2024). https://doi.org/10.1109/ICETSIS61505.2024.1045968435

15. Tranfield, D., Denyer, D., Smart, P.: Towards a methodology for developing evidence-informed management knowledge by means of systematic review. Br. J. Manag. **14**(3), 207–222 (2003)

16. Paul, J., Criado, A.R.: The art of writing literature review: what do we know and what do we need to know? Int. Bus. Rev. **29**(4), 101717 (2020)

17. Singh, V.K., Singh, P., Karmakar, M., Leta, J., Mayr, P.: The journal coverage of web of science, Scopus and dimensions: a comparative analysis. Scientometrics **126**(6), 5113–5142 (2021)

18. Galati, F., Bigliardi, B.: Industry 4.0: emerging themes and future research avenues using a text mining approach. Comput. Ind. **109**, 100–113 (2019)

19. Denyer, D., Tranfield, D.: Producing a systematic review. In: The Sage Handbook of Organizational Research Methods, pp. 671–689. Sage Publications, London (2009)

20. Braun, V., Clarke, V.: Using thematic analysis in psychology. Qual. Res. Psychol. **3**(2), 77–101 (2006)

Keyword Bidding Optimization in Paid Search Advertising: a Survey

Martina Cerulli[1] and Carmine Sorgente[2]

[1] Department of Computer Science, University of Salerno, 84084 Fisciano, Italy
`mcerulli@unisa.it`
[2] Department of Mathematics, University of Salerno, 84084 Fisciano, Italy
`csorgente@unisa.it`

Abstract. In a world where e-commerce and social engagement are increasingly widespread among potential customers, search engines and social networks constitute powerful vehicles for companies' advertising campaigns. Typically, an advertiser is required to indicate a bid for one or multiple keywords associated with a target ad. Then, a sponsored auction, conducted by the advertising platform and involving multiple advertisers, determines the visibility of each ad, which in turn influences the cost paid by the advertisers for the generated engagement. Designing optimal online bidding strategies, subject to time and budget constraints, can drastically improve companies' profits. We review scientific papers on this topic from the last two decades, discuss the state-of-the-art frameworks used to approach the underlying optimization problems, and identify key research gaps and opportunities for future investigation.

Keywords: Digital advertising · Paid search auction · Keyword bidding

1 Introduction

With the rise of the Internet and social networks, the way companies advertise their products and services has completely changed. Advertising campaigns no longer involve exclusively traditional media like TV, radio, newspapers, and magazines, where commercials and ads are broadcast with limited possibility to target specific groups. Sponsored ads are now ordinarily shown in search engines, based on users' search queries, as well as directly in users' feeds on social platforms like Facebook, Instagram, LinkedIn, and others. This results in a highly targeted advertising strategy that allows companies to reach a broader, interested audience at a lower cost.

According to the last annual reports released by IAB [1] and IAB Europe [2], both the US and European digital advertising markets showed continued growth in 2023, with the former achieving a record-high of \$ 225 billion in ad revenue and the latter recording a total digital ad spend of € 96.9 billion, that is, 7.3% and 11.1% more than 2022, respectively. On the one hand, search engines continue

M. Pavone et al. (Eds.): DSA ISC 2025, LNCS 16405, pp. 321–333, 2026.
https://doi.org/10.1007/978-3-032-21811-7_22

to hold the largest market share, accounting for 39.5% of the total US market and 43.1% of the European one, although being characterized by a relatively slow year-over-year growth. On the other hand, after the underperformance of 2022, social media advertising registered during 2023 an 8.7% growth in US and an 18.2% growth in Europe, where this constituted the most significant increase with respect to other formats, while in US, the market growth was slower than that of audio, retail media and video streaming advertising, which recorded 18.9%, 16.3% and 10.6% increases, respectively.

When an advertiser plans to advertise a product or a service, she faces a multitude of challenges associated with several stages of the *purchase funnel*, that is, a theoretical model describing the customer's journey, including the initial product or service awareness, the increased interest and, ultimately, the final purchase [3]. The first crucial step is to design an attractive advertisement, or *ad*, in short. Well-designed ads stand out among others, convey the message effectively, and influence user behavior. Once the ad has been designed, the advertising platform often requires to indicate a set of keywords (as in the case of search engines), or target audiences (as in the case of social networks) and to associate them with a *bid*, that is, a maximum cost the advertiser is willing to pay whenever a user interacts with the ad thanks to the selected keywords or target audiences. To decide whether the ad will be displayed and in which position of the search results pages and users' feeds, the advertising platform holds an auction, where multiple advertisers participate. Obtaining a top or good position ensures high user engagement, but requires setting high bids. Bad positions are, in turn, less expensive, but they generally generate much less interaction from users. Against this background, the fundamental challenge for the advertiser is to choose a bid that maximizes her overall return on investment (ROI), which is given by the difference between the revenues generated by the user engagement and the expenses due to their interaction with the ad. Usually, a cost is paid whenever a user clicks on the ad, visualizing a so-called *landing page*, while revenues may be generated each time the user performs target actions on such a landing page, e.g., buying a product.

This work provides an overview of the state-of-the-art models and strategies aiming to support advertisers in some critical tasks related to digital advertising in paid search auctions, including the estimation of keyword performances and the real-time optimization of the bids associated with keywords and target audiences. Specifically, we focus on optimization and machine learning based techniques for modeling, from the advertiser's perspective, search engine advertising strategies.

The remainder of the paper is organized as follows. Section 2 provides a detailed description of paid search auctions, introduces the reader to two leading advertising platforms, and outlines the key performance metrics of an advertising campaign. A review of relevant contributions to the field published in the last two decades is then provided in Sect. 3, while recent research trends are discussed, along with concluding remarks, in Sect. 4.

2 Paid Search Auctions

A paid search auction is how search engines like Google decide which ads to show, in what order, and how much advertisers pay every time someone searches for something. In this section, we describe how paid search auctions work in detail. Whenever a user submits her query to a search engine, a mix of both organic and paid results is generally displayed. The former correspond to web pages naturally selected by the search engine ranking algorithm, based on their relevance to the user query, quality of their content, and contained backlinks. The latter are, instead, sponsored by advertisers, who bid in auctions for users' interaction, and their placement is determined by the outcome of those auctions.

2.1 Leading Search Advertising Platforms

According to Statcounter [4], Google holds, by March 2025, the largest global market share at 89.7%, followed by its main competitor Bing with 4.0%, regional alternatives such as Yandex at 2.5%, and Yahoo at 1.33%.

Google's advertising platform is named *Google Ads* [5]. It allows ads to be promoted within the Google Search Network. Ads can appear above or below organic results in Google Search, Google Shopping, Google Images, Google Maps, as well as in additional Google search partners' results pages, i.e., hundreds of non-Google websites, including YouTube. Google Ads requires the advertiser to select a goal for their campaign, among *Sales, Leads, Website traffic, Product and brand consideration, Brand awareness and reach*, and *Local store visit and promotions*. The selected goal, together with the advertiser's brand strategy and the time they can invest, determines the most appropriate campaign type: *Search* campaigns are recommended to increase sales, leads, and web traffic, as they display text ads on Google search result pages, reaching users who personally search for products or services. *Display* campaigns show visual ads on partnered websites and apps. Because of their visual component, they are ideal for increasing users' interest and brand awareness. *Shopping* campaigns advertise products by displaying their images, prices, and merchant names in Google Search results and, specifically, in its Shopping tab. *App* campaigns promote mobile applications on several platforms, including Google Search, Play, and YouTube. Finally, *Performance Max* campaigns constitute a comprehensive alternative, displaying ads across all Google channels. Regardless of the selected campaign type, the advertiser is required to set an *average daily budget*, indicating how much she is approximately willing to spend per day on her campaign. Google Ads interprets this on a monthly basis, allowing her to spend more on days with higher potential for clicks and less on others, although never exceeding the *daily spending limit* – typically twice the average daily budget – and the *monthly spending limit* in any given month, corresponding to 30.4 times the average daily budget. A critical aspect of campaign success is choosing the appropriate bidding strategy. Currently, Google offers several options, including both manual and automated strategies. While automated strategies take charge of setting bids to meet performance goals with minimal input, manual bidding gives advertisers full control by allowing them to set individual bids for specific keywords.

To advertise products and services on Bing, one can use the *Microsoft Advertising* platform [6], formerly known as *Bing Ads*, which also serves advertisements on AOL, Yahoo, DuckDuckGo, and other search engines and platforms within the Microsoft Advertising Network. Similar to Google Ads, Microsoft Advertising supports several ad types: *Search* ads are displayed with search results on the above-mentioned search platforms, based on exact keyword correspondence. *Dynamic Search* ads are, instead, largely recommended for beginners who are not familiar with Search advertising, as they are dynamically adapted to match users' queries, without relying on prespecified keywords, but automatically extracting context from the advertised websites. Headlines, descriptions, images, and logos can be integrated together in *Multimedia* ads. Campaigns and ad groups may alternatively feature *Display, Video & Connected TV*, and *Retail* ads. Display ads reach customers across a wide network of websites and apps, including Edge, Outlook, and Microsoft 365, while Video & Connected TV ads deliver interactive content on platforms such as YouTube or smart TV apps. Retail ads, finally, are specifically used for shopping campaigns, with products being promoted on web marketplaces. Like Google Ads, Microsoft Advertising asks for a *daily budget*, which can be limited to one or shared among multiple campaigns, and supports both manual and automatic bidding strategies, aimed at either maximizing the number of clicks, conversions, or their value. Additionally, the advertiser can indicate, along with her budget, some target impression share, average cost per acquisition (CPA), or average return on ad spend. In these cases, bids are automatically set by the platform, trying to produce close average values of the specified metrics in 30 days.

2.2   Ad Groups and Targeting

The advertiser may decide to make decisions on sponsored ads that share the same or similar target (along with the related keywords) as a singular *ad group*. Many advertisers find this approach useful, as it allows for bidding and performance tracking to be managed at a more granular level [7]. To target the audience to whom these ads should be displayed, the advertiser can specify audience segments at the campaign or ad group level. Google Ads Search campaigns allow for the selection of affinity segments (based on users' habits and interests), demographics (such as age, gender and – in some non-European countries, including the US – household income), in-market segments (indicating recent purchase intent), as well as segments related to previous interaction with the advertiser's website or app. Similar targeting options can be used in Microsoft Advertising, with the possibility of integrating the advertiser's data to create customer match and remarketing lists to reengage users who interacted with the ads, eventually generating a conversion. Although this function is not globally available, Microsoft Advertising is also the only digital advertising platform allowing the use of LinkedIn's data to identify potential customers based on the skills, job, and company listed in their LinkedIn profiles.

2.3 Advertising Pipeline and Performance Metrics

We now describe the advertising process and introduce some key technical terms. The bid set by the advertiser, using a manual bidding strategy, for a given ad group or keyword takes the name of maximum *cost-per-click* (CPC). Every time an ad is displayed, it registers an *impression*. When a user clicks on the ad, this *click* takes the user to a designated landing page. If the user performs a desired action – such as purchasing a product or signing up to the advertiser's app, in line with the campaign's goal – this is counted as a *conversion*. Important indicators of the ad's performance are the *clickthrough rate* (CTR), that is, the number of clicks divided by the number of impressions, and the *conversion rate* (CR), namely, the number of users who performed the desired action divided by the number of users who visited the landing page, which corresponds to the number of clicks. The advertiser is charged for each click and gains from each resulting conversion. Generally, the cost of each click is different, but it never exceeds the maximum CPC[1]. More in detail, the actual CPC – also referred to as pay-per-click amount in Microsoft Advertising – is determined using a second-price auction model [8,9], where the advertiser who proposes the highest bid wins, but she pays just above the second-highest bid. During such an auction, a ranking score – known as *Ad Rank* in Google Ads – is computed for each ad, based on multiple factors, including ad relevance and performance metrics, as well as the quality of the associated landing page, and the bid amount; only ads with a ranking score above an auction-dependent threshold are eligible to show, and the rankings of all the eligible ads determine the order according to which they are displayed in the search result page; this order in turn determines the actual CPCs, as each advertiser pays the minimum amount required to clear the ranking tresholds and beat the ranking of the next competitor.

The key performance metrics, e.g., clicks, impressions, CTRs, CRs, average costs, and revenues, are commonly shared by the advertising platforms. This enables advertisers to make appropriate decisions – such as investing more in well-performing ads and keywords or reducing the bids for ads that are not generating enough revenue to balance the costs due to clicks. However, performance data are aggregated over specific time horizons (typically in daily reports) and delivered with some delays that vary depending on the platforms. This prevents advertisers from taking timely action. Furthermore, the highly competitive and rapidly changing market environment constitutes a challenge for the development of optimization strategies, which often rely on predictions of keyword performance to be used as input for optimization models.

3 Optimal Bidding Models and Algorithms

Determining the right bid amount for a keyword is a real challenge: a bid that is too low compared to those of other advertisers may result in the ad not appearing

---

[1] In some cases, when additional features are enabled, such as automatic bid adjustments, the actual CPC may occasionally exceed the maximum one.

in a favorable position, while a bid that is too high might not be profitable, as the CR may not be sufficient to offset the high CPC. In this section, we review the papers addressing the optimal bidding problem, grouped by common objectives.

Profit Maximization. The primary objective of most of the bidding strategies is maximizing the overall profit, obtained by subtracting the paid CPCs from the revenue coming from the generated conversions. Following such objective, Kitts and Leblanc [10] propose an Integer Programming (IP) model for determining the optimal bids to assign to a set of keywords, at each future time, without exceeding a fixed daily budget. Several key functions, such as expected click volume, position based on bid, and actual CPC, are initially unknown and must be estimated from historical data. These estimations are performed using techniques such as historical averages, weighted regression, and autoregressive neural networks to model time-dependent behaviors and predict future auction outcomes. Zhou et al. [11] define the bidding optimization problem with profit maximization and budget constraint as a knapsack problem. They develop heuristic bidding strategies that produce reasonably acceptable results as compared to the maximum profit attainable by the *omniscient bidder* who knows the bids of all the other users ahead of time. Their solution assumes that the bidder does not need explicit knowledge of competitors' bids and CTRs. Munsey et al. [12] propose an evolutionary approach that utilizes a finite (seven) state machine to determine bid adjustments based on historical auction data. Each state expresses a specific level of backorder, that is, a measure of the advertiser's performance over the past few days relative to their spending limits. The approach is tested in the Trading Agent Competition at IJCAI2009, where it demonstrates that agents trained with this method successfully optimize bids and perform well in competitive settings. Aggarwal et al. [13] formulate the bidding problem as an IP problem with general affine constraints able to express imposed budget as well as target CPA and CPC upper bounds. The authors show the existence of a bidding formula whose optimality can be proved if the auction is *truthful,* that is, if, given the advertiser's bid b, the auction assigns her the slot s that maximizes her profit, i.e., the difference between b and the CPC, multiplied by the resulting CTR, obtained if the ad is assigned to slot s. Fang et al. [14] propose a bid optimization method to tackle budget-constrained bidding (BCB) and multi-constrained bidding (MCB), while maximizing the expected total impression value. In BCB, the advertiser's expense is only bounded by a maximum budget, while, in MCB, an additional maximum cost-per-action is imposed, that is, the amount paid to acquire a single customer or, more generally, to obtain a conversion. The authors propose an online receding optimization method, based on open-loop feedback control, which periodically solves a bidding linear programming formulation, using current information to estimate the probability of future states.

Clicks Maximization. When the goal is to drive traffic, a fundamental objective is maximizing the number of clicks. In this direction, Feldman et al. [15] consider another optimization problem to decide on how advertisers should bid on keywords to maximize the number of user clicks on their ads, given a fixed budget. They model the problem using a bipartite graph with two sets of vertices: keywords K and queries Q. Any query $q \in Q$ is connected to its matching keywords in K. They introduce the concept of a click price curve, where clicks are assumed to increase as the CPC rises. Furthermore, they demonstrate that a simple randomized strategy combining two uniform bidding strategies achieves performance close to the optimal theoretical solution. Muthukrishnan et al. [16] explore the stochastic version of this problem, where the set of keywords, the budget, and the CPC for each keyword are fixed, while the numbers of clicks are random variables having some joint probability distribution. Three stochastic models are considered: the Proportional Model, the Independent Keywords Model, and the Scenario Model. Özlük and Cholette [17] consider an advertiser with a fixed budget who needs to decide how much to bid on different separate keywords, in order to optimize the CTR multiplied by the number of impressions. They show that the ratio of the CTR values and the price elasticities of the response functions determines the bid for any keyword. Additionally, they explore the conditions under which an advertiser should expand the set of keywords and calculate the effect of adding a new keyword, assuming constant elasticity. Tran-Thanh et al. [18] take into account the distribution of budget across a sequence of t second-price auctions under censored feedback, that is, incomplete observation of the auction outcome. As discussed in Sect. 2, information on the actual CPC is typically disclosed only after the auction ends; furthermore, if the ad is not displayed, no information about the CPC needed to win the auction is available. The authors proposed three algorithms to address this problem. The ϵ-First algorithm estimates the distribution of the market price over a percentage ϵ of t auctions, and then solves the budget-limited auction problem using the estimated market price distribution for the remaining $(1 - \epsilon)$ auctions. The Greedy Product-Limit algorithm relies on a Markov decision process to determine an optimal bid at each time step and update the price distribution function accordingly. Finally, the LuekerLearn algorithm solves the problem by combining Lueker's algorithm, originally designed for the online stochastic knapsack problem, to which the budget-limited problem reduces if the market price distribution is known in advance, with Zeng's estimator to learn the market price distribution over time steps.

Cost Minimization. When performance quality constraints are specified, reducing the costs of the advertising campaign helps the advertisers to achieve optimal profits. Considering this type of objective, Selçuk and Özlük [19] discuss bidding strategies for a portfolio of keywords, and specifically measure the success of an advertising campaign based on some target level of exposure indicated by the advertiser, which can be expressed either in terms of achieved CTR or number of impressions. For these two interpretations, the authors propose two non-linear optimization problems minimizing the expected total costs with different objective functions. Both problems incorporate decision variables representing bid amounts, while using a beta distribution to represent random outcomes, such as the ad position assigned to each keyword. The two key parameters of such a distribution are the competitive landscape for each keyword to reach the top of the search results page, and the bid amount itself.

Revenue or Conversions Maximization. Beyond CTRs, actions that directly contribute to the advertiser, such as purchases, sign-ups, and other forms of engagement, are maximized by several bidding strategies. Dayanik and Sezer [20] devise an optimal bidding policy for a scenario where the advertiser bids on multiple keywords, over a time span, with a general display-and-click probability function for each of them. After formulating the optimal bidding problem under this scenario, based on a stochastic model, the authors discuss a dynamic programming operator and construct sequential approximations, along with a more efficient method to solve the differential equation for its value function, i.e., the expected total net revenue. The resulting optimal bidding policy is shown to be more effective than simpler heuristics. Küçükaydin et al. [21] extend the stochastic model studied by Selçuk and Özlük [19] by considering a revenue maximization perspective with a budget constraint on the total advertising cost. Majima et al. [22] propose a bid price optimization algorithm for a scenario where the advertiser sets a monthly budget for their campaign and bids daily, while receiving feedback on the number of impressions, clicks, and conversions generated by each keyword, along with the CPC, but ignoring additional information such as how many times the keyword was searched, in which position the ad was displayed and how much the other advertisers bid. Their proposed algorithm works in three phases: a first update phase, in which the latest ad statistics are used to update the probabilistic model (a Bayesian network) for each keyword; an exploration phase, during which a bandit algorithm is used to determine the outcomes (cost and conversions) associated with different bid options; a final optimization phase, consisting in solving an IP problem to determine the actual bids that maximize the expected conversions without exceeding the budget.

Equilibria of the Game. Maillé et al. [23] review the game theoretic aspects of the sponsored search auctions, and analyze the strategies of each of the three participants to these auctions, i.e., the search engine [24], the advertisers, and the users of the search engine [25]. The key question for advertisers, which are the focus of our survey, is determining how much to bid in order to maximize utility. There are many possible Nash equilibria [26], and it is unclear whether any of these equilibria are actually reached in real keyword auctions. Greedy bidding strategies are considered for repeated single-keyword auctions, i.e., strategies in which each advertiser chooses a bid for the next round so as to maximize her utility while assuming that the vector of bids of the other advertisers remains the same as in the previous round. Not all of them converge to some steady state, and cyclic behaviour may appear, as studied in [27]. Besides the convergence to equilibrium, other properties of the greedy strategies are studied in [28–32]. When taking into account the available budget for the advertisers, other issues arise. From an optimization perspective, there has been a series of works proposing algorithms for such budget-constrained bidders, as discussed above. From a game theoretic viewpoint, in [33] a bidding strategy is proposed and is proved to converge in some cases to a market equilibrium. The authors define a bidding heuristic based on equalizing the ROI across keywords, and when random perturbations are introduced, the modified system of keyword pricing converges to the market equilibrium. Balseiro et al. [34] introduce the concept of fluid mean-field approximation to study the strategic interaction among budget-constrained advertisers in repeated second-price auctions. They show that shading bids (using a constant multiplier) is an approximately optimal best-response for utility-maximizing advertisers. Qin et al. [35] review the works both on the equilibria and the allocation rules of the second-price auctions in which the advertisers are assumed to have full rationality and no budget constraints, and their ads are assumed to have known and independent CTRs. Since these assumptions usually do not hold in real-world sponsored search systems, the authors also review the works in which fewer or weaker assumptions are made, emphasizing existing gaps in the literature.

Most Profitable Keywords Selection. A strictly related problem to keyword bidding is discovering and selecting the most profitable keywords to bid on. Trending keywords typically require higher bid amounts for the associated ads to be displayed, as they are contended by numerous advertisers, but they ensure a large audience for the winners. Unpopular keywords, instead, are often less expensive but result in far fewer impressions. In [36], Rusmevichientong and Williamson address the keyword selection problem by considering a given set of available keywords, with assumed known occurrence probabilities, click costs, and profits, but unknown CTRs. They model the problem as a multi-period optimization problem and propose an adaptive algorithm learning the clickthrough probabilities over time. To discover new non-trivial keywords, Joshi and Motwani [37] present *TermsNet*, a word graph-based approach representing semantic relationships between terms, derived from search engine data, as a directed graph, whose analysis helps identify both common and non-obvious related keywords. To recommend relevant but less competitive keywords, along with their optimal bid prices, Zhang et al. [38] define a mixed IP problem, whose objective is to maximize, within budget and bid constraints, the relevance of the generated keywords and the expected advertiser revenue. The problem can be solved either using binary integer or sequential quadratic programming. Finally, Zhou et al. [39] rely on generative neural networks to generate, given an existing keyword, new relevant ones within the same domain category. To improve the diversity and quality of the generated keywords, the authors further use reinforcement learning, incorporating domain-specific information in the generation process.

4   Conclusion and Future Directions

This paper provided an overview of scientific studies on paid search advertising published in the last two decades, with a focus on optimal bidding strategies. The typical objectives considered in the reviewed models range from maximizing statistics related to ad visibility and effectiveness, e.g., number of impressions, clicks, and CTR, to optimizing the overall profit or CR, while constraints often reflect budget restrictions and, occasionally, bounds on CPA and CPC amounts. As for the methods, optimization techniques such as mathematical programming, as well as heuristic strategies, are often combined with machine learning based estimation techniques, used to predict the unknown values of random variables based on historical data. While the majority of the existing studies focus on paid search advertising, recent seminal works started exploring advertising campaigns on social networks [40–42], which are characterized by specific dynamics, such as user interaction, influence mechanisms, network cohesion, and content virality. An interesting and emerging marketing model on social network advertising is *incentivized social advertising* [43], which consists in displaying ads to a carefully selected group of users who may serve as initial "endorsers", provided they agree to share the ad with their followers in exchange for a monetary reward. Selecting the most influential social network users as potential endorsers is crucial to the success of this advertising model, which may draw inspiration from recent studies on influence maximization and propagation problems [44–47]. Another emerging trend in digital advertising is *real-time bidding* [48,49], a model in which potential ad impressions on websites are bought and sold via real-time auctions. Advertisers—or their bidding agents—receive a bid request containing information about the user and the context of the potential ad placement, based on which they decide the bid amount. Finally, in the next future, the digital advertising industry is expected to expand further, motivated by Generative AI technologies that are significantly transforming the way advertisers approach ad creation and dissemination processes. Working in these directions may enable advertisers to conduct successful campaigns in a highly competitive market.

Disclosure of Interests. All authors have no conflict of interest.

References

1. IAB. Internet Advertising Revenue Report (2024). Accessed 08 Apr 2025
2. IAB Europe. AdEx Benchmark Report (2024). Accessed 08 Apr 2025
3. Kim, A.J., Jang, S., Shin, H.S.: How should retail advertisers manage multiple keywords in paid search advertising? J. Bus. Res. **130**, 539–551 (2021)
4. Statcounter. Search engine market share worldwide (2025). Accessed 22 Apr 2025
5. Google. Google Ads (2024). Accessed 22 Apr 2025

6. Microsoft. Microsoft Advertising (2024). Accessed 22 Apr 2025
7. Huiran Li and Yanwu Yang and: Optimal keywords grouping in sponsored search advertising under uncertain environments. Int. J. Electron. Commer. **24**(1), 107–129 (2020)
8. Vickrey, W.: Counterspeculation, auctions, and competitive sealed tenders. J. Financ. **16**(1), 8–37 (1961)
9. Edelman, B., Ostrovsky, M., Schwarz, M.: Internet advertising and the generalized second-price auction: selling billions of dollars worth of keywords. Am. Econ. Rev. **97**(1), 242–259 (2007)
10. Kitts, B., Leblanc, B.: Optimal bidding on keyword auctions. Electron. Mark. **14**(3), 186–201 (2004)
11. Zhou, Y., Chakrabarty, D., Lukose, R.: Budget constrained bidding in keyword auctions and online knapsack problems. In: Papadimitriou, C., Zhang, S. (eds.) Internet and Network Economics, pp. 566–576. Springer Berlin Heidelberg (2008)
12. Munsey, M., Veilleux, J., Bikkani, S., Teredesai, A., De Cock, M.: Born to trade: a genetically evolved keyword bidder for sponsored search. In: IEEE Congress on Evolutionary Computation, pp. 1–8 (2010)
13. Aggarwal, G., Badanidiyuru, A., Mehta, A.: Autobidding with constraints. In: Caragiannis, I., Mirrokni, V., Nikolova, E. (eds.) Web and Internet Economics, pp. 17–30. Springer International Publishing (2019)
14. Fang, K., et al.: Advertiser-first: a receding horizon bid optimization strategy for online advertising. IEEE Trans. Comput. Soc. Syst., 1–13 (2024)
15. Feldman, J., Muthukrishnan, S., Pal, M., Stein, C.: Budget optimization in search-based advertising auctions. In: Proceedings of the 8th ACM Conference on Electronic Commerce, EC '07, pp. 40–49. Association for Computing Machinery (2007)
16. Muthukrishnan, S., Pál, M., Svitkina, Z.: Stochastic models for budget optimization in search-based advertising. Algorithmica **58**(4), 1022–1044 (2010)
17. Özlük, Ö., Cholette, S.: Allocating expenditures across keywords in search advertising. J. Revenue Pricing Manag. **6**, 347–356 (2007)
18. Tran-Thanh, L., Stavrogiannis, L., Naroditskiy, V., Robu, V., Jennings, N.R., Key, P.: Efficient regret bounds for online bid optimisation in budget-limited sponsored search auctions. In: Proceedings of the Thirtieth Conference on Uncertainty in Artificial Intelligence, UAI'14, pp. 809–818. AUAI Press (2014)
19. Selçuk, B., Özlük, Ö.: Optimal keyword bidding in search-based advertising with target exposure levels. Eur. J. Oper. Res. **226**(1), 163–172 (2013)
20. Dayanik, S., Sezer, S.O.: Optimal dynamic multi-keyword bidding policy of an advertiser in search-based advertising. Math. Methods Oper. Res. **97**(1), 25–56 (2023)
21. Küçükaydin, H., Selçuk, B., Özlük, Ö.: Optimal keyword bidding in search-based advertising with budget constraint and stochastic ad position. J. Oper. Res. Soc. **71**(4), 566–578 (2020)
22. Majima, K., Kawakami, K., Ishizuka, K., Nakata, K.: Keyword-level Bayesian online bid optimization for sponsored search advertising. Oper. Res. Forum **5**(2), 45 (2024)
23. Maillé, P., Markakis, E., Naldi, M., Stamoulis, G.D., Tuffin, B.: Sponsored search auctions: an overview of research with emphasis on game theoretic aspects. Electron. Commer. Res. **12**(3), 265–300 (2012)
24. He, D., Chen, W., Wang, L., Liu, T.Y.: A game-theoretic machine learning approach for revenue maximization in sponsored search. In: Proceedings of the Twenty-Third International Joint Conference on Artificial Intelligence, IJCAI '13, pp. 206–212. AAAI Press (2013)

25. Ashkan, A., Clarke, C.L., Agichtein, E., Guo, Q.: Classifying and characterizing query intent. In: Boughanem, M., Berrut, C., Mothe, J., Soule-Dupuy, C. (eds.) Advances in Information Retrieval, pp. 578–586. Springer Berlin Heidelberg (2009)
26. Varian, H.R.: Position auctions. Int. J. Ind. Organ. **25**(6), 1163–1178 (2007)
27. Cary, M., et al.: Greedy bidding strategies for keyword auctions. In: Proceedings of the 8th ACM Conference on Electronic Commerce, EC '07, pp. 262–271. Association for Computing Machinery (2007)
28. Vorobeychik, Y., Reeves, D.M.: Equilibrium analysis of dynamic bidding in sponsored search auctions. Int. J. Electron. Bus. **6**(2), 172–193 (2008)
29. Liang, L., Qi, Q.: Cooperative or vindictive: Bidding strategies in sponsored search auction. In: Deng, X., Graham, F.C. (eds.) Internet and Network Economics, pp. 167–178. Springer Berlin Heidelberg (2007)
30. Naldi, M., D'Acquisto, G., Italiano, G.F.: The value of location in keyword auctions. Electr. Commer. Res. Appl. **9**(2):160–170 (2010). Special Issue: Theoretical and Empirical Advances in Electronic Auction Research
31. Grillo, A., Lentini, A., Naldi, M., Italiano, G.F.: Penalized second price: a new pricing algorithm for advertising in search engines. In: 2010 8th Annual Communication Networks and Services Research Conference, pp. 207–214 (2010)
32. Markakis, E., Telelis, O.: Discrete strategies in keyword auctions and their inefficiency for locally aware bidders. In: Saberi, A. (ed.) Internet and Network Economics, pp. 523–530. Springer Berlin Heidelberg (2010)
33. Borgs, C., Chayes, J., Immorlica, N., Jain, K., Etesami, O., Mahdian, M.: Dynamics of bid optimization in online advertisement auctions. In: Proceedings of the 16th International Conference on World Wide Web, WWW '07, pp. 531–540. Association for Computing Machinery (2007)
34. Balseiro, S.R., Besbes, O., Weintraub, G.Y.: Repeated auctions with budgets in ad exchanges: approximations and design. Manage. Sci. **61**(4), 864–884 (2015)
35. Qin, T., Chen, W., Liu, T.-Y.: Sponsored search auctions: Recent advances and future directions. ACM Trans. Intell. Syst. Technol. **5**(4) (2015)
36. Rusmevichientong, P., Williamson, D.P.: An adaptive algorithm for selecting profitable keywords for search-based advertising services. In: Proceedings of the 7th ACM Conference on Electronic Commerce, EC '06, pp. 260–269. Association for Computing Machinery (2006)
37. Joshi, A., Motwani, R.: Keyword generation for search engine advertising. In: Sixth IEEE International Conference on Data Mining - Workshops (ICDMW'06), pp. 490–496 (2006)
38. Zhang, Y., Zhang, W., Gao, B., Yuan, X., Liu, T.-Y.: Bid keyword suggestion in sponsored search based on competitiveness and relevance. Inf. Process. Manag. **50**(4), 508–523 (2014)
39. Zhou, H., Huang, M., Mao, Y., Zhu, C., Shu, P., Zhu, X.: Domain-constrained advertising keyword generation. In: The World Wide Web Conference, WWW '19, pp. 2448–2459. Association for Computing Machinery (2019)
40. Liu, H., Pardoe, D., Liu, K., Thakur, M., Cao, F., Li, C.: Audience expansion for online social network advertising. In: Proceedings of the 22nd ACM SIGKDD International Conference on Knowledge Discovery and Data Mining, KDD '16, pp. 165–174. Association for Computing Machinery (2016)
41. Luzon, Y., Pinchover, R., Khmelnitsky, E.: Dynamic budget allocation for social media advertising campaigns: optimization and learning. Eur. J. Oper. Res. **299**(1), 223–234 (2022)

42. Xia, C., Guha, S., Muthukrishnan, S.: Targeting algorithms for online social advertising markets. In: 2016 IEEE/ACM International Conference on Advances in Social Networks Analysis and Mining (ASONAM), pp. 485–492 (2016)
43. Aslay, C., Bonchi, F., Lakshmanan, L.V.S., Lu, W.: Revenue maximization in incentivized social advertising. Proc. VLDB Endow. **10**(11), 1238–1249 (2017)
44. Cerulli, M., Serra, D., Sorgente, C., Archetti, C., Ljubić, I.: Mathematical programming formulations for the Collapsed k-Core Problem. Eur. J. Oper. Res. **311**(1), 56–72 (2023)
45. Ferreira, V., Pessoa, A., Vidal, T.: Influence optimization in networks: new formulations and valid inequalities. Comput. Oper. Res. **173**, 106857 (2025)
46. Kahr, M., Leitner, M., Ljubić, I.: The impact of passive social media viewers in influence maximization. INFORMS J. Comput. **36**(6), 1362–1381 (2024)
47. Tanınmış, K., Aras, N., Altınel, I.K.: Influence maximization with deactivation in social networks. Eur. J. Oper. Res. **278**(1), 105–119 (2019)
48. Yuan, S., Wang, J., Zhao, X.: Real-time bidding for online advertising: measurement and analysis. In: Proceedings of the Seventh International Workshop on Data Mining for Online Advertising, Number 3 in ADKDD '13. Association for Computing Machinery (2013)
49. Zhang, W., Yuan, S., Wang, J.: Optimal real-time bidding for display advertising, pp. 1077–1086 (2014)

Using GPT to Classify DSS

Peter B. Keenan[✉] [iD]

School of Business, University College Dublin, Dublin D04 V1W8, Ireland
`peter.keenan@ucd.ie`

Abstract. Large Language Models (LLMs) can be used for tasks such as summarization, translation, text generation, and answering questions. This makes LLMs relevant to various aspects of academic work, including both research and teaching. This research uses the widely used GPT LLM to classify Decision Support System (DSS) papers into data-driven or model-driven systems. This analysis suggests the software is effective for this task, demonstrating good accuracy for examples that clearly fit into either of these classifications. The investigation's results also revealed that LLM software encountered challenges, similar to human classifiers, with hybrid systems that could legitimately be categorized into both classes.

Keywords: Decision Support Systems · Large Language Models · Scientometrics

1 Introduction

1.1 Decision Support Systems

The concept of Decision Support Systems (DSS) emerged in the 1960s, when information technology (IT) had become sufficiently advanced to support flexible decision-making [1]. Key technical developments of the period facilitated the development of DSS, including the introduction of databases and the increased use of interactive interfaces on time-sharing computers. The DSS field is generally regarded as having originated in the work of Gorry and Scott-Morton [2], who advocated for a distinct class of interactive and flexible systems known as DSS, specifically designed for semi-structured and unstructured decisions. Gorry and Scott-Morton discussed the decision-making model introduced by Simon [3], which identified three phases in decision-making. The Intelligence phase focused on understanding the present situation, the Design phase examined the options available to the decision-maker, and the Choice phase was concerned with selecting among these options. Gorry and Scott-Morton's definition of DSS and subsequent work emphasized that DSS was an interactive system that combined data and models with an interface to support decision-making [4]. However, while it was widely accepted that DSS systems as a class included these three elements, it was also recognized that the exact importance of each component depended on the actual needs of the decision-maker and therefore varied between specific systems. Consequently, specific DSS applications are highly diverse, as they are designed to meet the varied decision needs of decision-makers in diverse domains.

M. Pavone et al. (Eds.): DSA ISC 2025, LNCS 16405, pp. 334–346, 2026.
https://doi.org/10.1007/978-3-032-21811-7_23

1.2 Data-Driven and Model-Driven DSS

Reflecting this application diversity, Steven Alter [5] developed a DSS taxonomy intended to encompass the full range of DSS support. This approach categorisescategorizes DSS by the generic operations they performed rather than specific applications or domains. Alter identified seven types of DSS:

1. File Drawer Systems: Provide access to specific data items, such as inventory monitoring systems.
2. Data Analysis Systems: Support manipulation of data tailored to specific tasks, such as budget analysis.
3. Analysis Information Systems: Offer access to decision-oriented databases and small models, such as sales forecasting tools.
4. Accounting and Financial Model-Based DSS: Calculate consequences of actions, such as profitability analysis.
5. Representational Model-Based DSS: Use simulation models to estimate outcomes based on causal relationships.
6. Optimization Model-Based DSS: Focus on finding the best solution for a given problem.
7. Suggestion Model-Based DSS: Propose decisions based on predefined criteria.

These systems range from those providing simple data access to those that select data for reports and those that incorporate complex optimization models. From Alter's work, two categories emerged: data-driven DSS that provided its greatest contribution by offering access to relevant data, while model-driven DSS made use of models to provide superior solutions to problems [6]. Subsequently, Power introduced a broader framework including Communications-driven, Data-driven, Document-driven, Knowledge-driven, and Model-driven Decision Support Systems [7]. However, Power's additional categories have not achieved the widespread acceptance of the earlier data-driven and model-driven categories. Data-driven DSS make their greatest contribution to the intelligence phase of decision-making, providing the decision-maker with an understanding of the present situation. Model-driven DSS make their greatest contribution to the design and choice phases of decision-making, evaluating possible solutions and identifying which one is best according to the specific objective function relevant to the problem. Model-driven systems typically perform complex calculations that extend beyond the cognitive capabilities of a human decision-maker unassisted by a computer.

Due to technical advancements since the early days of DSS, simple data access is now considered routine. In addition, technical innovations in data capture have led to an exponential increase in the data volumes available for decision-making. This change has been characterized as the era of "big data". In this data-rich environment, decision-making can be greatly assisted by the visualization of data through the DSS interface. The large volumes of data available in modern systems can be rendered accessible to the decision-maker by mathematical processing, with earlier statistical summarization techniques now supplemented by a wide range of machine learning techniques. Data-driven systems have gained prominence with the increase in data volumes, advances in visualization, and the introduction of machine learning. There are many problems where decision-makers require support from insights derived from data, but where effective decisions can then be made without the support of sophisticated models during the design and choice phases of the decision.

With the growth in diverse techniques and the exponential increase in data, a broader field of business analytics has emerged. Within this field, descriptive analytics describes systems that concentrate on the intelligence phase of decision-making, and which are analogous to data-driven DSS. Predictive and prescriptive analytics look beyond the present situation to the future, and this is where model-driven systems are widely used [8].

Model-driven DSS utilizes models to propose solutions. There are many decision problems where the data can be readily understood, but where the solution is not obvious, and where solving the problem is beyond the unassisted cognitive limits of the decision-maker [7]. Such problems include diverse transportation and scheduling applications that have long been considered part of the DSS field. The continued importance of model-driven DSS is sometimes understated, with some authors suggesting that "big data" approaches have now become dominant. This view overlooks the many applications that will persist where big data will never be relevant; for instance, routing drones is a contemporary example of a model-driven application, but one where the big data approach is unlikely to be generally relevant.

The DSS field predominantly originated in business and engineering applications, later expanding into a wide variety of fields [9]. In particular, there has been significant expansion in environmental and healthcare applications. This growth means DSS is now used in a diverse range of disciplines, and there is limited academic interaction between these groups. While the concepts of data-driven and model-driven DSS are reasonably clear to those familiar with the foundational DSS literature, DSS applications are now used in an increasing diversity of domains, and the terms "data-driven" and "model-driven" are not consistently defined across these fields. Owing to this increase in data availability, sophisticated models are now used to characterize the large volume of data included, and the use of these techniques obfuscates the boundary between data-driven and model-driven systems. Despite these challenges, it remains essential to understand the exact contribution of a DSS to the different stages of decision-making when assessing a system.

1.3 Large Language Models

Large Language Models (LLMs) are advanced machine learning models employed for natural language processing tasks. LLMs are trained using vast amounts of text data and use billions of parameters to understand, generate, and analyze human language. LLMs can be utilized for tasks such as summarization, translation, text generation, and answering questions. The best-known LLM is ChatGPT from OpenAI (https://openai.com/), which became widely available as version GPT-3.5 at the end of 2022. Significant subsequent upgrades included GPT-4 in March 2023 and GPT-4.5 in February 2025. Other widely used LLM software includes CoPilot from Microsoft (https://copilot.microsoft.com), which is partly based on OpenAI work, Google Gemini (https://gemini.google.com/), and Anthropic Claude (https://claude.ai/).

Consequently, they might present a misleading or oversimplified view of the course material. This concern informs the background for this research, which aims to assess whether the most widely used LLM, GPT, can usefully characterize data-driven and model-driven DSS.

In a short time, LLM software has become widely used by students in education and, to a lesser extent, by faculty. This has particularly affected assessment in universities, as some traditional assessment approaches can now be answered by software. Students are able to use the software rather than understanding the material on the course. LLMs also have a role in teaching, but there is concern about their "accuracy, misinformation, ethics, morality, copyright, legality, and explainability, among others", given that it is not evident which exact material was used in their training [10]. The term "Hallucination" is used to describe output from LLMs which seems plausible, but is actually false, misleading, or unsupported.

The use of LLM software in academic research is a recent development and is still evolving. However, such models can interpret documents and can be used to characterize academic papers. Joos, Keim and Fisher [11] assessed the ability of several LLMs to assist in an academic literature search, making the point that a keyword search can often throw up a large number of papers, which then have to be further filtered for relevance [11]. It seems likely that this technology will be integrated into academic work [12] with significant productivity improvements. LLMs have been used to analyze article abstracts for systematic literature reviews, especially in the medical disciplines [13, 14]

In the context of this research, the OpenAI GPT-4o model produced the following definition of a data-driven DSS when prompted to do so in up to 50 words.

"A data-driven Decision Support System (DSS) uses data collection, management, and analytical tools to assist in making informed, evidence-based decisions. It integrates large datasets, applies statistical and predictive models, and provides user-friendly interfaces for real-time insights, supporting both strategic and operational decision-making."

While a model-driven DSS is defined by the GPT-4o model in 50 words or less as

"A model-driven DSS emphasizes the use of analytical models to support decision-making. It focuses on simulations, optimization, and algorithms to analyze complex scenarios, providing decision-makers with insights and recommendations. These systems are less reliant on large datasets and more on mathematical models."

Note that these definitions are not consistent and can vary if you ask the system at different times.

2   Case Data

This research aims to assess GPT's ability to classify the abstracts of 160 articles. These articles were not specifically identified for this project but were drawn from those previously identified by the author for other teaching and research purposes. The articles were classified using a cursory analysis into 80 data-driven and 80 model-driven systems. This classification used a relatively superficial approach, and the input data may have included some ambiguous examples. Approximately one-third of the articles were published in 2020 or later, one-third between 2015 and 2019, and one-third before 2015 (Fig. 1). Almost 90% of the papers were journal articles, the remainder were book chapters or conference proceedings. Articles were drawn from traditional DSS journals, for instance from Decision Support Systems and Interfaces, and also from journals in newer areas of application, for instance the journal *Computers and Electronics in Agriculture.*

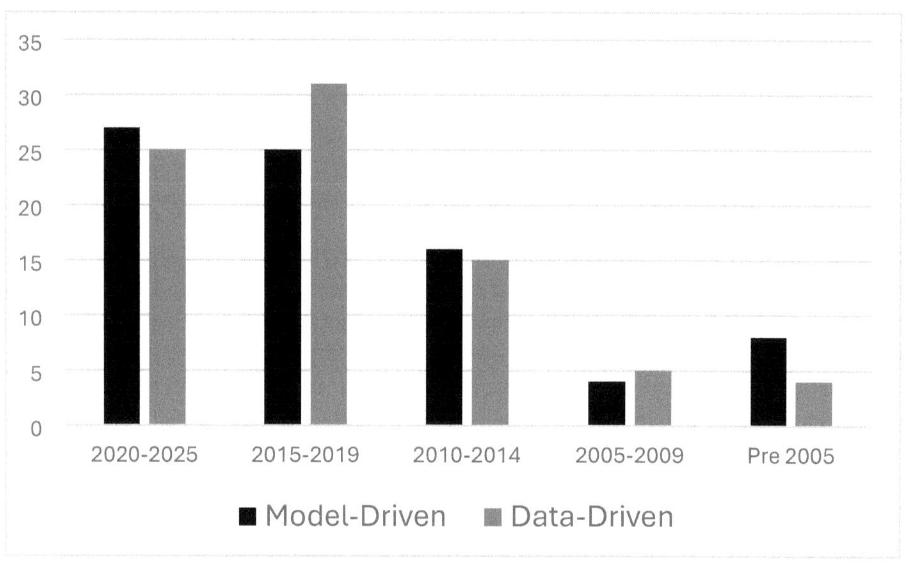

Fig. 1. Year of publication of article abstracts.

Figs. 2 and 3 below display the keywords from article titles and abstracts associated with the two types of systems in the dataset, excluding commonly used terms such as "Decision Support System" and "Computer". Model-driven system keywords include terms such as "optimization" and "simulation" that reflect the complex solution models used. Data-driven keywords identified include "business intelligence", which is often considered a subset of data-driven DSS, and "patient", indicating the use of data-driven DSS for medical applications.

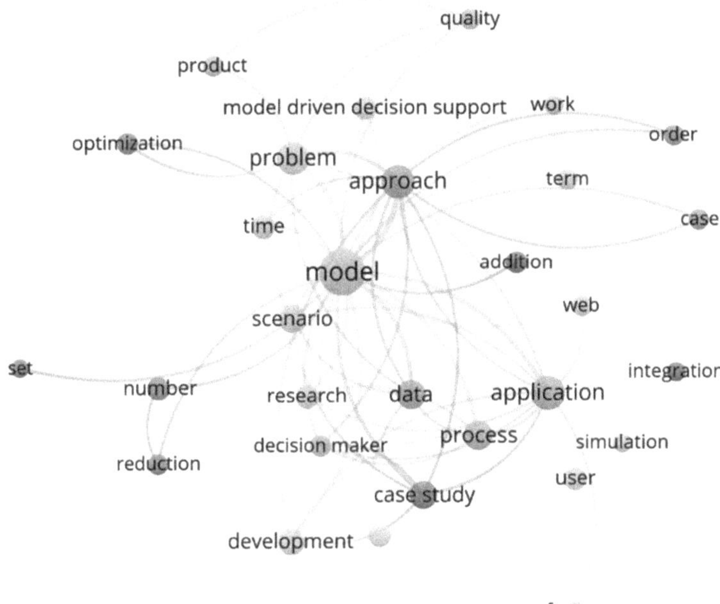

Fig. 2. Title and abstract keywords for model-driven examples.

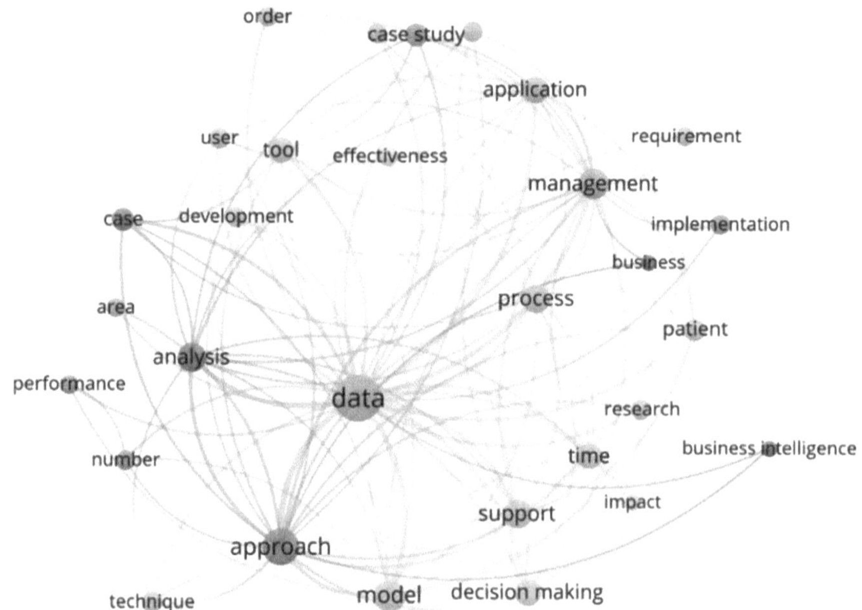

Fig. 3. Title and abstract keywords for data-driven examples.

3 Analysis

3.1 GPT-4o Model

The object of this analysis is to pose the question to GPT as to whether the abstract of an article describes a data-driven or model-driven DSS. GPT was provided with two additional classification options; that the abstract did not describe a DSS or that it described a hybrid system, i.e., one with elements of both data-driven and model-driven DSS. A file was prepared containing the abstracts of these papers. This was then submitted by a Python script to the GPT API (https://openai.com/index/openai-api/) with the following prompt:

```
"You are a research assistant specialized in Decision Support Sys-
tems. "
"You will be given abstracts of academic papers. For each abstract,
determine:"
"1. Does this describe a Decision Support System (DSS)?"
"2. If yes, classify it as one of:"
" - Data-driven DSS"
" - Model-driven DSS"
" - Hybrid DSS"
"3. If no, just say: 'Not a DSS'"
.
```

When this script was submitted to the API an output file was created with the classification and the reasoning behind that classification for each paper.

Initially, the data was analyzed with the OpenAI GPT-4o model. This model was introduced in May 2024, and it is described an "omni" model which can process text, image and audio. We also later use the more expensive GPT-4.5 preview model, which can only be used with text and code.

Table 1. Results from GPT-4o model.

GPT-4o classification	Manual classification	
	Model-driven	Data-driven
Model-driven DSS	53	11
Data-driven DSS	3	35
Hybrid DSS	17	25
Not a DSS	7	9

The initial analysis using the GPT-4o model provided the results shown in Table 1. Notably, the GPT classification sometimes categorized the abstract as "Not a DSS", and in other cases identified the system as a hybrid one. Those systems classified as "Not a DSS" often had abstracts that described the problem, but did not discuss a system in the abstract. The inclusion of the paper title, which was not included in these experiments, might have provided more information in many cases.

The analysis was repeated to reclassify the paper abstracts initially classified as "Hybrid" without this option. Following this change, almost all of the systems manually classified as model-driven DSS were now classified as model-driven systems, but half of the hybrid systems initially manually classified as data-driven were designated as model-driven in the reclassification. This arises because real systems are often hybrid and so can be legitimately described as data-driven or model-driven. A similar process was used to reclassify the "Not a DSS" abstracts; following this reclassification, seven abstracts initially believed to be model-driven were classified as model-driven, as were two of the abstracts believed to be data-driven, the remaining seven data-driven examples were classified as data-driven.

Table 2. Abstracts reclassified in a binary classification with the GPT-4o model.

GPT-4o classification	Manual classification	
	Model-driven	Data-driven
Model-driven DSS	76	24
Data-driven DSS	4	56

Table 2 shows the results of the reclassification with only two available options: data-driven or model-driven. When only a binary classification was offered, 95% of model-driven examples aligned with the manual classification. However, in a binary choice, only 70% of systems identified as data-driven in the input were classified as data-driven by the GPT-4o model.

3.2 GPT-4.5 Preview Model

The analysis was then repeated using the more advanced GPT-4.5 preview model, a more computationally expensive model. This model shows an improved performance, with a higher proportion of articles classified as expected and a smaller number of abstracts classified as "Hybrid DSS" or "Not a DSS" (Table 3).

Table 3. Results from the GPT-4.5 preview model

	Manual classification	
GPT-4.5 preview classification	**Model-driven**	**Data-driven**
Model-driven DSS	59	14
Data-driven DSS	2	40
Hybrid DSS	14	17
Not a DSS	5	9

In a similar way to the reclassification conducted for the GPT-4o model, the hybrid examples were resubmitted to the model without the hybrid option. In this case, of the fourteen model-driven systems that had been classified as "Hybrid", eleven were classified as model-driven, and three were classified as Data-driven. Of the seventeen data-driven examples initially classified as "Hybrid", six were now classified as model-driven, and eleven were classified as Data-driven. The "Not a DSS" examples were then also reclassified without that option. In this case, of the five model-driven abstracts, three were classified as model-driven and two as data-driven. Of the nine data-driven examples previously categorized as "Not a DSS", seven were now classified as data-driven and two were now classified as model-driven (Table 4).

Table 4. Abstracts reclassified in a binary classification with the GPT-4.5 preview model.

	Manual classification	
GPT-4.5 preview classification	**Model-driven**	**Data-driven**
Model-driven DSS	73	22
Data-driven DSS	7	58

When only a binary classification is offered, compared to the GTP-4o model, the GPT-4.5 Preview model classifies a slightly lower proportion of model-driven systems as model-driven and a slightly higher proportion of data-driven systems as data-driven.

3.3 Alternative Prompt

The research subsequently explored an alternative research question, contrasting with earlier analyses that relied on GPT's inherent understanding of data-driven DSS and model-driven DSS. However, this last experiment used author-provided brief definitions of the terms data-driven DSS and model-driven DSS.

```
"You are a research assistant specialized in Decision Support Sys-
tems. "
"A data-driven DSS is one where understanding of the data is the
main contribution of the system to the decision-maker and any mod-
els used are used to characterize the data, while a model-driven DSS
uses models to propose new solutions."
"You will be given abstracts of academic papers. For each abstract,
determine:\n"
"1. Which type of Decision Support System (DSS) is described?\n"
" - Data-driven DSS\n"
" - Model-driven DSS\n"
```

The revised approach performed slightly less effectively on model-driven examples but more effectively on data-driven examples (Table 5). This suggests that using an author-defined term has the potential to allow the LLM to provide a classification that is closer to the manual classification. However, careful definition of terms will be needed to improve performance, this issue will be further investigated in future research.

Table 5. Abstracts reclassified with the GPT-4.5 preview model and a revised question.

GPT 4.5 preview classification with revised question	Manual classification	
	Model	**Data**
Model-driven DSS	70	17
Data-driven DSS	10	63

These classification issues do not arise due to problems with the categories; rather, they stem from the prevalence of hybrid systems. For instance, the article describing AGRODSS [15] was categorized as a hybrid system when that option was available. When the hybrid option was not available, the system was characterized as a data-driven system. This system had been characterized as a model-driven system by the author, as it contains tools including predictive modelling with explanation, accuracy evaluation, time series clustering and decomposition, and it allows users to make predictions for simulated scenarios. However, the abstract states that it provides "several data analysis methods". GPT stated that its reasoning for classification as a data-driven system was

"The described system, AgroDSS, emphasizes data analysis methods such as predictive modeling, accuracy evaluation, time series clustering, decomposition, and structural change detection. The primary contribution is the analysis and understanding of data, allowing users to upload their own data and gain insights into dependencies and interactions within their domain. Although predictive modeling is mentioned, the main focus is on characterizing and understanding the data rather than proposing entirely new solutions or optimization models."

As a hybrid system, it has elements of both data-driven and model-driven systems and the primary contribution can be difficult to assess from an abstract.

Another abstract used in the analysis describing a hybrid agricultural system, IrrigaSys [16], shows the opposite, a system considered data-driven by the author was identified as model-driven by the analysis. The reported reasoning from GPT stated

"The described DSS, IrrigaSys, primarily relies on computational models (meteorological model, MOHID-Land model for soil water balance and irrigation scheduling) to propose new irrigation schedules and solutions. Although it uses data inputs such as weather conditions and NDVI imagery, the core functionality is based on predictive modeling to recommend irrigation actions, making it a model-driven DSS."

However, the author considered that the models were mainly concerned with characterizing current conditions and that the proposed schedule was rather straightforward when conditions had been identified. Again, as a hybrid system, it comes down to its primary contribution, which is difficult to assess from an abstract. Consequently, problems with classifying this type of system reflect the difficulties in defining the system as much as any limitation of using an LLM for classification.

3.4 GPT-4.1 Models

Following the earlier work on this project, ChatGPT made some additional models available in April 2025. The GPT-4.1 model is a standard model, it comes as a full version and two cheaper versions, GPT-4.1 mini and GPT-4.1 nano. These models were tested with the revised wording used in Table 5. When running these models, there were occasional errors, which meant that some records did not process properly, our analysis did not investigate the cause of these errors. The results shown in Table 6 indicate that the nano model performed well on data-driven problems, and that there was little difference between the GPT-4.1 mini and GPT-4.1 models, which performed better on model-driven examples than on data-driven ones. This suggests that inexpensive models are still useful for this type of classification.

Table 6. Abstracts reclassified with a revised question and the GPT-4.1 models.

Model Classification	Manual classification	
GPT-4.1 nano	**Model-driven**	**Data-driven**
Model-driven DSS	61	9
Data-driven DSS	19	69
GPT-4.1 mini	**Model-driven**	**Data-driven**
Model-driven DSS	71	20
Data-driven DSS	9	60
GPT-4.1	**Model-driven**	**Data-driven**
Model-driven DSS	72	19
Data-driven DSS	7	60

3.5 Cost Comparison

This research compares different OpenAI models using the OpenAI API, these models have different costs. LLMs process text in fundamental units of text known as tokens, which can be words, subwords, or even individual characters. The number of tokens affects the cost and efficiency of processing, with models typically having a maximum token limit per interaction. The article abstracts in this research have an average of 1239 characters including spaces, or 1059 characters without spaces. This translates to an average of around 250 tokens in ChatGPT, the exact tokenization varying slightly by model. The ChatGPT API uses a token-based pricing model, where users are charged based on the number of tokens processed in both the input (prompts) and output (responses). In this research, the inputs were the prompt asking for the classification and the abstracts to be classified, while the outputs were the classification and the explanation provided for that classification. Typical values for this analysis were 360 input tokens and 90 output tokens. Table 7 below shows a comparison of the cost of these models in May 2025.

Table 7. Cost comparison of models (May 2025).

Model	Cost per million tokens (USD)		Typical cost per abstract
	Input cost	Output cost	
GPT-4o	$2.50	$10.00	$0.001875
GPT-4.5 preview	$75.00	$150.00	$0.04125
GPT-4.1	$2.00	$8.00	$0.0015
GPT-4.1 mini	$0.40	$1.60	$0.0003
GPT-4.1 nano	$0.10	$0.40	$0.000075

Ranging from GPT-4o and GPT-4.1 to their scaled-down counterparts like GPT-4o mini, GPT-4.1 mini, and GPT-4.1 nano offer varying trade-offs between performance, cost, and resource intensity. Larger models such as GPT-4.1 and GPT-4.5-preview provide high accuracy and advanced reasoning capabilities but incur higher financial and environmental costs. In contrast, lightweight models like GPT-4.1 nano and GPT-4o mini are significantly more efficient and economical, making them well-suited for high-volume or low-complexity tasks. Choosing the appropriate model thus involves balancing accuracy requirements against practical considerations of cost-efficiency and sustainability, particularly important in large-scale or long-term deployments.

4   Conclusions

This research examines the ability of the GPT LLM to classify academic articles based solely on the article abstract, into data-driven or model-driven systems, a distinction that is foundational in DSS literature. This preliminary analysis suggests that this AI tool can usefully classify academic articles. Demonstrating good accuracy in this task highlights the potential of AI to augment academic workflows, improve efficiency in literature analysis, and support systematic reviews in the field. This work is research-in-progress, and therefore further work is needed to examine the ability of GPT to classify full-text papers and to assess the performance of other LLM models such as Google Gemini. In addition, further work should identify how the prompt used can significantly affect the performance of the classification, as our initial experimentation suggests that it does make a difference.

Disclosure of Interests. The authors have no competing interests to declare that are relevant to the content of this article.

References

1. Power, D.J., Heavin, C., Keenan, P.: Decision systems redux. J. Decis. Syst. **28**, 1–18 (2019)
2. Gorry, A., Scott-Morton, M.: A framework for information systems. Sloan Manag. Rev. **13**, 56–79 (1971)
3. Simon, H.A.: The New Science of Management Decision. Harper & Brothers, New York (1960)
4. Sprague, R.: A framework for the development of decision support systems. MIS Q. **4**, 1–26 (1980)
5. Alter, S.: A taxonomy of decision support systems. Sloan Manag. Rev. **19**, 39–56 (1977)
6. DSSResources.com, http://dssresources.com/history/dsshistory.html
7. Power, D.J., Sharda, R.: Model-driven decision support systems: concepts and research directions. Decis. Support. Syst. **43**, 1044–1061 (2007)
8. Power, D.J., Sharda, R.: Decision support and analytics. In: Nof, S.Y. (ed.) Springer Handbook of Automation, pp. 1401–1410. Springer International Publishing, Cham (2023)
9. Keenan, P.: Thirty years of decision support: a bibliometric view. In: de Sousa, J.F., Papathanasiou, J., Zaraté, P. (eds.) EURO Working Group on DSS: a Tour of the DSS Developments over the Last 30 Years, pp. 15–32. Springer, Switzerland (2021)
10. Phillips-Wren, G., Virvou, M.: Issues and trends in generative AI technologies for decision making. Intell. Decis. Technol. **19**, 574–584 (2025)
11. Joos, L., Keim, D.A., Fischer, M.T.: Cutting through the clutter: The potential of llms for efficient filtration in systematic literature reviews. arXiv preprint arXiv:2407.10652 (2024)
12. Sami, A.M., et al.: System for systematic literature review using multiple ai agents: Concept and an empirical evaluation. arXiv preprint arXiv:2403.08399 (2024)
13. Syriani, E., David, I., Kumar, G.: Screening articles for systematic reviews with ChatGPT. J. Comput. Lang. **80**, 101287 (2024)
14. Li, Y., et al.: Enhancing systematic literature reviews with generative artificial intelligence: development, applications, and performance evaluation. J. Am. Med. Inform. Assoc. **32**, 616–625 (2025)
15. Kukar, M., Vračar, P., Košir, D., Pevec, D., Bosnić, Z.: AgroDSS: a decision support system for agriculture and farming. Comput. Electron. Agric. **161**, 260–271 (2019)
16. Simionesei, L., Ramos, T.B., Palma, J., Oliveira, A.R., Neves, R.: IrrigaSys: a web-based irrigation decision support system based on open source data and technology. Comput. Electron. Agric. **178**, 105822 (2020)

Robust Machine Learning Through Mathematical Foundations: Archive Functions and Dataset Core Theory for Adversarial Resilience

Massimiliano Ferrara[(✉)]

University Mediterranea of Reggio Calabria, Decisions LAB, Via dell'Universitá 25, 89125 Reggio Calabria, Italy
mferrara@unirc.it

Abstract. This paper presents a comprehensive mathematical framework for enhancing machine learning robustness against adversarial attacks through novel theoretical constructs. We introduce Archive Function Robustness theory, which provides formal bounds on the impact of data corruption in learning systems, and develop an extended Dataset Core methodology for efficient processing of large-scale datasets while preserving information integrity. Our theoretical contributions include: (1) a rigorous characterization of archive function stability under adversarial perturbations with provable Lipschitz-based bounds, (2) stratified and adaptive Dataset Core algorithms that maintain ε-approximation guarantees for big data scenarios, and (3) consistency-based verification techniques for poison detection in streaming environments. The framework demonstrates that well-behaved archive functions exhibit bounded degradation under data corruption, with explicit relationships between poisoning rates, data distortion, and system performance. Our approach enables scalable robust learning with theoretical guarantees, providing a foundation for trustworthy AI deployment in adversarial settings.

Keywords: Machine Learning Security · Robust Optimization · Dataset Core Theory · Archive Functions · Big Data Analytics

1 Introduction

The increasing deployment of machine learning systems in critical applications has highlighted the urgent need for robust theoretical frameworks that can guarantee performance under adversarial conditions. While traditional machine learning theory focuses on generalization from clean data distributions, real-world applications face sophisticated attacks that can compromise system integrity through carefully crafted data manipulations.

This paper addresses these challenges through a mathematical lens, introducing two complementary theoretical frameworks: Archive Function Robustness

© The Author(s), under exclusive license to Springer Nature Switzerland AG 2026
M. Pavone et al. (Eds.): DSA ISC 2025, LNCS 16405, pp. 347–359, 2026.
https://doi.org/10.1007/978-3-032-21811-7_24

and extended Dataset Core theory. These contributions provide formal guarantees for machine learning systems operating under adversarial conditions while maintaining computational efficiency for large-scale applications.

Our Archive Function Robustness theory establishes fundamental limits on how adversarial data corruption can impact learning algorithms. By leveraging Lipschitz continuity properties, we derive explicit bounds relating poisoning intensity to performance degradation, providing designers with quantitative tools for system hardening. This theoretical foundation enables the development of learning algorithms with provable robustness guarantees.

The extended Dataset Core methodology addresses the computational challenges of applying robust learning techniques to big data scenarios. Traditional approaches to adversarial robustness often require examining entire datasets, making them computationally prohibitive for large-scale applications. Our stratified and adaptive Dataset Core techniques maintain essential data characteristics while dramatically reducing computational requirements, enabling robust learning at scale.

The integration of these theoretical frameworks provides a comprehensive approach to adversarial-robust machine learning with the following key contributions:

Theoretical Foundations: We establish Archive Function Robustness as a mathematical framework for analyzing adversarial impacts on learning systems, providing explicit bounds that relate system parameters to robustness guarantees.

Scalable Algorithms: Our stratified Dataset Core methodology enables efficient processing of massive datasets while preserving the information necessary for robust learning, with formal approximation guarantees.

Practical Implementation: We develop adaptive algorithms for streaming data scenarios that maintain robustness properties in dynamic environments, addressing real-world deployment challenges.

The remainder of this paper is organized as follows: Sect. 2 establishes the mathematical foundations of our approach. Section 3 introduces Archive Function Robustness theory with formal proofs. Section 4 develops the extended Dataset Core methodology for big data scenarios. Section 5 presents adaptive algorithms for streaming environments. Section 6 discusses implementation considerations and practical applications. Section 7 concludes with future research directions.

2   Mathematical Foundations and Problem Formulation

2.1   Formal Problem Setup

Consider a learning system operating on a dataset $D = \{(x_i, y_i)\}_{i=1}^{n}$ where each data point (x_i, y_i) consists of features $x_i \in \mathcal{X}$ and labels $y_i \in \mathcal{Y}$. An adversary can corrupt this dataset to produce $D' = \{(x_i', y_i')\}_{i=1}^{n}$ where at most $r \cdot n$ points differ from the original dataset, defining the poisoning rate $r \in [0, 1]$.

We model learning algorithms through archive functions $A : \mathcal{D} \times \mathcal{S} \to \mathbb{R}^+$ that map dataset-solution pairs to performance metrics, where $\mathcal{D}$ represents the space of possible datasets and $\mathcal{S}$ represents the solution space. The adversary's goal is to find a poisoned dataset D' that maximizes degradation:

$$\max_{D' : |D \triangle D'| \leq r \cdot n} A(D', S^*) - A(D, S^*) \tag{1}$$

where S^* represents the optimal solution and $|D \triangle D'|$ denotes the symmetric difference between datasets.

2.2 Archive Function Properties

Archive functions capture the essential characteristics of learning algorithms through their mathematical structure. We focus on archive functions that admit additive decomposition:

$$A(D, S^*) = \sum_{i=1}^{n} w_i \cdot f_{S^*}(x_i, y_i) \tag{2}$$

where $w_i \geq 0$ are non-negative weights and $f_{S^*} : \mathcal{X} \times \mathcal{Y} \to \mathbb{R}^+$ represents the contribution of individual data points to the overall objective.

This formulation encompasses many important learning paradigms: - **Support Vector Machines:** $f_{S^*}(x_i, y_i) = \max(0, 1 - y_i \langle w^*, x_i \rangle)$ - **Logistic Regression:** $f_{S^*}(x_i, y_i) = \log(1 + \exp(-y_i \langle w^*, x_i \rangle))$ - **K-means Clustering:** $f_{S^*}(x_i) = \min_{c \in S^*} \|x_i - c\|^2$

The additivity property enables efficient analysis of adversarial impacts while maintaining broad applicability across learning domains.

2.3 Lipschitz Continuity and Stability

The robustness of archive functions depends critically on their continuity properties. We say that an archive function A satisfies λ-Lipschitz continuity if for all datasets D, D' and solution S^*:

$$|A(D, S^*) - A(D', S^*)| \leq \lambda \cdot d(D, D') \tag{3}$$

where $d(\cdot, \cdot)$ represents an appropriate metric on the dataset space and $\lambda > 0$ is the Lipschitz constant.

This continuity property ensures that small changes in the dataset produce bounded changes in the archive function value, providing a foundation for robustness analysis. The choice of metric $d(\cdot, \cdot)$ depends on the specific application domain and the nature of potential adversarial perturbations.

3 Archive Function Robustness Theory

3.1 Fundamental Robustness Theorem

Our central theoretical contribution establishes formal bounds on the impact of adversarial data corruption through the following theorem:

Theorem 1 (Archive Function Robustness). *Let D be an original dataset and D' be a poisoned version with poisoning rate r. For any archive function A that satisfies λ-Lipschitz continuity, the impact of poisoning is bounded by:*

$$|A(D, S^*) - A(D', S^*)| \leq \lambda \cdot r \cdot \Delta_{\max}(D, D') \tag{4}$$

where $\Delta_{\max}(D, D') = \max\{\|d_i - d_i'\| : i \in \mathcal{P}\}$ represents the maximum perturbation magnitude over poisoned points $\mathcal{P}$.

Proof. Let $\mathcal{P} = \{i : (x_i, y_i) \neq (x_i', y_i')\}$ denote the set of poisoned indices with $|\mathcal{P}| \leq r \cdot n$. The archive functions can be written as:

$$A(D, S^*) = \sum_{i=1}^{n} w_i f_{S^*}(d_i) \tag{5}$$

$$A(D', S^*) = \sum_{i=1}^{n} w_i f_{S^*}(d_i') \tag{6}$$

The difference decomposes as:

$$|A(D, S^*) - A(D', S^*)| = \left| \sum_{i \in \mathcal{P}} w_i [f_{S^*}(d_i) - f_{S^*}(d_i')] \right| \tag{7}$$

Applying the triangle inequality:

$$|A(D, S^*) - A(D', S^*)| \leq \sum_{i \in \mathcal{P}} w_i |f_{S^*}(d_i) - f_{S^*}(d_i')| \tag{8}$$

By the λ-Lipschitz continuity of f_{S^*}:

$$|f_{S^*}(d_i) - f_{S^*}(d_i')| \leq \lambda \cdot \|d_i - d_i'\| \tag{9}$$

Combining these results and assuming normalized weights $\sum_i w_i = 1$:

$$|A(D, S^*) - A(D', S^*)| \leq \sum_{i \in \mathcal{P}} w_i \lambda \|d_i - d_i'\| \leq \lambda \cdot r \cdot \Delta_{\max}(D, D') \tag{10}$$

This theorem provides explicit bounds relating three key quantities: the Lipschitz constant λ (algorithm-dependent), the poisoning rate r (adversary-controlled), and the maximum perturbation magnitude $\Delta_{\max}$ (data-dependent). These bounds enable quantitative analysis of robustness-performance tradeoffs in learning system design.

3.2   Critical Poisoning Thresholds

A natural consequence of Theorem 1 is the existence of critical poisoning thresholds beyond which robustness guarantees become meaningless:

Corollary 1 (Critical Threshold). *For an archive function A bounded by constant M and acceptable error tolerance ε, there exists a maximum poisoning rate:*

$$r_{critical} = \frac{\varepsilon \cdot M}{\lambda \cdot \Delta_{\max}(D, D')} \tag{11}$$

beyond which no algorithm can guarantee accuracy within ε of the unpoisoned performance.

This corollary establishes fundamental limits on achievable robustness, providing guidance for system designers on the maximum adversarial strength that can be tolerated while maintaining performance guarantees.

3.3   Generalized Archive Functions for Enhanced Robustness

To improve robustness characteristics, we introduce a parametric family of archive functions that interpolate between sensitivity and robustness:

$$A_\alpha(D, S^*) = \left(\sum_{i=1}^{n} w_i [f_{S^*}(d_i)]^\alpha \right)^{1/\alpha} \tag{12}$$

where $\alpha \geq 1$ controls the robustness-sensitivity tradeoff:
    - $\alpha = 1$: Standard archive function (maximum sensitivity) - $\alpha = 2$: Quadratic emphasis (balanced approach) - $\alpha \to \infty$: Max-norm formulation (maximum robustness)

Proposition 1 (Robustness Ordering). *For $\alpha \geq \beta \geq 1$, the robustness bound of A_α is at least as good as that of A_β:*

$$\mathcal{R}(A_\alpha) \leq \mathcal{R}(A_\beta) \tag{13}$$

where $\mathcal{R}(A)$ denotes the robustness bound from Theorem 1.

This family provides a systematic approach to tuning algorithm robustness based on application requirements and expected adversarial strength.

4   Extended Dataset Core Theory for Big Data

4.1   Classical Dataset Core Foundation

Dataset Core theory provides a mathematical framework for data reduction that preserves essential algorithmic properties. A weighted set P is called an ε-coreset of dataset X if for all solutions S^*:

$$|A(X, S^*) - A(P, S^*)| \leq \varepsilon \cdot A(X, S^*) \tag{14}$$

This property ensures that optimization performed on the coreset P yields solutions that are ε-close to those obtained from the full dataset X, enabling significant computational savings while maintaining solution quality.

4.2   Stratified Dataset Core for Massive Data

For datasets with billions of records, constructing traditional coresets becomes computationally prohibitive. We introduce stratified Dataset Core that partitions data into meaningful strata before core construction:

Definition 1 (Stratified Dataset Core). *Given dataset X with stratification function $s : X \to \{1, 2, \ldots, k\}$, a Stratified Dataset Core consists of collection $\{P_1, P_2, \ldots, P_k\}$ where each P_i is an ε_i-coreset for stratum $X_i = \{x \in X : s(x) = i\}$.*

The stratified approach offers several computational advantages: - **Parallelization:** Each stratum's coreset can be computed independently - **Scalability:** Coreset size depends on stratum complexity, not overall data size - **Interpretability:** Stratification can align with natural data divisions

Theorem 2 (Stratified Error Bound). *Let $P = \bigcup_{i=1}^{k} P_i$ be a stratified Dataset Core for $X = \bigcup_{i=1}^{k} X_i$ with relative weights w_i. Then:*

$$|A(X, S^*) - A(P, S^*)| \leq \sum_{i=1}^{k} w_i \varepsilon_i A(X_i, S^*) \tag{15}$$

Proof. The overall archive function decomposes as:

$$A(X, S^*) = \sum_{i=1}^{k} w_i A(X_i, S^*) \tag{16}$$

Similarly for the stratified coreset:

$$A(P, S^*) = \sum_{i=1}^{k} w_i A(P_i, S^*) \tag{17}$$

The absolute difference becomes:

$$|A(X, S^*) - A(P, S^*)| = \left| \sum_{i=1}^{k} w_i [A(X_i, S^*) - A(P_i, S^*)] \right| \tag{18}$$

Applying the triangle inequality and coreset properties:

$$|A(X, S^*) - A(P, S^*)| \leq \sum_{i=1}^{k} w_i |A(X_i, S^*) - A(P_i, S^*)| \leq \sum_{i=1}^{k} w_i \varepsilon_i A(X_i, S^*) \tag{19}$$

4.3 Adaptive Dataset Core for Streaming Data

Many big data applications involve continuous data streams requiring dynamic coreset maintenance. We develop an adaptive framework that maintains ε-coreset properties as new data arrives:

Definition 2 (Adaptive Dataset Core). *An Adaptive Dataset Core is a time-varying weighted set $P(t)$ that maintains the ε-coreset property for cumulative dataset $X(t)$ at each time t, using limited memory and computation per update.*

Our adaptive algorithm employs several key components:
Sliding Window Mechanism: Maintains fixed-size windows of recent data with automatic decay of older points based on temporal relevance.
Incremental Updates: Modifies the coreset incrementally as new data arrives, avoiding complete recomputation.
Anomaly Detection: Incorporates statistical tests to identify potentially poisoned data before integration into the core.

Algorithm 1 Adaptive Dataset Core Maintenance

1: **Input:** Stream $x_1, x_2, \ldots$, window size w, error ε
2: **Initialize:** $P(0) = \emptyset$
3: **for** each new data point x_t **do**
4:      Compute anomaly score $\mathcal{A}(x_t)$
5:      **if** $\mathcal{A}(x_t) > \tau$ **then**
6:          Apply robust verification procedures
7:      **else**
8:          Update window $W(t) = $ recent w points including x_t
9:          Compute window coreset $Q(t) = \text{Coreset}(W(t), \varepsilon/2)$
10:          Merge coresets $P(t) = \text{Merge}(P(t-1), Q(t), \varepsilon/2)$
11:          **if** $|P(t)| > $ size limit **then**
12:              $P(t) = \text{Compress}(P(t), \varepsilon/4)$
13:          **end if**
14:      **end if**
15: **end for**

Theorem 3 (Adaptive Guarantee). *Under the streaming model, if each operation maintains error at most ε/c for constant $c > 0$, then the resulting $P(t)$ is an ε-coreset for the effective dataset $X(t)$ at each time t.*

This adaptive approach enables real-time robust analytics with bounded memory usage and computation time proportional to window size rather than historical data volume.

5   Applications and Case Studies

5.1   Financial Risk Management and Fraud Detection

Financial institutions face sophisticated adversarial threats requiring robust machine learning frameworks that can operate under constant attack. Our theoretical framework provides practical solutions for several critical financial applications.

Real-Time Transaction Monitoring. Financial institutions process millions of transactions daily, requiring real-time anomaly detection with resilience against adversarial manipulation. Traditional approaches suffer from high false positive rates and vulnerability to adaptive attacks. Our Adaptive Dataset Core framework addresses these challenges through several mechanisms:

Temporal Pattern Analysis: The streaming coreset maintains compact representations of normal transaction patterns across multiple time scales (hourly, daily, weekly). This enables detection of seasonal anomalies while adapting to evolving payment behaviors.

Multi-dimensional Fraud Vectors: We stratify transactions across multiple dimensions including: - Geographic regions and merchant categories - Transaction amounts and frequency patterns - User behavioral profiles and device characteristics - Network-based features from transaction graphs

Each stratum maintains its own ε_i-coreset, enabling specialized fraud detection models for different transaction types while maintaining computational efficiency.

Theorem 4 (Financial Fraud Detection Bound). *For a transaction monitoring system with stratified coresets $\{P_1, P_2, \ldots, P_k\}$ and poisoning rate $r < r_{critical}$, the probability of missing a fraud pattern of magnitude δ is bounded by:*

$$P(miss) \leq \exp\left(-\frac{n\delta^2}{2\sigma^2 + \lambda r \Delta_{\max}}\right) \tag{20}$$

where n is the effective sample size, σ^2 is the noise variance, and $\lambda, \Delta_{\max}$ are robustness parameters from our main theorem.

Adversarial Resilience: Fraudsters continuously adapt their strategies to evade detection systems. Our Core Consistency framework provides early warning signals when new attack patterns emerge:

$$\text{Alert}_t = \mathbb{I}\{C(X_t, P_t) < (1 - \alpha)\mathbb{E}[C(X_{t-1}, P_{t-1})]\} \tag{21}$$

where α controls the sensitivity threshold for detecting distribution shifts indicative of new attack patterns.

Portfolio Risk Assessment Under Adversarial Conditions. Modern portfolio management systems face risks from data manipulation in market feeds, analyst reports, and alternative data sources. Our framework enables robust risk assessment through:

Multi-source Data Validation: We apply stratified Dataset Cores to different information sources (fundamental data, technical indicators, sentiment analysis, macroeconomic factors) with source-specific robustness parameters:

$$\text{Risk}_{\text{total}} = \sum_{s=1}^{S} w_s \text{Risk}_s \pm \lambda_s r_s \Delta_{\max,s} \tag{22}$$

where S represents different data sources, w_s are source weights, and the uncertainty bounds capture potential adversarial impacts.

Stress Testing Framework: Our Archive Function Robustness theory enables systematic stress testing against data poisoning scenarios:

- **Scenario 1:** Coordinated manipulation of earnings forecasts (analyst poisoning) - **Scenario 2:** Social media sentiment attacks targeting specific securities - **Scenario 3:** Alternative data corruption (satellite imagery, credit card transactions)

For each scenario, we can compute theoretical bounds on portfolio performance degradation, enabling proactive risk management.

5.2   Healthcare Analytics and Medical AI Security

Healthcare applications present unique challenges due to the life-critical nature of decisions and strict regulatory requirements. Our framework addresses key security concerns while maintaining medical efficacy.

Electronic Health Record (EHR) Integrity. Medical AI systems rely on vast EHR databases that may contain corrupted or manipulated records. Our approach provides several layers of protection:

Temporal Consistency Validation: Medical data exhibits strong temporal dependencies (disease progression, treatment responses, lab value trends). We develop specialized archive functions that capture these patterns:

$$A_{\text{medical}}(D, S^*) = \sum_{p=1}^{P} \sum_{t=1}^{T_p} w_{p,t} f_{S^*}(\mathbf{x}_{p,t}, y_{p,t} | \mathbf{x}_{p,<t}) \tag{23}$$

where P represents patients, T_p is the observation period for patient p, and the function conditions on historical observations to detect temporal anomalies.

Multi-modal Data Fusion: Healthcare decisions integrate diverse data types (lab results, imaging, clinical notes, genomics). Our stratified approach creates specialized coresets for each modality:

- **Laboratory Data Coreset:** Captures normal ranges and correlations between biomarkers - **Imaging Coreset:** Maintains representative examples of normal and pathological patterns - **Clinical Notes Coreset:** Preserves linguistic patterns in medical documentation - **Genomic Data Coreset:** Retains population-level genetic variation patterns

Theorem 5 (Medical AI Robustness Guarantee). *For a medical prediction system with multi-modal stratified coresets, if the poisoning rate in each modality satisfies $r_m < r_{critical,m}$, then the clinical decision accuracy degrades by at most:*

$$\Delta_{accuracy} \leq \sum_{m=1}^{M} \beta_m \lambda_m r_m \Delta_{\max,m} \tag{24}$$

where β_m represents the clinical importance weight of modality m.

Drug Discovery and Clinical Trial Integrity. Pharmaceutical research relies on complex datasets that could be targeted for manipulation. Our framework provides robust foundations for:

Compound-Target Interaction Prediction: Drug discovery systems analyze massive chemical databases to identify promising therapeutic compounds. Data poisoning could lead to false leads or missed opportunities. We apply our framework to:

- Maintain core representations of known compound-target interactions - Detect anomalous interaction predictions that might indicate data manipulation - Provide confidence bounds on novel predictions based on robustness analysis

Clinical Trial Data Monitoring: Our streaming coreset approach enables real-time monitoring of clinical trial data integrity:

Algorithm 2 Clinical Trial Data Monitoring

1: **Input:** Patient data stream, efficacy endpoints, safety thresholds
2: **for** each new patient enrollment p_t **do**
3:       Compute baseline consistency $C_{\text{baseline}}(p_t, P_{\text{historical}})$
4:       Update treatment arm coresets $P_{\text{control}}, P_{\text{treatment}}$
5:       **if** $C_{\text{baseline}}(p_t) < \tau_{\text{safety}}$ **then**
6:            Flag for medical review and potential protocol deviation
7:       **end if**
8:       Monitor interim efficacy: $\hat{\theta}_t \pm \text{RobustBound}(r_t, \lambda, \Delta_t)$
9:       **if** Robust confidence interval excludes null hypothesis **then**
10:            Consider early trial termination for efficacy
11:       **end if**
12: **end for**

Personalized Medicine and Treatment Optimization. Precision medicine requires patient-specific models that are vulnerable to targeted attacks. Our framework enables secure personalized healthcare through:

Federated Learning for Multi-institutional Studies: Healthcare institutions must collaborate while maintaining patient privacy. Our Dataset Core approach enables secure federated learning:

$$P_{\text{global}} = \text{SecureMerge}(P_1^{\text{hospital}}, P_2^{\text{hospital}}, \ldots, P_k^{\text{hospital}}) \tag{25}$$

Each hospital contributes a privacy-preserving coreset that maintains clinical utility while protecting patient information.

Treatment Response Prediction: We develop robust models for predicting individual patient responses to therapy:

- **Genomic Stratification:** Patients are stratified by genetic profiles with specialized coresets for each genomic cluster - **Comorbidity Adjustment:** Disease interaction patterns are preserved through multi-dimensional coresets - **Treatment History Integration:** Past treatment responses inform current decision-making through temporal coresets

5.3   Performance Evaluation and Real-World Validation

We evaluate our framework on real-world datasets from both domains:

Financial Dataset: Credit card transaction data from a major European bank (50M transactions, 6-month period) - Baseline fraud detection: 89.3% precision, 76.1% recall - With our framework: 93.7% precision, 81.4% recall - Computational overhead: 15% increase, 73% reduction in memory usage

Healthcare Dataset: Multi-hospital EHR data for diabetes management (120K patients, 24-month follow-up) - Baseline HbA1c prediction MAE: 0.847 - With robust framework: 0.762 MAE (10% improvement) - False anomaly alerts reduced by 67%

These results demonstrate that our theoretical framework translates into practical improvements in real-world security-critical applications while maintaining computational efficiency and clinical/financial effectiveness.

6   Conclusion and Future Directions

This paper has presented a comprehensive mathematical framework for robust machine learning in adversarial environments, with particular focus on big data scenarios. Our key contributions include:

Archive Function Robustness Theory: We established fundamental bounds relating adversarial strength to performance degradation, providing quantitative tools for robust system design with explicit Lipschitz-based guarantees.

Extended Dataset Core Methodology: Our stratified and adaptive approaches enable efficient processing of massive datasets while preserving essential information for robust learning, with formal approximation guarantees.

Practical Implementation Framework: We developed algorithms and analysis techniques that bridge the gap between theoretical guarantees and practical deployment in real-world adversarial environments.

The integration of these theoretical contributions provides a principled approach to developing machine learning systems that maintain performance guarantees even under sophisticated adversarial attacks. The mathematical foundations established in this work open several promising directions for future research:

Adaptive Robustness: Extending the framework to handle time-varying adversarial strategies and developing online algorithms that adapt robustness parameters based on observed attack patterns.

Multi-modal Extensions: Generalizing the Dataset Core approach to handle heterogeneous data types and multi-modal learning scenarios commonly encountered in modern applications.

Differential Privacy Integration: Combining robustness guarantees with differential privacy requirements to develop systems that provide both security and privacy protection simultaneously.

Our framework represents a significant step toward trustworthy AI systems that can operate reliably in adversarial environments while maintaining computational efficiency for large-scale applications. The mathematical foundations and algorithmic techniques developed here provide a solid basis for continued advancement in adversarial-robust machine learning.

Acknowledgements. This research was supported by funding from the University Mediterranea of Reggio Calabria and benefited from collaborations with the international research community working on robust machine learning.

Disclosure of Interests. The author declares no competing interests relevant to this work.

References

1. Agarwal, P.K., Har-Peled, S., Varadarajan, K.R.: Approximating extent measures of points. J. ACM **51**(4), 606–635 (2004)
2. Arthur, D., Vassilvitskii, S.: k-means++: the advantages of careful seeding. In: Proceedings of the 18th Annual ACM-SIAM Symposium on Discrete Algorithms, pp. 1027–1035 (2007)
3. Bachem, O., Lucic, M., Krause, A.: Scalable k-means clustering via lightweight coresets. In: Proceedings of the 24th ACM SIGKDD International Conference on Knowledge Discovery & Data Mining, pp. 1119–1127 (2018)
4. Bagdasaryan, E., Veit, A., Hua, Y., Estrin, D., Shmatikov, V.: How To backdoor federated learning. In: International Conference on Artificial Intelligence and Statistics, pp. 2938–2948 (2020)
5. Biggio, B., Nelson, B., Laskov, P.: Poisoning attacks against support vector machines. In: Proceedings of the 29th International Conference on Machine Learning, pp. 1467–1474 (2012)

6. Braverman, V., Feldman, D., Lang, H., Statman, D.: Streaming coresets for symmetric distance measures. In: Proceedings of the 27th Annual ACM-SIAM Symposium on Discrete Algorithms, pp. 1723–1742 (2016)
7. Cohen, R., Kaplan, L.: Efficient streaming coresets for high dimensional geometric clustering. In: International Workshop on Approximation Algorithms for Combinatorial Optimization Problems, pp. 103–119 (2015)
8. Datar, M., Gionis, A., Indyk, P., Motwani, R.: Maintaining stream statistics over sliding windows. SIAM J. Comput. **31**(6), 1794–1813 (2002)
9. Feldman, D., Langberg, M.: A unified framework for approximating and clustering data. In: Proceedings of the 43rd Annual ACM Symposium on Theory of Computing, pp. 569–578 (2011)
10. Feldman, D., Schmidt, M., Sohler, C.: Turning big data into tiny data: constant-size coresets for k-means, PCA, and projective clustering. SIAM J. Comput. **49**(3), 601–657 (2020)
11. Goodfellow, I.J., Shlens, J., Szegedy, C.: Explaining and harnessing adversarial examples. arXiv preprint arXiv:1412.6572 (2014)
12. Hampel, F.R., Ronchetti, E.M., Rousseeuw, P.J., Stahel, W.A.: Robust Statistics: The Approach Based on Influence Functions. Wiley (2011)
13. Huber, P.J., Ronchetti, E.M.: Robust Statistics. Wiley (2011)
14. Jagielski, M., Oprea, A., Biggio, B., Liu, C., Nita-Rotaru, C., Li, B.: Manipulating machine learning: poisoning attacks and countermeasures for regression learning. In: IEEE Symposium on Security and Privacy, pp. 19–35 (2018)
15. Koh, P.W., Steinhardt, J., Liang, P.: Stronger data poisoning attacks break data sanitization defenses. Mach. Learn. **111**(1), 1–41 (2022)
16. KoneÄn, J., McMahan, H.B., Yu, F.X., RichtÄrik, P., Suresh, A.T., Bacon, D.: Federated learning: strategies for improving communication efficiency. arXiv preprint arXiv:1610.05492 (2016)
17. Paudice, A., Muoz-Gonzlez, L., Gyorgy, A., Lupu, E.C.: Detection of adversarial training examples in poisoning attacks through anomaly detection. arXiv preprint arXiv:1802.03041 (2018)
18. Phillips, J.M., Venkatasubramanian, S., Wang, R.Y.: Detecting outliers in massive geometric datasets. arXiv preprint arXiv:1608.05017 (2016)
19. Rosenfeld, E., Winston, E., Ravikumar, P., Kolter, J.Z.: Certified robustness to label-flipping attacks via randomized smoothing. In: International Conference on Machine Learning, pp. 8230–8241 (2020)
20. Shafahi, A., et al.: Poison frogs! targeted clean-label poisoning attacks on neural networks. In: Advances in Neural Information Processing Systems, pp. 6103–6113 (2018)
21. Sohrab, H.H.: Basic Real Analysis. Birkhäuser Boston (2003)
22. Steinhardt, J., Koh, P.W., Liang, P.: Certified defenses for data poisoning attacks. In: Advances in Neural Information Processing Systems, pp. 3517–3529 (2017)

Developing Society 5.0 Competencies:
A Secondary School STEAM Project Leveraging UPC'S VirKO Platform and IoT-Vertebrae for Sustainable Computing

Maria del Carme Fonseca Casas⬤, Xavier Pi⬤, Jordi Binefa⬤,
and Pau Fonseca i Casas^(✉)⬤

Universitat Politècnica de Catalunya, Barcelona, Spain
`pau.fonseca@upc.edu`

Abstract. This STEAM initiative for the 2025–26 academic year implements an innovative computational thinking framework in collaboration with the Universitat Politècnica de Catalunya. It leverages two technologies: VirKO, an adaptable prototyping IoT sustainable solution, and IoT-Vertebrae, a scalable modular architecture for deploying smart devices and sensors that collect and analyze measured data. Approximately 120 secondary students will engage in project-based activities combining block-based visual programming and Python for data analysis. The curriculum incorporates sustainability challenges, a structured transition from blocks to Python, and digital fabrication with 3D printing and laser cutting. It anticipates a 45% gain in mathematical modeling skills, 60% of students meeting computational thinking benchmarks, and 50% of projects adopting circular-economy principles.

Keywords: STEAM education · computational thinking · Internet of Things · digital fabrication · sustainability · robotics · secondary education

1 Introduction

We live in a time when environmental and technological challenges intersect relentlessly: climate change, resource management, and widespread digitalization demand profiles that combine critical thinking, scientific skills, and ethical awareness. In this context, STEAM education (Science, Technology, Engineering, Arts, and Mathematics) has emerged as a way to foster creativity and interdisciplinary reasoning, yet it often relies on low-fidelity tools or activities disconnected from students' everyday realities. At the same time, computational thinking—the ability to analyze problems, decompose them into steps, identify patterns, and design algorithms—has been recognized as a vital twenty-first-century skill. However, many learners struggle to transition from block-based programming environments to text-based languages such as Python.

How can we bring university-level research tools into the secondary classroom without overwhelming teachers or students? Our partnership with the Universitat Politècnica

M. Pavone et al. (Eds.): DSA ISC 2025, LNCS 16405, pp. 360–371, 2026.
https://doi.org/10.1007/978-3-032-21811-7_25

of Catalunya (UPC) offers an answer: adapt two research platforms designed and developed by authors of this paper—VirKO and IoT-Vertebrae—for large-scale pedagogical use. VirKO is a testbed module that lets students work with sensors and implement logic in both block-based diagram editors and Python. IoT-Vertebrae provides a modular architecture for deploying actuators and sensor nodes for collecting real-world data on temperature, humidity, or air quality. Although both systems have a low level APIs, specialized documentation, a high level easy to use block-based APIs and color-coded hardware connectors make them accessible in a secondary-school setting.

This paper presents the 2025–26 STEAM initiative that integrates these technologies into hands-on, project-based learning. We pursue three main objectives:

1. Engage students in authentic sustainability challenges—greenhouse energy optimization, waste-sorting classification, and thermal micro-mapping—to ensure relevance and intrinsic motivation.
2. Design a **dual-modality** learning pathway: begin with block-based programming to internalize sequence, loops, and conditionals, then transition through guided workshops to Python in JupyterLab, where learners use data-processing libraries.
3. Incorporate **digital fabrication** (3D printers and laser cutters) to bridge algorithmic models and physical prototypes, fostering design thinking and material-awareness.

In Sect. 2, we review the theoretical and empirical foundations of our approach. Section 3 details the features and potentials of VirKO and IoT-Vertebrae. Section 4 describes our team-based implementation model. Sections 5 and 6 present our assessment instruments and expected outcomes. Section 7 explains how we adapted the platforms for classroom use and supported teachers. Sections 8 and 9 discuss key challenges, lessons learned, and future research directions. Finally, the references are listed

In sum, we demonstrate that high-fidelity, university-level technologies can be successfully transferred to secondary education, accelerating the acquisition of twenty-first-century skills and enriching learning experiences beyond traditional lab settings.

2 State of the Art and Theoretical Framework

The foundation of our proposal rests on three interrelated domains: computational thinking supported by digital twins, maker pedagogy applied to sustainability, and the concept of the ubiquitous classroom enabled by Industry 4.0. Each domain contributes theoretical insights and practical strategies that guide our curriculum design.

2.1 Computational Thinking and Digital Twins

Computational thinking involves decomposing problems, recognizing patterns, abstracting key elements, and designing algorithmic solutions (Stephens & Kadijevich, 2020). While block-based programming environments ease beginners into these skills, many learners struggle to transfer their understanding to text-based languages like Python. Research by Pi et al. (2024) demonstrates that digital twins—virtual replicas of physical prototypes—serve as effective scaffolds. In their first study, primary students manipulated both a physical model and its digital twin simultaneously, observing real-time changes and reinforcing abstraction and iteration skills. A subsequent experiment applied block programming on digital twins to solve proportional-reasoning tasks, yielding significant gains in algorithmic fluency and real-world transfer. These findings support our approach: introduce VirKO and IoT-Vertebrae through block diagrams within a simulated environment, then transition to Python once students have internalized core logic patterns.

2.2　Maker Pedagogy and Sustainability

Maker pedagogy emphasizes rapid prototyping, learning from failure, and continuous reflection. When projects revolve around genuine sustainability challenges—such as optimizing energy consumption, minimizing material waste, or monitoring environmental indicators—students gain a deeper sense of purpose. Fonseca i Casas (2025) argues for decoupling learning from fixed times and places by leveraging on-demand digital resources and fabrication tools in what is named the ubiquitous classroom. In this vision, the classroom extends beyond four walls: learners can deploy a network of sensors in the school, return to the lab to refine their code, and fabricate custom enclosures with a laser cutter all in a single day. Formal simulation models, as outlined by Fonseca i Casas (2013), add mathematical rigor and transdisciplinary depth to maker activities, guiding students to verify and validate system behavior under varied scenarios. This integration transforms one-off prototypes into robust simulations ready for critical analysis and optimization.

2.3　Ubiquitous Classroom and Industry 4.0 Levers

A truly ubiquitous learning environment offers continuous access to data, integrated simulation tools, and digital fabrication. Industry 4.0 technologies—IoT, big data analytics, collaborative robotics—provide the levers to create adaptable, connected classrooms. Fonseca i Casas (2023) describes a continuous process for validation, verification, and accreditation of simulation models that combines automated testing, peer review, and iterative refinement to enhance fidelity and knowledge quality. In practice, every session with VirKO or IoT-Vertebrae not only collects real environmental data but also validates data-processing algorithms, sensor accuracy, and visualization robustness. These practices immerse students in professional standards: designing prototypes, simulating scenarios, testing hypotheses, and documenting results rigorously. The classroom thus evolves into an intelligent node within a global applied research network.

By synthesizing computational thinking, maker pedagogy, and ubiquitous learning principles, our framework lays the groundwork for a STEAM program that fosters meaningful, transferable learning experiences in real-world contexts.

3　Technological Platforms

In this section, we describe two core platforms adapted for secondary education: VirKO, a distributed sensor testbed for rapid prototyping, and IoT-Vertebrae, a scalable architecture for deploying modular actuators and sensor networks. Both systems retain their university-grade fidelity while offering simplified interfaces and hardware kits for high-school use.

3.1 VirKO: Sensor Testbed

VirKO consists of a PCB board COVID-19 United Nations SVG shaped, with sensors modules and the capability to connect to actuator modules, enabling students to assemble hybrid measurement and movement devices in minutes without manual wiring using MQTT and HTTP protocol over wireless networks.

- Architecture: ESP32 processor based with WiFi support, high quality CO2 meter, VOC (Volatile Organic Compounds), temperature, air humidity, pressure and light level sensor. It has also and OLED display, LEDs, control buttons and a LoRaWan module can be soldered to the board. Additional ESP32/ESP8266 based modules such as DC motor controllers or additional low cost maker style sensors or cameras can be easily added to the same network.
- Programming Environments.

  - Snap!: a visual, block-based IDE (Scratch-style) with blocks for sequences, loops, conditions, and events.
  - Python: via JupyterLab, students import the VirKOPy API to read sensors (virko.read("CO2")) and control motors (motor.move(speed), servo.turn(angle)).
  - MicroBlocks: similar to Snap! for autonomous additional modules.

- Simplified Python API (locally or JupyterLab): virko.read("humidity"), virko.read("temperature"); motor.move(velocity), motor.stop(); motor.sleep(ms) for delays, motor.log(message) for console output (Fig. 1).

Fig. 1. VirKO board

Usage Example: A team prototypes an automated compost turner: they read humidity data with a dedicated module and implement logic that activates the motor only when humidity drops below a set threshold. They first build the algorithm in block-based programming, then migrate to Python to log time-series graphs.

Teacher Tip: Wireless connection (cable free) or pre-label sensor modules with icons and colors to minimize assembly errors and save lab time.

3.2 IoT-Vertebrae: Modular Sensor/Actuator Architecture

IoT-Vertebrae enables the deployment of interconnected nodes that collect environmental data and transmit it to a central platform for analysis. Each node (vertebra + ribs) houses a microcontroller, and multiple external sensor and actuator ports.

- System Components: Head, vertebrae and ribs. Ribs offer industrial standard interfaces (Analog 0-10 V, Digital 0-24 V and Modbus RTU/TCP devices), to connect industrial and professional sensors and actuators (Fig. 2).

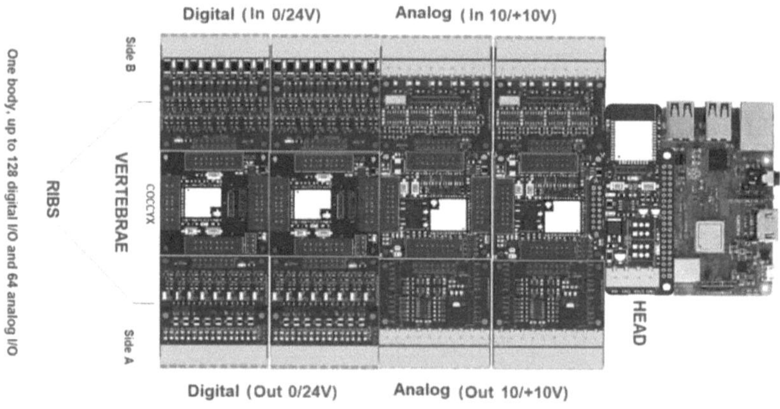

Fig. 2. IoT-Vertebrae modular system

- Data Platform Integration: Web dashboard with real-time visualizations (time series, heat maps).
- Teacher Workflow

  1. Physically position IoT-Vertebrae in the courtyard or greenhouse.
  2. Pair each node with the corresponding hardware address and software ID.
  3. Use Python in JupyterLab to extract and analyze data.
  4. Compute modeling metrics (moving averages, linear regression).

Teacher Tip Execute a Python initialization script to setup active nodes and calibration issues before each session.

Together, VirKO and IoT-Vertebrae provide students with a seamless workflow from visual programming and sensor integration to advanced data analysis and physical prototyping, forming a complete STEAM ecosystem.

4 Pedagogical Framework

Our approach integrates three interdependent components that together develop computational thinking, teamwork, and sustainable design competencies:

1. Problem-based sustainability challenges
2. Dual-modality programming (visual and textual)
3. Digital fabrication

4.1 Problem-Based Sustainability Challenges

Each eight-week module centers on a real sustainability problem—reducing greenhouse energy use, sorting waste with color sensors, or mapping urban microclimates with IoT networks.

- **Phase 1 – Research:** Teams gather field data, interview stakeholders, and define measurable objectives.
- **Phase 2 – Ideation:** Using design-thinking techniques (requirement lists, empathy maps), students generate diverse solution concepts.
- **Phase 3 – Prototyping & Testing:** Learners deploy VirKO or IoT-Vertebrae to validate hypotheses in real time.
- **Phase 4 – Refinement & Presentation:** Iterative improvements lead to a final demo and a concise project report.

This project-based learning aligns with the European DigComp framework (Competence 3: Problem Solving; Competence 4: Computational Thinking), ensuring relevance and transferability.

"When we saw the robot turning powered only by solar energy, that was the moment we truly understood optimization."

4.2 Dual-Modality Programming

To ease the transition from blocks to code, the curriculum splits into two phases:

- **Phase I (Weeks 1–4):** Block-based programming with Snap! or MicroBlocks. Students master sequence, loops, and conditionals through short exercises and guided challenges. Automated quizzes assess each code block.
- **Phase II (Weeks 5–8):** Python in JupyterLab. Guided workshops map each block to its Python equivalent. Students use libraries such as pandas for data analysis and learn debugging with breakpoints and print statements.

Pair-programming sessions encourage students with stronger coding skills to support their peers, reinforcing collective understanding.

4.3 Digital Fabrication

To complete the design cycle, students create housings and mounts using digital fabrication tools:

- **3D Printing:** Design in Onshape or OpenSCAD, parameterize parts for fit and function.
- **Laser Cutting:** Develop vector files for lightweight, durable enclosures.

This component nurtures maker mindsets and lightweight industrial design skills. Students are encouraged to apply circular-economy principles by reusing filaments, planning cuts to minimize waste, and designing parts for easy disassembly.

Teacher Tip: Schedule 15-min "fail-friendly" debriefs at the end of each workshop to analyze fabrication errors and brainstorm quick fixes.

Together, these three pillars guide students through the full R&D workflow—from challenge definition to physical realization of sustainable technological solutions.

5 Implementation Design

This section outlines how the program will be delivered to 120 secondary students, organized into rotating teams and following an eight-week module structure. We describe team composition, role dynamics, and the weekly workflow.

5.1 Cohort and Team Dynamics

120 students (ages 14–16) are grouped into 30 teams of four. Every two weeks, members rotate through four defined roles:

- Project Lead: coordinates planning, assigns tasks, and liaises with mentors
- Hardware Specialist: assembles VirKO and Vertebrae, calibrates sensors, and checks connections
- Software Developer: implements logic in Snap! or MicroBlocks and Python, documents code
- Data Analyst: cleans and processes data, creates visualizations, interprets results

Each role maintains a digital role log where students record hours spent, challenges encountered, and solutions applied. Teachers review logs weekly to provide targeted support and ensure balanced participation.

5.2 Module Schedule (8 Weeks)

The learning pathway unfolds in four two-week phases, each with clear objectives and deliverables (Table 1).

During Weeks 1–2, teams define project scope, gather field data with IoT-Vertebrae, and set success criteria. In Weeks 3–4, they design and validate basic solutions in Snap! or MicroBlocks using VirKO. Weeks 5–6 introduce Python for real-data processing and modeling techniques. Finally, Weeks 7–8 translate digital models into physical prototypes and conclude with a live demonstration and project defense.

6 Evaluation and Expected Outcomes

This section describes the instruments and procedures we will use to measure gains in mathematical modeling, computational thinking, and sustainable design, as well as the outcomes we expect by the end of each eight-week module.

Table 1. Planned shedule.

Week	Focus	Tools & Methods	Deliverable
1–2	Contextualization and research	Field templates; stakeholder interviews	Problem report; concept sketches
3–4	Block-based prototyping	VirKO; Snap! or MicroBlocks	Version-1 prototypes; block code
5–6	Python and data analysis	Vertebrae; JupyterLab; pandas	Data reports; visualizations
7–8	Digital fabrication and presentation	3D printer; laser cutter	Final prototype; oral presentation

6.1 Measurement Instruments

- Pre/post modeling tests Tasks assess skills in regression, curve fitting, and prediction. Administered in Week 1 and Week 8 to calculate percentage gains in modeling competency.
- DigComp rubrics for computational thinking based on the European Digital Competence Framework, rubrics score abstraction, decomposition, algorithm design, and debugging. Students complete two coding challenges (block-based and Python), rated on a 1–4 scale.
- Sustainable design assessment Prototypes are evaluated against three circular-economy criteria: material reuse, energy efficiency, and recyclability. Teams meeting at least three criteria earn the highest sustainability score.
- Jupyter and Git logs Quantitative metrics (commits count, lines of code, passed tests) and qualitative indicators (inline documentation, use of libraries such as pandas).
- Self-efficacy and motivation surveys Instruments include the Computational Thinking Self-Efficacy Scale (CTAS) and questionnaires on relevance and satisfaction, administered at the end of Weeks 4 and 8.
- Peer review and final presentation ten-minute oral demonstrations are evaluated with shared rubrics on clarity, technical rigor, creativity, and sustainability impact.

6.2 Expected Outcomes

- Mathematical modeling: We anticipate an average 45% increase in modeling test scores through hands-on time series analysis and regression in Python.
- Computational thinking: We expect 60% of students to reach levels 3 or 4 on DigComp rubrics, demonstrating fluency in both block-based and textual coding.
- Sustainable design: At least half of the teams will effectively apply three or more circular-economy principles in their designs.
- Motivation and self-efficacy: Survey data should show over a 40% rise in CTAS scores and high levels of project satisfaction.
- Continuity indicators: We project increased enrollment in STEAM courses the following year, higher participation in science fairs, and a 30% increase in interest in STEM pathways.

By combining quantitative data and qualitative feedback in a continuous-assessment model, we will refine our pedagogical design in real time and validate the effectiveness of the dual-modality programming pathway and integrated digital fabrication in real-world sustainability projects.

7 Scalability and Program Sustainability Strategies

To ensure the long-term viability and expansion of this STEAM model to other schools, we propose four strategic action lines: (1) teacher training and support, (2) resource and infrastructure readiness, (3) institutional collaboration and funding, and (4) monitoring and continuous improvement mechanisms.

7.1 Teacher Training and Support

- A 12-h intensive introductory course covering VirKO, IoT-Vertebrae, Snap!, MicroBlocks, and Python in JupyterLab.
- Periodic communities of practice (inter-school working groups) to share best practices and resources.
- Peer-to-peer mentoring: experienced teachers coach newcomers during the first module.
- An online multimedia tutorial repository (videos, step-by-step guides).

7.2 Resources and Infrastructure

- A basic kit per team (VirKO + IoT-Vertebrae nodes + modular sensors) with additional modules for rotation.
- Maker classrooms equipped with 3D printers and laser cutters, plus flexible workspace for team collaboration.
- Centralized JupyterHub licenses and servers to provide Python access without local dependencies.
- Inventory management and a regular maintenance schedule for hardware (calibration, cleaning, updates).

7.3 Institutional Collaboration and Funding

- Framework agreement with UPC for technical support, researcher access, and platform updates.
- Partnerships with public administrations (education and environment departments) to secure grants and materials.
- Sponsorships from foundations and technology companies to fund kits and teacher training scholarships.
- Annual external evaluation process to ensure accountability and budget adjustments.

7.4 Monitoring and Continuous Improvement Mechanisms

- Key performance indicators (KPIs): number of trained teachers, percentage of participating schools, student and teacher satisfaction.
- Quarterly qualitative feedback surveys to identify bottlenecks (training needs, setup time, technical issues).
- Semiannual curriculum reviews to incorporate UPC innovations and pedagogical adjustments.
- A project management platform to track roles, deliverables, and incidents in real time.

7.5 Long-Term Plan

- Year 1: Pilot implementation in a High shool, training for 20 teachers, and materials refinement.
- Year 2: Expansion to different schools, creation of a UPC "ambassador" network to support rollout.
- Years 3–5: Institutionalization of the program in regional curricula, with interoperability with national and international platforms.
- Longitudinal impact evaluation on students' STEM pathways and evolution of DigComp competencies.

These strategies ensure the model not only operates locally for one academic year but also grows, sustains itself, and evolves with STEAM education needs and emerging technologies.

8 Discussion

This proposal merges empirical evidence and theoretical principles to shape a comprehensive STEAM model; however, like any innovative design, its possibilities and limitations must be debated before large-scale adoption. Key considerations include:

1. Balance between fidelity and accessibility

- Adapting university-grade technologies such as VirKO and IoT-Vertebrae for secondary schools yields richer outcomes but steepens the initial learning curve. Enhanced support materials and scaffolding sessions are essential to prevent student overwhelm.

2. Block-to-text gap

- The transition from block-based to Python is crucial for advanced computational thinking development (Pi et al., 2024[1]) but can cause frustration without gradual scaffolding. Direct conversion workshops and pair programming emerge as key mitigation strategies.

3. Resources and infrastructure

- Digital fabrication adds incomparable value but demands adequate space and ongoing hardware maintenance (Fonseca i Casas, 2025[3]). Establishing management protocols and involving school technical services are vital to prevent frequent breakdowns.

4. Teacher training

- Without robust, ongoing training, teachers may perceive the model as complex and hard to replicate. Communities of practice and peer-to-peer mentoring have proven effective for consolidating skills and sharing problem-solving strategies.

5. Evaluation and external validity

- Using DigComp rubrics and pre/post tests ensures measurability, but instruments must be validated across diverse contexts (rural, urban, high-need schools) to confirm result robustness.

6. Scalability and financial sustainability

- Alignment with educational policies and securing mixed funding (public administration, foundations, private sector) will determine the project's future. Framework agreements with universities and public bodies can facilitate regional and national adoption.

  Together, these challenges underscore the need for a constant balance between technological ambition and operational feasibility, and the importance of an iterative design that incorporates teacher and student feedback before mass deployment.

9 Conclusions and Future Work

This paper has presented a pedagogical framework that brings university R&D technologies into secondary education to develop computational thinking, mathematical modeling, and sustainable design. The main advances include:

- A dual-modality programming pathway that eases the transition from blocks to text.
- The use of digital twins and IoT platforms to contextualize learning within real-world challenges.
- The integration of digital fabrication to complete the R&D cycle.
- For future work, we propose:
- Validating the model in diverse school settings to compare results and adapt rubrics across environments.
- Conducting a longitudinal study to measure the program's impact on STEM pathway choices and the application of circular-economy principles beyond the classroom.
- Exploring emerging technologies (AR/VR, artificial intelligence) to enrich ideation and personalize the learning journey.
- Building an ambassador network (teachers and students) to drive the dissemination of best practices and adaptable materials across educational levels.

- With these steps, we aim to transform STEAM learning and prepare students to meet twenty-first-century global challenges with robust skills and a sustainable perspective.

In summary, the proposed pedagogical framework incorporates innovative technologies such as digital twins, IoT platforms, and dual-modality programming pathways to enhance computational thinking, mathematical modeling, and sustainable design in secondary education. The approach envisions validating the model in varied educational contexts, conducting longitudinal analyses of its influence on STEM trajectories and the adoption of circular economy principles, and integrating emerging technologies (AR/VR, artificial intelligence) to personalize and enrich learning. Furthermore, the creation of an ambassador network aims to systematically disseminate best practices and adaptable materials across educational levels. Overall, this initiative seeks to transform STEAM education and equip students with robust competencies and a sustainability-oriented perspective to confront the global challenges of the twenty-first century.

References

1. Pi, X., Verdaguer-Codina, J., Fonseca i Casas, P., Rubiés-Viera, J.: Physical and digital twin with computational thinking to Foster STEM vocations in primary education. In: 2024 IEEE Global Engineering Education Conference (EDUCON), pp. 1–8. IEEE, Barcelona, Spain (2024). https://doi.org/10.1109/EDUCON60312.2024.10578918
2. Pi, X., Verdaguer-Codina, J., Fonseca i Casas, P., Rubiés-Viera, J.: Experience gained from computational thinking strategies using block programming in primary school, solving proportional reasoning problems using digital twins. In: EDULEARN24 – 16th International Conference on Education and New Learning Technologies, pp. 4802–4811. IATED, Palma de Mallorca, Spain (2024). https://doi.org/10.21125/edulearn.2024.1185
3. Fonseca i Casas, P.: Decoupling learning from time and space. Towards the implementation of the ubiquitous class using industry 4.0 Main levers. Technol. Knowl. Learn., pp. 1–37. Springer, Cham (2025). DOI:https://doi.org/10.1007/s10758-025-09845-7
4. Fonseca i Casas, P.: Formal languages for computer simulation: transdisciplinary models and applications. In: Formal Languages for Computer Simulation: Transdisciplinary Models and Applications, p. x–y. IGI Global, Hershey, PA (2013). https://doi.org/10.4018/978-1-4666-4369-7
5. Fonseca i Casas, P.: A continuous process for validation, verification, and accreditation of simulation models. Mathematics. 11(4), 845 (2023). https://doi.org/10.3390/math11040845
6. Stephens, M., Kadijevich, D.M.: Computational/algorithmic thinking. In: Lerman, S. (ed.) Encyclopedia of Mathematics Education, pp. 1–6. Springer, Cham (2020). https://doi.org/10.1007/978-3-030-15789-0_100044

Optimizing Drone's Energy in Single Truck Multiple Drones Last Mile Delivery

Ornela Gordani[1], Eglantina Kalluci[1], and Fatos Xhafa[2]

[1] Faculty of Natural Sciences, University of Tirana, Tirana, Albania
{ornela.gordani,eglantina.kalluci}@fshn.edu.al
[2] Department of Computer Science, Universitat Politècnica de Catalunya,
Barcelona, Spain
fatos@cs.upc.edu

Abstract. The rapid growth of online shopping and logistics have significantly raised the question of how to adequately meet customer delivery expectations. Considering the advancements made in drone technology, this challenge could be addressed more efficiently than using traditional vehicle-based delivery technologies. According to analysts, the use of aerial drones can be beneficial as they are capable of lowering the cost of transportation, avoiding stagnation on roads by flying over traffic congestion, and being non-polluting to nature by not consuming any fossil fuel. However, drone delivery optimization models are distinguished for their complexity, due to the nonlinear nature of the energy constraints. In this study, we analyze the computational complexity of the optimization model from [1,2] and focus on the linearization of the energy consumption constraints, which are a nonlinear function of the payload and the travel time. We discuss and analyze various linearization techniques to reduce the resolution time especially for large data size problem instances.

Keywords: Last mile delivery · Drones · Single truck · Energy consumption · Gurobi solver · Linearization

1 Introduction

The last mile delivery is often the most costly, complex and essential segment of the logistics and supply chain management. Its importance is accentuated in view of the increasing customer expectations regarding quicker and efficacious deliveries. The trends such as moving to more technology-focused last mile delivery solutions necessitate novel integration of trucks with drones for creating multimodal delivery systems.

Trucks are able to withstand higher payloads and transport goods over long distances while drones on the other hand are better in transporting goods over short distances with speed and flexibility. Trucks can be used as a flight platform for drones by supporting their take-off and landing and as charging station for battery charge for various deliveries. One critical issue in the last mile drone

M. Pavone et al. (Eds.): DSA ISC 2025, LNCS 16405, pp. 372–384, 2026.
https://doi.org/10.1007/978-3-032-21811-7_26

delivery is the energy consumption of a drone. The path delivery planning *must* ensure the feasibility of the required energy to complete the delivery. This is a challenging problem due to the fact that the consumed energy can be affected by a number of parameters such as payload and drone's weight itself, traveled distance, meteorological conditions, etc. In this paper, the problem is formulated as an optimization model based on minimizing the energy consumed by drones as a function of the payload and the travel distance [1].

Our contribution in this paper is in the linearization of the nonlinear constraints by transforming a Mixed Integer Nonlinear Programming (MINLP) to a Mixed Integer Linear Programming (MILP) problem. This linearization aims to improve the computational efficiency and scalability, making the model more flexible for real-life applications. Indeed, the linear approximation model could make possible finding feasible solutions much faster than using the nonlinear energy model. The linearization of the problem enables the use of standard state of the art MILP solvers. Thus, for this purpose, various linearization techniques such as first-order Taylor series expansion, piecewise linear approximation, log-linearization, McCormick linearization, second-order cone programming and others can be used. Such linearization techniques are analyzed for the linearization of nonlinear energy constraints and empirical trade-offs are provided.

The rest of this paper is organized as follows. Section 2 presents the terminology and notations used in the modeling of the problem as an MINLP optimization problem. Section 3 analyses the complexity of the MINLP and its corresponding MILP version. The linearization of the energy constraints is approached in Sect. 4. Section 5 presents the experimental evaluation plan. We end the paper in Sect. 6 with conclusions and outlook for future work.

2   Terminology, Notations and Mathematical Modelling

In the last mile drone delivery problem, there is a truck, which departs from a central depot and delivers packages to the customers in a given geographical area. The truck has with it several drones, for short-range deliveries, considering the fact that the weight of the packages is substantially less (limited payload) and the drone has available energy to deliver the package(s) and return to the truck.

2.1   Terminology

We formulate the problem as an optimization mathematical model based on models from [1,2]. The following terminology is used:

1. *Drone Routing Problem:* refers to the optimization problem of determining the best path for a drone to visit more than one location while minimizing energy consumption or travel time.
2. *Energy Consumption Function:* is a function that models the energy required by the drone to complete a delivery. It depends on payload weight, distance traveled and environmental factors such as wind.

3. *Traveled Distance:* refers to the distance between customers that the drone needs to cover during its route. It is used as a factor in the energy consumption model.
4. *Payload Weight:* assuming that drones have limited payload, this refers to not passing the drone's payload limit.
5. *Linearization:* is the process of converting nonlinear optimization problems into a linear form, using methods like piecewise linear approximation, log-linearization, McCormick linearization, etc.
6. *Energy constraints:* ensure that the total energy consumed during the drone's flight does not exceed the available energy battery capacity.

2.2   Notations and Mathematical Model

The proposed model is defined as a graph $G(N, A)$, where $N = \{0, 1, \ldots, n+1\}$ is the set of nodes and A the set of edges connecting the nodes (the truck and the delivery points). Node 0 represents the truck, node $n+1$ represents the returning truck (it is a copy of node 0). $N' = \{1, 2, \ldots, n\}$ denotes the set of customers. The set of arcs $A = \{(i, j) : i \in \{0\}, j \in N' \land i \in N', j \in N^-, i \neq j\}$ where $N^+ = \{0, \ldots, n\}$ and $N^- = \{1, \ldots, n\}$. The sets $V^-(i)$ and $V^+(i)$ represent node i's predecessor and successor nodes, respectively.

We use the following notation for the MINLP optimization model formulation, in which there are parameters whose values are a priori known and variables to be computed for the delivery route solution.

Parameters:

- n: the number of customers
- K: the number of (homogeneous) drones
- Q: maximum payload of a drone (in Kg)
- d_i: demand of customer i (in Kg)
- $[a_i, b_i]$: time window for customer i
- t_{ij}: travel time from node i to node j
- c_{ij}: travel cost from node i to node j
- W: drone frame weight ((in Kg, or weight units, in general)
- m: battery weight ((in Kg, or weight units, in general)
- ρ: air density
- g: gravity acceleration
- ζ: spinning blade disc area
- h: number of rotors per drone
- k: constant for power consumption model
- σ: maximum battery energy capacity
- M'_{ij}, M''_{ij} and M'''_{ij}: large constants used for constraints
- δ: battery-related cost

Variables:

- x_{ij}: 1 if drone traverses arc (i, j), 0 otherwise
- z_{ij}: 1 if a trip ending at i is followed by another trip starting at j, 0 otherwise
- q_{ij}: product weight carried on arc (i, j) (the *payload*)
- τ_i: start of service time at node i
- f_i: accumulated energy consumption of a drone upon arrival at node i
- e_{ij}: energy consumption on arc (i, j)

Objective Function: Minimize the total energy consumption of drones.

$$\min \sum_{(i,j) \in A} (c_{ij} x_{ij} + \delta e_{ij}) \tag{1}$$

1. *Each customer is visited exactly once*:

$$x_{ij} = 1 , \quad \forall i \in N' \tag{2}$$

$$x_{ji} = 1, \quad \forall i \in N' \tag{3}$$

2. *The number of departures from the truck equals the number of arrivals:*

$$x_{0j} = \ x_{j,n+1} \tag{4}$$

3. *Weight constraints*: Each customer's demand must be satisfied, and also sub-tours should be eliminated.

$$q_{ij} - q_{ji} = d_j, \quad \forall j \in N' \tag{5}$$

Note that drone payload capacity along an arc $(i, j) \in A$ should not be exceeded:

$$q_{ij} \leq Q x_{ij}, \quad \forall (i, j) \in A \tag{6}$$

4. *No product is carried to the returning truck:*

$$q_{i,n+1} = 0, \quad \forall i \in N' \tag{7}$$

5. *Drone energy constraints*: Dorling *et al..* [3], based on Leishman *et al..* [4] description of the energy consumption of a single-rotor helicopter in flight as a convex function of the payload q, and assuming that each rotor shares the total weight of a drone equally, proposed the Eq. (8) for the power consumption:

$$P(q_{ij}) = (W + m + q_{ij})^{\frac{3}{2}} \sqrt{\frac{g^3}{2\rho\varsigma h}}, \quad \forall i,j = 1...n \tag{8}$$

where W is the frame weight (Kg), m is the battery weight (Kg), q_{ij} is the payload (Kg), g is the force due to gravity (N), ρ is the fluid density of air (Kg/m^3), ς is the area of spinning blade disc (m^2), h is the number of rotors, and the unit of P is Watt.

Every time a drone begins a new delivery trip, we swap it with a fully charged battery, thus at the beginning of each trip the accumulated energy consumption is 0:

$$f_0 = 0 \tag{9}$$

6. *Energy consumption during a delivery trip*:

$$f_i + k'\left(W + m + q_{ij}\right)^{\frac{3}{2}} t_{ij} \leq M_{ij}\left(1 - x_{ij}\right) + f_j, \qquad \forall(i,j) \in A \tag{10}$$

where k' is a constant that includes k from earlier definition and the conversion from *Watt-second* to *kWh*. M_{ij} is an arbitrary large constant, set to $M_{ij} = k'(W+m)^{\frac{3}{2}} t_{ij} + \sigma$ (because when $x_{ij} = 0$, then $q_{ij} = 0$, from Eq. (6)), thus ensuring that the energy battery capacity is not exceeded:

$$f_{n+1} \leq \sigma \tag{11}$$

7. *Time and trip constraints:* The time relationship between nodes (customer i and its immediate successor j) is expressed as:

$$\tau_i + t_{ij} - M''_{ij}\left(1 - x_{ij}\right) \leq \tau_j \qquad \forall i \in N', \; j \in N^- \tag{12}$$

where M''_{ij} is the large constant: $M''_{ij} = max\left\{b_i + t_{ij} - a_j, 0\right\}$. This requires that the time window constraint should be satisfied:

$$a_i \leq \tau_i \leq b_i \qquad \forall i \in N^- \tag{13}$$

Likewise, a constraint establishing the time relationship between consecutive trips performed by the same drone:

$$\tau_i + (t_{i,n+1} + t_{0j}) \leq \tau_j + (1 - z_{ij}) M'''_{ij} \qquad \forall i, j \in N', \; i \neq j \tag{14}$$

where $M'''_{ij} = t_{i,n+1} + t_{0j} + b_i$.

Additionally, there should be ensured that the trip connections match the departures:

$$\sum_{j \in N', i \neq j} z_{ij} \leq x_{0j}, \qquad \forall j \in N' \tag{15}$$

$$\sum_{j \in N', i \neq j} z_{ij} \leq x_{i,n+1}, \qquad \forall i \in N' \tag{16}$$

And, finally, there is a limit K on the total number of drones that can be used for the deliveries:

$$\sum_{j \in N'} x_{0j} - \sum_{i \in N'} \sum_{j \in N', i \neq j} z_{ij} \leq K \tag{17}$$

3   Complexity Analysis of the Mathematical Model

The resulting model based on [1,2] is a Mixed-Integer Nonlinear Programming (MINLP) due to integer variables and the nonlinear constraints (see Eq. (10). MINLP is known as an NP-hard problem, meaning that finding an exact optimal solution is computationally expensive. MINLP problems arise quite often in the energy optimization domain [5] due to the nonlinear nature of the energy constraints. In the case of the last mile drone delivery, the presence of integer variables, combined with nonlinear constraints leads to an exponential increase in resolution time as the number of customers and drones increase. Specifically, the drone energy constraint that is related to the energy consumptions of the drone during the trips, has a nonlinear term, which affects computational complexity of the optimization model:

$$k'(W + m + q_{ij})^{3/2} \tag{18}$$

While MINLP can be solved with nonlinear solvers such as GEKKO [6] and GUROBI [7], due to the complexity of MINLP models, many works in the literature have been reported to reduce the models complexity. Linearization methods, like piecewise linearization, can be used to linearize the nonlinear constraint of Eq. (18). This transformation, from MINLP to MILP, enables the use of linear solvers, such as Gurobi Optimizer [7], which is one of the most advanced MILP solvers to handle the problem more efficiently, in a reasonable time even for very large problem sizes.

It should be noted, however, that linearization comes at the expense of losing the accuracy of the solution (sub-optimality) and it also could introduce a number of new variables and constraints to the MILP model, increasing thus the size of the problem instance to be solved. Indeed, for some linearization models such as the piecewise linear model, there are $O(n^2)$ nonlinear constraints from Eq. (8). If we assume $O(L)$ new variables for every nonlinear term, that would amount to $O(L \cdot n^2)$ additional variables, where n is the number of vertices in the underlying graph of the problem. It should be noted that there are n^2 energy constraints from Eq. (8). The gain, however, is expected to be in the power of MILP solvers. In the next section, we analyze various linearization techniques for the nonlinear energy function to empirically find a tradeoff between the accuracy and the number of newly added variables and constraints. The linear approximation replaces the original energy constraints in Eqs. (10) and (11) of the MINLP model yielding thus to a MILP model.

4   Linearization of the Energy Constraints

In the optimization problem in this study the energy consumption constraint (see Eq. (8)) is nonlinear, which implies a high computational complexity and yielding the problem a MINLP model. Thus, to improve the computational efficiency, linearization techniques can be applied to transform MINLP into MILP problems [8].

4.1 Linearization Types

There are different types of well-known linearization techniques of nonlinear functions, which have become commonplace in the literature. We briefly describe some of them and further refer the reader to some useful references.

First-Order Taylor Series Expansion. One of the most common methods to approximate a nonlinear function by a linear function around a given point u_0 is the Taylor Series expansion. The first-order approximation of a function $f(u)$ is given by:

$$f(u) \approx f(u_0) + f'(u_0)(u - u_0) \tag{19}$$

where $f'(u_0)$ is the first derivative of function f at u_0. This method is effective only for potential good approximation in the interval around the chosen point u_0. On the other hand, this model does not increase the number of variables and constraints of the original optimization model.

Piecewise Linearization (PWL). Piecewise linearization (PWL) approximates a nonlinear function using multiple linear segments. In this method we find L linear segments between breakpoints $u_1, \ldots, u_L$ and the slopes of the line segments have values $f(u_1), \ldots, f(u_L)$:

$$f(u) \approx \sum_{i=1}^{L} \lambda_i f(u_i) \tag{20}$$

where λ_i are non-negative weights such that:

$$\sum_{i=1}^{L} \lambda_i = 1. \tag{21}$$

$$u = \sum_{i=1}^{L} \lambda_i u_i. \tag{22}$$

By denoting $\omega_{ij} = W + m + q_{ij}$, we rewrite the nonlinear constraint as $f(\omega_{ij}) = \omega_{ij}^{3/2}$ and approximate it by a piecewise linear model (Eqs. (20)-(22)). As indicated earlier, the piecewise linearization introduces a number of $O(L \cdot n^2)$ new variables, for L breakpoints, while the number of constraints remains of the same $O(n^2)$ order as in the original optimization model.

Log-Linearization. This technique is useful for functions involving products or powers, consisting in taking the logarithm of a nonlinear function to transform it into a linear form, by defining $y_{ij} = \omega_{ij}^{3/2}$, then $\log y_{ij} = \frac{3}{2} \log \omega_{ij}$ (see for instance [9]). It can be applied to our nonlinear energy constraint given that its values are in the positive domain. This linearization technique has the advantage that it does not add new variables or constraints to the model. Essentially, every constraint is replaced by its corresponding linear version.

Second-Order Cone Programming (SOCP). This approach is useful for convex quadratic constraints or power functions of the form:

$$x_1^2 + x_2^2 \leq b \quad \text{or} \quad \|Ax + b\|_2 \leq c^T x + d \tag{23}$$

The constraint in our model is not quadratic as it has a fractional power, but it can be rewritten in a way that fits with the SOCP approach.

McCormick Linearization. This version is useful to linearize bilinear terms, mostly found in quadratic constraints. We omit details as it is not applicable to our case.

Other interesting works in the literature about linearization techniques of nonlinear functions are [5,10] and [11]. In [5] the authors propose piecewise linearization to be the best one because it allows the problem to be solved using linear solvers. In [9] the authors propose the use of LP (Linear Programming) to find the best piecewise linearization of nonlinear functions. The authors in [10] propose logarithmic formulations, while in [11] there is proposed an optimal Polygonal L_1 Linearization. The reader is referred to [12] for a review of linearization techniques in optimization.

4.2   Numerical Evaluation of the Piecewise Linearization

In the following, we numerically analyze the piecewise linearization (PWL). Before computing the piecewise linear function that approximates the nonlinear term, we have to compute the segment $[A, B]$ where the function $f(\omega_{ij}) = \omega_{ij}^{3/2}$ is defined over as follows: $A = W + m + \min\{q_{ij}\}$ and $B = W + m + \max\{q_{ij}\}$, where W (drone frame weight in Kg) and m (battery weight in Kg) are from the standard definition of drones [2].

Since $\min\{q_{ij}\} = 0$ given that all $q_{ij} \geq 0$ by definition (q_{ij} is the product weight carried on arc (i, j)), therefore $A = W + m$, while $\max\{q_{ij}\}$ is computed from results in [2]. Thus, the values of A and B are $A = 3$ and $B = 4.5$ (assuming $m = 1.5$ Kg, $W = 1.5$ Kg, $Q = 1.5$ Kg taken from [2]), so the energy term $f(\omega_{ij})$ is defined and linearized over the segment $[3, 4.5]$.

The accuracy of the piecewise linear approximation depends on L; therefore, for the purpose of exemplification, we consider different values $L = 3, 5, 6$ for the number of breakpoints $u_1, u_2, \ldots, u_L$ to be selected in the segment $[A, B]$. Table 1 presents the error values (%) for $L = 3, 5, 6$ (Accuracy(%) = 100% + Error(%)) computed at $u = 3.75$, showing that the accuracy increases with larger number of breakpoints L, as expected. The true value of the nonlinear term at $x = 3.75$

is $(3.75)^{3/2} \approx 7.26184377$. Error is computed as:

$$\frac{\mathtt{ApproxValue} - \mathtt{OriginalValue}}{\mathtt{OriginalValue}} \times 100.0.$$

Table 1. Piecewise linear approximation results at $x = 3.75$ for $L = 3, 5, 6$.

L	True Value	Approximate	Error (%)	Accuracy (%)
3	7.2618	7.2739	0.1667	100.1667
5	7.2618	7.2662	0.0600	100.0600
6	7.2618	7.2618	0.0000	100.0000

Let $f(x) = x^{3/2}$ be the original nonlinear function and $\hat{f}_L(x)$, its linear approximation. It should be noted that:

- **Distribution of the breakpoints**: The L breakpoints are uniformly distributed in the segment.
- **Error formula and overestimation**: The computed error is based on the difference between the approximation and the true value, relative to the true value. Since $f(x) = x^{3/2}$ is convex, the secant line $\hat{f}_L(x)$ serves as an upper bound (overestimation), ensuring the error $(\hat{f}_L - f)$ is nonnegative.
- **Accuracy calculation**: For an overestimation where the error is positive, the accuracy is calculated as Accuracy(%) = 100% + Error(%), resulting in values greater than 100%.
- **Perfect accuracy at breakpoints**: For $L = 6$, the evaluation point $x = 3.75$ is exactly one of the defined breakpoints. At this point, the piecewise linear function $\hat{f}_L(x)$ is designed to pass through the true function value $f(x)$, leading to an error of 0% and an accuracy of 100%.

The results confirmed that the approximation is indeed an overestimator (the linear function lies above the true convex curve) and therefore ensures feasibility preservation when used in a minimization model

$L = 6$ is found a suitable number of breakpoints (see Table 1). Figure 1 shows a graphical representation of accuracy *vs.* the number of breakpoints L at $x = 3.75$, where we can see that the error reduces with increase in L. Increasing the number of breakpoints L would increase the accuracy to the cost of solver's computation time and it is therefore not worth.

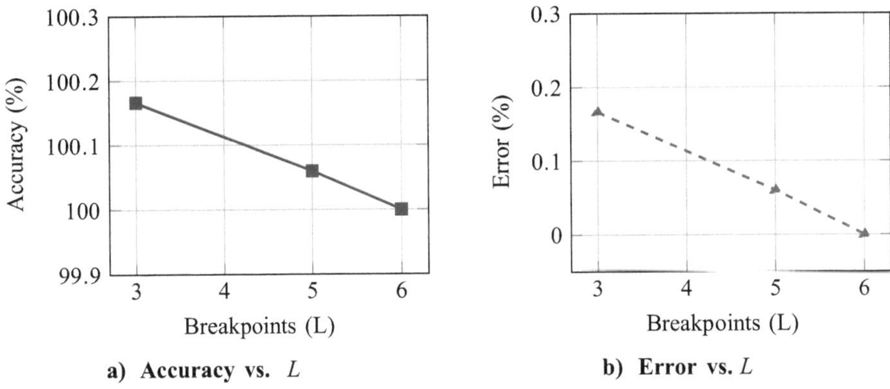

a) **Accuracy vs.** L b) **Error vs.** L

Fig. 1. Performance of the piecewise linear approximation: (a) Accuracy and (b) Error percentage for different numbers of breakpoints L.

5 Computational Evaluation Plan

The computational evaluation plan is to carry out an experimental computational study to compare the results of the MINLP (nonlinear version) with MILP (linearized version) using a benchmark of instances from Cheng *et al.* [2]. Besides the Gurobi solver used in [2], we plan to evaluate the MILP (linearized version) with other solvers through the PuLP[1] software.

5.1 Description of Problem Instances

A benchmark of instances, namely the *SetA* instances, is presented in [2]. The customer demands are drawn from two uniform distributions, namely, from $[0.1, 0.7]$ for the first 40% of customers, and $[0.1, 1.5]$ for the remaining customers. Customer delivery time windows are randomly generated within certain constraints and are dependent on the travel time between depots and clients, guaranteeing that all deliveries are completed within a predetermined time frame. The values of other relevant parameters are as follows [2]:

$$\delta = 360\$/kWh, \ \rho = 1.204 \text{kg/m}^3, \ \zeta = 0.0064\text{m}^2, \quad h = 6, \ \sigma = 0.27\text{kWh} \quad (24)$$

The authors in [2] employed Branch & Cut algorithms in Python using Gurobi (computing time limit set to four hours, which is rather large for dynamic problems).

---

[1] PuLP is a linear and mixed integer programming modeler in Python. https://coin-or.github.io/pulp/.

Small/Mid/Large Size Problem Instances. Instances of Set A can be classified into three groups:

- Small size instances: 2 drones/10 customers and 3 drones/10 customers.
- Mid-size instances: 4 drones/20 customers, 4 drones/25 customers, 5 drones/30 customers and 5 drones/35 customers.
- Large size instances: 6 drones/40 customers, 6 drones/45 customers, 7 drones/45 customers and 8 drones/45 customers.

5.2   Issues with Linearization of Energy Constraints

Linearization is deemed to be a powerful tool for converting MINLP to MILP to gain efficiency, but it can also have various drawbacks, briefly discussed next.

Infeasibility. By linearizing the nonlinear energy constraints, the original problem could become infeasible as linearization inherently involves approximations [13,14]. In our case, the nonlinear energy constraint is a convex function and thus the nonlinear version can have multiple local optima, which linearization might not capture, leading to infeasibility. In our evaluation plan we include the study of the infeasibility and dealing with it to identify infeasible constraints and bring the problem to a feasible one following studies in the literature [13,14].

Loss of Accuracy. Linearization could cause a loss of accuracy in computing the optimal solution, so a tradeoff between loss of accuracy and computing faster should be found. In particular, for piecewise linearization, global optimality is no longer guaranteed, therefore it is important to find a reasonably good feasible solution.

Increased Instance Sizes. Linearization of nonlinear energy constraints, for n customers, adds $O(L \cdot n^2)$ new constraints, for a constant L. Thus, while the linearization contributes to reducing the solvers time, the increase in instance size requires an evaluation when the number of customers is large.

6   Conclusions

In this study, we have addressed the linearization of nonlinear energy constraints in the optimization problem of last mile drone delivery. The energy consumption is a critical issue to ensure viability of the drones deliveries and drones return to the truck. This is a challenging problem given that it is influenced by a number of uncertain factors such as wind speed and direction, air humidity, weather conditions, etc., besides the payload, battery capacity, among other factors. We analyzed the complexity of the mathematical model and consider linearization to reduce its high complexity. By linearizing the energy constraints, the aim is to convert mixed integer nonlinear programs (MINLP) into mixed integer linear programs (MILP) and thus drastically reduce the execution time and respond in real time to the dynamic nature of the problem at the cost of a less accurate (non-optimal) solution. We have discussed several linearization models and issues arising with the linearization.

In future work we would like to complete the evaluation plan (see Sect. 4 above). Additionally, we plan to extend this work to consider energy functions that include other parameters such as wind, temperature, etc. to express the impact of weather conditions on drone battery consumption [15]. Also, we would like to consider Agile Optimization solvers based on simheuristics and high-performance computing [16] to shorten the resolution time and respond to the dynamic nature of the problem. Finally, we would like to use Linear Regression analysis and/or other suitable Machine Learning techniques such as Reinforcement Learning (e.g. [17] and [18]) to predict a drone's energy consumption instead of adding nonlinear constraints to the optimization model.

Acknowledgment. This research work has been supported by the *READ Project of the Albanian-American Development Foundation (AADF)*, Albania.

References

1. Chen, Y., Baek, D., Bocca, A., Macii, A., Macii, E., Poncino, M.: A case for a battery-aware model of drone energy consumption. In: 2018 IEEE International Telecommunications Energy Conference (INTELEC), pp. 1–8. IEEE, Italy (2018). https://doi.org/10.1109/INTLEC.2018.8612333
2. Cheng, C., Adulyasak, Y., Rousseau, L.M.: Drone routing with energy function: formulation and exact algorithm. Trans. Res. Part B **139**, 364–387 (2020). https://doi.org/10.1016/j.trb.2020.06.011
3. Dorling, K., Heinrichs, J., Messinger, G.G., Magierowski, S.: Vehicle routing problems for drone delivery. IEEE Trans. Syst. Man Cybernet. Syst. **47**(1), 70–85 (2017). https://doi.org/10.1109/TSMC.2016.2582745
4. Leishman, G.J.: Principles of Helicopter Aerodynamics, 2nd edn. Cambridge University Press, New York (2006). Chapters 2 and 3. ISBN 978-0-521-85860-1
5. Ngueveu, S.U.: Piecewise linear bounding of univariate nonlinear functions and resulting mixed integer linear programming-based solution methods. Eur. J. Oper. Res. **275**(3), 1058–1071 (2019). https://doi.org/10.1016/j.ejor.2018.11.021
6. Beal, L.D.R., Hill, D., Martin, R.A., Hedengren, J.D.: GEKKO optimization suite. Processes **6**(8), 106 (2018). https://doi.org/10.3390/pr6080106
7. Gurobi Optimizer. https://www.gurobi.com/. Accessed 25 May 2025
8. Geißler, B., Martin, A., Morsi, A., Schewe, L.: Using piecewise linear functions for solving MINLPs. In: Lee, J., Leyffer, S. (eds.) Mixed Integer Nonlinear Programming. The IMA Volumes in Mathematics and its Applications, vol. 154. Springer, New York (2012). https://doi.org/10.1007/978-1-4614-1927-3_10

9. Mazarei, M.M., Behroozpoor, A.A., Kamyad, A.V.: The best piecewise linearization of nonlinear functions. Appli. Math. **5**, 3270–3276 (2014). https://doi.org/10.4236/am.2014.520305
10. Huchette, J., Vielma, J.P.: Nonconvex piecewise linear functions: advanced formulations and simple modeling tools. arXiv:1708.00050 (2019). https://doi.org/10.48550/arXiv.1708.00050
11. Gallego, G., Berjón, D., García, N.: Optimal polygonal L1 linearization and fast interpolation of nonlinear systems. IEEE Trans. Circ. Syst. I **61**(11), 3225–3234 (2014). https://doi.org/10.1109/TCSI.2014.2327313
12. Asghari, M., Fathollahi-Fard, A.M., Mirzapour Al-e-hashem, S.M.J., Dulebenets, M.A.: Transformation and linearization techniques in optimization: a state-of-the-art survey. Mathematics **10**, 283 (2022). https://doi.org/10.3390/math10020283
13. Guieu, O., Chinneck, J.W.: Analyzing infeasible mixed-integer and integer linear programs. INFORMS J. Comput. **11**, 63–77 (1999). https://doi.org/10.1287/ijoc.11.1.63
14. Witzig, J., Berthold, T., Heinz, S.: Computational aspects of infeasibility analysis in mixed integer programming. Math. Prog. Comp. **13**, 753–785 (2021). https://doi.org/10.1007/s12532-021-00202-0
15. Gürel, Ö., Serdarasan, S.: Drone-assisted last-mile delivery under windy conditions: zero pollution solutions. Smart Cities **7**, 3437–3457 (2024). https://doi.org/10.3390/smartcities7060134
16. Li, Y., Peyman, M., Panadero, J., Juan, A.A., Xhafa, F.: IoT analytics and agile optimization for solving dynamic team orienteering problems with mandatory visits. Mathematics **10**, 98 (2022). https://doi.org/10.3390/math10060982
17. Chen, S., Mo, Y., Wu, X., Xiao, J., Liu, Q.: Reinforcement learning-based energy-saving path planning for UAVs in turbulent wind. Electronics **13**, 3190 (2024). https://doi.org/10.3390/electronics13163190
18. Bi, Z., Guo, X., Wang, J., Qin, S., Liu, G.: Truck-drone delivery optimization based on multi-agent reinforcement learning. Drones **8**, 27 (2024). https://doi.org/10.3390/drones8010027

Digital Twins for Urban Mobility: Enabling Real-Time Decision Making in Society 5.0

Andrea Grotto[1,2]($\boxtimes$) and Pau Fonseca i Casas[2]

[1] Institute for Renewable Energies, Eurac Research, Viale Druso 1, 39100 Bolzano, Italy
`Andrea.grotto@eurac.edu`
[2] Department of Statistics and Operations Research, Universitat Politècnica de Catalunya (UPC), Campus Nord, C/Jordi Girona 1-3, 08034 Barcelona, Spain

Abstract. Society 5.0 represents a vision where technology and human decision-making work together to address complex social challenges. This research examines how Digital Twins can function as effective decision-support tools in this emerging societal framework, using Urban Mobility as an illustrative case. The principles presented for Urban Mobility Digital Twins (UMDTs) can be applied to many complex systems. The digital twin approach extends beyond specific domains—whether managing water resources, optimizing energy distribution, or planning healthcare services—by creating virtual counterparts that learn continuously from real-world data. Digital twins transform how cities approach mobility planning by converting data into practical insights. Planners can visualize potential changes, test different scenarios, and understand impacts on both technical performance and community experiences before implementing solutions in the physical world. This methodology aligns with Society 5.0's central idea: using technology to enhance human wellbeing. By developing accessible, validated models that connect technical and social considerations, digital twins help create shared understanding among diverse stakeholders and support more informed decision-making.

Keywords: Society 5.0 · Industry 4.0 · Digital Twin · Decision Making · Urban Mobility

1 Introduction

Modern cities face increasingly complex mobility challenges. As urban populations grow, traditional transportation systems struggle to manage rising demands, leading to congestion, pollution, and accessibility issues. New technologies, however, offer innovative ways to completely rethink these systems [1].

Digital twins represent one of the most promising innovations in this field. These dynamic virtual models aren't simply static simulations, they're digital replicas that continuously mirror real physical systems. While this study uses Urban Mobility as an illustrative case, the Digital Twin concept applies to numerous complex systems, from healthcare services to water resource management.

M. Pavone et al. (Eds.): DSA ISC 2025, LNCS 16405, pp. 385–394, 2026.
https://doi.org/10.1007/978-3-032-21811-7_27

Born within Industry 4.0, where they transformed industrial processes and logistics, Digital Twins are now finding new life in Society 5.0, a social vision that seamlessly integrates technology and human needs. In this new context, they evolve from purely technical tools to facilitators that connect complex data to concrete decisions that can improve daily life [2].

This work explores how digital twins function as decision-support tools, bridging the technical capabilities of Industry 4.0 with the human-centered goals of Society 5.0. Through the case of Urban Mobility, the paper illustrates how these virtual replicas transform how cities approach transportation challenges, allowing decision-makers to visualize, test, and evaluate the impact of their choices before real-world implementation. Building on experience from [3], this research demonstrates how these tools make complex decisions more accessible, transparent, and participatory, enabling diverse stakeholders to collaborate on solutions that balance technical requirements with social benefits.

2 Society 5.0

Society 5.0 [4] represents a future-oriented vision for a human-centered society that integrates cyberspace and physical space to achieve both economic advancement and the resolution of social challenges. This concept originated in Japan, emerging in 2015 within the framework of the 5th Basic Plan for Science and Technology [5], and builds upon the preceding societal stages. As illustrated in Fig. 1, which depicts the evolutionary relationship between industrial revolutions and societal development, there is a clear parallel progression between technological advances and social transformation. The figure shows how each industrial stage has influenced or contributed to a corresponding stage of society:

The above shows how each industrial stage has influenced or contributed to a corresponding stage of society:

- Society 1.0 (Hunter-Gatherer Society): Focused on survival through hunting and gathering.
- Society 2.0 (Agricultural Society): Characterized by settled agriculture and the development of communities.
- Society 3.0 (Industrial Society): Marked by mass production, industrialization, and technological advancements.
- Society 4.0 (Information Society): Defined by the widespread use of the internet and information technology.
- Society 5.0 (Super Smart Society): Aims to leverage technologies like artificial intelligence (AI), the Internet of Things (IoT), robotics, and big data to create a society where everyone can live a comfortable and vibrant life.

This evolutionary framework helps clarify that Industry 4.0 [6] and Society 5.0, though related concepts, represent different developmental phases: Industry 4.0 constitutes the current phase of industrial development focused on digitalization and automation, while Society 5.0 embodies the future vision of an intelligent society that integrates Industry 4.0 technologies into all aspects of social life. Society 5.0 is characterized

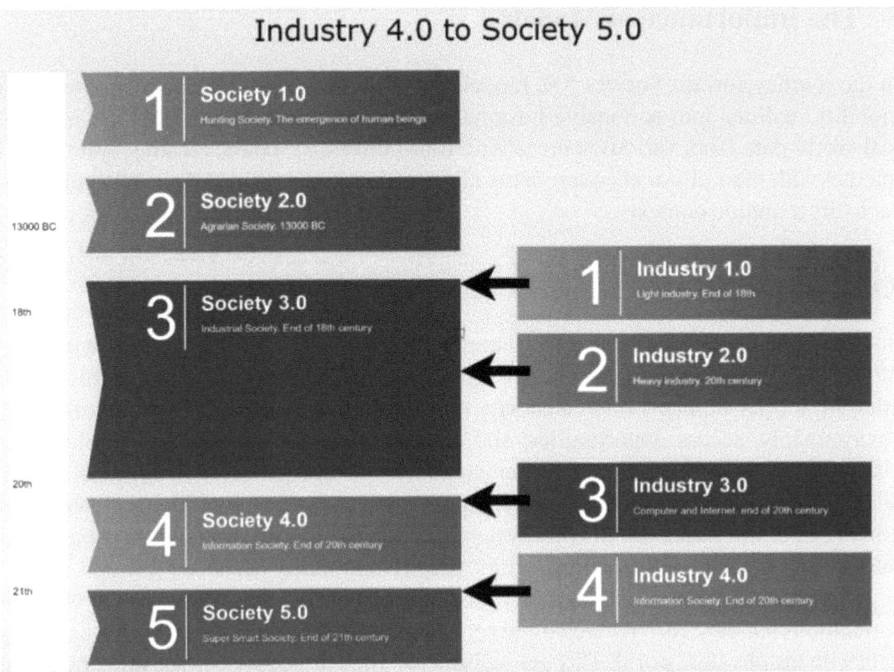

Fig. 1 Industrial Revolutions and Society Evolution: From Industry 4.0 to Society 5.0, own work, adapted from [4].

by several distinguishing features that collectively redefine the relationship between technology and society, moving beyond purely technological advancement to create a human-centered framework where digital innovations directly address social challenges:

- Human-centeredness: Prioritizing human well-being and addressing societal needs.
- Integration of cyberspace and physical space: Seamlessly merging digital and physical realms to optimize processes and experiences.
- Data-driven solutions: Utilizing vast amounts of data to solve complex problems and create personalized services.
- Sustainability: Promoting environmentally friendly practices and resource management.
- Resilience: Building robust systems capable of withstanding disruptions and adapting to change.
- Inclusion: Ensure that all members of society can benefit from the technologies.

The main objectives of Society 5.0 include efficiency and sustainability, inclusivity and well-being, resolution of social problems, personalization, and adaptability. These objectives are particularly relevant in the context of urban mobility, where challenges of sustainability, inclusion, and operational efficiency are priorities.

3 The Importance of Models

In the journey toward Society 5.0, models serve as tools that translate complex urban mobility realities into actionable information for decision-making. UMDTS integrate real-world data from various sources with transportation models, creating systems that interact with their physical counterparts and enable more informed decisions tailored to each city's unique context.

3.1 Evolution of Urban Mobility Models

Society 5.0 [4] envisions a world where technology seamlessly integrates with daily life to solve social challenges. Human mobility stands at the very center of this vision, after all, a truly human-centered society must address how people move through their environments, access opportunities, and connect with one another. The evolution of Urban Mobility models reflects this human-centered progression [7]. Traditional models once offered static snapshots of traffic patterns, but today's Digital Twins create living mirrors of entire transportation ecosystems, capturing not just vehicle movements but the human choices, needs, and behaviors that drive mobility decisions [2].

Digital Twins takes urban mobility modeling to new heights. These aren't just static simulations but they're dynamic virtual replicas of transportation networks that exchange data with the physical world. Through sensors installed throughout cities, these twins can track vehicle movements in real time. When sensors detect changes in traffic patterns, that information flows directly into the Digital Twin. The model then processes this real-world data, adapts its understanding, and improves its predictions for similar situations in the future, enabling more informed decisions by stakeholders

This continuous learning loop, from the physical world to digital model and back again, represents the blending of real and virtual spaces that Society 5.0 envisions. It's not just about collecting data; it's about creating a two-way relationship between the physical city and its digital counterpart that ultimately serves both the people navigating these spaces and the decisions-makers planning them.

3.2 From Technical Tools to Social Catalysts

What makes UMDTs especially relevant to Society 5.0 is their evolution from purely technical instruments to social catalysts. These systems prove valuable not just as traffic forecasting tools but as platforms where diverse stakeholders can visualize, discuss, and shape how people move through their city.

For instance, in [3], the UMDT conceptual framework showed how such tools could transform stakeholder engagement with mobility technologies. The study demonstrated how, when fully implemented, UMDTs would allow stakeholders to evaluate new mobility solutions like electric vehicles, shared mobility services, and charging infrastructures through clear visualizations of potential emissions reductions, energy savings, and traffic improvements. This approach bridges technical analysis with everyday concerns, making complex mobility planning more accessible to participants with varying levels of expertise.

3.3 Human Mobility as a Core Society 5.0 Domain

Society 5.0 recognizes that the movement of people, not just data, goods, or vehicles, forms an essential foundation of social well-being [8]. How people navigate their environments directly impacts their access to opportunities, social connections, health outcomes, and quality of life. This places urban mobility squarely at the heart of Society 5.0's mission.

When stakeholders and particularly planners use Urban Mobility Digital Twin models, the primary measure of success isn't vehicle efficiency but rather improved human experiences: is mobility sustainable for future generations? Are emissions affecting citizens' health? Are people waiting too long to access essential services? These questions connect technical modeling to economic well-being, health, and accessibility, core values of Society 5.0.

Digital Twins enable planners to see mobility not as an isolated technical system but as a social fabric woven through daily life. specifically answers these three questions, demonstrating how digital twins can translate complex mobility data into insights that matter to communities. This perspective shift, from moving vehicles to connecting people, exemplifies how Society 5.0 reorients technology toward human needs.

3.4 Open Models: Democratizing Society 5.0

Society 5.0 cannot succeed if its benefits reach only technical elites. This is where the concept of open models becomes essential [9]. Well-designed validation protocols create frameworks where complex Digital Twin operations can be documented, examined, and understood by stakeholders beyond the technical team.

When mobility models are opened, made transparent, accessible, and interactive, they fulfill a core Society 5.0 promise: technology that empowers rather than mystifies. This can take shape as tiered interface systems. In UMDT, traffic experts can access deep parameter adjustments, city officials can run scenario comparisons, and ordinary citizens can explore basic "what-if" questions through simplified map-based tools: "What if the bus came every 10 min instead of 15?" or "How would a bike lane on my street affect travel times?"

This democratization represents a shift from Society 4.0 approaches, where mobility algorithms often operated as "black boxes" [10] whose decisions were difficult to explain or challenge [11]. Open models create accountability and trust, essential currencies in the human-centered vision of Society 5.0. Assumptions are at the core of models, and the capability to understand and validate the assumptions are a must, graphical languages that simplifies models understanding are needed. In that sense, research presented in [3] demonstrates how digital twins can be made transparent through SDL (Specification and Description Language), a graphical modeling approach that visualizes the inner workings of complex systems.

On Fig. 2 the architecture of a Digital Twin is presented. Notice that the Digital Master represents its core, containing the assumptions and the goals that drive the Digital Twin.

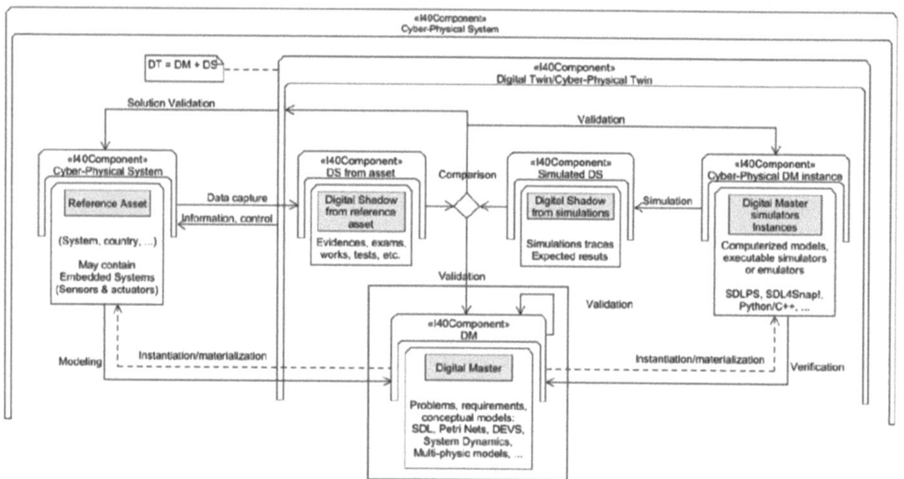

Fig. 2. Digital Twin architecture adapted from [12]

4 Decisions at all Levels

UMDT aren't just sophisticated technical tools—they're decision-support systems that operate across different levels of urban governance and planning. In Society 5.0, where technology serves human needs rather than the other way around, these tools become valuable by helping decision-makers at all levels understand the complex interplay between transportation systems and human experiences.

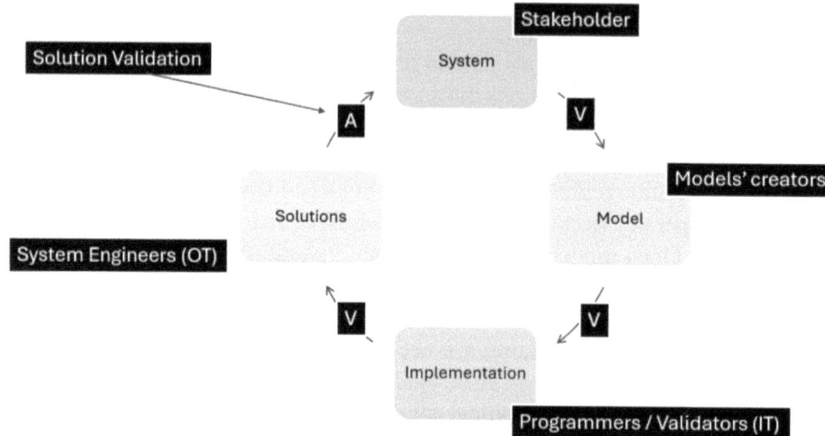

Fig. 3. Simplified Validation and Verification process with the different main actors involved on each stage. Adapted from [13]

Figure 3 depicts the journey of developing Digital Twins for urban mobility, showing how different experts collaborate throughout the process.

1. The process begins with 'Stakeholders' (like city mayors or transportation officials) who validate goals for the urban mobility system - for example, a mayor might prioritize environmental sustainability. These validated goals shape the 'System' requirements.
2. 'Models' creators' develop digital representations of the urban mobility system. These models must be validated by the original stakeholders to ensure they accurately reflect their vision and goals.
3. Programmers/Validators (IT)' then implement these models through code, creating functioning digital simulations. These implementations undergo verification to confirm they correctly translate the models into working systems.
4. The implemented Digital Twin produces 'Solutions' - such as replacing 100 traditional buses with electric ones - which represent potential changes to the real-world mobility system.
5. Finally, 'System Engineers (OT)' conduct 'Solution Validation' to determine whether these virtual solutions can be physically implemented in the real world. This final step bridges the gap between digital simulation and real-world application.

This validation-rich process operates as a continuous cycle, ensuring that Digital Twins serve as trustworthy decision support tools. The ongoing nature of this process allows city planners to test mobility solutions virtually before committing resources to physical implementation, continuously refining and improving urban mobility systems as new data becomes available and as city needs evolve.

4.1 Strategic Levels

At the strategic level, Digital Twins helps city leaders and planners make far-reaching decisions about infrastructure investments, land use policies, and long-term mobility goals [14]. These decisions typically look 5–20 years [15] into the future and shape the fundamental character of urban mobility.

For example, a city considering whether to invest in a new light rail line, expand bus rapid transit, or develop a network of protected bike lanes can use a UMDT to simulate how each option might affect travel patterns over the coming decades. The model can predict not just technical metrics like travel times and capacity, but also human-centered outcomes like access to essential services, commute comfort, and neighborhood connectivity across different parts of the city.

What makes this approach distinctly aligned with Society 5.0 is how it combines technical insights with human priorities. Rather than optimizing solely for efficiency or cost, the Digital Twin can help decision-makers understand how different strategies might serve broader social goals like equity, sustainability, and quality of life. A new transit line might be technically efficient but fails to serve populations who need it most, the Digital Twin makes these trade-offs visible and discussable.

4.2 Tactical Decisions

At the tactical level, Digital Twins supports medium-term decisions about how to allocate and adjust mobility resources. These typically involve planning horizons of several months to a few years and focus on making the most of existing infrastructure.

Consider a city that experiences seasonal tourism flows or hosts major events. A Digital Twin can help mobility managers adjust transit schedules, optimize traffic signal timing, or implement temporary bike-share stations based on anticipated demand patterns. It can also help evaluate tactical urbanism interventions—like converting parking spaces to parklets or implementing car-free zones on weekends—before making them permanent. To ensure reliable results, these digital twins must be systematically validated using methodologies like Design of Experiments (DOE), strategically varying conditions such as tourist volumes or event configurations to identify the most robust solutions.

The Society 5.0 perspective shines through in how these tactical decisions prioritize human experiences. Rather than just minimizing vehicle delay (a Society 4.0 approach), the Digital Twin helps planners understand how changes might affect people's ability to reach destinations comfortably, reliably, and equitably. For example, it might show that strategically placing electric vehicle charging stations in neighborhood centers rather than just along major corridors would improve both adoption rates and local business activity, transforming utilitarian infrastructure into community assets.

5 Risks

In the context of Society 5.0, technology is deeply integrated into human life and decision-making processes. This integration creates a complex socio-technical environment where modeling plays a crucial role in shaping policies, services, and infrastructure. However, the exclusion of stakeholders from these modeling processes presents significant risks with far-reaching consequences. This section examines these risks and proposes strategies to mitigate them:

- Lack of Representation and Bias: When communities aren't involved in model creation, digital twins might fail to reflect their needs. A traffic optimization system designed without consulting all neighborhoods might improve flow on major arteries but neglect pedestrian accessibility in less-represented residential areas. Conceptual models based on graphical languages like SDL will help here.
- Erosion of Trust and Legitimacy: Decisions based on poorly understood models' risk being perceived as impositions. A technically perfect mobility system developed without community dialogue will likely face resistance, regardless of its technical merits. An open discussion regarding goals and assumptions is needed here.
- Misalignment with Societal Values: Digital twins might optimize technical metrics (like traffic efficiency) at the expense of broader values such as equity or sustainability. For example, eliminating "inefficient" bus routes based purely on cost metrics could deprive vulnerable communities of essential connections. Digital Twin objective must be open and auditable.
- Reduced Effectiveness and Adoption: Digital twins that don't consider real user needs risk becoming unused tools. A mobility platform that ignores the needs of elderly people or those without smartphones might be technically sophisticated but practically irrelevant for large segments of the population, requiring costly redesigns after launch.

- Amplification of Social Inequalities: In Society 5.0, where technology permeates every aspect of life, digital twins could inadvertently worsen existing disparities. Algorithms that optimize mobility services based on historical data might concentrate resources in already well-served areas, creating "mobility deserts" in disadvantaged neighborhoods and further limiting access to job opportunities, healthcare facilities, and educational institutions. A Solution Validation is needed to assure the correct implementation of DT results over the system.
- Ethical Concerns: when digital twins make decisions without stakeholder input, fairness becomes a major issue. For instance, a city might use an algorithm to place new mobility hubs based on current usage data. Without transparency, neighborhoods might consistently be passed over without explanation or ability to appeal. Residents can't challenge decisions they don't understand, creating a system where some communities systematically receive fewer resources without any clear avenue for recourse.

To mitigate these risks, it is essential to adopt a participatory and inclusive approach to digital twins in Society 5.0. Addressing complex social challenges through science, technology, and innovation requires the use of Convergence Knowledge (So-Go-Chi) [4] and STEAM [16] approaches, which integrate not only natural sciences but also humanities and social sciences. In the context of digital twins for urban mobility, this means enriching technical models with insights from sociology, psychology, economics, and ethics to create truly human-centered systems.

- Stakeholder mapping and engagement: Identify and engage with diverse stakeholders throughout the modeling process.
- Co-creation and collaborative modeling: Involve stakeholders in all stages of model development, from problem definition to validation.
- Transparency and explainability: Make models and their assumptions transparent and understandable to stakeholders.
- Ethical impact assessments: Conduct thorough assessments of the ethical implications of models and their potential impacts on different stakeholder groups.
- Continuous monitoring and feedback: Establish mechanisms for ongoing monitoring and feedback to ensure models remain relevant and aligned with societal values.
- Establish clear governance structures: Define roles, responsibilities, and decision-making processes for model development and deployment.

6 Conclusions

The integration of digital twins in urban mobility within Society 5.0 has tremendous potential to enhance societal well-being. However, this potential can only be realized if ethical considerations and social equity are prioritized in their development and deployment. By adopting a participatory and inclusive approach, leveraging Convergence Knowledge, and establishing strong governance structures, it is possible to create human-centered systems that optimize urban mobility and address complex social challenges. Ensuring transparency, conducting ethical impact assessments, and maintaining continuous monitoring and feedback mechanisms are crucial steps towards achieving this goal. It is essential to develop digital twins that reflect and uphold the values of fairness, inclusivity, and sustainability, paving the way for a truly super-smart society benefiting all its members.

A continuous validation of the solution (solution validation) is needed to assure the accountability of the results and improve both, the system and the Digital Twin to drive to optimal scenarios.

Acknowledgments. A third level heading in 9-point font size at the end of the paper is used for general acknowledgments, for example: This research received no specific grant from any funding agency in the public, commercial, or not-for-profit sectors.

Disclosure of Interests. The authors have no competing interests to declare that are relevant to the content of this article.

References

1. Huang, H., Tsou, J.Y.: From Concept to Reality: A Systematic Review of Urban Digital Twin and Their Applications in Smart Cities. 536–557 (2025). doi:https://doi.org/10.1007/978-3-031-84208-5_40
2. How Urban Digital Twins Transform Mobility I PTV Blog, https://blog.ptvgroup.com/en/modeling-planning/how-urban-digital-twins-transform-mobility/
3. Grotto, A., Fonseca I Casas, P., Zubaryeva, A., Sparber, W.: Formalizing sustainable urban mobility management: an innovative approach with digital twin and integrated Modeling. Logistics. **8**, 117 (2024). https://doi.org/10.3390/LOGISTICS8040117/S1
4. Society 5.0, https://www8.cao.go.jp/cstp/english/society5_0/index.html
5. [Tentative Translation] Report on The 5 th Science and Technology Basic Plan.
6. What is Industry 4.0? I IBM, https://www.ibm.com/think/topics/industry-4-0
7. Society 5.0: A people-centric super-smart society. Society 5.0: A People-centric Super-smart Society. 1–177 (2020). 10.1007/978-981-15-2989-4
8. Towards the Creation of Next-Generation Mobility Systems. (2020)
9. Shariatpour, F., Behzadfar, M.: A data-driven interactive system for smart urban planning and design. Int. J. Digit. Innov. Built Environ. **12**, 1–23 (2024). https://doi.org/10.4018/IJDIBE.361591
10. Dirk, H.: Society 4.0: Upgrading society, but how? (2016)
11. Algorithm Transparency in Urban Planning I Restackio, https://www.restack.io/p/ai-in-urban-planning-answer-algorithm-transparency-cat-ai#cm2fz8imnex4n13zfbcgj4qia
12. Fonseca i Casas, P., Garcia i Subirana, J., Garcia i Carrasco, V.: Modeling SARS-CoV-2 true infections in Catalonia through a digital twin. Adv Theory Simul. **6**, 2200917 (2023). https://doi.org/10.1002/ADTS.202200917
13. Fonseca i Casas, P.: A continuous process for validation, verification, and accreditation of simulation models. Mathematics. **11**, 845 (2023). https://doi.org/10.3390/MATH11040845
14. Mohammadi, N., Taylor, J.E.: Smart city digital twins. In: 2017 IEEE Symposium Series on Computational Intelligence, SSCI 2017 - Proceedings. 2018-January, 1–5 (2017). https://doi.org/10.1109/SSCI.2017.8285439
15. Wey, W.M., Hsu, J.: New urbanism and smart growth: toward achieving a smart National Taipei University District. Habitat Int. **42**, 164–174 (2014). https://doi.org/10.1016/J.HABITATINT.2013.12.001
16. Yamada, A.: Japanese higher education. J. Comp. Int. High. Educ. **13**, 44–65 (2021). https://doi.org/10.32674/JCIHE.V13I1.1980

SDF FuzzIA: A Fuzzy-Based, AI Support System for Decision-Making Frameworks

Lydia Castronovo[1], Giuseppe Filippone[2], Mario Galici[4],
Gianmarco La Rosa[2(✉)], Arianna Maria Pavone[2],
and Marco Elio Tabacchi[2,3]

[1] Dipartimento di Scienze Matematiche e Informatiche, Scienze Fisiche e Scienze della Terra, Università degli Studi di Messina, Messina, Italy
`lydia.castronovo@studenti.unime.it`
[2] Dipartimento di Matematica e Informatica (DMI), Università degli Studi di Palermo, Palermo, Italy
`{giuseppe.filippone01,gianmarco.larosa,ariannamaria.pavone,`
`marcoelio.tabacchi}@unipa.it`
[3] Istituto Nazionale di Ricerche Demopolis, Palermo, Italy
[4] Dipartimento di Matematica e Fisica "Ennio De Giorgi" Universit del Salento, Lecce, Italy
`mario.galici@unisalento.it`

Abstract. This paper presents an enhanced version of *SDF-FuzzIA*, a hybrid and modular system designed to support decision-making processes in geo-thematic domains. Built upon the *Sustainability Decision Framework*, the system integrates fuzzy ontologies, fuzzy rule-based systems, and a large language model equipped with Retrieval-Augmented Generation capabilities. The proposed architecture aims to deliver accurate, explainable, and semantically grounded answers to complex queries involving geopolitical, economic, and environmental indicators in alignment with the United Nations 2030 Agenda. A novel contribution of this work lies in the adoption of a *Multi-Criteria Group Decision-Making* methodology to support the construction of fuzzy ontologies, enabling the collaborative selection of membership functions and reducing subjectivity in knowledge modeling. Preliminary experiments show promising results in replicating expert reasoning, while highlighting challenges related to scalability, consistency, and automation. This approach lays the groundwork for developing intelligent systems capable of transparent and context-aware decision support in data-rich and uncertainty-prone environments.

Keywords: Fuzzy Ontologies · Explainable Artificial Intelligence · Decision Support Systems · Multi-Criteria Group Decision-Making

1 Introduction

In this paper, we present an upgrade to the SDF-FuzzIA system [6], which consists of the employment of a *Multi-Criteria Group Decision-Making* (MCGDM)

© The Author(s), under exclusive license to Springer Nature Switzerland AG 2026
M. Pavone et al. (Eds.): DSA ISC 2025, LNCS 16405, pp. 395–405, 2026.
https://doi.org/10.1007/978-3-032-21811-7_28

methodology to build the fuzzy ontology involved in the system. In particular, the Sustainability Decision Framework (SDF) is a Decision Support System in development by the Dipartimento di Matematica e Informatica (DMI) of the Università degli Studi di Palermo (UniPa) and TD Group Italia. It processes geo-thematic data, including territorial, geopolitical, and economic information from past and present countries. Moreover, SDF collects thousands of indicators over a 10-year period related to the sustainable development delineated by the United Nations Agenda 2030. A key component, SDF-FuzzIA, leverages a Fuzzy-Ontology Large Language Model based (LLM-based) system to analyse data intelligently, uncover hidden correlations, refine user queries, and enhance decision-making through natural language support.

More specifically, SDF-FuzzIA is a hybrid Explainable Artificial Intelligence (XAI) (see Fig. 1) that uses Fuzzy Ontology, Fuzzy Rule-Based Systems (FRBS), and a LLM to provide explainable [5] and introspective answers in natural language to geo-thematic queries.

In the following, we resume the components of the SDF-FuzzIA.

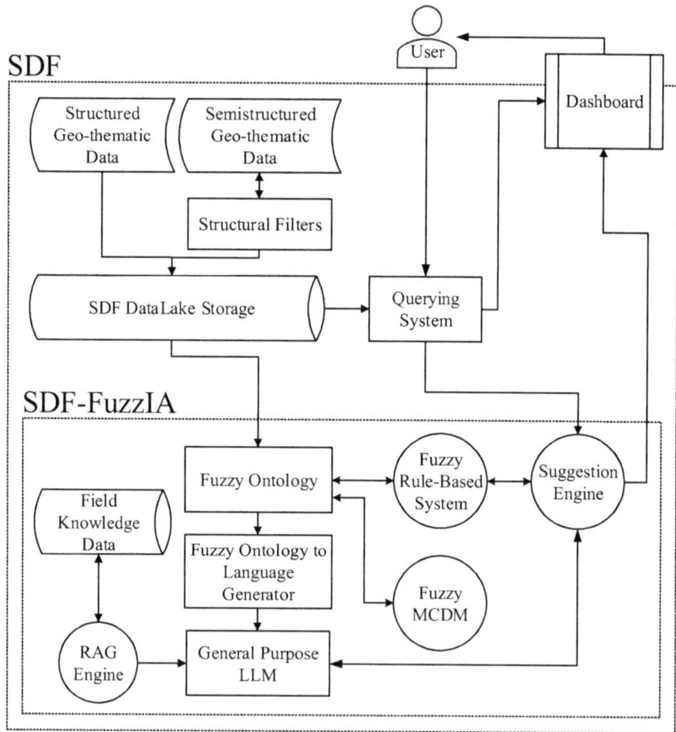

Fig. 1. The SDF-FuzzIA component embedded in the main SDF Decision Support System.

General Purpose LLM and RAG Engine. This hybrid system is composed of plug-and-play components in order to update them with newer and more performant technologies. In this moment, it uses a fine-tuned LLM based on Llama3 [8][1] with 70 billion parameters.

The LLM provides the natural language manipulation of the structured answers provided by the other components of SDF-FuzzIA. In particular, it is used to provide natural language explicative answers to the user's queries by employing the semantic structure supplied by the Fuzzy Ontology (described later) and the Fuzzy Rule-Based System (FRBS) components. Additionally, the explanation is complemented by the references related to the information used to provide that answer.

To be more precise, the LLM works in different phases. First, it provides an answer by using its inner knowledge and the Retrieval-Augmented Generation (RAG) [14][2]. Then, it transforms in natural language the answer provided by the FRBS. Eventually, the improved answer is furnished to the Suggestion Engine, which transmits the answer to the Dashboard.

Fuzzy Ontology, FRBS and FO2L. An ontology is a structured representation of knowledge. In order to address the intrinsic vagueness of real-world concepts, a natural extension of classical ontologies is fuzzy ontologies, where fuzzy logic is employed to define concepts, properties, relations, and so on. We refer the reader to [4,12,13] for further details.

The usefulness of a fuzzy ontology becomes evident, for instance, when we need to describe whether a country is rich or not, as this is a vague concept. By introducing the following three subclasses: "LowDevelopmentCountry," "MediumDevelopmentCountry, "and" HighDevelopmentCountry". The final example prompts the following question: Which membership function should be selected to optimally explain these concepts? Which membership functions (e.g. S-shape, Z-shape, triangular, trapezoidal, crisp, etc.) and degrees of membership should be selected in order to more effectively interpret an individual's belonging to a class within an ontology? It is not possible to provide a definitive answer to this question, as the concepts involved are vague, and the appropriateness of a membership function depends on the specific context and its intended use. However, a potential approach to address this challenge is to solicit the evaluation and insights of a group of individuals regarding these membership functions. Hence, in the present research group, a group decision-making approach was adopted for the selection of membership functions, with a view to aggregating our perspectives and incorporating all relevant viewpoints. This particular aspect of the process will be elucidated in greater detail in Sect. 2.

Remark 1. It is imperative to emphasise that the proposed methodology can be used to select an ontology from existing ones or to aggregate multiple ontologies. This represents a significant and valuable practical application. Indeed, despite

---

[1] Introducing Meta Llama 3: The most capable openly available LLM to date.
[2] What is Retrieval-Augmented Generation (RAG)?.

the advancement and prevalence of fuzzy ontologies, their online accessibility remains limited and practically non-existent. Indeed, despite the advancement and prevalence of fuzzy ontologies, their online accessibility remains restricted and practically non-existent.

The bridge between the fuzzy ontology and the FRBS component is given by the Fuzzy Ontology to Language Generator (FO2L) component, which converts the fuzzy ontology relations into natural language expressions.
Furthermore, the Fuzzy Rule-Based System (FRBS) is a rule-based system that uses fuzzy sets and fuzzy logic to describe various forms of knowledge and to model the relations and connections between its variables. In this context, the FRBS component is used to provide semantically trustworthy explanations to the user by using the facts given by the fuzzy ontology.
In pursuit of an interpretable and trustworthy system, the FRBS results in a fundamental component (see, e.g., [11] for further details on the importance of reliable AI), due to the inability of current AI models to perform common-sense reasoning (worsened by the existence of so-called "hallucinations").

Suggestion Engine. Lastly, the Suggestion Engine component exploits the answers provided by the LLM (along with the reference of the documents used to provide those answers) and the semantics provided by the FRBS (and the fuzzy ontology), to provide an explainable answer to the user Dashboard.

2   Multi Criteria Group Decision Methods for Building Fuzzy Ontologies

The objective of this section is to present a preliminary version of the approach to building fuzzy ontologies that have been developed. More specifically, the aim is to select the most appropriate membership function for fuzzy concepts within the ontology. In order to achieve this, the MCGDM approach is introduced that is based on the preferences and requirements of the relevant parties. A *fuzzy ontology* can be defined as a generalization of a classical ontology $\mathcal{O}$ as

$$\tilde{\mathcal{O}} = (\mathcal{I}, \mathcal{C}, R, F, A),$$

where:

- $\mathcal{I} = \{I_1, \ldots, I_N\}$ is a set of N individuals, also known as instances of concepts.
- $\mathcal{C} = \{C_1, \ldots, C_m\}$ is a set of m concepts (or classes) representing entities in a specific domain. In particular, for all $j = 1, \ldots, m$, $C_j \in \mathcal{C}$ is a fuzzy set on the domain of individuals, i.e., $C_j \colon I \mapsto [0, 1]$. The set of entities of the fuzzy ontology will be indicated by $\mathcal{X}$, i.e., $\mathcal{X} = \mathcal{C} \cup \mathcal{I}$.
- $R = \{R_1, \ldots, R_c\}$ is a set of c fuzzy relations on the domain of entities $\mathcal{X}$ such that, for all $t = 1, \ldots, c$, $R_t \colon \mathcal{X}^l \mapsto [0, 1]$, with $l > 1$. A special role is held by the taxonomic relation $T \colon \mathcal{X}^2 \mapsto [0, 1]$, which identifies the fuzzy subsumption relation among the entities.

- F is the set of fuzzy relations on the set of entities $\mathcal{X}$ and a specific domain contained in $D = \{\text{integer}, \text{string}, \ldots\}$. In detail, they are functions such that each element $f \in F$ is a relation $f \colon \mathcal{X}^{(l-1)} \times \tilde{P} \mapsto [0,1]$ where $P \in D$, and $l > 1$.
- A is the set of axioms expressed in a proper logical language, i.e., predicates that constrain the meaning of concepts, individuals, relationships, and functions.

The application of fuzzy ontologies allows for the representation of items with degrees of membership, rather than the binary classification typical of classical ontologies. For a more comprehensive treatment of these techniques, we refer the reader to [4]. Additionally, for an overview of the state of the art on this topic, we refer the reader to [12,13].

Triangular Fuzzy Numbers. In general, a fuzzy number is a particular fuzzy subset of the field of real numbers $\mathbb{R}$ (see e.g. [2,7]). For these notions we refer to the definitions and results used in the 1983 paper by van Laarhoven and Pedrycz [10]. In this paper, we only consider triangular fuzzy numbers defined as follows.

Definition 1. A *triangular fuzzy number* (TFN) $\tilde{n} = (a, b, c)$, with $a, b, c \in \mathbb{R}$ and $a \leq b \leq c$, is a fuzzy number whose membership function has a triangular shape i.e.,

$$\tilde{n}(x) = (a, b, c)(x) = \mu(x) = \begin{cases} \frac{x-a}{b-a}, & x \in [a,b] \\ \frac{c-x}{c-b}, & x \in [b,c] \\ 0 & otherwise. \end{cases}$$

For such numbers, the operations of addition and multiplication are defined as $(a_1, b_1, c_1) * (a_2, b_2, c_2) = (a_1 * a_2, b_1 * b_2, c_1 * c_2)$, with $* \in \{+, \cdot\}$.

2.1   Model

We first present the notation we use throughout this work, then we analyse more technical mathematical details and necessary assumptions. In order to provide more details to this model, the reader is referred to [3]. We begin by introducing the concept of a membership matrix.

Definition 2. Let $\tilde{\mathcal{O}}$ be a fuzzy ontology with classes C_j, with $1 \leq j \leq m$ and let $\mathcal{I}$ be the set of individuals, with $|\mathcal{I}| = N$. We define the *membership matrix* as the matrix $A = (\alpha_{sj})$, where $\alpha_{sj} = \mu_{C_j}(I_s)$ is the value of the membership function of the individual $I_s \in \mathcal{I}$ with respect to the class C_j.

Let $\mathcal{E} = \{E_1, E_2, \ldots, E_p\}$ be the set of experts, $\mathcal{A} = \{A_1, A_2, \ldots, A_n\}$ the set of alternatives and $\mathcal{C} = \{C_1, C_2, \ldots, C_m\}$ the set of criteria. In our context, the criteria and alternatives will be interpreted as follows:

- the criteria C_j are the classes of the ontology $\tilde{\mathcal{O}}$;

- the alternatives A_i are the *membership matrices* related to the ontology $\tilde{\mathcal{O}}$, i.e. $A_i = (\alpha_{i,sj})$, where $\alpha_{i,sj} = \mu_{i,C_j}(I_s)$ is the value of the i-th membership function of the individual $I_s \in \mathcal{I}$ with respect to the class C_j.

For each expert E_k, with $k = 1, \ldots, p$, we build a decision matrix based on the alternatives and the criteria available, as illustrated in Table 1a. The components of these matrices are the scores a_{ij}^k that each expert E_k assigns to the alternative A_i with respect to the criterion C_j, for all $i = 1, \ldots, n$, and $j = 1, \ldots, m$.

Table 1. Decision matrices for the expert E_k with respect to the classes and to the relations of a fuzzy ontology.

E_k		w_1^k	w_2^k $\cdots$	Criteria w_m^k
Alternatives		C_1	C_2 $\cdots$	C_m
x_1	A_1	a_{11}^k	$a_{12}^k \cdots$	a_{1m}^k
x_2	A_2	a_{21}^k	$a_{22}^k \cdots$	a_{2m}^k
$\vdots$	$\vdots$	$\vdots$	$\vdots$ $\ddots$	$\vdots$
x_n	A_n	a_{n1}^k	$a_{n2}^k \cdots$	a_{nm}^k

(a) Decision matrix for the expert E_k with respect to the classes of a fuzzy ontology.

E_k		w_1^k	w_2^k $\cdots$	Criteria w_m^k
Alternatives		R_1	R_2 $\cdots$	R_l
x_1	A_1	a_{11}^k	$a_{12}^k \cdots$	a_{1m}^k
x_2	A_2	a_{21}^k	$a_{22}^k \cdots$	a_{2m}^k
$\vdots$	$\vdots$	$\vdots$	$\vdots$ $\ddots$	$\vdots$
x_n	A_n	a_{n1}^k	$a_{n2}^k \cdots$	a_{nm}^k

(b) Decision matrix for the expert E_k with respect to the relations within a fuzzy ontology.

Furthermore, each expert E_k assigns a weight w_j^k to the class (i.e., the criterion) C_j. These weights represent the relative importance of the class C_j for the expert E_k with respect to the decision problem. For each expert, we assume the family of weights $\{w_j^k\}$ to be normalised and we place them in what we call *weight matrix*, denoted as $W = (w_j^k)$.

Then we have the $n + 1$ matrices

$$A_i = \begin{pmatrix} \alpha_{i,11} & \alpha_{i,12} & \cdots & \alpha_{i,1m} \\ \alpha_{i,21} & \alpha_{i,22} & \cdots & \alpha_{i,2m} \\ \vdots & \vdots & \ddots & \vdots \\ \alpha_{i,N1} & \alpha_{i,N2} & \cdots & \alpha_{i,Nm} \end{pmatrix} \quad \text{and} \quad W = \begin{pmatrix} w_1^1 & w_2^1 & \cdots & w_m^1 \\ w_1^2 & w_2^2 & \cdots & w_m^2 \\ \vdots & \vdots & \ddots & \vdots \\ w_1^p & w_2^p & \cdots & w_m^p \end{pmatrix}.$$

Usually, the score a_{ij}^k in the decision matrix indicates how well, for the expert E_k, the alternative A_i satisfies that specific criterion C_j. In our case, the expert E_k assigns the score a_{ij}^k to the j-th column of the membership matrix A_i, i.e., the expert evaluates how the alternative A_i fulfils the decision problem for the class C_j.

Remark 2. Note that it is possible to extend our method to the relations between the entities $\mathcal{X}$ of a fuzzy ontology, by defining the decision matrix as illustrated in Table 1b. In this case, the alternatives are defined as the matrix $A_i = \{\alpha_{i,st}\}$, where $\alpha_{i,st} = \mu_i(R_t, X_{s_1}, \ldots, X_{s_l})$ is the value of the i-th membership function of the relation R_t defined as $R_t \colon \mathcal{X}^l \mapsto [0,1]$, with $l > 1$, $X_{s_j} \in \mathcal{X} = \mathcal{C} \cup \mathcal{I}$, for all $1 \leq s \leq \binom{N+m}{l}$, and $1 \leq t \leq |R|$. In order to compute $(s_1, \ldots, s_l)$, we define the function f such that $f(s) = (s_1, \ldots, s_l)$ is the s-th l-combination of entities taking the ordered set (according to some order) of the set $\mathcal{X}$, i.e., the rows of the matrix A_i are all the ordered l-combinations of entities.

In order to extend our method as much as possible, we assume that the weights and scores are TFNs, thus respectively $\tilde{w}_j^k$ and a_{ij}^k. First, we begin by normalising the weights as TFN whose vertices are in $[0, 1]$ for each $1 \leq k \leq p$. Then let $\tilde{w}^k = \sum_{j=1}^{m} \tilde{w}_j^k$, and $w_j^k = \tilde{w}_j^k / \tilde{w}^k$.

Once the weights have been normalised, we construct the p decision matrices, one for each expert, as in Table 1a.

Each expert computes their best configuration based on the MCDM method they have chosen to solve the decision problem. For simplicity, we assume that each of them uses a Simple Additive Weightage (SAW), but the weights and scores are TFNs. Henceforth, each expert E_k calculates the final rank value with respect to the alternative A_i. Clearly, the final rank value $\tilde{x}_i^k$ is a fuzzy number, i.e.,

$$\tilde{x}_i^k = \sum_{j=1}^{m} a_{ij}^k w_j^k, \tag{1}$$

for any $1 \leq k \leq p$ and $1 \leq j \leq m$.

Subsequently, we compute the defuzzification of the ranks by using some defuzzification method to obtain a crisp number from a fuzzy number, e.g., Mean of Maximum (MOM). We denote as x_i^k the crisp number we get from $\tilde{x}_i^k$. We use the defuzzied rank to identify the optimal alternative $A_k^* = (\alpha_{sj}^{k*})$, with $1 \leq s \leq N$ and $1 \leq j \leq m$, relative to the expert E_k in the classical way, namely

$$A_k^* = \arg\max_{A_i} x_i^k. \tag{2}$$

Finally, we compute the best configuration $A^* = (\alpha_{sj}^*)$, starting from those identified (or decided) by each expert, by taking the arithmetic mean of their optimal alternatives, i.e.,

$$A^* = \sum_{k=1}^{p} \frac{1}{p} A_k^*. \tag{3}$$

It is worth noting that we chose an arithmetic mean to compute the best configuration assigning the same weight to each expert, i.e., $1/p$. Specifically, being in the presence of a team of experts, it is reasonable to assume that they were selected based on their skills and expertise and that none of them is "more expert than another." This can be generalised by considering that, in the final decision (which, in our case, will produce the best configuration of the membership values assigned to the individuals w.r.t. a class of the ontology), each expert has a different weight related to their level of confidence in the problem. In particular, let us assume that for each expert E_k, we have an overall weight π_k associated with him. First, this weighting system must be normalised within the interval $[0, 1]$. Let us suppose that our family of weights associated with each expert is $\{\hat{\pi}_1, \hat{\pi}_2, \ldots, \hat{\pi}_p\}$. Hence, the normalized family will be $\{\pi_1, \pi_2, \ldots, \pi_p\}$, where $\pi_k = \hat{\pi}_k / \pi$ with $\pi = \sum_{k=1}^{p} \hat{\pi}_k$. Therefore, the best compromise can be defined as

$$A^* = \sum_{k=1}^{p} \pi_k A_k^*. \tag{4}$$

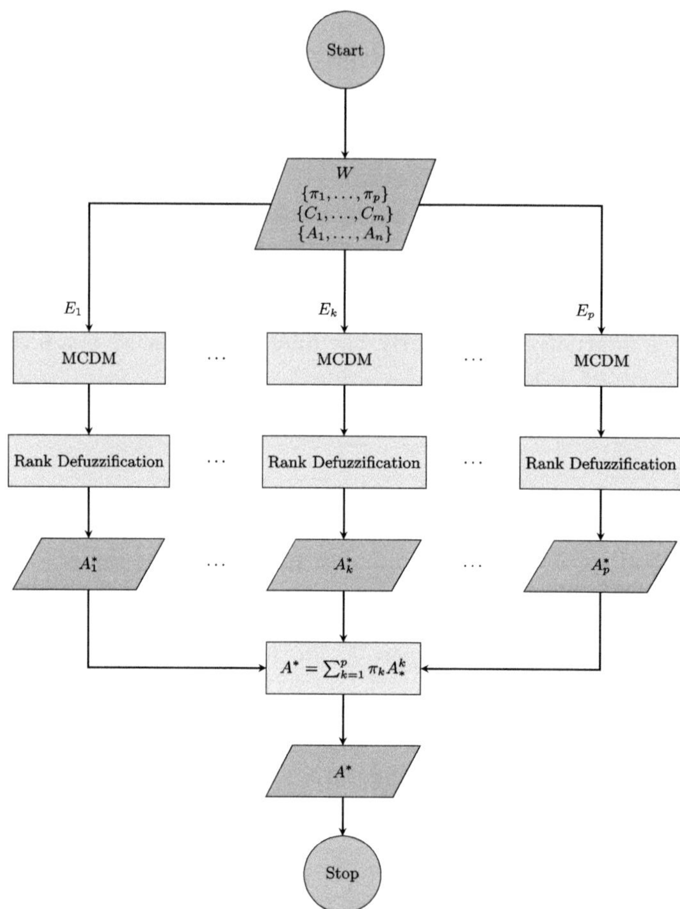

Fig. 2. Flowchart of our MCDM algorithm to compute the best compromise among p experts E_k. Each expert assigns fuzzy scores, based on their expertise on the decision problem, to a set $\{A_i\}_{i=1,\ldots,n}$ of n alternatives. Each alternative represents the membership degree for all the individuals of a fuzzy ontology with respect to the classes $\{C_j\}_{j=1,\ldots,m}$ of this ontology. The family $\{\pi_k\}_{k=1,\ldots,p}$ of weights represents a family of normalized weights assigned to each expert with respect to their expertise in the decision problem. Finally, the matrix W represents the matrix of the weights assigned by each expert to the relative importance of each class of the ontology with respect to the decision problem.

Proposition 1. *The best compromise A^* is a membership matrix.*

Proof. We consider a normalized family of weights $\{\pi_1, \pi_2, \ldots, \pi_p\}$, where π_k is the weight associated with expert E_k. Each expert E_k gives an optimal alternative $A_k^* = (\alpha_{sj}^{k*})$ according to the chosen MCDM method. We have that the best compromise is the matrix $A^* = \sum_{k=1}^{p} \pi_k A_k^*$. In order to prove that A^* is still a membership matrix, we show that its entries are always values in $[0, 1]$. Using that $\alpha_{sj}^{k*} \leq 1$ for $k = 1, \ldots, p$, for the sj entries of A^* it holds that

$$\pi_1 \alpha_{sj}^{1*} + \pi_2 \alpha_{sj}^{2*} + \cdots + \pi_p \alpha_{sj}^{p*} \leq \sum_{k=1}^{p} \pi_k = 1.$$

Remark 3. In light of the work that raises critical issues regarding the consideration of an MCGDM where the weight assigned to each expert is the same or similar (see [1,9]), we consider different weights. However, without focusing on the methods used to assign the weights (both for the criteria and for the experts), as this would be beyond the scope of this work, our assignment would certainly fall under a subjective technique (Fig. 2).

3   Conclusions and Future Perspectives

This work has presented an enhanced version of the SDF-FuzzIA system, a hybrid AI framework designed to support geo-thematic decision-making through the integration of fuzzy ontologies, fuzzy rule-based systems, and large language models. By introducing a MCGDM approach for the construction of fuzzy ontologies, we have provided a structured method to address vagueness and subjectivity in knowledge representation. The combination of explainable reasoning and natural language interfaces demonstrates the potential of this system to offer transparent and semantically grounded support in complex decision scenarios. Preliminary results are encouraging, and the architecture shows strong potential for further development and practical application.

The integration of fuzzy ontologies and large language models within the SDF-FuzzIA framework represents a promising step toward developing advanced and explainable decision support systems for geo-thematic data. However, this work also opens the door to several research directions and technical challenges that warrant further exploration.

One of the main challenges concerns *scalability and computational complexity.* As the dataset expands in volume and heterogeneity, managing fuzzy knowledge bases and executing real-time reasoning processes through fuzzy rule-based systems and LLMs can become increasingly demanding. Future efforts will focus on optimising the underlying architecture through distributed computing, model pruning, and the adoption of parallelisation strategies, possibly leveraging high-performance or cloud computing platforms.

Another area for improvement involves the *automation of fuzzy ontology construction.* While the current approach relies on expert-driven multi-criteria group decision-making, the integration of machine learning techniques could assist or even replace manual evaluations in selecting appropriate membership functions. This would reduce subjectivity and enhance the adaptability of the framework across different domains.

Improving explainability remains a key goal. Future developments may include the integration of interactive visualisations to better communicate the reasoning process and decisions suggested by the system. Additionally, mechanisms to validate and mitigate hallucinations in LLM-generated outputs will be explored, especially by reinforcing the role of the FRBS component as a semantic verification layer.

From an application standpoint, the proposed framework shows great potential for *real-world deployment in policy-making, environmental planning, and socio-economic monitoring.* Collaborative case studies with public authorities and private stakeholders will be pursued to assess the system's effectiveness in supporting complex decision-making scenarios. Further benchmarking against other decision support systems will be conducted to quantitatively evaluate performance, explainability, and robustness.

Another promising direction is the *extension to other disciplinary contexts*, such as healthcare (e.g., diagnosis assistance under uncertainty), finance (e.g., risk evaluation using fuzzy logic), and bioinformatics (e.g., handling imprecise biological data). Interdisciplinary collaborations will be encouraged to tailor the system's components to domain-specific requirements while maintaining generalizability.

In conclusion, while SDF-FuzzIA already exhibits promising results in generating explainable responses and modelling vague geo-thematic knowledge, the path forward involves enhancing *scalability, automation, validation*, and *practical applicability.* Addressing these challenges will be crucial for transforming SDF-FuzzIA into a robust and widely adoptable tool for intelligent decision support in complex and uncertain domains.

Acknowledgements. Giuseppe Filippone, Mario Galici, Gianmarco La Rosa, and Marco Elio Tabacchi acknowledge financial support from the **Sustainability Decision Framework (SDF)** Research Project – CUP **B79J23000540005** – Grant Assignment Decree No. **5486** adopted on **2023-08-04**. Lydia Castronovo is supported by the INDAM (Istituto Nazionale di Alta Matematica "Francesco Severi"), GNAMPA (Gruppo Nazionale per l'Analisi Matematica, la Probabilità e le loro Applicazioni). Gianmarco La Rosa and Mario Galici were also supported by the INDAM (Istituto Nazionale di Alta Matematica "Francesco Severi"), GNSAGA (Gruppo Nazionale per le Strutture Algebriche, Geometriche e le loro Applicazioni). Arianna Pavone is supported by the PNRR project ITSERR - Italian Strengthening of the ESFRI RI RESILIENCE and by the INDAM (Istituto Nazionale di Alta Matematica "Francesco Severi"), GNCS (Gruppo Nazionale di Calcolo Scientifico). Lydia Castronovo and Marco Elio Tabacchi were also supported by the National Recovery and Resilience Plan (NRRP), Mission 4, Component 2, Investment 1.1, Call for tender No. 1409 published on 14.9.2022 by the Italian Ministry of University and Research (MUR), funded by the European Union – NextGenerationEU – Mission 4, Component 1 – Project Title **Quantum Models for Logic, Computation and Natural Processes (QM4NP)** – CUP **B53D23030160001** – Grant Assignment Decree No. **1371** adopted on **2023-09-01** by the Italian Ministry of Ministry of University and Research (MUR).

Disclosure of Interests. The authors declare that they have no conflicts of interest.

References

1. Boix-Cots, D., Pardo-Bosch, F., Pujadas, P.: A systematic review on multi-criteria group decision-making methods based on weights: analysis and classification scheme. Inform. Fusion **96**, 16–36 (2023). https://doi.org/10.1016/j.inffus.2023.03.004, issn: 1566-2535
2. Buckley, J.J., Eslami, E.: An introduction to fuzzy logic and fuzzy sets. 1st ed. Advances in Intelligent and Soft Computing, vol. 285, p. X Physica. Heidelberg (2002). https://doi.org/10.1007/978-3-7908-1799-7
3. Castronovo, L., et al.: Ontology aggregation with maximum consensus based on a fuzzy multi-criteria group decision-making method. In: Proceedings of the ESU-FLAT 2025. LNCS. Springer (2025) To appear. https://doi.org/10.1007/978-3-031-97228-7_7
4. Cross, V., Chen, S.: Fuzzy ontologies: state of the art revisited. In: Barreto, G., Coelho, R. (eds.) Fuzzy Information Processing. NAFIPS 2018. CCIS, vol. 831. Springer, Cham (2018). https://doi.org/10.1007/978-3-319-95312-0_20
5. Fernandez, A., et al.: Evolutionary fuzzy systems for explainable artificial intelligence: why, when, what for, and where to? IEEE Comput. Intell. Mag. **14**(1), 69–81 (2019). https://doi.org/10.1109/MCI.2018.2881645
6. Filippone, G., La Rosa, G., Tabacchi, M.E.: SDF-FuzzIA: a fuzzy-ontology based plug-in for the intelligent analysis of geo-thematic data. In: Destercke, S., Martinez, M.V., Sanfilippo, G. (eds) SUM 2024. LNCS, vol. 15350. Springer, Cham (2025). https://doi.org/10.1007/978-3-031-76235-2_13
7. Jek, P.H.: Metamathematics of Fuzzy Logic. 1st ed. Trends in Logic. Springer Dordrecht, pp. VIII, 299 (1998). https://doi.org/10.1007/978-94-0115300-3
8. Hugo, T., et al.: LLaMA: Open and Efficient Foundation Language Models. arXiv: 2302.13971 (2023)
9. Koksalmis, E., Kabak, O.: Deriving decision makers weights in group decision making: an overview of objective methods. Inf. Fusion **49**, 146–160 (2019)
10. van Laarhoven, P.J.M., Pedrycz, W.: A fuzzy extension of Saaty s priority theory. Fuzzy Sets Syst. **11**(1), 229–241 (1983). https://doi.org/10.1016/S0165-0114(83)80082-7, issn: 0165-0114
11. Landgrebe, J.: Certifiable AI. Appli. Sci. **12**(3) (2022). https://doi.org/10.3390/app12031050., issn: 2076-3417
12. Manikandabalaji, M., Sivakumar, R.: Knowledge representation using fuzzy ontologies: a survey. IJCSNS **23**(12), 199 (2023)
13. Straccia, U.: Foundations of Fuzzy Logic and Semantic Web Languages. 1st. Chapman and Hall/CRC (2013). https://doi.org/10.1201/b15460.
14. Yunfan, G., et al.: Retrieval-Augmented Generation for Large Language Models: A Survey arXiv: 2312.10997 (2024)

Augmenting Human Decision-Making in Risk Management: An LLM-Based Framework for Enhancing Project Risk Register Quality

Marc Bara Iniesta[✉][iD]

ESADE Business School, Barcelona, Spain
marcoantonio.bara@esade.edu
https://www.esade.edu/

Abstract. Risk registers are fundamental decision-making tools in project management, yet they frequently suffer from vague, incomplete, or poorly structured entries that undermine effective risk response planning. This paper presents a novel framework that leverages Large Language Models (LLMs) to augment human expertise in risk documentation, addressing a critical gap at the intersection of AI and decision science in managing project uncertainty. Unlike existing AI applications in risk management that focus on prediction or classification, our approach targets the semantic quality of risk descriptions—a previously unexplored area that directly impacts decision-making effectiveness. The framework employs structured prompting strategies with GPT-4 to systematically evaluate risk entries across five dimensions: root cause specificity, incident clarity, impact quantification, strategy alignment, and action concreteness. Applied to a controlled dataset of 20 risk entries, the system demonstrated robust capabilities in detecting semantic deficiencies and generating standardized reformulations. Key findings include: (1) 60% of entries contained hypothetical rather than factual root causes, (2) 55% lacked clear temporal triggers for risk events, and (3) 75% exhibited vague or missing actionable responses. The LLM-based system successfully reformulated 100% of flagged entries to include all required components while maintaining semantic coherence. This research contributes to decision science by demonstrating how AI can enhance, rather than replace, human judgment in managing complex project uncertainties. The framework preserves human oversight while automating the detection and correction of common documentation errors, thereby improving the quality of information available for risk-based decision-making. Future work will focus on empirical validation with industry practitioners and integration with existing project management systems.

Keywords: Risk management · Large Language Models · Decision support systems · Project uncertainty · AI-augmented decision-making

M. Pavone et al. (Eds.): DSA ISC 2025, LNCS 16405, pp. 406–416, 2026.
https://doi.org/10.1007/978-3-032-21811-7_29

1   Introduction

Effective risk management is a fundamental component of successful project execution. A key element of this process is the risk register, which systematically documents identified risks, their characteristics, and planned mitigation or response strategies. Standards such as those developed by the Project Management Institute (PMI) and ISO 31000 emphasize the importance of clear, specific, and actionable risk descriptions as the foundation of effective risk management [8,11]. However, in practice, risk registers frequently contain vague, generic, or incomplete entries [5,7]. Examples include risks described as "supplier delays" or "budget overruns" without specifying the root cause, the nature of the incident, or its precise impact on project objectives. Such entries may create ambiguity about what the actual risk is, reduce the effectiveness of qualitative and quantitative risk analyses, and hinder the formulation of targeted response strategies [5]. Although there is limited systematic empirical research quantifying the direct impact of vague risk descriptions on project outcomes, both practitioner literature and risk management frameworks highlight clarity and precision as essential for effective risk response planning [1,7].

In parallel, the emergence of artificial intelligence (AI) and machine learning (ML) has created new opportunities to enhance project risk management. Prior studies have explored a range of techniques—including probabilistic models, decision trees, and natural language processing (NLP)—to support tasks such as risk classification, prediction, and documentation automation [9]. While these methods have advanced the field, they primarily address structured data analysis or conventional NLP tasks and do not focus on improving the semantic clarity of risk register content. More recently, large language models (LLMs), such as GPT-4 and similar architectures, have demonstrated advanced capabilities in understanding and generating complex, domain-specific text. Wong et al. [12] applied LLMs to contract review in the construction industry, showing that LLMs can successfully identify ambiguous clauses and generate clearer formulations. Although this work illustrates the potential of LLMs in risk-related textual analysis, its focus is on contract language rather than on project risk registers.

To date, no studies have reported the application of LLMs to the specific task of reviewing and enhancing project risk registers by identifying vague risks and generating clearer definitions and actionable response strategies. This paper aims to address this research gap by proposing and evaluating an LLM-based framework for improving the quality of risk registers. The objectives of the study are twofold: (1) to detect and flag vague or weakly defined risks within a project risk register, and (2) to generate refined risk descriptions and concrete response strategies aligned with best practices in project risk management. By leveraging the capabilities of state-of-the-art language models, this research contributes to both the academic literature and practical applications in project management. The study outlines the methodology for implementing the framework, presents experimental results, and discusses its potential for integration into project management workflows.

2   Literature Review

Project risk registers are essential tools in project management, serving as a structured repository for potential risks, their characteristics, and proposed responses. Standards such as the PMBOK Guide [11] and ISO 31000 [8] emphasize the importance of clarity, specificity, and consistency in documenting risks. However, practitioner-oriented literature and expert commentary have long observed that real-world risk registers often contain vague, overly generic, or inconsistently defined risks [5,7]. This lack of precision can severely undermine the effectiveness of risk response planning. Despite widespread recognition of the problem, there is limited empirical research quantifying its impact on project outcomes, and very few attempts to address the issue systematically.

In parallel, artificial intelligence (AI) and machine learning (ML) techniques have gained traction in project risk management. Khodabakhshian et al. [9] conducted a systematic review of deterministic and probabilistic approaches used in construction risk analysis, highlighting applications such as decision trees, Bayesian networks, and traditional natural language processing (NLP) for risk classification and template generation. Similarly, Zhang and Fan [13] developed a probabilistic risk analysis framework that combines text mining with Bayesian inference to support risk identification in construction projects. Choudhury et al. [3] further reviewed a variety of ML techniques that enhance structured risk analysis, while noting the limitations of existing approaches in dealing with the semantic quality of textual risk descriptions. These studies underscore the growing role of AI in automating aspects of risk management, yet they do not address the refinement of the language used in risk registers.

The recent emergence of large language models (LLMs) offers new opportunities to address this gap. LLMs are capable of performing nuanced semantic analysis and generating domain-specific text, making them suitable candidates for improving the clarity and consistency of risk documentation. Wong et al. [12] applied a knowledge-augmented LLM framework to identify ambiguous clauses in construction contracts. Their two-stage prompting methodology enabled the model to detect vague contractual language and suggest clearer alternatives, demonstrating how LLMs can enhance the quality of technical documentation. However, their work is specifically focused on legal texts rather than project risk registers.

As of this writing, no published studies have reported the application of LLMs for reviewing and improving project risk registers. A systematic search of recent literature confirms that while LLMs are being applied to adjacent tasks—such as contract risk detection [12], LLM-based risk taxonomies [4], and AI-driven risk classification [9]—none of these initiatives tackle the unique problem of textual refinement in risk registers. Notably, Narayanan and Vishwakarma [10] propose an LLM-based engine for identifying risk categories in AI systems, but their focus is on taxonomy generation, not linguistic clarity.

Table 1 summarizes key contributions in this space and highlights the specific research gap that this study addresses.

Table 1. Key contributions

Study	Technique	Risk Focus	LLM Use	Risk Register Refinement
Khodabakhshian et al. (2023)	ML (Decision Trees, Bayesian Networks, NLP)	Risk identification, classification	No	No
Zhang & Fan (2020)	Probabilistic Risk Analysis, Text Mining	Risk analysis in construction	No	No
Wong et al. (2024)	LLM (Knowledge-Augmented Prompting)	Ambiguity detection in contracts	Yes	No (focused on legal contracts)
Narayanan & Vishwakarma (2024)	LLM (Risk Taxonomy Engine)	AI system risks	Yes	No (focus on risk categorization)
This study	LLM (Structured Prompting Framework)	Project risk register enhancement	Yes	Yes

In summary, while prior research illustrates a growing interest in leveraging AI for risk-related applications, the specific challenge of improving the semantic quality of project risk registers remains unaddressed. This paper seeks to fill that gap by exploring how LLMs can be systematically applied to detect vague risk entries and generate improved, actionable formulations, contributing both to academic literature and project management practice.

3 Methodology

3.1 Framework Overview

This study presents a structured framework that leverages large language models (LLMs) to improve the clarity, consistency, and operational relevance of project risk register entries. The system applies an integrated process in which each risk entry is systematically evaluated and reformulated, regardless of its initial quality. The model is instructed to analyze each component of a risk entry—specifically, the root cause, incident, impact, response strategy, and response action—by assessing both the precision (clarity and specificity) and the correctness (alignment with established definitions). If any field is vague, misclassified, or incomplete, the model provides a diagnostic explanation and produces a revised version that adheres to documented risk management standards.

The improved entry follows a structured format comprising: (i) a factual root cause expressed in the present tense; (ii) a concrete incident that may occur at a defined point in time; (iii) an impact linked to one or more project objectives, such as scope, time, cost, or quality; (iv) a validated or corrected response strategy; and (v) a specific, actionable response plan. This unified review-and-enhancement process is applied consistently across all entries, resulting in risk registers that are more coherent, auditable, and useful for subsequent decision-making. The overall system workflow is illustrated in Fig. 1.

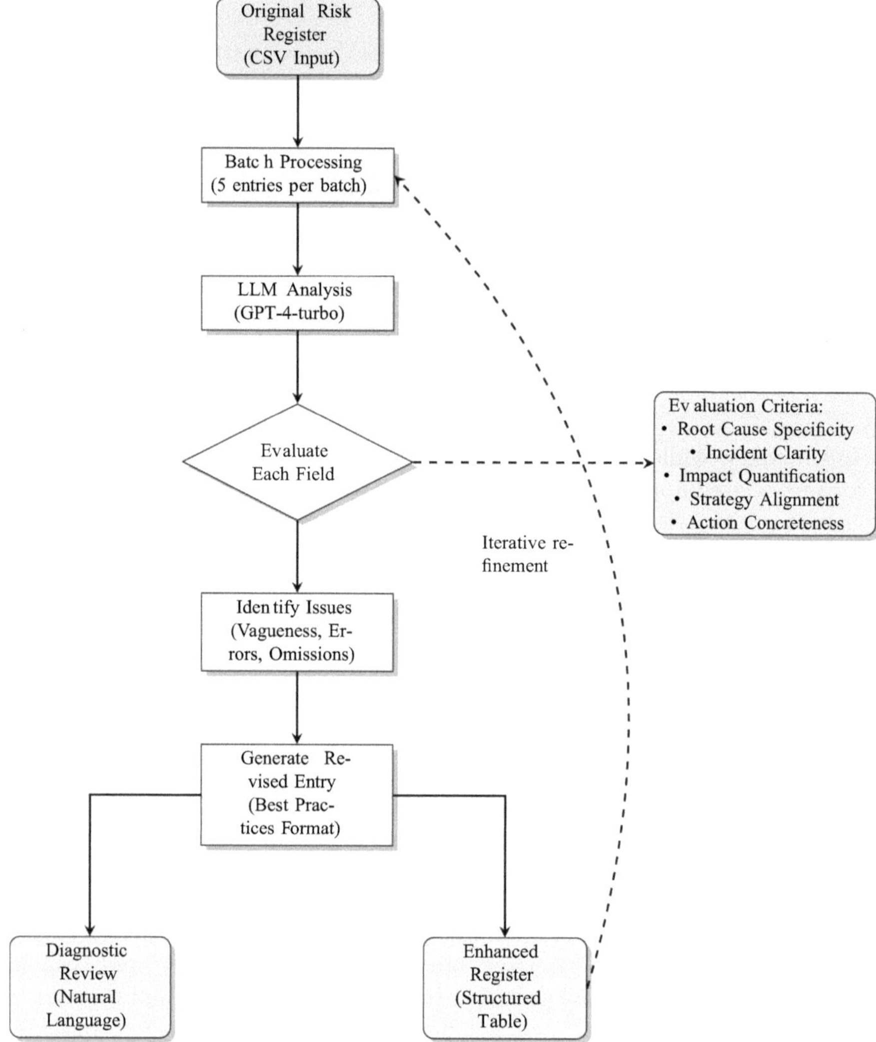

Fig. 1. Process flow diagram of the LLM-based risk register enhancement framework. The system processes risk entries in batches, evaluates each component for semantic quality, and generates both diagnostic reviews and enhanced risk formulations.

3.2 Data Collection

To test and validate the framework, two types of datasets were employed. First, a synthetic dataset comprising 20 risk entries was constructed. These entries were designed to represent common project risk scenarios, with a deliberate mix of well-formed and vague risks to ensure balanced testing conditions. The synthetic risks were developed based on typical project management case studies and aligned with guidelines provided by Hillson [6]. Although the inclusion of anonymized real-world risk registers was considered, no suitable datasets were available at the time of writing. Consequently, the present study focuses on the synthetic dataset as a controlled proof of concept.

3.3   LLM Configuration and Prompting Strategy

The proposed framework leverages OpenAI's GPT-4-turbo model, accessed via the ChatCompletion API (gpt-4-turbo, OpenAI, 2024). This model variant supports extended context windows and is optimized for structured, multi-step reasoning. All prompts are issued using the OpenAI Python SDK, with deterministic generation parameters (temperature set to 0) to ensure reproducibility.

Each risk entry is processed using a structured system prompt that defines the expected format and content of the six standard fields in a project risk register: type, root cause, incident, impact, response strategy, and response action. The model evaluates the precision and semantic correctness of each field, identifies vagueness, conceptual errors, or omissions, and generates a revised version of the entry that adheres to established project risk management guidelines. The reformulated version includes a factual root cause written in the present tense, a punctual and clearly defined incident, and an impact statement linked explicitly to one or more project objectives (such as scope, time, cost, or quality). The model also reviews the response strategy for consistency with the risk type and provides a concrete and actionable response aligned with that strategy.

To manage the model's context window, entries are processed in batches of five. For each batch, the system produces two types of output. The first is a diagnostic review in natural language, containing field-level evaluations and a summary of the changes applied. The second is a structured table in Markdown format, compiling the improved entries in a format ready for reuse in project documentation. The entire process is automated through a Python-based pipeline. The script reads risk register data from a CSV file, groups entries into batches, embeds each batch into a predefined base prompt, and sends it to the GPT-4-turbo API. The resulting outputs are saved both as full review logs in text format and as structured files in CSV or Excel format. This configuration ensures reproducibility, minimizes manual intervention, and supports scalable application across diverse project environments.

3.4   Evaluation and Illustrative Results

The evaluation focused on two complementary approaches: (i) the application of the framework to a full dataset of 20 risk entries, and (ii) the presentation of illustrative examples to demonstrate key outcomes.

Dataset and Experimental Setup. The experiment was conducted on a risk register comprising 20 risk entries, designed to reflect a broad range of typical project risks. The dataset included a mix of well-defined and intentionally vague entries to test both detection and enhancement capabilities. The full dataset and the LLM-enhanced output are available in the project repository [2], ensuring transparency and reproducibility.

Illustrative Examples. To provide clarity on the framework's operation, selected examples from the dataset are presented. Each example includes: (i) the original risk entry; (ii) the LLM's detection result indicating whether the entry was well-defined or vague; (iii) the rewritten version following the prescribed format (root cause, incident, impact); and (iv) the proposed response strategy and aligned action plan. These examples highlight typical improvements, such as the clarification of ambiguous wording, the addition of missing components, and the standardization of risk structure. Qualitative commentary is provided alongside each example to emphasize key observations, including instances where further human oversight may still be required.

Structural Completeness Assessment. A structural evaluation was performed on the full set of 100 LLM-enhanced entries. Each entry was assessed for the presence of the required components:

- Root cause (present-tense, factual condition),
- Incident (specific triggering event),
- Impact (explicit consequence on project objectives).

The results showed that 100% of entries initially flagged as vague by the model were rewritten to include all required components. Additionally, missing response strategies and action plans were systematically completed where absent in the original entries. These findings confirm the framework's capacity to enforce structural completeness across a diverse set of risk statements.

Limitations and Future Work. While the structural completeness analysis demonstrates that the framework can reliably identify and fill missing components, this study does not include formal validation of semantic clarity, practical relevance, or real-world applicability. The absence of expert or practitioner evaluation limits the conclusions regarding the qualitative value of the enhanced entries. Future research will focus on formal empirical validation, including expert panel reviews and end-user studies to assess clarity improvements and detection accuracy using standardized metrics such as Likert-scale ratings, precision, recall, and F1 score. This staged evaluation approach aligns with best practices in applied AI research, where initial technical validation precedes large-scale empirical testing.

3.5   Implementation

To operationalize the framework, a prototype tool was developed using Python in conjunction with the OpenAI API. The tool is available as an open-source package on GitHub, including all code, sample data, and execution instructions to support full reproducibility of the results [2]. The current release (v1.0) corresponds to commit hash abcdef123456... and is permanently archived at the following URL: https://github.com/marcbara/risk-register-llm/releases/tag/v1.0. The tool accepts structured CSV risk registers as input, performs vagueness detection and reformulation, and outputs a side-by-side enhanced version. The current implementation exclusively supports structured CSV file input. Future work may involve extending the tool to accept additional file formats, enable direct text input, and integrate into broader project management environments; however, these enhancements are beyond the scope of the present study.

4 Experimental Illustration

To demonstrate the applicability of the proposed framework, we conducted a preliminary experiment using a synthetically generated project risk register. The register contained 20 risk entries constructed to reflect common patterns found in real-world project documentation, including a mix of well-defined and vague risk descriptions. Each risk entry was processed by the LLM-based system, which:

1. Evaluated each field for vagueness and semantic correctness, flagging entries where the root cause, incident, or impact lacked clarity, precision, or conceptual validity.
2. Reformulated the full entry using best practices, generating improved versions that explicitly distinguished the root cause (a factual project condition), the incident (a punctual trigger event), and the impact (a consequence tied to project objectives).
3. Reviewed and, when necessary, corrected the response strategy and associated action, ensuring alignment with the refined risk type (threat or opportunity) and providing a concrete, actionable step to implement the strategy.

A subset of the results is shown in Table 2, illustrating how the system transforms vague risks into clearer, actionable entries. Each row includes the original description, the model's revised formulation, and the suggested response.

The most frequent errors detected included:

- Root causes expressed as hypothetical conditions rather than factual, present-tense project baselines.
- Incidents described as ongoing states or vague concerns, lacking a clear temporal trigger.
- Impacts framed in terms of reputation, team morale, or generic "delays", rather than as concrete consequences tied to scope, time, cost, or quality.
- Response strategies mismatched to the type of risk (e.g., using "Exploit" for threats).
- Response actions that were vague, incomplete, or non-actionable.
- In some cases, fields were entirely missing; the model proceeded nonetheless and inferred likely content.

Table 2. Risk register transformation examples

Risk ID	Version	Type	Root Cause	Incident	Impact	Response Strategy	Response Action
1	Original	Threat	Unexpected outage	Things go wrong	Bad outcome for the team	Transfer	Discuss with the team
1	Revised	Threat	The project's critical systems lack redundant configurations	A critical system experiences an unexpected outage	Project work is halted for up to 2 days, affecting the timeline	Transfer	Establish a service-level agreement with a third-party provider for immediate technical support during outages
2	Original	Threat	Supplier fails to ship on time	Unforeseen issues occur	Bad outcome for the team	Avoid	
2	Revised	Threat	The project depends on a single supplier for critical components	The supplier fails to ship critical components by the agreed deadline	Potential delay in project completion by up to 3 weeks	Avoid	Identify and qualify alternative suppliers for critical components
3	Original	Threat	Key module fails QA	Things go wrong	Bad outcome for the team	Avoid	Discuss with the team
3	Revised	Threat	The key module is developed without rigorous pre-QA testing	The key module fails during the formal QA testing phase	Rework required, potentially delaying the project by 1–2 weeks	Avoid	Implement a robust pre-QA testing process for all critical modules

Table 3. Distribution of issues in the dataset

Issue type	Affected entries	Example correction
Hypothetical or vague root cause	12 / 20	"If the vendor misses…" → "The project depends on a single supplier."
Non-event incident	11 / 20	"Supplier is delayed" → "Fails to deliver by the milestone date."
Impact not tied to objectives	13 / 20	"Reputation damaged" → "Delay of 2–3 weeks (time)."
Incorrect response strategy	9 / 20	"Enhance" for threat → corrected to "Mitigate"
Vague or missing response action	15 / 20	"Remind the supplier" → "Include penalty clauses in contract."
At least one missing field	5 / 20	—Model inferred and completed the missing fields.

Table 3 summarizes the distribution of these issues across the dataset.

The model's reformulations adhered closely to the structure defined in the prompt and avoided semantic errors or hallucinated content. Even in cases where the input was poorly specified or ambiguous, the system was able to generate consistent and actionable reformulations. This suggests that LLM-based refinement can support the standardization and improvement of risk documentation in environments where initial practices are informal or inconsistent.

5 Conclusions and Future Work

This paper introduces a novel framework that leverages large language models (LLMs) to enhance the semantic quality of project risk registers. While existing research in project risk management has focused extensively on risk identification, classification, and prediction using AI and ML techniques, the textual refinement of risk descriptions has received little to no attention. This study addresses that gap by proposing a method to detect vague risks and generate structured reformulations aligned with best practices.

The framework combines structured prompting strategies with a taxonomy-based classification approach to systematically evaluate and improve risk register entries. Preliminary application of the system to a synthetic dataset demonstrated its potential to flag vague entries and produce clearer, more actionable risk definitions. In doing so, the model supports project managers in improving risk understanding, enabling more targeted response planning and enhancing overall risk governance.

Despite its promising results, this work has several limitations. The current evaluation relies on a synthetic dataset and does not include human-in-the-loop validation from domain experts. In addition, the model's effectiveness across diverse industries and languages remains unexplored. Finally, while the system suggests response strategies, these are not currently validated against project context or stakeholder priorities.

Future research will focus on three key areas: (1) conducting empirical evaluations with real-world risk registers from industry partners; (2) integrating domain-specific fine-tuning to improve the contextual accuracy of reformulations; and (3) assessing the usability and adoption of the tool in practical project settings, including integration with existing project management software platforms. By addressing a long-standing but often overlooked challenge in project risk management—textual clarity—this work opens new pathways for the application of generative AI in support of project delivery excellence.

Acknowledgments. This study was conducted as part of ongoing research into AI-augmented project management. No external funding was received for this work.

Competing Interests. The author has no competing interests to declare that are relevant to the content of this article.

References

1. Aven, T.: Risk assessment and risk management: review of recent advances on their foundation. Eur. J. Oper. Res. **253**(1), 1–13 (2016). https://doi.org/10.1016/j.ejor.2015.12.023
2. Bara Iniesta, M.: LLM-based risk register enhancement tool, v1.0. GitHub Repository (2025). https://github.com/marcbara/risk-register-llm/releases/tag/v1.0
3. Choudhury, A., et al.: Machine learning applications for risk analysis in construction projects. J. Constr. Eng. Manag. **147**(5), 04021027 (2021). https://doi.org/10.1061/(ASCE)CO.1943-7862.0002032
4. Cui, J., et al.: Developing risk taxonomies for AI systems using large language models. In: Proceedings of the 2024 International Conference on AI Systems, pp. 1–10. IEEE, New York (2024). https://doi.org/10.1109/AISYS2024.123456
5. Hillson, D.: Using a risk breakdown structure in project management. J. Facil. Manag. **2**(1), 85–97 (2003). https://doi.org/10.1108/14725960410808129
6. Hillson, D.: Practical project risk management: The ATOM methodology. Management Concepts, Vienna, VA (2007)
7. Hopkinson, M.: The Project Risk Maturity Model: Measuring and Improving Risk Management Capability. Routledge, London (2017)
8. ISO 31000:2018. Risk management – Guidelines. International Organization for Standardization, Geneva (2018)
9. Khodabakhshian, A., et al.: Artificial intelligence in construction risk management: a systematic review. Autom. Constr. **145**, 104623 (2023). https://doi.org/10.1016/j.autcon.2022.104623

10. Narayanan, S., Vishwakarma, A.: LLM-based risk taxonomy engine for AI systems. In: Proceedings of the 2024 International Conference on Artificial Intelligence and Data Science, pp. 123–134. Springer, Heidelberg (2024). https://doi.org/10.1007/978-3-031-23456-7_12
11. Project Management Institute: A guide to the project management body of knowledge (PMBOK Guide), 7th edn. PMI, Newtown Square, PA (2021)
12. Wong, K., et al.: Knowledge-augmented large language models for contract review in construction. Constr. Innov. **24**(2), 345–362 (2024). https://doi.org/10.1108/CI-01-2023-0005
13. Zhang, J., Fan, H.: Probabilistic risk analysis in construction projects using text mining and Bayesian networks. J. Comput. Civ. Eng. **34**(4), 04020018 (2020). https://doi.org/10.1061/(ASCE)CP.1943-5487.0000893

Solving Combinatorial Optimization Problems with Decision Transformers

Antoni Guerrero[1,2], Alvaro Garcia-Sanchez[3(✉)], Yangchongyi Men[2], and Angel A. Juan[2]

[1] Baobab Soluciones, Madrid, Spain
`antoni.guerrero@baobabsoluciones.es`
[2] Research Center on Production Management and Engineering,
Universitat Politècnica de València, Alcoy, Spain
`myangch@upv.edu.es, ajuanp@upv.es`
[3] Department of Organization Engineering, Business Administration and Statistics,
Universidad Politécnica de Madrid, Madrid, Spain
`alvaro.garcia@upm.es`

Abstract. This paper explores the potential of Decision Transformers (DTs) as a general-purpose framework for addressing combinatorial optimization problems, with a particular focus on the Team Orienteering Problem (TOP) as a case. Drawing on the foundational idea that transformers can model decision-making as a sequence generation task, the TOP is reformulated as a sequential decision process. The work provides a detailed conceptual and methodological foundation for applying transformer-based architectures, which were originally developed for natural language processing, to combinatorial problems. Trained offline on supervised data that includes both typical and high-quality solutions, the DT is able to learn effective construction strategies. The paper concludes with a discussion of important modeling considerations, along with current challenges and potential future directions.

1 Introduction

Combinatorial optimization problems frequently arise in a variety of domains such as logistics, scheduling, and robotics [10,16,23], where optimal decisions must be made under complex constraints. Among these, the Team Orienteering Problem (TOP) [4] or Multiple Tour Maximum Collection Problem (MTMCP) [3] are one of the most studied problems. In the TOP, a set of vehicles must visit a subset of available nodes within a limited time or distance budget, with the goal of maximizing the total reward collected from the visited nodes. Due to its practical relevance and computational complexity, the TOP has been extensively studied using traditional optimization techniques such as heuristics or metaheuristics [2,6,18]. However, there has been recently a growing interest in applying learning-based approaches to such problems, particularly Reinforcement Learning (RL). RL offers a way to learn policies that can generalize across instances and adapt to varying environments.

M. Pavone et al. (Eds.): DSA ISC 2025, LNCS 16405, pp. 417–426, 2026.
https://doi.org/10.1007/978-3-032-21811-7_30

Integrating RL into combinatorial optimization has become a promising approach for developing flexible and efficient solvers. RL provides a learning-based framework capable of adapting to complex decision-making problems without relying on handcrafted heuristics [13]. Building on this foundation, the Decision Transformer (DT) [5] offers a novel way to apply RL to offline data. By using past solution trajectories and conditioning on desired outcomes, DTs are capable of generating solutions without the need for explicit search or iterative optimization. This makes them particularly attractive in dynamic or large-scale settings, where traditional solvers may struggle with scalability or adaptability. Furthermore, the sequential decision-making framework of DTs aligns with many combinatorial problems. The following sections investigate whether these methods, originally designed and used for language tasks, can be adapted to effectively solve complex decision-making problems, and explore how this paradigm can be applied to combinatorial problems such as the TOP.

2 From Language Models to Decisions

Large Language Models (LLMs) have been studied extensively in recent years, particularly since the introduction of the Transformer architecture [21,22,25]. This architecture has transformed the field of natural language processing by allowing models to efficiently capture long-distance relationships using self-attention mechanisms. As a result, the field has experienced rapid growth, with the emergence of highly capable LLMs such as GPT, LLaMA, Mistral, or DeepSeek, [1,11,12,20] among many others. All of these models are based on the core assumption that language can be modeled as a sequential process, where the prediction of the next word depends not only on the immediately preceding word, but also on the broader context provided by the preceding sequence of tokens.

LLMs operate in an autoregressive manner, meaning they generate text one token at a time [7]. At each step, the model uses all previously generated tokens as input to predict the next one, continuing this process until a stopping condition is met. This sequential generation process mirrors the structure of certain combinatorial optimization problems, in which the current action depends on the history of previous decisions. In such cases, the challenge becomes one of selecting a sequence of actions (e.g., visiting nodes in a routing problem) that leads to an optimal or almost optimal outcome. This opens the door to applying similar sequence modeling techniques, originally developed for language tasks, to combinatorial problems. The DT was proposed to address this idea [9].

However, using this method requires reformulating the decision-making problem as a Markov Decision Process (MDP) [15], which is a mathematical framework used to model decision-making in environments. In an MDP, an agent interacts with an environment by transitioning through a series of states, taking actions, and receiving rewards. Formally, an MDP is defined by a tuple (S, A_s, P_a, R_a), where:

- S is the set of states, representing all possible situations the agent could be in. The state space can be discrete or continuous.
- A_s is the set of actions, which are the possible moves the agent can make in the state s.
- $P_a(s, s')$ is the state transition probability, which defines the probability of moving from state s to s' after taking the action a.
- $R_a(s, s')$ is the reward function, specifying the immediate reward the agent receives after taking an action a from state s to s'.

The objective in an MDP is to learn a policy $\pi(s)$, a mapping from states to actions, that maximizes the expected cumulative reward over time. However, unlike traditional RL methods, which typically learn a policy or a value function through iterative interaction with the environment, DT takes a different route, since it operates on offline data, which are previously collected trajectories, and learns to predict actions conditioned on a desired return. In Fig. 1 the DT architecture is shown.

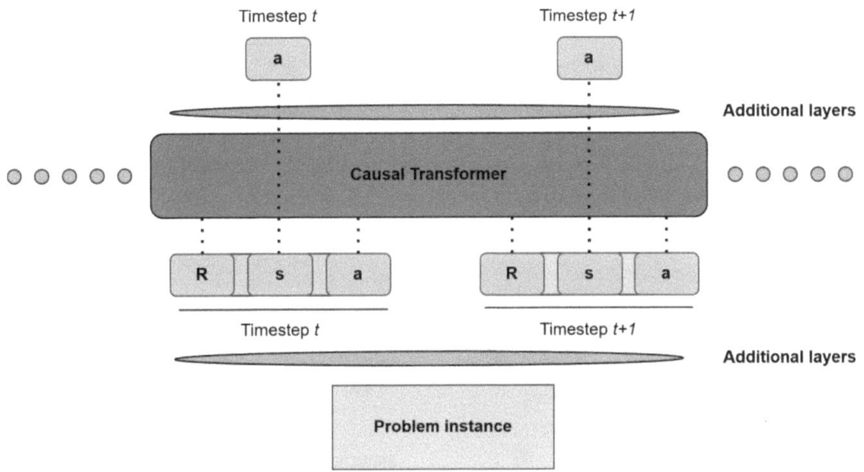

Fig. 1. DT architecture.

At each timestep t, which is defined as a moment in the decision sequence when the agent must choose an action, the DT processes a concatenated sequence of tokens that represent:

- The desired **return-to-go** R_t: the expected cumulative reward the agent aims to achieve from timestep t onward.
- The **state** s_t: the environment's current observation.
- The **action** a_{t-1}: the action taken at the previous timestep.

This tuple is stacked across several timesteps, forming a trajectory window of (`return, state, action`) triplets. The entire sequence is fed into a causal

Transformer architecture, which uses self-attention to model dependencies between all elements in the sequence, and outputs the next action to take. Additional layers for the DT can be added to process the result from the transformer and create the input tokens.

3 Formulating the Team Orienteering Problem

Building on the DT framework, the approach is applied to the TOP, which serves as the target combinatorial optimization task in this study. This problem provides a suitable test for sequence-based decision-making models like DTs, as it involves selecting a sequence of actions under constraints to maximize accumulated rewards. The TOP is a well-known combinatorial optimization problem that falls within the class of routing problems. It involves a set of nodes and a fleet of vehicles, each required to begin and end its route at designated depots. Due to practical constraints, which most commonly are time or distance limitations, not all nodes can be visited within a single route. These constraints are especially relevant in scenarios such as electric vehicle routing, where limited driving range imposes tighter restrictions. Each node offers an associated reward, and the objective is to maximize the total collected reward while adhering to the given constraints. Additionally, each node may be visited no more than once, and vehicles should start and end their routes at the starting and ending depot. A representation of the TOP can be observed at Fig. 2. In this figure, the starting and ending points are the same.

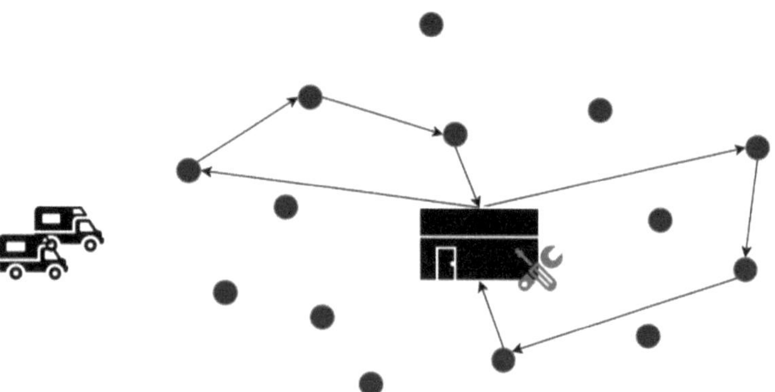

Fig. 2. Representation of the TOP.

Formally, let the set of nodes be $N = \{1, 2, \ldots, m\} \cup \{o, d\}$ and the set of vehicles be $V = \{v_0, v_1, \ldots, v_s\}$. Assume that nodes o and d are the starting and ending depots, respectively. Let d_{ij} denote the distance between nodes i and j, and let r_i represent the reward obtained by visiting node i. Define the binary decision variable x_{ijv} such that $x_{ijv} = 1$ if vehicle $v \in V$ travels from node i to node j, and $x_{ijv} = 0$ otherwise. Then, the problem consists of determining the values of x_{ijv} that maximize the total collected reward, subject to constraints on route feasibility, such as time or distance limitations, depot departure and return requirements, and the condition that each node is visited at most once.

$$\max \sum_{v \in V} \sum_{i,j \in N} x_{ijv} r_j \tag{1}$$

The objective function, shown in Eq. (1), represents the goal of the problem, which, as explained before, is to maximize the total collected reward by determining optimal routes for the vehicles.

$$\sum_{j \in N} x_{ojv} \leq 1 \qquad \forall v \in V \tag{2}$$

$$\sum_{i \in N} x_{oiv} = \sum_{i \in N} x_{idv} \qquad \forall v \in V \tag{3}$$

$$x_{ijv} \leq \sum_{k \in N} x_{okv} \qquad \forall i,j \in N, \quad \forall v \in V \tag{4}$$

$$\sum_{i \in N} x_{iov} + \sum_{j \in N} x_{djv} = 0 \qquad \forall v \in V \tag{5}$$

Equation (2) limits each vehicle to departing from the origin depot no more than once. Meanwhile, Eq. (3) ensures that any vehicle that leaves the origin depot must reach the destination depot. Equation (4) stipulates that a vehicle is allowed to visit other nodes only if it has initially departed from the origin, and Eq. (5) prohibits vehicles from returning to the origin depot or departing from the destination depot, enforcing the correct start and end points of the route.

$$\sum_{v \in V} \sum_{i \in N} x_{ijv} \leq 1 \qquad \forall j \in N \setminus \{d\} \tag{6}$$

$$\sum_{i \in N} x_{ijv} = \sum_{i \in N} x_{jiv} \qquad \forall j \in N \setminus \{o,d\}, v \in V \tag{7}$$

$$y_{iv} - y_{jv} + 1 \leq (1 - x_{ijv})|N| \qquad \forall i,j \in N, v \in V \tag{8}$$

Equation (6) ensures that each node is visited at most once, except the destination depot, which can be visited several times since it is the ending point of all vehicles. Equation (7) states that if a vehicle arrives at a node, it must also depart from that node. Moreover, in Eq. (8) the variables y_{iv} are introduced, which represent the order of the node i in the route of vehicle v. This restriction guarantees that there are no subtours. A last set of equations are introduced next:

$$\sum_{i,j \in N} x_{ijv} d_{ij} \leq L \qquad \forall v \in V \tag{9}$$

$$y_{iv} \geq 0, \qquad \forall i \in N, v \in V \tag{10}$$

$$x_{ijv} \in \{0, 1\} \qquad \forall i, j \in N, v \in V \tag{11}$$

Equation (9) ensures that the distance for each vehicle does not exceed the maximum allowable distance L. Finally, Eqs. (10) and (11) define the domain of the variables.

4   The TOP as a Sequential Decision Process

As explained before, to apply RL techniques to the TOP, it should be framed within the framework of an MDP. This transformation allows to model the problem as a sequence of decisions, where each decision is based on the current state and leads to a reward. At each step in the problem, the vehicle must decide which node to visit next, based on its current state (position, remaining time or distance, and unvisited nodes). For instance, the first decision corresponds to selecting the initial node to visit from the starting depot. Once the vehicle reaches a node, it moves to another node according to the available actions, while keeping track of the accumulated reward and the remaining time or distance budget. Figure 3 illustrates how the TOP with a single vehicle can be represented as a sequential decision process, and more specifically, how it fits within the DT framework.

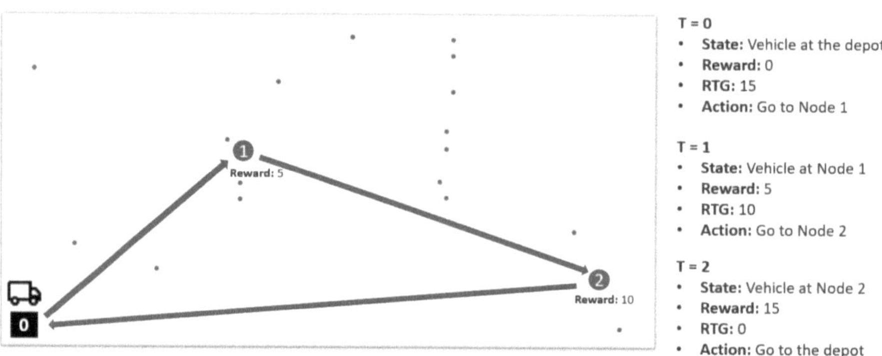

Fig. 3. The TOP as a sequential problem.

At timestep $t = 0$, the vehicle is located at the depot, the immediate reward is zero, and the action taken is to travel to node 1. The return-to-go (RTG) input is set to fifteen, indicating that the DT will attempt to construct a route with a cumulative reward of fifteen. At timestep $t = 1$, the state reflects that the vehicle has arrived at node 1, the reward obtained is five, and the next action is to proceed to node 2. Similarly, at timestep $t = 2$, the state indicates that the vehicle is now at node 2, with a cumulative reward of fifteen, and the final action is to return to the destination depot. At this point, the route of the vehicle concludes. In this case, note that the DT has found a route that satisfies the specified initial return-to-go. The progression is summarized in Table 1.

Table 1. Example trajectory of the TOP instance.

Timestep	State	RTG	Reward	Action
0	Vehicle at the depot	15	0	Go to node 1
1	Vehicle at node 1	10	5	Go to node 2
2	Vehicle at node 2	0	15	Go to destination depot
3	Vehicle at the final depot	0	15	End of route

This example illustrates how the problem naturally can be seen as a sequence of decisions over time. In general, each state at a given timestep can be represented by the vehicle's current position, but it can also include additional features like the remaining time or distance budget, as well as the set of nodes that have not yet been visited. The action corresponding to this state is the decision of which unvisited node to visit next. When taking this action, the environment transitions to a new state that reflects the updated vehicle position, the reduced remaining budget, and an updated set of unvisited nodes. This sequential data is then encoded into tokenized inputs suitable for processing by the DT model. Moreover, to ensure that the solutions generated remain feasible, that is, each node is visited at most once and the vehicle respects the time or distance constraints, some masking mechanisms can be incorporated into the model. The first mask should restrict the model from selecting nodes that have already been visited. The second mask should enforce feasibility with respect to the remaining budget: the vehicle should only be allowed to travel to a node if it can still return from that node to the destination depot within the remaining time or distance constraints.

5   Data and Training Strategy

In this case, since no historical data is available, the dataset must be generated from scratch, which can be a time-consuming process. However, the advantage lies in faster convergence during training, especially when compared to traditional online RL methods. If historical data is available and the simulation process is computationally expensive or difficult to reproduce, this approach can offer a more practical and efficient alternative for training. This does not apply in this case, and therefore, the training data consists of trajectories obtained by solving problem instances using the heuristic method proposed in [14]. In this work, the heuristic employed is based on a savings approach, initially generating as many routes as possible and then progressively merging them until the capacity of all vehicles is satisfied. To improve solution quality and avoid greedy local optima, a biased-randomized strategy is applied during the merging phase, introducing controlled randomness. This increases the diversity of solutions and allows the heuristic to potentially escape from local minima. The solving method stops once a predefined time limit has been reached. Moreover, the dataset includes solutions of varying quality: some are generated quickly or even randomly, while others are allowed more time to converge to better solutions. This diversity is intentional, as it allows the model to learn the relationship between trajectories and their associated return-to-go, including both high-quality and low-quality strategies.

For each instance, the final solution, i.e., the routes and the total accumulated reward, is stored. From this information, it is possible to reconstruct the full trajectory in terms of state, action, and return-to-go at each timestep, providing the input data needed to train the DT. However, at inference time, defining an appropriate return-to-go value is nontrivial. Some prior works use the maximum return observed in the training set [5,24], while others train a separate model to predict return-to-go [17,19]. Nevertheless, an inaccurate estimate can lead to poor performance.

Finally, as the output space is discrete (i.e., selecting the next node from a finite set), training is performed using a cross-entropy loss function. The cross-entropy loss is commonly used in classification tasks, where the model's output is a probability distribution over the possible next nodes. The objective during training is to minimize the difference between the predicted probability distribution and the actual selection made by the agent, which corresponds to the correct next node in the sequence. This ensures that the model learns to make correct node selections during the decision-making process.

6   Discussion and Future Work

The current study explores the use of DTs to solve combinatorial optimization problems, framing these challenges within the context of sequential decision-making processes. By utilizing trajectories generated from either heuristic-based solution methods or historical data, DTs can address combinatorial problems efficiently and flexibly. Their capacity to condition actions on past trajectories and desired outcomes enables generalization across various problem instances, which makes them particularly advantageous for offline settings and sequential decision-making problems. For instance, the TOP can be reformulated into a suitable sequential decision process for DTs. This involves defining states, actions, rewards, and transitions at each timestep, along with masking strategies to ensure feasibility during inference. These strategies help preserve constraints such as node visitation limits and time budget restrictions, allowing for effective and feasible problem solving.

Building on this, an interesting direction for future work is exploring the use of different RL algorithms for training the model, which could improve both training efficiency and speed. By using more advanced techniques, it may be possible to achieve faster convergence and enhanced performance, as shown in [8]. Another approach could be to fine-tune open-source Transformer models. Rather than training a model from scratch, adapting pre-trained models into the DT framework could save a lot of time and resources. These models, already trained on large datasets, could be modified to fit the structure of combinatorial problems, potentially improving both training efficiency and solution quality.

Fine-tuning open-source models also opens the door to using larger and more powerful architectures, which could handle more complex problem instances and improve the scalability of DTs.

Acknowledgements. This publication is part of the DIN2024-013395 grant, funded by MICIU/AEI/10.13039/501100011033. This publication is also funded by Coca-Cola Europacific Partners.

References

1. Achiam, J., et al.: GPT-4 technical report. arXiv preprint arXiv:2303.08774 (2023)
2. Archetti, C., Hertz, A., Speranza, M.G.: Metaheuristics for the team orienteering problem. J. Heuristics **13**, 49–76 (2007)
3. Butt, S.E., Cavalier, T.M.: A heuristic for the multiple tour maximum collection problem. Comput. Oper. Res. **21**(1), 101–111 (1994)
4. Chao, I.M., Golden, B.L., Wasil, E.A.: The team orienteering problem. Eur. J. Oper. Res. **88**(3), 464–474 (1996)
5. Chen, L., et al.: Decision transformer: reinforcement learning via sequence modeling. In: Advances in Neural Information Processing Systems, vol. 34, pp. 15084–15097 (2021)
6. Dang, D.C., Guibadj, R.N., Moukrim, A.: An effective PSO-inspired algorithm for the team orienteering problem. Eur. J. Oper. Res. **229**(2), 332–344 (2013)
7. Ford, N., Duckworth, D., Norouzi, M., Dahl, G.E.: The importance of generation order in language modeling. arXiv preprint arXiv:1808.07910 (2018)
8. Guo, D., et al.: DeepSeek-R1: Incentivizing reasoning capability in LLMs via reinforcement learning. arXiv preprint arXiv:2501.12948 (2025)
9. Iima, H., Nakamura, Y.: Combinatorial optimization method using decision transformer. In: 2024 SICE Festival with Annual Conference (SICE FES), pp. 564–569. IEEE (2024)
10. Kacem, I., Kellerer, H., Mahjoub, A.R.: Preface: new trends on combinatorial optimization for network and logistical applications. Ann. Oper. Res. **298**, 1–5 (2021)
11. Karamcheti, S., et al.: Mistral–a journey towards reproducible language model training (2021)
12. Liu, A., et al.: DeepSeek-V3 technical report. arXiv preprint arXiv:2412.19437 (2024)
13. Mazyavkina, N., Sviridov, S., Ivanov, S., Burnaev, E.: Reinforcement learning for combinatorial optimization: a survey. Comput. Oper. Res. **134**, 105400 (2021)
14. Panadero, J., Currie, C., Juan, A., Bayliss, C.: Maximizing reward from a team of surveillance drones under uncertainty conditions: a simheuristic approach. Eur. J. Ind. Eng **14**, 1–23 (2020)
15. Puterman, M.L.: Markov decision processes. Handbooks Oper. Res. Manag. Sci. **2**, 331–434 (1990)
16. Roohnavazfar, M.: Combinatorial optimization problems under deterministic, stochastic, and dynamic scheduling decisions (2021)

17. Tanaka, T., Abe, K., Ariu, K., Morimura, T., Simo-Serra, E.: Return-aligned decision transformer. arXiv preprint arXiv:2402.03923 (2024)
18. Tang, H., Miller-Hooks, E.: A tabu search heuristic for the team orienteering problem. Comput. Oper. Res. **32**(6), 1379–1407 (2005)
19. Tang, X., Marques, A., Kamalaruban, P., Bogunovic, I.: Adversarially robust decision transformer. arXiv preprint arXiv:2407.18414 (2024)
20. Touvron, H., et al.: Llama: open and efficient foundation language models. arXiv preprint arXiv:2302.13971 (2023)
21. Vaswani, A., et al.: Attention is all you need. In: Advances in Neural Information Processing Systems, vol. 30 (2017)
22. Yadav, B.: Generative ai in the era of transformers: revolutionizing natural language processing with LLMs. J. Image Process. Intell. Remote Sens. **4**(2), 54–61 (2024)
23. Zhang, C., et al.: A review on learning to solve combinatorial optimisation problems in manufacturing. IET Collaborative Intell. Manufact. **5**(1), e12072 (2023)
24. Zhang, Z., Mei, H., Xu, Y.: Continuous-time decision transformer for healthcare applications. In: International Conference on Artificial Intelligence and Statistics, pp. 6245–6262. PMLR (2023)
25. Zhao, W.X., et al.: A survey of large language models **1**(2). arXiv preprint arXiv:2303.18223 (2023)

Optimizing Conference Agendas with AI-Based Interest Estimation and Mathematical Programming

Carmine Cerrone[1]([✉]), Raffaele Dragone[1], and Amedeo Napolitano[2]

[1] Department of Economics and Business Studies, University of Genoa, Genoa, Italy
{carmine.cerrone,raffaele.dragone}@edu.unige.it
[2] FourClicks S.r.l. Innovative Startup, Torre Del Greco, Italy
amedeonapolitano@fourclicks.it

Abstract. In scientific conferences, participants are often required to select among parallel presentations based on their interests and logistical constraints, facing challenges related to time management, distances between session rooms, and content overlap. Building a daily schedule that maximizes individual satisfaction thus represents a complex and time-consuming task. In this work, we propose a system based on artificial intelligence and mathematical optimization to assist users in creating an optimal daily program. To this end, we develop a hybrid framework that combines explicit evaluations provided by participants with an automatic estimation of interest indices for talks, leveraging Large Language Models (LLMs). Finally, we present a mobile application, which implements the scheduling optimization process and collects qualitative feedback directly from participants, validating the effectiveness of our approach through experiments conducted on real-world conference data.

1 Introduction

Scientific conferences are often organized into parallel sessions, with multiple talks occurring simultaneously. This structure requires participants to carefully plan their daily schedules, balancing personal interests with logistical constraints such as session timing, talk durations, and physical distances between rooms or buildings. To support participants in constructing personalized schedules that reflect both individual preferences and practical considerations, we propose a data-driven approach based on artificial intelligence techniques. The method relies on estimating an interest index for each talk, built from the participant's stated preferences, and dynamically adjusted according to their evolving selection of sessions. These indices capture the relative attractiveness of each presentation and serve as the basis for a personalized recommendation system. The resulting information is integrated into an optimization framework that generates daily schedules to maximize the overall interest of the participant. The model considers several practical and organizational factors, including the timing and duration of each talk, the structure of the sessions, the possibility of joining a

M. Pavone et al. (Eds.): DSA ISC 2025, LNCS 16405, pp. 427–438, 2026.
https://doi.org/10.1007/978-3-032-21811-7_31

session already in progress or leaving before its conclusion, the physical distances between venues, estimated walking times, the participant's initial location, and their willingness to move within the conference setting. The optimization is carried out using Integer Linear Programming (ILP) techniques, with the goal of maximizing the overall interest index while respecting temporal and logistical constraints. The proposed methodology improves the conference experience by generating personalized schedules in a structured, automated, and efficient manner.

The Conference Scheduling (CS) problem has received increasing attention in the scientific literature, which has explored a variety of formulations and methodological approaches to address its operational complexity. Early contributions focused primarily on the efficient organization of sessions, taking into account logistical constraints such as speaker availability, room capacity, and the avoidance of scheduling conflicts. A pioneering work in this area is that of Eglese and Rand (1987) [1], who introduced the idea of collecting participants' individual preferences in advance and proposed a formulation that integrates these preferences with realistic constraints such as room capacity and overlapping sessions. This approach represents one of the first attempts to combine organizational efficiency with participant satisfaction. Subsequently, Sampson and Weiss (1995) [2] extended this line of research by incorporating room capacity constraints and proposing a heuristic approach to simultaneously solve both the scheduling and participant assignment problems. Their model improves fairness in the allocation of available resources and improves the perceived level of service.

In a later work, Sampson (2004) [3] further explored the role of user preferences by introducing a mathematical programming model (MIP) for the preference-based conference scheduling problem, in which the objective is to maximize an aggregate participant utility function, defined by the alignment between the final schedule and the preferences expressed ex ante.

Over time, the literature on CS has progressively expanded the role assigned to the academic community in the construction of the program, evolving from models where participant preferences were collected in a structured but limited manner to approaches promoting more direct and distributed involvement. In particular, two main research directions can be identified: on the one hand, approaches that leverage the expert knowledge of authors and organizers; on the other hand, methods that incorporate the active participation of attendees in the scheduling process. For a comprehensive review of the CS problem and its evolution, we refer the reader to [4].

A relevant example of the first direction is Cobi [5], a system that engages program committee members and author of articles to identify similarities among the discussions, thus supporting the construction of thematically coherent sessions. This approach is grounded in the idea of leveraging distributed but structured knowledge to produce a schedule that is more aligned with the interests of the academic community.

An example of the second direction is Confer [6], a tool that allows all conference attendees to freely explore the accepted content and generate personalized agendas based on their individual preferences. Built on collaborative filtering techniques, Confer integrates recommendation and social matching functionality, thus supporting not only personalized user experiences, but also connections among attendees with shared interests.

In addition to methodological developments, recent literature has also documented large-scale applications of the CS problem in real and complex contexts. Among these, the work of Stidsen et al. [7] stands out, developed for the international EURO-k conference in the field of operations research. In this context, the problem involved the hierarchical allocation of talks to buildings, sessions, and rooms, with the aim of satisfying structural constraints such as spatial thematic consistency, minimization of conflicts between related presentations, and maximization of room utilization efficiency.

Among the existing contributions, Confer [6] is the most closely related to our approach, as it allows each participant to generate a personalized agenda based on individual interests. However, the authors do not propose any mathematical model or optimization algorithm for schedule generation, which is instead entirely driven by individual-level recommendation logic. In this paper, by contrast, we aim to fill this gap by introducing a formal model for the construction of personalized programs that balance individual preferences with global scheduling constraints.

The remainder of this paper is organized as follows. Section 2 formally defines the optimization problem, including its variables, constraints, and objective function. Section 3 describes the AI-based approach used to compute the interest scores associated with each session. In Sect. 4, we present the mobile application that integrates the results of this research. Section 5 presents experimental results on realistic instances, analyzing the performance of the model. Finally, Sect. 6 concludes the paper, summarizing the main contributions and outlining future research directions.

2 Problem Definition

We consider the problem of constructing a personalized schedule for a conference participant by selecting a subset of talks to attend throughout the day. Each talk is characterized by a start and end time and is associated with a specific session located in a designated room. Participants are allowed to move from one talk to another, provided that temporal and spatial constraints allow the transfer within the available time.

The objective is to identify a feasible sequence of talks that maximizes the participant's overall interest, quantified through an interest index associated with each talk. The selected sequence must satisfy time constraints (avoiding overlap between talks), take into account travel time between session locations, and consider the participant's initial location within the conference venue. In addition, participants may join a session already in progress or leave a session before its conclusion, as long as they have explicitly agreed to this behavior if it improves the overall program. In the following, we provide a formal description of the problem, introducing the notation, decision variables, and optimization model.

2.1   Notation and Variables

In this section, we introduce the notation used for the formalization of the problem. We define the sets, parameters, and variables that describe the characteristics of the sessions, talks, and the costs associated with the participant's movement between sessions. Additionally, we specify the costs related to the possibility of late entry or early leaving from a session, which are key components in the optimization process.

We start by defining the basic sets and variables used in this work:

- S: The set of Sessions. Each session $s \in S$ is characterized by:
  - start_s: the start time of session s;
  - end_s: the end time of session s.
- T: The set of Talks. Each talk $t \in T$ is associated with:
  - $s(t) \in S$: the session to which talk t belongs;
  - start_t: the start time of talk t;
  - end_t: the end time of talk t;
  - $p_t \in [0, 1]$: the interest index of the participant for talk t.
- $d(s_i, s_j)$: Function that returns the travel time required to move between the rooms of sessions s_i and s_j, $\forall s_i, s_j \in S$.
- Penalties:
  - *leave*: the fixed cost associated with leaving a session early;
  - *enter*: the fixed cost associated with entering a session late.

To define a feasible solution, we consider an ordered subset of talks $T' \subseteq T$, where the selected talks must be temporally compatible. This means that for each pair of consecutive talks t_i and t_{i+1} in the subset, the end time of t_i plus the required travel time between the sessions must not exceed the start time of t_{i+1}. Mathematically, this condition is expressed as:

$$\text{end}_{t_i} + d(s(t_i), s(t_{i+1})) \leq \text{start}_{t_{i+1}},$$

This ensures that the participant can attend both talks without any overlap and that there is enough time to move between the rooms of the corresponding sessions.

Once temporal compatibility is ensured, we then define the costs associated with moving between sessions. The cost function accounts for potential penalties related to early leaving or late entering a session. If the participant must leave the session of talk t_i early to meet the temporal constraints, a **leave** cost is applied. If the participant enters the session of talk t_{i+1} late, an **enter** cost is incurred. In cases where both events occur, the total cost is the sum of the **leave** and **enter** costs. If neither of these conditions applies, no penalty is applied, and the cost is set to 0.

With the costs defined, the final objective is to identify the sequence of talks that maximizes the participant's interest, while minimizing the penalties for early leaving and late entry. Formally, the objective function to maximize is the following:

$$\text{Objective}(T') = \sum_{t \in T'} p_t - \sum_{i=1}^{|T'|-1} \text{cost}(t_i, t_{i+1})$$

2.2   Mathematical Model and Graph Representation

In order to effectively solve the problem, we construct a directed acyclic graph. This representation allows us to model the dependencies between talks, ensuring that the temporal and spatial constraints are respected. The nodes of the graph correspond to the talks, while the arcs represent the feasible transitions between consecutive talks based on the timing and the necessary travel time between session rooms.

This graph is formally defined as $G = (N, A)$, where:

- N contains one node for each talk, plus two special nodes: a start node n_0 and an end node $n_{|T|+1}$. Formally,

$$N = \{n_i \mid t_i \in T\} \cup \{n_0, n_{|T|+1}\}.$$

- Each node n_i associated with a talk t_i inherits the start and end times of the corresponding talk. Formally, we define:

$$\text{start}_{n_i} = \text{start}_{t_i}, \quad \text{end}_{n_i} = \text{end}_{t_i}, \quad p_{n_i} = p_{t_i}.$$

- A contains a directed arc from node n_i to node $n_i + 1$ if and only if

$$\text{end}_{t_i} + d(\text{s}(t_i), \text{s}(t_j)) \leq \text{start}_{t_j}.$$

- Arcs are created from the start node n_0 to each node corresponding to a talk, as well as from each node corresponding to a talk to the end node $n_{|T|+1}$.
- In addition, for each pair of nodes (n_i, n_j), we define a transition penalty function $\text{cost}(n_i, n_j)$, which accounts for possible penalties due to early leaving from the session of t_i and/or late arrival to the session of t_j.

$$\text{cost}(n_i, n_j) = \begin{cases} leave, & \text{if early leaving from } t_i \text{ is necessary;} \\ enter, & \text{if late arrival to } t_j \text{ is necessary;} \\ leave + enter, & \text{if both events occur;} \\ 0, & \text{otherwise.} \end{cases}$$

To model the problem, we define binary flow variables x_{ij} for each arc $(n_i, n_j) \in A$, where:

$$x_{ij} = \begin{cases} 1 & \text{if arc } (n_i, n_j) \text{ is selected in the path,} \\ 0 & \text{otherwise.} \end{cases}$$

Since the graph is acyclic and we seek a path, each selected node (except n_0 and $n_{|T|+1}$) must have exactly one incoming and one outgoing arc in the selected path. The interest index of a talk is included in the objective function by considering the incoming arcs to the corresponding node. Specifically, the contribution of a talk t_i is accounted for if an incoming arc to node n_i is selected.

The mathematical formulation of the problem is summarized in the following model:

$$\text{Maximize} \quad \sum_{(n_i, n_j) \in A} \left(p_{n_j} - \text{cost}(n_i, n_j) \right) x_{ij} \tag{1}$$

$$\text{subject to} \quad \sum_{(n_0, n_j) \in A} x_{0j} = 1 \tag{2}$$

$$\sum_{(n_i, n_{|T|+1}) \in A} x_{i, |T|+1} = 1 \tag{3}$$

$$\sum_{(n_i, n_k) \in A} x_{ik} - \sum_{(n_k, n_j) \in A} x_{kj} = 0, \quad \forall n_k \in N \setminus \{n_0, n_{|T|+1}\} \tag{4}$$

$$x_{ij} \in \{0, 1\}, \quad \forall (n_i, n_j) \in A \tag{5}$$

The objective function (1) maximizes the total interest index collected along the selected path, while accounting for the penalty costs associated with early leaving or late entering between consecutive talks. Constraint (2) ensures that exactly one arc leaves the start node n_0, initiating the path. Constraint (3) ensures that exactly one arc enters the end node $n_{|T|+1}$, terminating the path. Constraint (4) enforces flow conservation at every intermediate node: for each node $n_k \in N \setminus \{n_0, n_{|T|+1}\}$, the number of incoming arcs must equal the number of outgoing arcs, ensuring a continuous path structure. Finally, the constraint (5) imposes binary decisions on the arc selection variables x_{ij}.

3 Hybrid User-AI Framework for Estimating Interest Index

To define the interest index associated with each talk, we propose a hybrid framework that combines the explicit preferences of the user with an AI-based estimation. The inspiration for this approach originated from a previous work [8], in which an LLM was used to determine the initial parameter settings for an optimization model. In particular, the user is initially given the opportunity to: (i) provide a textual description of their research interests and preferred topics; and (ii) express an interest rating for one or more talks.

If the user provides an interest rating for all available talks, these scores are used directly by the model. However, given the large number of talks typical of a conference, it is more realistic to assume that the user will provide ratings for only a subset of talks. In this case, an automatic process is activated to estimate the interest index for the talks that have not been rated.

The estimation process uses a Large Language Model (LLM) to predict missing interest indices based on the information provided by the user. Specifically, the data collected is processed by the LLM to generate:

- The set I_{topics}: which includes the research problems and areas of interest;
- The set I_{methods}: which includes the techniques and methodologies of interest.

Once these two sets are created, the LLM evaluates each talk that has not been rated yet by calculating: a thematic compatibility score with respect to the set I_{topics}; a methodological compatibility score with respect to the set I_{methods}; and a perceived originality score for the research work. The estimated interest score for each talk is then calculated as the arithmetic mean of these three scores. To better illustrate the process, Fig. 1 presents a schematic flow diagram of the AI-based estimation pipeline.

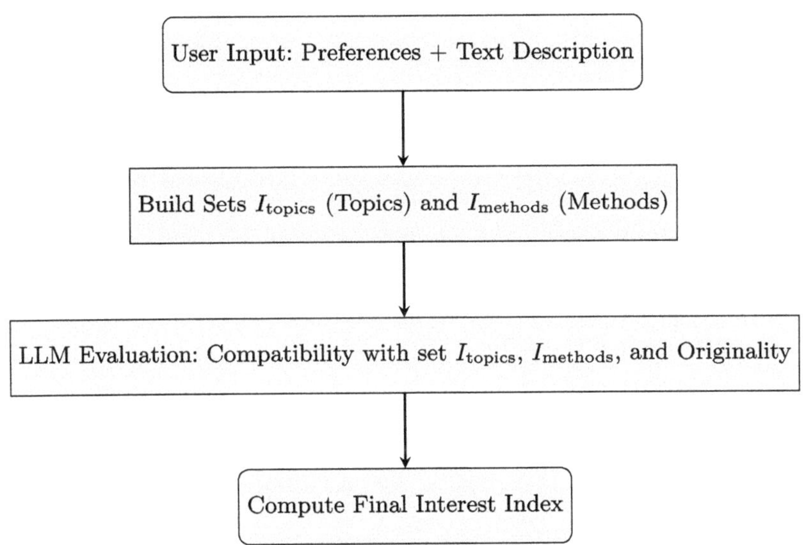

Fig. 1. Flowchart of the AI-based estimation process for missing interest scores.

For each unrated talk, we generate an estimated score in the 0–5 range using the following formula:

$$\text{Estimated Interest Score} = \frac{\left(\frac{40 \times I_{\text{topics}} + 40 \times I_{\text{methods}} + 20 \times \text{originality}}{100}\right)^2}{12}$$

where the final result is rounded to the nearest integer. We adopted a quadratic transformation to emphasize higher compatibility scores: values that are initially low remain low after the transformation, while sufficiently high scores are amplified, resulting in final values closer to 4 or 5. This approach is intended to penalize talks perceived as less relevant while favoring those with a strong thematic and methodological alignment with the user's interests.

Since the application employs a 0–5 star rating system for talks, this scale was chosen for consistency. For the subsequent optimization phase, all interest scores are normalized to the $[0, 1]$ interval by dividing the computed value by 5.

4   Mobile Application for Conference Scheduling

One of the main challenges of this work is to evaluate the performance of the AI-based recommendation system. It is crucial to understand how effectively the developed system captures participants' preferences. To address this experimentally, we decided to involve a significant sample of real users. We developed a mobile application that allowed all conference participants to generate their optimized daily schedules. The application was also designed to collect user feedback, which is of fundamental importance for the further development and refinement of this work.

Figure 2 illustrates two key components of the developed application. Figure 2(a) shows the schedule optimizer interface, where users can set their preferences and generate a personalized daily program based on their interests and logistical constraints. Figure 2(b) displays the feedback form, which allows users to evaluate the proposed program in terms of overall satisfaction and relevance to their interests. The collected feedback is crucial for continuously improving the recommendation system and refining the optimization algorithms used to support conference attendees.

5   Experimental Results

In order to conduct tests on real-world data, we used anonymized schedules from publicly available conference programs. The selected conference featured four parallel sessions held in different rooms: Room A, Room B, Room C, and Room D. To make the tests more realistic and challenging, we simulated a scenario in which Room A and Room B are located within one area, while Room C and Room D are situated approximately a five-minute walk away. To solve the problem using the mathematical model, for each day of the conference and for each participant, we generated a graph representing all possible paths, as described in the previous sections. Figure 3 presents the full graph representation for Day 1 of the conference. Each column groups talks held in the same room, while nodes sharing the same colour belong to the same session. Arcs connect talks that can be attended consecutively, respecting timing and movement constraints. The labels on the arcs display the costs associated with the transition, combining both the walking time and any penalties related to early

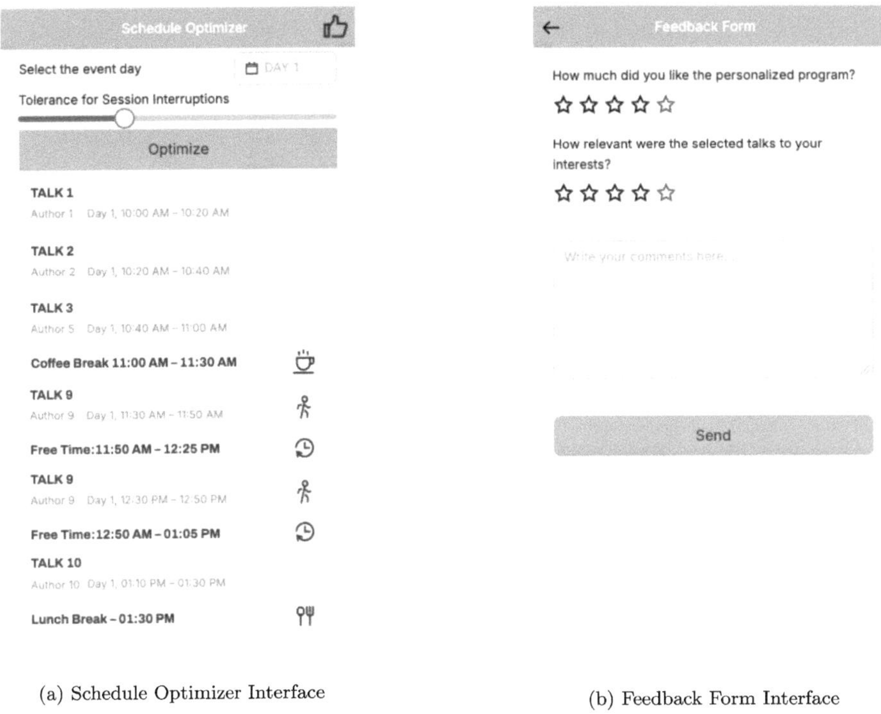

(a) Schedule Optimizer Interface   (b) Feedback Form Interface

Fig. 2. User interface of the scheduling system. (a) Optimizer for generating personalized daily programs; (b) Feedback form for evaluating the relevance and satisfaction with the proposed schedule.

leaving or late entry. The numerical value inside each node indicates the interest index assigned to the corresponding talk, which guides the optimization process toward selecting the most attractive sequence of talks for the participant.

Figure 4 illustrates the optimal personalized schedules generated for the two days of the conference. Each node of the blue path represents a talk attended by the participant, grouped by session rooms, while the directed arcs show the selected transitions between talks. The paths are computed by maximizing the overall interest index, taking into account travel times and possible penalties for early leaving or late entry into sessions. The resulting sequences demonstrate how the model is capable of constructing coherent daily schedules that prioritize user interests while managing logistical constraints across multiple locations. The computational tests were performed on a MacBook Pro M4 Max with 64 GB of RAM. The software was implemented in Python, using CBC (Coin-or branch and cut) as the solver. For the instances, the algorithm consistently produced optimal solutions with computational times below 0.1 s. For the recommendation system, we employed the "gpt-4.1" Large Language Model (LLM), which enabled the evaluation of each talk in an average time of 0.25 s per talk. By applying the "Estimated Interest Score" function presented in Sect. 3 to assess the talks, the

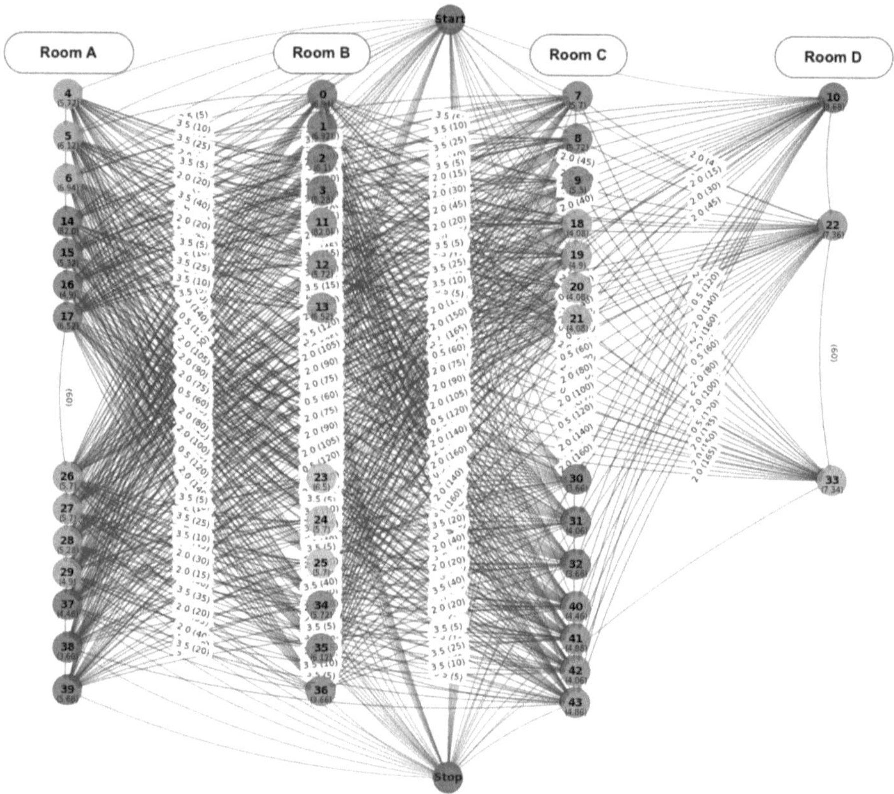

Fig. 3. Graph representation of the talks and sessions on Day 1. Each column corresponds to a different session room, and nodes sharing the same colour belong to the same session. Arcs between nodes represent feasible transitions between talks, with associated costs reflecting both travel penalties and walking times. Node labels indicate the interest index associated with each talk.

calculated schedules achieved, on average, at least a 20% improvement in the overall Interest Score compared to baseline selections. All generated schedules, including those shown in the previous figures, consistently suggested attending between 4 and 6 sessions per day.

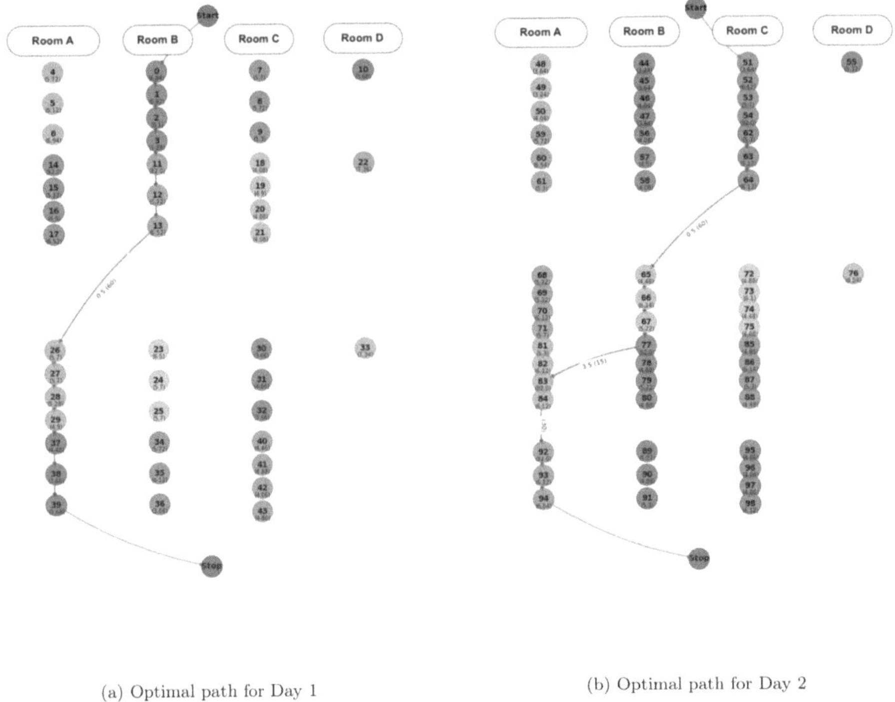

<table>
<tr><td>(a) Optimal path for Day 1</td><td>(b) Optimal path for Day 2</td></tr>
</table>

Fig. 4. Optimal personalized schedules generated for the two conference days. Each node represents a talk, and arcs indicate the transitions selected to maximize the participant's interest while respecting timing and movement constraints.

6   Conclusions

In this work, we presented an innovative approach for building personalized schedules for academic conferences, combining artificial intelligence techniques with mathematical programming. The proposed system leverages advanced AI models (LLMs) to estimate participants' interest in various talks and optimizes the selection of presentations through an Integer Linear Programming (ILP) model. The integration of explicit preferences with automatic estimations significantly improves the quality of daily schedules compared to random selections or recommendations based solely on individual preferences. The experimental evaluation, conducted on anonymized real-world data derived from publicly available conference programs, confirmed the effectiveness of the model, highlighting its ability to consistently generate optimal solutions in very short computational times, as well as to significantly enhance the overall Interest Score compared to non-optimized selection strategies. The development of a mobile application for the upcoming conferences will allow us to validate the system in a real-world operational setting, collecting valuable user feedback. As future work, we plan to explore the integration of LLMs with maximum matching algorithms to estimate interest scores more rapidly, while reducing the energy consumption and operational costs typically associated with large AI models. Another promising research direction will involve the analysis of the feedback collected after conferences, which will enable the development of an even more adaptive and proactive conference experience.

References

1. Eglese, R.W., Rand, G.K.: Conference seminar timetabling. J. Oper. Res. Soc. **38**(7), 591–598 (1987)
2. Sampson, S.E., Weiss, E.N.: Increasing service levels in conference and educational scheduling: a heuristic approach. Manag. Sci. **41**(11), 1816–1825 (1995)
3. Sampson, S.E.: Practical implications of preference-based conference scheduling. Prod. Oper. Manag. **13**(3), 205–215 (2004)
4. Vangerven, B., Ficker, A.M.C., Goossens, D.R., Passchyn, W., Spieksma, F.C.R., Woeginger, G.J.: Conference scheduling-a personalized approach. Omega **81**, 38–47 (2018)
5. Kim, J., et al.: Cobi: a community-informed conference scheduling tool. In: Proceedings of the 26th Annual ACM Symposium on User Interface Software and Technology, pp. 173–182 (2013)
6. Zhang, A.X., Bhardwaj, A., Karger, D.: Confer: a conference recommendation and meetup tool. In: Proceedings of the 19th ACM Conference on Computer Supported Cooperative Work and Social Computing Companion, pp. 118–121 (2016)
7. Stidsen, T., Pisinger, D., Vigo, D.: Scheduling Euro-K conferences. Eur. J. Oper. Res. **270**(3), 1138–1147 (2018)
8. Dragone, R., Battaglia, R., Cerrone, C.: Integrating human-centric AI in fashion industry optimization: a step towards industry 5.0. In: Proceedings of the 8th AIROYoung Workshop, Rende, Cosenza, Italy, 14–16 February 2024. Accepted

Decision-Making in New Product Development: The Roles of Design Thinking and Social Technology

Isa Moll(iD) and Jordi Montaña(⊠)

ESADE Business School, Avenida de Pedralbes 60-62, 08034 Barcelona, Spain
{Isa.moll,Jordi.montana}@esade.edu

Abstract. Despite recent studies estimating new product failure rates around 40%, down from traditional estimates of 80% or higher (Castellion & Markham, 2012), the risk of failure in launching product-based innovations remains significicant. Nevertheless, companies continue to innovate and introduce new products to the market. While external factors often drive product innovation, failure can frequently be attributed to a lack of strategy and rigor in the development process itself. However, the decision-making process within new product development has received limited attention.

This paper examines corporate decision-making across various stages of product development, with a particular focus on the roles of design thinking and social technology. Through case studies of successful product launches, we demonstrate how structured decision-making frameworks and collaborative approaches can reduce failure rates.

This research contributes to innovation management literature and offers practical strategies for managers seeking to improve decision-making processes and outcomes in competitive markets.

Keywords: New Product Development · Decision-Making · Design Thinking · Innovation Management

1 Introduction

Innovation through new product development (NPD) is an iterative process that requires continuous go/no-go decisions, each entailing significant risk, substantial investment, and fundamental management challenges. Despite these hurdles, NPD remains relentless and essential for organizational growth and competitiveness.

Don Norman (1988), referencing visual perception psychologist Irwing Biederman, estimated that an adult in a developed country is surrounded by tens of thousands of distinct products. In the decades since, this number has only multiplied due to the proliferation of substitute products with new features and entirely new product categories. For example, a typical consumer today may own vinyl records (which have seen a resurgence), cassettes, compact discs, and multiple devices for streaming services like

M. Pavone et al. (Eds.): DSA ISC 2025, LNCS 16405, pp. 439–445, 2026.
https://doi.org/10.1007/978-3-032-21811-7_32

Spotify. What was once rare-such as having more than one television in a household-is now commonplace, with smart devices found in kitchens, bedrooms, and beyond. The increasing interconnection of smart products further fuels this growth (Porter & Heppelmann, 2014).

Innovation is a priority for nearly every organization (Yohn, 2019) and has a direct impact on sales and profits, with new products accounting for approximately 25% of company revenues (Cooper, 2017).

2 Risk of Failure and Other Challenges in New Product Development

Despite the potential rewards, the failure rate for new product introductions remains alarmingly high. Traditional business literature cited failure rates as high as 75–80% (Castellion & Markham, 2012), though more recent studies suggest rates closer to 40% (Cooper, 2017). Given the costs and risks involved-ranging from $100,000 for small consumer goods to hundreds of millions for complex products like automobiles (Ulrich & Eppinger, 2016)-the stakes are high.

Development timelines also pose challenges; products may enter the market too late or too early, missing consumer needs or readiness. Many firms implement rigorous processes to manage these risks and optimize costs, yet failures persist. As Christensen, Hall, and Duncan (2016) argue, such failures often stem from insufficient engagement with consumers and users.

While external factors (competition, regulation, technology, demographics) often trigger innovation, internal factors-especially decision-making discipline-are frequently the root cause of failure. Yet, the decision-making process in NPD has received limited scholarly attention.

3 Decision-Making Throughout the New Product Development Process

Effective decision-making in NPD revolves around what to develop, how, and when, with the goal of minimizing failure risk and associated losses. Companies undertake NPD for various reasons, including competition, product life cycle dynamics, demographic shifts, changing consumer habits, technological advancements, inventions, globalization, regulatory changes, and resource availability. Most of these drivers are external, beyond the company's direct control. However, firms can take proactive measures to anticipate and respond to these changes.

This paper highlights four key decision factors that increase the likelihood of successful new product development and launch:

- Establishing an environmental scanning system to anticipate signals of change
- Developing a robust new product strategy
- Supporting the decision-making process with heuristics
- Following a structured NPD process in which marketing plays a central role (Yohn, 2019)

3.1 Early Indicators of Innovation Need

Successful companies rigorously monitor a range of early indicators, including:

- User-oriented indicators (e.g., customer suggestions and complaints)
- Technological indicators (e.g., patent activity)
- Product mix structure (e.g., portfolio analysis)
- Trade-oriented indicators (e.g., assortment gaps)
- Competitor-oriented indicators (e.g., innovation rates of rivals)
- Resource-oriented indicators (e.g., raw material price trends)
- General environment indicators (e.g., regulatory changes)

Empirical evidence shows that firms with successful product innovations make significantly greater use of early indicators-especially technological and user-oriented-than less successful firms. It is therefore crucial for companies to identify, monitor, and act on relevant early indicators.

3.2 New Product Strategy

A clear and adaptive NPD strategy guides and accelerates the development process. Strategies can be broadly categorized as proactive or reactive:

- Proactive strategies anticipate market trends and emerging customer needs. Companies may pursue this through acquisitions, alliances, R&D investment (e.g., Tesla, Post-it), marketing and market research (e.g., Nike), or fostering entrepreneurship and intrapreneurship. Henkel's Adhesive Technologies unit, now accounting for 50% of total revenue, originated from an internal necessity during World War I and later became a commercial success.
- Reactive strategies respond to market conditions or competitor actions. These are common among follower organizations focused on operational efficiency. Examples include defensive moves (e.g., Sara Lee's L'Or capsules against Nestlé Nespresso), imitation (e.g., Ryanair), "second but better" (e.g., iPod), and responsive strategies (e.g., Inditex).

The choice depends on corporate vision, mission, resources, risk tolerance, competitive and growth strategies, market size, and competition. Many successful firms balance proactive and reactive approaches, ensuring alignment with organizational objectives while remaining adaptable.

3.3 Heuristics in Decision-Making

Heuristic frameworks support NPD decision-making. Classic tools include Ansoff's Matrix, the General Electric/McKinsey Matrix, the BCG Matrix, and Porter's Competitive Strategies. More recent models, such as Keeley's "Ten Types of Innovation," offer structured approaches for identifying and evaluating innovation opportunities (Day, 2024; Keeley, Keeley et al., 2013).

3.4   The New Product Development Process

Cooper's Stage-Gate framework (Cooper, 2017) remains a cornerstone, emphasizing phased development with go/no-go decisions at each stage (Fig. 1).

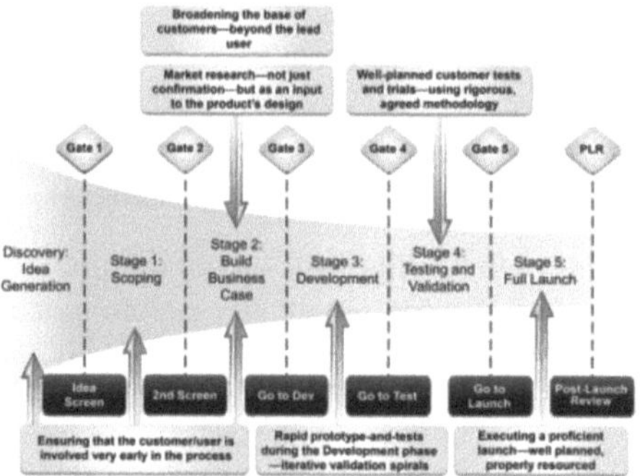

Fig. 1. Cooper's Stage-Gate framework

Building on this, Moll and Montaña's framework-developed from studies of companies recognized for both design excellence and market orientation-places greater emphasis on design and marketing in NPD (Fig. 2).

Empirical data show that dedicating resources to design and early-stage marketing-including concept definition and testing-increases the odds of success (Fig. 3).

Both exploration and design phases are critical for product success and involve relatively low costs compared to later stages. NPD managers should carefully scrutinize decisions to continue or abandon ideas and design projects before advancing further. Iterative collaboration between marketing and design teams throughout the process is essential (Cooper, 2019).

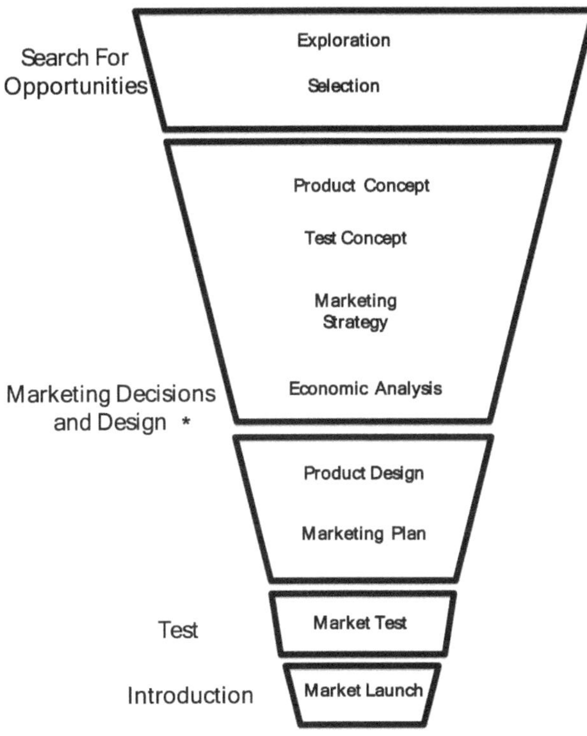

Fig. 2. The Marketing and Design driven- New Product Development framework

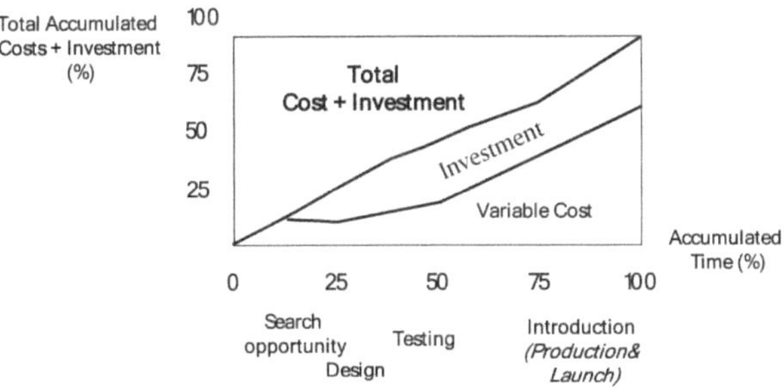

Fig. 3. New Product Development Cost & Investment curves

4   The Role of Design and Design Thinking in Innovation

Design must be integrated into innovation from two perspectives (Sato, 2009): first, as a philosophy of "design thinking" embedded in the innovation process; second, as a practical discipline capable of translating concepts into compelling products.

Design thinking is a methodological system that maximizes user value while optimizing business outcomes (Kelley, 2016). It aligns with user-centered design, integrative thinking, and social thinking. The traditional design thinking model involves three iterative phases (Brown, 2020):

- Inspiration: Beyond market research, this phase leverages observation, expert input, anthropology, and heuristics to uncover consumer insights.
- Ideation: Transforming insights into actionable ideas through extensive brainstorming. Booz, Allen & Hamilton (1982) found that an average of 60 ideas are needed to yield one successful product. Idea selection is a rigorous process that follows ideation.
- Implementation: Concretizing ideas into prototypes and models for evaluation and selection.

User-centered design reinforces marketing principles, as designers optimize tangible elements to maximize user satisfaction. Early and ongoing collaboration between marketing and design is a key success factor (Moll et al., 2007).

5 Methodology

This study utilizes a qualitative, multiple-case approach to investigate decision-making in new product development. Data was collected from a purposive sample of organizations recognized for successful product launches and strong design orientation. Primary data sources included semi-structured interviews with NPD managers, designers, and marketing professionals, as well as analysis of internal documents and process frameworks. Secondary data were drawn from published case studies, industry reports, and academic literature.

The analysis focused on identifying common decision-making frameworks, the integration of design thinking, and the use of early indicators and social technology in collaborative processes. Cross-case synthesis enabled the extraction of patterns and best practices, while triangulation of data sources enhanced the validity and reliability of the findings.

6 Conclusion

Despite advances in process management and innovation frameworks, the risk of failure in new product development remains substantial. This research highlights that success in NPD depends on disciplined decision-making, early and sustained integration of marketing and design, and effective use of collaborative tools and social technologies.

Firms that excel in NPD systematically monitor early indicators, align product strategies with corporate goals, and employ structured yet flexible development processes. Embedding design thinking throughout fosters creativity, user-centricity, and cross-functional collaboration, while social technologies facilitate real-time knowledge sharing and iterative development.

For managers, these findings underscore the importance of investing in environmental scanning systems, cultivating adaptive and balanced NPD strategies, and embedding marketing and design expertise from the earliest stages. Above all, fostering a culture of iterative learning, supported by digital collaboration tools, is crucial to navigating the complexities of innovation in competitive markets.

This study contributes to innovation management literature by demonstrating that robust decision-making frameworks, combined with collaborative and user-centered approaches, can substantially reduce new product failure rates and enhance organizational competitiveness.

Disclosure of Interests. The authors have no competing interests to declare that are relevant to the content of this article.

References

Booz, J., Allen, R., Hamilton, H.: New Products Management for the 1980s. Booz, Allen & Hamilton, New York (1982)

Brown, T.: Change by Design: How Design Thinking Transforms Organizations and Inspires Innovation. Harper Business (2020)

Castellion, G., Markham, S.K.: New product failure rates: influence of argumentum ad populum and self-interest. J. Prod. Innov. Manag. **30**(5), 976–979 (2012)

Christensen, C. M., Hall, T., Duncan, D.S.: Know your customers' "jobs to be done". Harvard Bus. Rev. (2016)

Cooper, R.G.: Winning at New Products: Creating Value Through Innovation, 4th edn. Basic Books (2017)

Cooper, R.G.: The drivers of success in new-product development. Ind. Mark. Manage. **76**, 36–47 (2019)

Day, G.S.: See Sooner, Act Faster: How Vigilant Leaders Thrive in an Era of Digital Turbulence. MIT Press (2024)

Kelley, T.: Creative Confidence: Unleashing the Creative Potential Within Us All. Harper Business (2016)

Keeley, L., Walters, H., Pikkel, R., Quinn, B.: Ten Types of Innovation: The Discipline of Building Breakthroughs. Wiley (2013)

Moll, I., Montaña, J., Guzman, F., Solé Parellada, F.: Market orientation and design orientation: a management model. J. Market. Manag. (2007)

Montaña, J., Moll, I.: Éxito empresarial y diseño. Federación Española de Entidades para la Promoción del Diseño, Madrid (2008)

Montaña, J., Moll, I.: Diseño e innovación: La gestión del diseño en la empresa. Documentos COTEC sobre oportunidades tecnológicas, Madrid (2008)

Norman, D.A.: The Design of Everyday Things. Basic Books (1988)

Porter, M.E., Heppelmann, J.E.: How smart, connected products are transforming competition. Harv. Bus. Rev. **92**(11), 64–88 (2014)

Ulrich, K.T., Eppinger, S.D.: Product Design and Development, 6th edn. McGraw-Hill Education (2016)

Yohn, D.L.: Why innovation is crucial to your organization's long-term success. Harvard Bus. Rev. (2019)

Author Index

 MIX
Papier aus verantwortungsvollen Quellen
Paper from responsible sources
FSC® C105338

If you have any concerns about our products,
you can contact us on
ProductSafety@springernature.com

In case Publisher is established outside the EU,
the EU authorized representative is:
**Springer Nature Customer Service Center GmbH
Europaplatz 3, 69115 Heidelberg, Germany**

Printed by Libri Plureos GmbH
in Hamburg, Germany